Aleksandr A. Samarskii
Evgenii S. Nikolaev

Numerical Methods for Grid Equations

Volume II
Iterative Methods

Translated from the Russian
by Stephen G. Nash

1989 Birkhäuser Verlag
Basel · Boston · Berlin

Authors' address:
Aleksandr A. Samarskii
Evgenii S. Nikolaev
Department of Computational
Mathematics and Cybernetics
Moscow University
Moscow 117234
USSR

Originally published as
Metody resheniya setochnykh uravnenii
by Nauka, Moscow 1978.

CIP-Kurztitelaufnahme der Deutschen Bibliothek

Samarskij, Aleksandr A.:
Numerical methods for grid equations / Aleksandr A. Samarskii
; Evgenii S. Nikolaev. Transl. from the Russ. by Stephen G.
Nash. – Basel ; Boston ; Berlin : Birkhäuser.
Einheitssacht.: Metody rešenija setočnych uravnenij <engl.>

NE: Nikolaev, Evgenij S.:

Vol. 2. Iterative methods. – 1989

Softcover reprint of the hardcover 1st edition 1989

Typesetting and Layout: *mathScreen online*, CH-4056 Basel

ISBN-13: 978-3-0348-9923-9 e-ISBN-13: 978-3-0348-9142-4
DOI: 10.1007/978-3-0348-9142-4

Volume II
Table of Contents

Volume I
Table of Contents

Chapter 5

The Mathematical Theory of Iterative Methods

The current chapter contains results and basic concepts from the theory of iterative methods; these methods will be studied in the succeding chapters. In Section 5.1 we state the simplest concepts of functional analysis, give the basic properties of linear and non-linear operators in a Hilbert space, and also give several theorems on the solubility of operator equations. In Section 5.2, we give a systematic treatment of difference schemes as operator equations in an abstract space and indicate the properties of the corresponding operators. In Section 5.3, we look at the basic definitions and concepts from the theory of iterative processes, examine a canonical form for iterative schemes, and also the concepts of convergence and number of iterations.

5.1 Several results from functional analysis

5.1.1 Linear spaces. In the preceding chapters we studied the basic direct method for solving the simplest difference equations. Those methods were characterized by the property that it is in principle possible to obtain with their aid a precise solution to the difference problem after a finite number of operations. Naturally it is assumed that the data for the problem are exact, and that all the computations are performed without rounding error.

These methods are so effective because they take into account the structure of the matrix of the system being solved. The requirement that the matrix satisfy special properties reduces the applicability of these methods, limiting them to the simplest problems.

To solve complex and, in particular, non-linear difference problems, iterative methods are most commonly used. The essence of iterative methods consists in constructing a sequence of approximations converging to the solution, starting with some initial guess. After a finite number of steps, the approximate solution is taken to be the solution of the problem.

Iterative methods are more universal in that they allow us to solve not one concrete problem, but a class of problems possessing definite properties. These properties are defined not by the structure of the grid equations, but by more general functional properties. Since in the majority of iterative methods the concrete structure of the equations is not used, the theory of iterative methods can be constructed from a single point of view, taking as our goal the investigation of a first-kind equation

$$Au = f,$$

where A is an operator, f is given, and u is the desired elements of some space H.

Before going on to construct and investigate iterative methods, we give here a short list of results from functional analysis (without proof).

A *linear space* over the field K of real or complex numbers is a set H together with the operations of addition and scalar multiplication which satisfies the following axioms (x, y, z are elements of H, λ and μ are scalars from K):

1) both operations are closed in H;
2) $x+y=y+x$, $x+(y+z)=(x+y)+z$ (commutativity and associativity of addition);
3) $\lambda(\mu x)=(\lambda\mu)x$ (associativity of multiplication);
4) $\lambda(x+y)=\lambda x+\lambda y, (\lambda+\mu)x=\lambda x+\mu x$ (distribution of multiplication over addition);
5) there exists an identity element 0 for which $x+0=x$ for any $x \in H$;
6) for any $x \in H$ there exists an additive inverse $(-x) \in H$ such that $x+(-x)=0$;
7) $1 \cdot x = x$.

Depending on whether the field K is real or complex, we obtain a *real* or *complex linear space* H.

In linear spaces, it is possible to introduce the concept of linear dependence and linear independence of elements. Elements $x_1, x_2, \ldots, x_n$ of the linear space H are called *linearly independent* if

$$\lambda_1 x_1 + \lambda_2 x_2 + \cdots + \lambda_n x_n = 0 \tag{1}$$

implies that $\lambda_1 = \lambda_2 = \ldots = \lambda_n = 0$. Conversely, if there exist $\lambda_1, \lambda_2, \ldots, \lambda_n$ not all zero satisfying (1), then the elements $x_1, x_2, \ldots, x_n$ are called *linearly dependent*.

The space H is called ***n*-*dimensional*** if in H there exist n linearly independent elements, and any $(n+1)$-st element is linearly dependent.

A non-empty closed set H_1 of elements of the linearly space H is called a ***subspace*** if $x_1, x_2, \ldots, x_n \in H_1$ implies that any linear combination $\lambda_1 x_1 + \lambda_2 x_2 + \ldots + \lambda_n x_n$ of these elements is also in H_1.

The sum of a finite number of subspaces $H_1, H_2, \ldots, H_n$ is the set of elements of the form

$$x = x_1 + x_2 + \cdots + x_n, \qquad x_i \in H_i, \quad i = 1, 2, \ldots, n. \tag{2}$$

Suppose that $H_1, H_2, \ldots, H_n$ are subspaces belonging to the linear space H. If each element $x \in H$ is uniquely representable in the form (2), then we say that H is the ***direct sum of the subspaces*** $H_1, H_2, \ldots, H_n$, and the expression (2) is called the ***expansion of the element*** x ***in elements from*** $H_1, H_2, \ldots, H_n$.

We will have in this case

$$H = H_1 \oplus H_2 \oplus \cdots \oplus H_n.$$

It is not difficult to show that if $H = H_1 \oplus H_2$, then H_1 and H_2 have in common only the zero element of the space. Conversely, if any element $x \in H$ can be represented in the form $x = x_1 + x_2, x_1 \in H_1, x_2 \in H_2$ and $H_1 \cap H_2 = 0$, then $H = H_1 \oplus H_2$.

H is called a ***normed linear space*** if, for any element $x \in H$, there is defined a real number $\| x \|$, called the ***norm***, which satisfies the conditions:

1) $\| x \| \geq 0$, and $\| x \| = 0$ if $x = 0$;
2) $\| x + y \| \leq \| x \| + \| y \|$ (the triangle inequality);
3) $\| \lambda x \| = |\lambda| \; \| x \|$, λ a scalar.

A sequence $\{x_n\}$ of elements of the linear normed space H is said to ***converge*** to an element $x \in H$ if $\| x - x_n \| \to 0$ as $n \to \infty$. If $\| x_n - x_m \| \to 0$ as $n, m \to \infty$, then the sequence $\{x_n\}$ is called a ***Cauchy sequence.***

A linear normed space H is called ***complete*** if any Cauchy sequence $\{x_n\}$ from this space converges to some element $x \in H$. A complete linear space is called a ***Banach space.*** Any finite-dimensional linear normed space is complete. Subspaces of a normed linear space are normed in a natural way.

One and the same linear space can be normed in an infinite number of ways. Suppose that the linear space has been normed in two different ways by the norms $\| x \|_1$ and $\| x \|_2$. If there exist constants $0 < m < M$ such that for any $x \in H$

$$m \| x \|_1 \leq \| x \|_2 \leq M \| x \|_1,$$

then the norms are called ***equivalent.*** Notice that in a finite-dimensional space any two norms are equivalent.

If two equivalent norms are introduced into a linear space, then the convergence of some sequence $\{x_n\}$ in one norm implies convergence in the other.

Suppose that H is a real (complex) linear space, and suppose that for any elements x, y of H there is an associated real (complex) number (x, y) for which:

1) $(x, y) = \overline{(y, x)}$ (symmetry);
2) $(x + y, z) = (x, z) + (y, z)$ (distributive law);
3) $(\lambda x, y) = \lambda(x, y)$ (homogeneity);
4) $(x, x) \geq 0$ for any $x \in H$, and $(x, x) = 0$ if and only if $x = 0$.

The number (x, y) is called the *inner product* of the elements x and y. The overline denotes the complex conjugate of a number.

A linear normed space H in which the norm is induced by the scalar product $\| x \| = \sqrt{(x, x)}$ is called a *unitary space*. A complete unitary space is called a *Hilbert space*. A finite-dimensional unitary space is complete.

The inner product satisfies the Cauchy-Schwartz-Bunyakovskij inequality $|(x, y)| \leq \| x \| \cdot \| y \|$. Elements x and y of a unitary space are called *orthogonal* if $(x, y) = 0$. An element $x \in H$ is called *orthogonal to the subspace H_1 of the space H* if x is orthogonal to any element $y \in H_1$. The set H_2 of all elements $x \in H$ orthogonal to the subspace H_1 of the space H is called the *orthogonal complement* of the subspace H_1. Notice that the orthogonal complement is itself a subspace of the space H.

Suppose that H_1 is an arbitrary subspace of the space H, and H_2 is its orthogonal complement. Then H is the direct sum of H_1 and H_2, $H = H_1 \oplus H_2$. Consequently, any element $x \in H$ is uniquely representable in the form $x = x_1 + x_2$, $x_\alpha \in H_\alpha$, $\alpha = 1, 2$, and $(x_1, x_2) = 0$.

A system $x_1, x_2, \ldots, x_n, \ldots$ of elements of the space H is called an *orthogonal system*, if $(x_m, x_n) = \delta_{mn}, m, n = 1, 2, \ldots$, where δ_{mn} is the Kronecker delta, equal to 1 for $m = n$ and equal to 0 for $m \neq n$.

If there does not exist a non-zero element $x \in H$ orthogonal to all the elements of the orthonormal system $\{x_n\}$, then this system is called *complete*. The Fourier series $\sum_{k=1}^{\infty} c_k x_k$, where $c_k = (x, x_k), k = 1, 2, \ldots$ can be constructed for any $x \in H$ from a complete orthonormal system $\{x_n\}$, it converges to this element, and for any $x \in H$ we have

$$\| x \|^2 = (x, x) = \sum_{k=1}^{\infty} c_k^2.$$

5.1.2 Operators in linear normed spaces. Suppose that X and Y are linear normed spaces. We say that the operator A is defined on the set $\mathcal{D} \subset X$ with values in Y (i.e., that A is an operator mapping from $\mathcal{D}$ into Y) if for any element $x \in \mathcal{D}$ there exists a corresponding element $y = Ax \in Y$. The set $\mathcal{D}$ is called the *domain of the operator* A and is denoted by $\mathcal{D}(A)$. The union of all the elements $y \in Y$ which can be represented in the form $y = Ax$ ($x \in \mathcal{D}(A)$), is called the *range of the operator* A and is denoted by im A. If $\mathcal{D}(A) = X$, im $A = X$, i.e. the operator A maps X onto itself, then we say that A is an operator on X.

The operator A is called *linear* if $\mathcal{D}(A)$ is a linear manifold in X and if for any $x_1, x_2 \in \mathcal{D}(A)$

$$A(\lambda_1 x_1 + \lambda_2 x_2) = \lambda_1 A x_1 + \lambda_2 A x_2,$$

where λ_1 and λ_2 are scalars from the field K.

A linear operator A is called *bounded* if there exists a constant $M > 0$ such that for any $x \in \mathcal{D}(A)$

$$\| Ax \|_2 \leq M \| x \|_1, \tag{3}$$

where $\| \cdot \|_1$ is the norm in X, and $\| \cdot \|_2$ is the norm in Y. An abitrary non-linear operator A is called *bounded* on $\mathcal{D}(A)$ if

$$\sup_{x \in \mathcal{D}(A)} \| Ax \|_2 < \infty.$$

For a linear operator A, the smallest constant M satisfying (3) is called the *norm* of the operator and is denoted by $\| A \|$. From the definition of the norm it follows that

$$\| A \| = \sup_{\|x\|_1 = 1} \| Ax \|_2 \qquad \text{or} \qquad \| A \| = \sup_{x \neq 0} \frac{\| Ax \|_2}{\| x \|_1}.$$

Notice that in a finite-dimensional space any linear operator is bounded. Suppose that A is an arbitrary operator mapping from X into Y. The operator A is called *continuous at the point* $x \in X$ if $\| x_n - x \|_1 \to 0$ ($x_n \in X$) implies that $\| Ax_n - Ax \|_2 \to 0$ as $n \to \infty$. A linear bounded operator is continuous.

An arbitrary operator A satisfies a *Lipschitz condition with constant* q if

$$\| Ax_1 - Ax_2 \|_2 \leq q \| x_1 - x_2 \|_1, \qquad x_1, x_2 \in \mathcal{D}(A). \tag{4}$$

Any linear bounded operator A satisfies a Lipschitz condition (4) with $q = \| A \|$.

Suppose that A is an arbitrary operator mapping from X into Y. The linear bounded operator $A'(x)$ is called the *Gateaux derivative of A at the point x* of the space X if for any $z \in X$

$$\lim_{t \to 0} \left\| \frac{A(x+tz) - Ax}{t} - A'(x)z \right\|_2 = 0.$$

Here the range of the operator A' belongs to Y.

If the operator A has a Gateaux derivative at every point of the space X, then (4) is valid for any $x_1, x_2 \in X$ with

$$q = \sup_{0 \leq t \leq 1} \| A'(x_1 + t(x_2 - x_1)) \| .$$

If A is a linear operator, then $A' = A$.

The set of all bounded operators mapping from X into Y forms a *linear normed space*, since the norm $\| A \|$ of an operator A satisfies all the axioms of a norm. Let us look at the set of all linear bounded operators mapping from X into X. On this set, it is possible to define the product AB of the operators A and B in the following way: $(AB)x = A(Bx)$. It is clear that AB is a linear bounded operator: $\| AB \| \leq \| A \| \cdot \| B \|$.

If $(AB)x = (BA)x$ for all $x \in X$, then the operators A and B are called *abelian or commutative*: in thie case we write $AB = BA$.

In connection with the solution of equations of the form $Ax = y$, we introduce the concept of an *inverse* operator A^{-1}. Suppose that A is an operator from X onto Y. If for each $y \in Y$ there corresponds only one $x \in X$ for which $Ax = y$, then using this correspondence we define the operator A^{-1}, called the *inverse* of A, and having domain Y and range X.

For any $x \in X$ and $y \in Y$ we have the identities $A^{-1}(Ax) = x$, $A(A^{-1})y = y$. It is not difficult to show that if A is linear, then so is A^{-1} (if it exists).

Lemma 1. *The linear operator A mapping from X onto Y has an inverse if and only if $Ax = 0$ implies $x = 0$.* □

Theorem 1. *Suppose that A is a linear operator from X onto Y. The inverse operator A^{-1} exists and is bounded (as an operator from Y onto X) if and only if there exists a constant $\delta > 0$ such that for any $x \in X$*

$$\| Ax \|_2 \geq \delta \| x \|_1 .$$

We also have the estimate $\| A^{-1} \| \leq 1/\delta$. Here $\| \cdot \|_1$ is the norm in X and $\| \cdot \|_2$ is the norm in Y. □

In other words, the inverse operator A^{-1} exists if and only if the homogeneous equation $Ax = 0$ only has the trivial solution.

Suppose that A and B are invertible linear bounded operators mapping into X. Then $(AB)^{-1} = B^{-1}A^{-1}$.

If the operator A is invertible, then it makes sense to talk about the powers A^k for any integers (and not just non-negative ones). Namely, we define $A^{-k} = (A^{-1})^k, i = 1, 2 \ldots$. Powers of the same operator commute.

We now introduce the concept of the *null* space of a linear operator A. The *null space of the linear operator* A is the set of all those elements x from the space X for which $Ax = 0$. The null space of the linear operator A is denoted by the symbol ker A.

The condition ker $A = 0$ is necessary and sufficient for the operator A to have an inverse.

A subspace X_1 of the space X is called an *invariant* subspace for the operator A mapping into X if A maps X_1 into itself, i.e. $Ax \in X_1$ for $x \in X_1$.

If the subspace X_1 is invariant relative to an invertible operator A, then it is invariant relative to the operator A^{-1}.

Ker A and im A are examples of invariant subspaces of the operator A. Notice that, if the operators A and B commute, then the subspaces ker B and im B are invariant relative to the operator A.

The number

$$\rho(A) = \lim_{k\to\infty} = \sqrt[k]{\| A^k \|}$$

is called the *spectral radius of the linear operator* A. It does not depend on the definition of the norm, and $\rho(A) = \inf_{\|\cdot\|} \| A \|$.

For any bounded linear operator A we have

$$\rho(A) \leq \| A \|, \quad \rho(A) \leq \sqrt[k]{\| A^k \|}, \qquad k = 2, 3, \ldots$$

Lemma 2. $\| A \| = \rho(A)$ *if and only if* $\| A^k \| = \| A \|^k$, $i = 2, 3, \ldots$ □

Notice here one more property of the spectral radius. If the operators A and B commute, then

$$\rho(AB) \leq \rho(A)\rho(B), \quad \rho(A + B) \leq \rho(A) + \rho(B).$$

5.1.3 Operators in a Hilbert space. Assume that the bounded linear operator A acts in the unitary space H. According to the general definition of an operator norm we have

$$\| A \| = \sup_{\|x\|=1} \| Ax \| = \sup_{x \in H} \sqrt{\frac{(Ax, Ax)}{(x, x)}}$$

and consequently, for any $x \in H$

$$(Ax, Ax) \leq \| A \|^2 (x, x).$$

Using the Cauchy-Schwartz-Bunyakovskij inequality, from this we obtain

$$|(Ax, x)| \leq \| Ax \| \| x \| \leq \| A \| (x, x). \tag{5}$$

From now on, we will only look at bounded operators.

The operator A^* is called the ***adjoint operator*** to A if, for any $x \in H$, it satisfies

$$(Ax, y) = (x, A^* y).$$

For any bounded linear operator A with domain $\mathcal{D}(A) = H$ there exists a unique operator A^* with domain $\mathcal{D}(A^*) = H$. The operator A^* is linear and bounded, $\| A^* \| = \| A \|$.

We now give the basic properties of the adjoint operator: $(A^*)^* = A$, $(A + B)^* = A^* + B^*$, $(AB)^* = B^* A^*$, $(\lambda A)^* = \bar{\lambda} A^*$. If the operators A and B commute, then the adjoint operators A^* and B^* also commute. If A has an inverse, then $(A^{-1})^* = (A^*)^{-1}$, i.e. the operations of inverting and taking the adjoint of an operator commute.

Lemma 3. *Let A be a linear operator in H. The space H can be represented in the form of the direct sum of the orthogonal subspaces*

$$H = \ker A \oplus \operatorname{im} A^*, \quad H = \ker A^* \oplus \operatorname{im} A.$$

Proof. In fact, suppose that H_1 is the orthogonal complement of im A^* in the space H, i.e.

$$H = H_1 \oplus \operatorname{im} A^*, \quad (x_1, x_2) = 0, \quad x_1 \in H_1, \quad x_2 \in \operatorname{im} A^*.$$

We will show that $H_1 = \ker A$. Let $x_1 \in \ker A$, then for any $x \in H$ we have $A^*x \in \text{im } A^*$ and

$$(x_1, A^*x) = (Ax_1, x) = 0.$$

Consequently, x_1 is orthogonal to im A^*, and therefore $x_1 \in H_1$. On the other hand, suppose $x_1 \in H_1$ (consequently, x_1 is orthogonal to im A^*). Then for any $x \in H$

$$0 = (x_1, A^*x) = (Ax_1, x).$$

Since x is an arbitrary element of H, $Ax_1 = 0$ and thus $x_1 \in \ker A$. The first assertion of the lemma is proved. The second is proved analogously. □

A linear operator A is called *self-adjoint* in H if $A = A^*$. For a self-adjoint operator $(Ax, y) = (x, Ay)$ for any $x, y \in H$.

An operator A is called *normal* if it commutes with its adjoint, $A^*A = AA^*$, and *skew-symmetric* if $A^* = -A$. Self-adjoint and skew-symmetric operators are normal.

It is known that if A and B are self-adjoint operators, then the operator AB is self-adjoint if and only if A and B commute.

If A is a linear operator, then A^*A and AA^* are self-adjoint operators, $\| A^*A \| = \| AA^* \| = \| A \|^2$, and

$$\ker A^*A = \ker A, \quad \text{im } A^*A = \text{im } A^*,$$
$$\ker AA^* = \ker A^*, \quad \text{im } AA^* = \text{im } A.$$

Any operator A can be represented as the sum of a self-adjoint operator A_0 and a skew-symmetric operator A_1

$$A = A_0 + A_1,$$

where $A_0 = 0.5(A + A^*)$, $A_1 = 0.5(A - A^*)$. If H is a real space, then it follows that

$$(Ax, x) = (A_0x, x), \qquad (A_1x, x) = 0.$$

In a complex space H we have a Cartesian representation of the operator A:

$$A = A_0 + iA_1,$$

where $A_0 = \text{Re } A = 1/2(A + A^*)$, $A_1 = \frac{1}{2i}(A - A^*)$ are self-adjoint operators in H. Here for any $x \in H$ we have

$$\text{Re } (Ax, x) = (A_0x, x), \qquad \text{Im } (Ax, x) = (A_1x, x).$$

If A is a self-adjoint operator in H, then we have the formula

$$\| A \| = \sup_{x \neq 0} \frac{|(Ax, x)|}{(x, x)}, \qquad x \in H.$$

Lemma 4. *If A is a bounded self-adjoint operator in H, then for any integer n, $\| A^n \| = \| A \|^n$.* □

Lemma 4 is also valid for normal operators.

From lemmas 2 and 4 it follows that for a normal (and, in particular, a self-adjoint) operator A, $\rho(A) = \| A \|$.

Lemma 5. *Suppose that the inner product of the elements x and y has been defined in two ways in the space H: $(x, y)_1$ and $(x, y)_2$. If the operator A is self-adjoint with respect to both inner products, then $\| A \|_1 = \| A \|_2 = \rho(A)$.* □

The spectral radius gives a lower bound for any operator norm. We now define the numerical radius of an operator, which allows us to obtain a two-sided estimate for a norm.

The *numerical radius of the operator A*, which is real in a complex space H, is defined by

$$\bar{\rho}(A) = \sup_{\|x\|=1} |(Ax, x)| \qquad x \in H.$$

For any bounded linear operator A we have: $\mu(A) \| A \| \leq \bar{\rho}(A) \leq \| A \|$, $\mu(A) \geq 1/2$ and, in addition, $\bar{\rho}(A^n) \leq [\bar{\rho}(A)]^n$ for any integer n. If the operator A is self-adjoint, then $\bar{\rho}(A) = \| A \|$. Notice also a number of other interesting properties of the numerical radius. For example, $\bar{\rho}(A^*) = \bar{\rho}(A)$, $\bar{\rho}(A^* A) = \| A \|^2$. Also, $\rho(A) \leq \bar{\rho}(A)$, where $\rho(A)$ is the spectral radius of an operator defined above.

A linear operator A, operating in the Hilbert space H, is called *positive* ($A > 0$) if $(Ax, x) > 0$ for all non-zero $x \in H$. For complex spaces H, the concept of positivity is only defined for self-adjoint operators, since in this case positivity of an operator implies self-adjointness.

Analogously we introduce the definitions of *non-negativity* of an operator A (for any $x \in H$, $(Ax, x) \geq 0$) and *positive definiteness* (for any $x \in H$ $(Ax, x) \geq \delta(x, x)$, where $\delta > 0$).

A non-linear operator A acting in H is called *monotonic* if

$$(Ax - Ay, x - y) \geq 0, \quad x, y \in H,$$

strictly monotonic if

$$(Ax - Ay, x - y) > 0, \quad x, y \in H, \quad x \neq y,$$

and *strongly monotonic* if for any $x, y \in H$ we have

$$(Ax - Ay, x - y) \geq \delta \parallel x - y \parallel^2, \qquad \delta > 0.$$

Theorem 2. *Assume that the non-linear operator A has a continuous Gateaux derivative at every point $x \in H$. Then the operator A is strongly monotonic on H if and only if there exists a $\delta > 0$ such that*

$$(A'(x)y, y) \geq \delta(y, y), \qquad y \in H. \square$$

Let A be a non-negative linear operator. The number (Ax, x) is called the *energy of the operator*. We will compare operators A and B using the energy. If $((A - B)x, x) \geq 0$ for any $x \in H$, then we will write $A \geq B$.

If there exist constants $\gamma_1 \geq \gamma_2 > 0$ such that $\gamma_1 B \leq A \leq \gamma_2 B$ for operators A and B, then we will call A and B *energy equivalent operators* (en. eq.), and γ_1 and γ_2 are the constants of energy equivalence for the operators A and B. Suppose that

$$\delta = \inf_{\|x\|=1} (Ax, x) \quad \text{and} \quad \Delta = \sup_{\|x\|=1} (Ax, x).$$

The numbers δ and Δ are called the ***bounds*** of the operator A (which is self-adjoint when H is a complex space). Obviously we have

$$\delta(x, x) \leq (Ax, x) \leq \Delta(x, x), \qquad x \in H$$

or

$$\delta\, E \,\leq\, A \,\leq\, \Delta\, E,$$

where E is the identity operator, $Ex = x$.

It is not difficult to verify that the inequality relation introduced onto the set of linear operators acting in H possesses the following properties:

1) $A \geq B$ and $C \geq D$ implies $A + C \geq B + D$,
2) $A \geq 0$ and $\lambda \geq 0$ implies $\lambda A \geq 0$,

3) $A \geq B$ and $B \geq C$ implies $A \geq C$,

4) if $A > 0$ and A^{-1} exists, then $A^{-1} > 0$.

Further it is clear that A^*A and AA^* are non-negative operators for any linear operator A. These operators will be positive if A is a positive operator.

Theorem 3. *The product AB of two commuting non-negative operators A and B, one of which is self-adjoint, is also a non-negative operator.* □

For any self-adjoint non-negative operator A, we have a generalized Cauchy-Schwartz-Bunyakovskij inequality

$$|(Ax, y)| \leq \sqrt{(Ax, x)}\sqrt{(Ay, y)}, \qquad x, y \in H.$$

Let D be a self-adjoint positive operator acting in H. Then it is possible to define the *energy space* H_D, consisting of elements of H, with the inner product $(x, y)_D = (Dx, y)$ and norm

$$\| x \|_D = \sqrt{(Dx, x)}.$$

Notice that if D is a self-adjoint, positive-definite, and bounded operator in H, then for any $x \in H$ we have the estimates (using the Cauchy-Schwartz-Bunyakovskij inequality)

$$\delta(x, x) \leq (Dx, x) \leq \| Dx \| \| x \| \leq \Delta(x, x), \quad \Delta = \| D \|, \quad \delta > 0.$$

These inequalities can be written in the form

$$\sqrt{\delta} \| x \| \leq \| x \|_D \leq \sqrt{\Delta} \| x \|,$$

from which it follows that the usual norm $\| \cdot \|$ and the energy norm $\| \cdot \|_D$ are equivalent.

We remark that a unitary energy space H_D can also be constructed from a non-self-adjoint positive operator D. For this, the inner product in H_D is defined as follows:

$$(x, y)_D = (D_0 x, y), \quad \text{where} \quad D_0 = 0.5(D + D^*).$$

We now give a series of lemmas which contain the basic inequalities that we will require later.

Lemma 6. *Suppose that $A \geq \delta E$, $\delta > 0$ for the linear operator A. Then for any $x \in H$ we have*

$$(Ax, Ax) \geq \delta(Ax, x).$$

If the condition $A \leq \Delta E$ is satisfied for the non-negative self-adjoint operator A, then for any $x \in H$ we have

$$(Ax, Ax) \leq \Delta(Ax, x). \square$$

Lemma 7. *From the condition $(Ax, Ax) \leq \Delta(Ax, x)$, $x \in H$, $\Delta > 0$ for a non-negative opeator A, it follows that*

$$A \leq \Delta E,$$

and from the condition $(Ax, Ax) \geq \delta(Ax, x)$, $\delta > 0$ for a non-negative self-adjoint operator A it follows that

$$A \geq \delta E. \square$$

Corollary 1. *Let A be a self-adjoint, positive-definite operator. Then from lemmas 6 and 7 it follows that*

$$\delta E \leq A \leq \Delta E, \qquad \delta > 0,$$

and

$$\delta(Ax, x) \leq (Ax, Ax) \leq \Delta(Ax, x), \qquad \delta > 0,$$

are equivalent.

Corollary 2. *From (5) and lemma 6 we obtain the estimate $(Ax, Ax) \leq \| A \| (Ax, x)$, $x \in H$, for a non-negative self-adjoint operator A in H.*

Lemma 8. *Suppose that A is a bounded, positive, self-adjoint operator in H, $A > 0$, $\| Ax \| \leq \Delta \| x \|$. Then the inverse operator A^{-1} is positive definite $A^{-1} \geq \frac{1}{\Delta} E$.* $\square$

Lemma 9. *Let A and B be self-adjoint positive-definite operators in H. Then*

$$\gamma_1 B \leq A \leq \gamma_2 B, \qquad \gamma_2 \geq \gamma_1 > 0$$

and

$$\gamma_1 A^{-1} \leq B^{-1} \leq \gamma_2 A^{-1}, \qquad \gamma_2 \geq \gamma_1 > 0$$

are equivalent. $\square$

Lemma 10. *If A is a positive-definite operator $A \geq \delta E$, $\delta > 0$, then the inverse operator A^{-1} exists and $\| A^{-1} \| \leq 1/\delta$.*

Proof. The proof follows from the inequality

$$\delta \| x \|^2 \leq (Ax, x) \leq \| Ax \| \, \| x \|, \qquad \delta > 0,$$

and from theorem 1. □

Remark. If A is a positive operator, then A^{-1} exists. In the case of a complex space H, the existence of the operator A^{-1} follows from the positivity of the real component $A_0 = 0.5(A + A^*)$ or the positivity of the imaginary component $A_1 = \frac{1}{2i}(A - A^*)$ of the operator A.

5.1.4 Functions of a bounded operator. In the theory of iterative methods, we are required to deal with functions of an operator. Let A be a bounded linear operator acting in the normed space X. If $f(\lambda)$ is an entire analytic function of the variable λ with the series expansion $\Sigma_{k=0}^{\infty} a_k \lambda^k$, then it is possible to define the *function $f(A)$ of the operator A* using the formula $f(A) = \Sigma_{k=0}^{\infty} a_k A^k$. The operator $f(A)$ will also be linear and bounded. As an example, we give here the exponential operator

$$e^A = \sum_{k=0}^{\infty} \frac{A^k}{k!}.$$

It is possible to extend the definition of a function of an operator to a wider class of functions and also to construct an operator calculus for bounded operators. We will give a more general definition only for self-adjoint bounded operators in a Hilbert space.

Let δ and Δ be lower and upper bounds on the self-adjoint operator A in H. Let $f(\lambda)$ be a continuous function on the interval $[\delta, \Delta]$. The operator $f(A)$ is called a *function of the self-adjoint operator A*.

The correspondence between functions of a real variable and functions of an operator possesses the following properties:

1) If $f(\lambda) = \alpha f_1(\lambda) + \beta f_2(\lambda)$, then $f(A) = \alpha f_1(A) + \beta f_2(A)$.
2) If $f(\lambda) = f_1(\lambda) f_2(\lambda)$, then $f(A) = f_1(A) f_2(A)$.
3) $AB = BA$ implies that $f(A)B = Bf(A)$ for any bounded linear operator B.
4) If $f_1(\lambda) \leq f(\lambda) \leq f_2(\lambda)$ for any $\lambda \in [\delta, \Delta]$, then $f_1(A) \leq f(A) \leq f_2(A)$.
5) $\| f(A) \| \leq \max\limits_{\delta \leq \lambda \leq \Delta} |f(\lambda)|$.
6) $\bar{f}(A) = [f(A)]^*$, where the overline denotes the complex conjugate of the function. If $f(\lambda)$ is a real-valued function, then it follows that the operator $f(A)$ is self-adjoint in H.

From property 4) it follows that if $f(\lambda) \geq 0$ on $[\delta, \Delta]$, then $f(A)$ is a non-negative operator.

An important example of a function of an operator is the square root of an operator. The operator B is called the *square root of the operator* A if $B^2 = A$.

Theorem 4. *For any non-negative self-adjoint operator* A, *there exists a unique non-negative self-adjoint square root which commutes with any operator that commutes with* A. □

The square root of an operator A will be denoted by $A^{1/2}$. Notice that $\| A \| = \| A^{1/2} \|^2$ if $A = A^* \geq 0$.

Theorem 5. *If* A *is a self-adjoint, positive-definite operator,* $A = A^* \geq \delta E$, $\delta > 0$, *then there exists a bounded self-adjoint operator*

$$A^{-1/2}, \quad \| A^{-1/2} \| \leq 1/\sqrt{\delta}\,.$$

Proof. The proof follows from the inequality

$$\delta(x,x) \leq (Ax,x) = (A^{1/2}x, A^{1/2}x) = \| A^{1/2}x \|^2$$

and from theorem 1. □

5.1.5 Operators in a finite-dimensional space. We will look at the n-dimensional unitary space H. Suppose that the elements $x_1, x_2, \ldots, x_n$ form an orthonormal basis in H. Using the definition of a finite-dimensional space, we can represent any element $x \in H$ uniquely in the form of a linear combination

$$x = c_1x_1 + c_2x_2 + \cdots + c_nx_n. \tag{6}$$

From the orthonormality of the system $x_1, x_2, \ldots, x_n$ it follows that $c_k = (x, x_k)$.

Thus, any element $x \in H$ can be put in correspondence with a vector $c = (c_1, c_2, \ldots, c_n)^T$, the components of which are the coefficients c_k from the expansion (6).

Let A be a linear operator defined on H. In the basis $x_1, x_2, \ldots, x_n$, it corresponds to a matrix $\mathcal{A} = (a_{ik})$ of dimension $n \times n$ where $a_{ik} = (Ax_k, x_i)$. Conversely, any matrix $\mathcal{A}$ of dimension $n \times n$ defines a linear operator in H.

For this, the element Ax is placed in correspondence with the vector

$$\left(\sum_{k=1}^{n} a_{1k}c_k,\ \sum_{k=1}^{n} a_{2k}c_k, \dots,\ \sum_{k=1}^{n} a_{nk}c_k\right)^T,$$

i.e. the vector $\mathcal{A}c$.

If the operator A is self-adjoint in H, then the matrix $\mathcal{A}$ corresponding to it is symmetric in any orthonormal basis. Notice that the self-adjoint operator A corresponds to a non-symmetric matrix in a non-orthonormal basis.

We look now at the properties of eigenvalues and eigenelements of a linear operator A. A number λ is called *eigenvalue of the operator* A if the equation

$$Ax = \lambda x \tag{7}$$

has a non-zero solution. An element $x \neq 0$ satisfying (7) is called an *eigenelement of the operator* A corresponding to the eigenvalue λ. In other words, the eigenvalues of the operator A are those values of λ for which $\ker(A - \lambda E) \neq 0$; the eigenelements corresponding to the eigenvalue λ are the non-zero elements of the subspace $\ker(A - \lambda E)$. This subspace is called the *eigensubspace* corresponding to the eigenvalue λ.

The set $\sigma(A)$ of eigenvalues of the operator A is called the *spectrum of the operator* A.

1. A self-adjoint operator A has n orthonormal eigenelements $x_1, x_2, \dots, x_n$. The corresponding eigenvalues $\lambda_k, k = 1, 2, \dots, n$ are real. If all the eigenvalues are distinct, then A is called an *operator with a simple spectrum.*

2. If A is a self-adjoint operator, then

$$\| A \| = \rho(A) = \max_{1 \leq k \leq n} |\lambda_k|,$$

where $\rho(A)$ is the *spectral radius of the operator* A. These results are also valid for normal operators A.

3. If $A = A^* \geq 0$, then all the eigenvalues of the operator A are non-negative. Then for any $x \in H$

$$\delta(x, x) \leq (Ax, x) \leq \Delta(x, x),$$

where $0 \leq \delta = \min_k \lambda_k$, $\Delta = \max_k \lambda_k$. For a self-adjoint operator A, the expression $(Ax, x)/(x, x)$ is called the *Rayleigh quotient.*

The largest and smallest eigenvalues of the operator A can be determined using the Rayleigh quotient as follows:

$$\delta = \min_{x \neq 0} \frac{(Ax, x)}{(x, x)}, \qquad \Delta = \max_{x \neq 0} \frac{(Ax, x)}{(x, x)}.$$

4. We will use $\lambda(A)$ to denote the eigenvalues of the operator A. Let $f(A)$ be a function of the self-adjoint operator A. Then $\lambda(f(A)) = f(\lambda(A))$ (the spectral mapping theorem).

5. If the self-adjoint operators A and B commute, $A = A^*$, $B = B^*$, $AB = BA$, then they have a common system of eigenelements. Also, the operators AB and $A + B$ have the same system of eigenelements as the operators A and B, and the eigenvalues satisfy

$$\lambda(AB) = \lambda(A)\lambda(B), \quad \lambda(A+B) = \lambda(A) + \lambda(B).$$

6. An arbitrary element $x \in H$ can be expanded in the eigenelements of a self-adjoint operator A

$$x = \sum_{k=1}^{n} c_k x_k, \quad c_k = (x, x_k), \quad \text{and} \quad \| x \|^2 = \sum_{k=1}^{n} c_k^2.$$

A number λ is called an *eigenvalue of the operator A relative to the operator B* if the equation

$$Ax = \lambda Bx \tag{8}$$

has a non-zero solution. An element $x \neq 0$ satisfying equation (8) is called an *eigenelement of the operator A relative to the operator B* corresponding to the number λ.

7. If the operators A and B are self-adjoint in H, and if in addition the operator B is positive definite, then there exist n eigenelements $x_1, x_2, \ldots, x_n$ which are orthonormal in the energy space H_B: $(x_k, x_i)_B = \delta_{ki}$, $k, i = 1, 2, \ldots, n$. The corresponding eigenvalues are real and satisfy the inequalities

$$\gamma_1(Bx, x) \le (Ax, x) \le \gamma_2(Bx, x),$$

where

$$\gamma_1 = \min_k \lambda_k = \min_{x \neq 0} \frac{(Ax, x)}{(Bx, x)},$$
$$\gamma_2 = \max_k \lambda_k = \max_{x \neq 0} \frac{(Ax, x)}{(Bx, x)}$$

Consequently, the constants of en. eq. for the self-adjoint operators A and B in the case where B is positive definite coincide with the smallest and largest eigenvalues of the generalized problem (8).

5.1.6 The solubility of operator equations. Suppose that we are required to find the solution of a first-kind operator equation

$$Au = f, \tag{9}$$

where A is a bounded linear operator in the Hilbert space H, f is given, and u is the desired elements of H. We will assume that H is finite-dimensional. We are interested in the solubility of the equation (9). We have

Theorem 6. *Equation* (9) *is soluble for any right-hand side f if and only if the corresponding homogeneous equation $Au = 0$ only has the trivial solution $u = 0$. In this case, the solution of equation* (9) *is unique.*

Proof. The proof of the theorem is based on lemma 1. □

The theorem can be formulated in another way: equation (9) has a unique solution for any $f \in H$ if and only if $\ker A = 0$ (see Section 5.2).

If $\ker A \neq 0$, then the equation is only soluble if further conditions are imposed on f. Recall that, by lemma 3, the space H is the direct sum of orthogonal subspaces: $H = \ker A \oplus \text{im } A^*$, $H = \ker A^* \oplus \text{im } A$.

Theorem 7. *Equation* (9) *is soluble if and only if the right-hand side f is orthogonal to the subspace $\ker A^*$. In this case, the solution is not unique and is only defined up to an arbitrary element of $\ker A$:*

$$u = \tilde{u} + \bar{u}, \quad \tilde{u} \in \ker A, \quad A\bar{u} = f, \quad \bar{u} \in \text{im} A^*. \ \square$$

Let f be orthogonal to $\ker A^*$. The solution of (9) having minimal norm will be called the *normal solution of equation* (9).

Lemma 11. *The normal solution is unique and belongs to the subspace* im A^* *(i.e. it is orthogonal to* $\ker A$*).*

Proof. Let $u = \tilde{u} + \bar{u}$, $\tilde{u} \in \ker A$, $\bar{u} \in \text{im } A^*$. Then $\| u \| = (u, u) = \| \tilde{u} \|^2 + \| \bar{u} \|^2 \geq \| \bar{u} \|^2$, since $\tilde{u}$ is an arbitrary element of the subspace $\ker A$. Consequently, $\| u \|$ will be minimal if $u = \bar{u} \in \text{im } A^*$. □

Suppose that f is not orthogonal to the subspace $\ker A^*$. Then the solution of equation (9) does not exist in the classical sense. Suppose

$$f = \tilde{f} + \bar{f}, \quad \tilde{f} \in \ker A^*, \quad \bar{f} \in \operatorname{im} A.$$

An element $u \in H$ for which $Au = \bar{f}$ is called a *generalized solution of equation* (9); a generalized solution minimizes the functional $\| Au - f \|$. In fact, since $(Au - \bar{f}) \in \operatorname{im} A$ for any $u \in H$,

$$\| Au - f \|^2 = \| Au - \bar{f} \|^2 + \| \tilde{f} \|^2 \geq \| \tilde{f} \|^2,$$

where equality is achieved if u is a generalized solution.

A generalized solution is determined up to an arbitrary element from the subspace $\ker A$. We will call the generalized solution of equation (9) having minimal norm the generalized normal solution. The normal solution is unique and belongs to $\operatorname{im} A^*$.

Clearly, the concept of normal solution introduced here is completely consistent with the one given above. Notice that, if the classical normal solution exists, then it coincides with the generalized normal solution.

We look now at equation (9) with an arbitrary non-linear operator A acting in the Hilbert space H. In this case it is necessary to use Banach's *contractive mapping principle* to prove the existence and uniqueness of a solution to equation (9).

Theorem 8. *Suppose that, in the Hilbert space H, we are given an operator B mapping the closed set T of the space H into itself. In addition, suppose that the operator B is uniformly contractive, i.e. it satisfies the Lipschitz condition*

$$\| Bx - By \| \leq q \| x - y \|, \qquad x, y \in T,$$

where $q < 1$ and does not depend on x and y. Then there exists one and only one point $x_* \in T$ such that $x_* = Bx_*$. □

The point x_* is called a *fixed point of the operator B*.

Corollary 1. *If the operator B has a Gateaux derivative in H which satisfies the condition $\| B'(x) \| \leq q < 1$ for any $x \in H$, then the equation $x = Bx$ has a unique solution in H.*

Corollary 2. *Suppose that the operator C maps the closed set T into itself and commutes with the operator B satisfying the contractive mapping condition. Then a fixed point of the operator B is a fixed point (possibly not unique) of the operator C. In particular, if some power B^n of the operator B is a contractive mapping, then a fixed point of the operator B^n is also a fixed point (unique) of the operator B.*

We turn now to the solution of equation (9) with a non-linear operator A. We have

Theorem 9. *Suppose that the operator A has a Gateaux derivative $A'(x)$ at each point $x \in H$ and that there exists a $\tau \neq 0$ such that $\| E - \tau A'(x) \| \leq q < 1$ for all $x \in H$. Then equation (9) has a unique solution in H.* □

Proof. Equation (9) can be written in the following form:

$$u = u - \tau A u + \tau f, \qquad \tau \neq 0. \tag{10}$$

We define the operator B: $Bx = x - \tau Ax + \tau f$. Clearly, the operator B has a Gateaux derivative equal to $B'(x) = E - \tau A'(x)$. From the conditions of the theorem we have that $\| B'(x) \| \leq q < 1$ for any $x \in H$. Therefore from corollary 1 of theorem 8 it follows that there exists a unique solution to equation (10) and consequently to equation (9). The theorem is proved. □

In Chapter 6 we will look at several ways of obtaining estimates for the norm of linear operations of the form $E - \tau C$ where τ is a scalar.

The contractive mapping principle does not exhaust all cases where the solution of the non-linear equation exists. To prove the solubility of the operator equation (9) it is also possible to use one of the variants of the fixed point theorem — *the Browder principle.*

Theorem 10. *Suppose that, in the finite-dimensional Hilbert space H, the continuous, monotonic (strictly monotonic) operator B satisfies the condition*

$$(Bx, x) \geq 0 \quad \text{for} \quad \| x \| = \rho > 0.$$

Then the equation $Bx = 0$ has at least one (unique) solution in the sphere $\| x \| \leq \rho$. □

We will use this theorem to formulate a condition which we can use to check if the operator equation (9) has a unique solution for any right-hand side f.

Theorem 11. *Suppose that we are given equation (9) in the finite-dimensional Hilbert space H with a continuous and strongly monotonic operator A,*

$$(Ax - Ay, x - y) \geq \delta \| x - y \|^2, \quad \delta > 0, \quad x, y \in H.$$

Then equation (9) has a unique solution in the sphere $\| u \| \leq \frac{1}{\delta} \| A0 - f \|$.

Proof. We write equation (4) in the following form:

$$Bu = Au - f = 0.$$

Clearly, the operator B is continuous and strongly monotonic. Using the condition of the theorem and the Cauchy-Schwartz-Bunyakovskij inequality, we obtain

$$\begin{aligned}(Bx, x) &= (Ax - f, x) = (Ax - A0, x - 0) - (f - A0, x) \\ &\geq \delta \parallel x \parallel^2 - \parallel f - A0 \parallel \parallel x \parallel = (\delta \parallel x \parallel - \parallel A0 - f \parallel) \parallel x \parallel .\end{aligned}$$

From this it follows that the operator B satisfies the condition $(Bx, x) \geq 0$ on the sphere $\parallel x \parallel = \frac{1}{\delta} \parallel A0 - f \parallel$. Therefore by theorem 10, the equation $Bu = 0$ (and together with it equation (9)) has a unique solution in this sphere. Theorem 11 is proved. □

Corollary 1. *If the operator A has a Gateaux derivative in H and is positive definite in H, then the conditions of theorem* 11 *are satisfied.*

In fact, since any linear operator is bounded in a finite-dimensional space, the Gateaux derivative is a continuous, bounded, and positive-definite operator in H. From theorem 2 it follows that A is a strongly-monotonic operator. In addition, from the boundedness of the Gateaux derivative it follows that the operator A satisfies a Lipschitz condition and is therefore continuous.

5.2 Difference schemes as operator equations

5.2.1 Examples of grid-function spaces. In Section 5.1.1, the basic concepts from the theory of difference schemes were introduced: grids, grid equations, grid functions, difference derivatives, etc. The theory formulates the general principles and laws for constructing difference schemes with a given property. A characteristic feature of this theory is the possibility of associating with each differential equation a whole class of difference schemes with the necessary properties. To construct a general theory, it is natural to get rid of the concrete structure and explicit form of difference equations. This leads us to define difference schemes as operator equations with operators acting in some functional space, namely in the space of grid functions.

The *space of grid functions* is taken to mean the set of functions defined on a certain grid. Since every grid function can be put in correspondence with a vector whose coordinates are the value of the grid function at the nodes of the grid, the operations of addition of functions and multiplication of a function by a scalar are defined as for vectors.

In the space of grid function, it is possible to introduce an inner product for functions, turning it into a Hilbert space. Various grid-function spaces can be distinguished from each other by the choice of the grid and norm. We give the following examples.

Example 1. Suppose we have defined the uniform grid $\bar{\omega} = \{x_i = ih, \ 0 \leq i \leq N, \ hN = l\}$ with step h on the interval $0 \leq x \leq l$. We denote by ω, ω^+, and ω^- the following parts of the grid $\bar{\omega}$:

$$\begin{aligned} \omega &= \{x_i \in \bar{\omega}, \quad 1 \leq i \leq N-1\}, \\ \omega^+ &= \{x_i \in \bar{\omega}, \quad 1 \leq i \leq N\}, \\ \omega^- &= \{x_i \in \bar{\omega}, \quad 0 \leq i \leq N-1\}. \end{aligned}$$

On the set H of real-valued grid functions defined on $\bar{\omega}$, we define an inner product and norm in the following way:

$$\begin{gathered} (u, v) = (u, v)_{\bar{\omega}} = \sum_{i=1}^{N-1} u_i v_i h + 0.5h(u_0 v_0 + u_N v_N), \\ \| u \| = \sqrt{(u, u)}, \quad u_i = u(x_i), \quad v_i = v(x_i). \end{gathered} \tag{1}$$

If u_i and v_i are considered as values of the functions $u(x)$ and $v(x)$ of the continuous argument $x \in [0, l]$ on the grid $\bar{\omega}$, then the inner product (1) is the trapezoid quadrature rule for the integral $\int_0^1 u(x)v(x)dx$. If the grid functions are defined in ω, ω^+, and ω^-, then the inner product of real-valued grid functions is correspondingly defined by the formulas

$$\begin{aligned} (u, v) &= \sum_{i=1}^{N-1} u_i v_i h, \quad u, v \in H(\omega), \\ (u, v) &= \sum_{i=1}^{N-1} u_i v_i h + 0.5h u_N v_N, \quad u, v \in H(\omega^+), \\ (u, v) &= \sum_{i=1}^{N-1} u_i v_i h + 0.5h u_0 v_0, \quad u, v \in H(\omega^-). \end{aligned}$$

It is easy to verify that these inner products satisfy all the axioms of an inner product, and therefore the spaces constructed are Hilbert spaces.

Example 2. Suppose now that we have introduced the arbitrary non-uniform grid

$$\bar{\omega} = \{x_i \in [0, l], \quad x_i = x_{i-1} + h_i, \quad 1 \le i \le N, \quad x_0 = 0, \quad x_N = l\}. \quad (2)$$

on the interval $0 \le x \le l$. Recall the definition of the average step $\hbar_i$ at the node x_i:

$$\hbar_i = 0.5(h_i + h_{i+1}), \quad 1 \le i \le N-1, \quad \hbar_0 = 0.5h_1, \quad \hbar_N = 0.5h_N. \quad (3)$$

Notice that the uniform grid is a special case of the non-uniform grid (2) with $h_i \equiv h$. Here we have $\hbar_i = h$, $1 \le i \le N-1$, $\hbar_0 = \hbar_N = 0.5h$.

As above, we denote by ω, ω^+, and ω^- the corresponding parts of the grid $\bar{\omega}$. By analogy with example 1, we define the inner products in real spaces of grid functions defined on the indicated grids by the formulas

$$(u, v) = \sum_{i=0}^{N} u_i v_i \hbar_i, \quad u, v \in H(\bar{\omega}), \quad (4)$$

$$(u, v) = \sum_{i=1}^{N-1} u_i v_i \hbar_i, \quad u, v \in H(\omega), \quad (5)$$

$$(u, v) = \sum_{i=1}^{N} u_i v_i \hbar_i, \quad u, v \in H(\omega^+),$$

$$(u, v) = \sum_{i=0}^{N-1} u_i v_i \hbar_i, \quad u, v \in H(\omega^-).$$

These spaces of grid functions are *Hilbert spaces* and have a *finite dimension* equal to the number of nodes of the corresponding grid.

It is convenient to write these inner products in the form

$$(u, v) = \sum_{x_i \in \Omega} u(x_i) v(x_i) \hbar(x_i), \qquad u, v \in H(\Omega),$$

where Ω can be taken to be $\bar{\omega}$, ω, ω^+, or ω^-. In addition to the indicated inner products, one often encounters sums of the form

$$(u, v)_{\omega^+} = \sum_{i=1}^{N} u_i v_i h_i, \qquad (u, v)_{\omega^-} = \sum_{i=0}^{N-1} u_i v_i h_{i+1}, \quad (6)$$

which can be used as inner products in the spaces $H(\omega^+)$ and $H(\omega^-)$. Clearly, the inner product (4) in the space $H(\bar{\omega})$ satisfies the identity

$$(u,v) = 0.5[(u,v)_{\omega^+} + (u,v)_{\omega^-}], \qquad u,v \in H(\bar{\omega}).$$

Example 3. Suppose that we have introduced an arbitrary non-uniform rectangular grid $\bar{\omega} = \bar{\omega}_1 \times \bar{\omega}_2$ into the rectangle $\bar{G} = \{0 \le x_\alpha \le l_\alpha,\ \alpha = 1,2\}$, where

$$\bar{\omega}_\alpha = \{x_\alpha(i_\alpha) \in [0, l_\alpha], \quad x_\alpha(i_\alpha) = x_\alpha(i_\alpha - 1) + h_\alpha(i_\alpha), \quad 1 \le i_\alpha \le N_\alpha,$$
$$x_\alpha(0) = 0, \quad x_\alpha(N_\alpha) = l_\alpha\}, \quad \alpha = 1,2.$$

Let $\hbar_\alpha(i_\alpha)$, $0 \le i_\alpha \le N_\alpha$, be the average step at the node $x_\alpha(i_\alpha)$ in the direction x_α:

$$\hbar_\alpha(i_\alpha) = 0.5[h_\alpha(i_\alpha) + h_\alpha(i_\alpha + 1)], \qquad 1 \le i_\alpha \le N_\alpha - 1,$$
$$\hbar_\alpha(0) = 0.5 h_\alpha(1), \qquad \hbar_\alpha(N_\alpha) = 0.5 h_\alpha(N_\alpha), \qquad \alpha = 1,2.$$

In the space $H(\Omega)$ of grid functions defined on Ω, where Ω is any part of the grid $\bar{\omega}$, we define the inner product by the formula

$$(u,v) = \sum_{x_i \in \Omega} u(x_i) v(x_i) \hbar_1 \hbar_2, \qquad x_i = (x_1(i_1), x_2(i_2)).$$

In particular, if the grid is uniform in any direction $h_\alpha(i_\alpha) \equiv h_\alpha$, $\alpha = 1,2$, and the grid functions are defined in ω (at the interior nodes of the grid $\bar{\omega}$), then the inner product can be written in the form

$$(u,v) = \sum_{i_1=1}^{N_1-1} \sum_{i_2=1}^{N_2-1} u(i_1,i_2) v(i_1,i_2) h_1 h_2, \qquad u,v \in H(\omega).$$

We limit ourselves here to these examples; other more complex examples will be considered in succeeding chapters when studying concrete difference problems.

5.2.2 Several difference identities. We move on now to consequences of the basic formulas, which we will use to transform expressions containing grid functions. We state these formulas in the case when the grid functions are given on the non-uniform grid defined in (2).

Recall the definition of the basic difference derivatives of a grid function:

$$y_{\bar{x},i} = \frac{y_i - y_{i-1}}{h_i}, \; y_{x,i} = y_{\bar{x},i+1} = \frac{y_{i+1} - y_i}{h_{i+1}}, \; y_{\check{x},i} = \frac{y_i - y_{i-1}}{\hbar_i},$$
$$y_{\hat{x},i} = \frac{y_{i+1} - y_i}{\hbar_i}, \; y_{\bar{x}\hat{x},i} = y_{x\check{x},i} = \frac{1}{\hbar_i}(y_{x,i} - y_{\bar{x},i}).$$

In Section 1.1.2, we obtained two formulas for summation by parts:

$$\sum_{i=m+1}^{n-1} u_{\hat{x},i} v_i \hbar_i = -\sum_{i=m+1}^{n} u_i v_{\bar{x},i} h_i + u_n v_n - u_{m+1} v_m, \tag{7}$$
$$\sum_{i=m+1}^{n-1} u_{\bar{x},i} v_i h_i = -\sum_{i=m}^{n-1} u_i v_{\hat{x},i} \hbar_i + u_{n-1} v_n - u_m v_m. \tag{8}$$

Substituting in these formulas the relations

$$h_i u_{\bar{x},i} = \hbar_i u_{\check{x},i}, \quad \hbar_i u_{\hat{x},i} = h_{i+1} u_{x,i},$$

we obtain, after a simple transformation, the formulas

$$\sum_{i=m+1}^{n-1} u_{\check{x},i} v_i \hbar_i = -\sum_{i=m}^{n-1} u_i v_{x,i} h_{i+1} + u_{n-1} v_n - u_m v_m, \tag{9}$$
$$\sum_{i=m+1}^{n-1} u_{x,i} v_i h_{i+1} = -\sum_{i=m+1}^{n} u_i v_{\check{x},i} \hbar_i + u_n v_n - u_{m+1} v_m, \tag{10}$$
$$\sum_{i=m+1}^{n} u_{\bar{x},i} v_i h_i = -\sum_{i=m}^{n-1} u_i v_{x,i} h_{i+1} + u_n v_n - u_m v_m. \tag{11}$$

We set $m = 0$ and $n = N$ in formulas (7), (9), (11), take into account definition (5) of the inner product in $H(\omega)$, and also the notation in (6). We obtain the identities

$$(u_x, v) = -(u, v_{\bar{x}})_{\omega^+} + u_N v_N - u_1 v_0, \tag{7'}$$
$$(u_{\check{x}}, v) = -(u, v_x)_{\omega^-} + u_{N-1} v_N - u_0 v_0, \tag{9'}$$
$$(u_{\bar{x}}, v)_{\omega^+} = -(u, v_x)_{\omega^-} + u_N v_N - u_0 v_0 \tag{11'}$$

for grid functions u_i and v_i defined on the grid $\bar{\omega}$. If we set $u_i = a_i y_{\bar{x},i}$ for $1 \le i \le N$ in (7′), then we obtain Green's first difference formula

$$((ay_{\bar{x}})_{\hat{x}}, v) = -(ay_{\bar{x}}, v_{\bar{x}})_{\omega^+} + a_N y_{\bar{x},N} v_N - a_1 y_{x,0} v_0. \qquad (12)$$

Analogously, setting $u_i = a_i y_{x,i}$ for $0 \le i \le N-1$ in (9′), we obtain

$$((ay_x)_{\hat{x}}, v) = -(ay_x, v_x)_{\omega^-} + a_{N-1} y_{\bar{x},N} v_N - a_0 y_{x,0} v_0.$$

If we subtract

$$(y, (av_{\bar{x}})_{\hat{x}}) = -(ay_{\bar{x}}, v_{\bar{x}})_{\omega^+} + a_N v_{\bar{x},N} y_N - a_1 v_{x,0} y_0,$$

from (12), then we obtain Green's second difference formula

$$((ay_{\bar{x}})_{\hat{x}}, v) - (y, (av_{\bar{x}})_{\hat{x}}) = a_N (y_{\bar{x}} v - v_{\bar{x}} y)_N - a_1 (y_x v - v_x y)_0. \qquad (13)$$

Notice that, for functions y_i and v_i which reduce to zero for $i = 0$ and $i = N$ ($y_0 = y_N = 0,\ v_0 = v_N = 0$), formula (12) has the form

$$((ay_{\bar{x}})_{\hat{x}}, v) = -(ay_{\bar{x}}, v_{\bar{x}})_{\omega^+},$$

and Green's second formula (13) has the form

$$((ay_{\bar{x}})_{\hat{x}}, v) = (y, (av_{\bar{x}})_{\hat{x}}).$$

In the general case of arbitrary grid functions defined on $\bar{\omega}$, the formulas (12) and (13) can be written in the form

$$(\Lambda y, v) = -(ay_{\bar{x}}, v_{\bar{x}})_{\omega^+}, \quad (\Lambda y, v) - (y, \Lambda v) = 0, \qquad (14)$$

where the difference operator Λ mapping $H(\bar{\omega})$ onto $H(\bar{\omega})$ is defined as follows:

$$\Lambda y_i = \begin{cases} \dfrac{1}{\hbar_0} a_1 y_{x,0}, & i = 0, \\ (ay_{\bar{x}})_{\hat{x},i}, & 1 \le i \le N-1, \\ -\dfrac{1}{\hbar_N} a_N y_{\bar{x},N}, & i = N. \end{cases}$$

Here the inner product in $H(\bar{\omega})$ is given by formula (4). Notice that (14) expresses the self-adjointness of the operator Λ in the space $H(\bar{\omega})$.

We examined the case when the grid functions take on real values on the grid. If they take on complex values on the grid $\bar{\omega}$, then we obtain a complex Hilbert space $H(\bar{\omega})$ with the inner product

$$(u, v) = \sum_{i=0}^{N} u_i \bar{v}_i \hbar_i, \qquad u, v \in H(\bar{\omega}), \tag{15}$$

where $\bar{v}_i$ is a scalar, the complex conjugate of v_i. Analogously, we define the inner product in $H(\omega)$

$$(u, v) = \sum_{i=1}^{N-1} u_i \bar{v}_i \hbar_i, \qquad u, v \in H(\omega), \tag{16}$$

and also in $H(\omega^+)$ and $H(\omega^-)$. Here the formulas for summation by parts (7′), (9′), and (11′) take the form

$$\begin{aligned}
(u_{\hat{x}}, v) &= -(u, v_{\bar{x}})_{\omega^+} + u_N \bar{v}_N - u_1 \bar{v}_0, \\
(u_{\check{x}}, v) &= -(u, v_x)_{\omega^-} + u_{N-1} \bar{v}_N - u_0 \bar{v}_0, \\
(u_{\bar{x}}, v) &= -(u, v_x)_{\omega^-} + u_N \bar{v}_N - u_0 \bar{v}_0,
\end{aligned}$$

and the Green difference formulas take the form:

$$\begin{aligned}
((ay_{\bar{x}})_{\hat{x}}, v) &= -(ay_{\bar{x}}, v_{\bar{x}})_{\omega^+} + a_N y_{\bar{x},N} \bar{v}_N - a_1 y_{x,0} \bar{v}_0, \\
((ay_{\bar{x}})_{\hat{x}}, v) - (y, (av_{\bar{x}})_{\hat{x}}) &= ((\bar{a} - a) y_{\bar{x}}, v_{\bar{x}})_{\omega^+} \\
&\quad + (a y_{\bar{x}} \bar{v} - \bar{a} y \bar{v}_{\bar{x}})_N - (a_1 y_{x,0} \bar{v}_0 - \bar{a}_1 y_0 \bar{v}_{x,0}).
\end{aligned}$$

Here the notation in (16) is used.

Using the operator Λ defined above and the notation in (15) for the inner product in $H(\omega)$, Green's second difference formula can be written in the form

$$(\Lambda y, v) - (y, \Lambda v) = ((\bar{a} - a) y_{\bar{x}}, v_{\bar{x}})_{\omega^+}.$$

From this it follows that, in a complex Hilbert space $H(\bar{\omega})$, the operator Λ is self-adjoint if all the a_i are real.

Relations analogous to Green's first and second difference formulas (12), (13) can also be obtained for the difference operator $(ay_{\bar{x}\hat{x}})_{\bar{x}\hat{x}}$. For example, we state here the analog to formula (12)

$$\begin{aligned}
\sum_{i=2}^{N-2} (ay_{\bar{x}\hat{x}})_{\bar{x}\hat{x},i} v_i \hbar_i &= \sum_{i=1}^{N-1} a_i y_{\bar{x}\hat{x},i} v_{\bar{x}\hat{x},i} \hbar_i \\
&\quad + [(ay_{\bar{x}\hat{x}})_{\bar{x}} v - a y_{\bar{x}\hat{x}} v_x]_{N-1} - [(ay_{\bar{x}\hat{x}})_x v - a y_{\bar{x}\hat{x}} v_{\bar{x}}]_1.
\end{aligned}$$

5.2.3 Bounds for the simplest difference operators. To study the properties of difference operators, we require inequalities which give estimates for operator bounds and for the energy equivalence constants for two operators acting in the space H of grid functions.

We first look at difference operators defined on the set of grid functions of one argument on the uniform grid $\bar{\omega} = \{x_i = ih \in [0, l],\ 0 \le i \le N,\ hN = l\}$. Below we will use the notation

$$(u, v) = \sum_{i=1}^{N-1} u_i v_i h + 0.5h(u_0 v_0 + u_N v_N), \quad (u, v)_{\omega^+} = \sum_{i=1}^{N} u_i v_i h.$$

We have

Lemma 12. *Any function $y_i = y(x_i)$ defined on the uniform grid $\bar{\omega}$ and reducing to zero for $i = 0$ and $i = N$ satisfies the inequalities*

$$\gamma_1(y, y) \le (y_{\bar{x}}^2, 1)_{\omega^+} \le \gamma_2(y, y), \tag{17}$$

where

$$\gamma_1 = \frac{4}{h^2}\sin^2\frac{\pi}{2N} \ge \frac{8}{l^2}, \quad \gamma_2 = \frac{4}{h^2}\cos^2\frac{\pi}{2N} < \frac{4}{h^2}.$$

Proof. In fact, let $\mu_k(i)$ be an orthonormal eigenfunction for the problem

$$(\mu_k)_{\bar{x}x} + \lambda_k \mu_k = 0, \qquad 1 \le i \le N-1,$$
$$\mu_k(0) = \mu_k(N) = 0. \tag{18}$$

In Section 1.5.1 it was noted that a grid function y_i satisfying the conditions of the lemma can be represented in the form of a sum

$$y_i = \sum_{k=1}^{N-1} c_k \mu_k(i), \qquad c_k = (y, \mu_k). \tag{19}$$

From (18) and (19) we find

$$y_{\bar{x}x,i} = \sum_{k=1}^{N-1} c_k (\mu_k)_{\bar{x}x,i} = -\sum_{k=1}^{N-1} \lambda_k c_k \mu_k(i), \quad 1 \le i \le N-1.$$

Using the orthonormality of the eigenfunctions μ_k we obtain

$$(y, y) = \sum_{k=1}^{N-1} c_k^2, \qquad -(y_{\bar{x}x}, y) = \sum_{k=1}^{N-1} \lambda_k c_k^2. \tag{20}$$

By Green's first difference formula (12), we have

$$-(y_{\bar{x}x}, y) = \left(y_{\bar{x}}^2, 1\right)_{\omega^+}. \tag{21}$$

The eigenvalues λ_k of problem (18) were found in Section 1.5.1:

$$\lambda_k = \frac{4}{h^2} \sin^2 \frac{k\pi h}{2l} = \frac{4}{h^2} \sin^2 \frac{k\pi}{2N}, \quad 1 \le k \le N-1,$$

where

$$\gamma_1 = \min_k \lambda_k = \lambda_1 = \frac{4}{h^2} \sin^2 \frac{\pi}{2N},$$
$$\gamma_2 = \max_k \lambda_k = \lambda_{N-1} = \frac{4}{h^2} \cos^2 \frac{\pi}{2N}.$$

From this and from (20), (21) follow the estimates (17) of lemma 12. □

Remark 1. The estimates (17) are precise in the sense that they become equalities if y_i is taken to be $\mu_1(i)$ and $\mu_{N-1}(i)$. Notice that $\gamma_1 = 8/l^2$ if $h = l/2$, i.e. for $N = 2$. For $N = 4$ we have $\gamma_1 = 32/(l^2(2+\sqrt{2})) > 8/l^2$.

Remark 2. If y_i reduces to zero only for $i = 0$ or for $i = N$, then in (17) we have

$$\gamma_1 = \frac{4}{h^2} \sin^2 \frac{\pi}{4N} \ge \frac{8}{l^2(2+\sqrt{2})}, \quad \gamma_2 = \frac{4}{h^2} \cos^2 \frac{\pi}{4N} < \frac{4}{h^2}.$$

If y_i is an arbitrary grid function on $\bar{\omega}$, then in (17) we have $\gamma_1 = 0$ and $\gamma_2 = 4/h^2$. To prove these assertions, we consider in place of problem (18) the corresponding eigenvalue problem studied in Section 5.1.5.

The inequalities (17) can be written in the form

$$\gamma_1(y, y) \le (-\Lambda y, y) \le \gamma_2(y, y), \tag{22}$$

if we introduce the difference operator Λ defined by the formula $\Lambda y_i = y_{\bar{x}x,i}$, $1 \le i \le N-1$ for functions y_i satisfying the conditions $y_0 = y_N = 0$. If the grid function y_i only reduces to zero on one end of the grid $\bar{\omega}$, then the operator Λ is defined using the formulas

$$\Lambda y_i = \begin{cases} y_{\bar{x}x,i}, & 1 \le i \le N-1, \\ -\dfrac{2}{h} y_{\bar{x},i}, & i = N, \quad \text{if} \quad y_0 = 0, \end{cases} \tag{23}$$

or

$$\Lambda y_i = \begin{cases} \dfrac{2}{h} y_{x,i}, & i = 0, \\ y_{\bar{x}x,i}, & 1 \le i \le N-1, \quad \text{if} \quad y_N = 0. \end{cases}$$

Taking into account that $(-\Lambda y, y) = (y_{\bar{x}}^2, 1)_{\omega^+}$ follows from Green's first difference formulas in each of these cases, we obtain the inequalities (22) where γ_1 and γ_2 are as indicated in remark 2, and y_i reduces to zero at the corresponding end of the grid $\bar{\omega}$.

If y_i is an arbitrary grid function, then the operator Λ is defined as:

$$\Lambda y_i = \begin{cases} \dfrac{2}{h} y_{x,0}, & i = 0, \\ y_{\bar{x}x,i}, & 1 \le i \le N-1, \\ -\dfrac{2}{h} y_{\bar{x},N}, & i = N. \end{cases}$$

In this case the inequalities (22) are also valid, and

$$(-\Lambda y, y) = -(y_{\bar{x}x}, y) + y_{\bar{x},N} y_N - y_{x,0} y_0 = (y_{\bar{x}}^2, 1)_{\omega^+}.$$

The constants γ_1 and γ_2 are as indicated in remark 2.

Thus, we have found bounds for the simplest difference operators. We shall show here that all of the operators Λ introduced in this section satisfy the inequality

$$|(-\Lambda u, v)| \le (-\Lambda u, u)^{1/2} (-\Lambda v, v)^{1/2}. \tag{24}$$

The idea behind (24) will be illustrated using as an example the operator $\Lambda y = y_{\bar{x}x}$. We introduce the space $H(\omega)$ of grid functions defined on ω with inner product

$$(u, v) = \sum_{i=1}^{N-1} u_i v_i h, \qquad u, v \in H(\omega).$$

The difference operator Λ in the space $H(\omega)$ corresponds to the linear operator A defined by

$$A y_i = -\Lambda \mathring{y}_i \qquad 1 \le i \le N-1,$$

where $y \in H(\omega)$, $y_i = \mathring{y}_i$ for $1 \le i \le N-1$ and $\mathring{y}_0 = \mathring{y}_N = 0$. The operator A maps $H(\omega)$ onto $H(\omega)$.

By $(u,v) = (\mathring{u}, \mathring{v})$ we have $(Au,v) = -(\Lambda\mathring{u}, \mathring{v})$, where $\mathring{u}_0 = \mathring{u}_N = 0$, $\mathring{v}_0 = \mathring{v}_N = 0$. From (22) it follows that $(Au,u) \geq \gamma_1(u,u)$, $\gamma_1 > 0$. Thus, the operator A is positive definite in $H(\omega)$.

We shall show that it is self-adjoint in $H(\omega)$. In fact, from Green's second difference formulas (13) we have

$$(Au,v) = -(\Lambda\mathring{u}, \mathring{v}) = -(\mathring{u}_{\bar{x}x}, \mathring{v}) = -(\mathring{u}, \mathring{v}_{\bar{x}x}) = (u, Av).$$

Since the Cauchy-Schwartz-Bunyakovskij inequality

$$|(Au,v)| \leq (Au,u)^{1/2}(Av,v)^{1/2}$$

is valid for non-negative self-adjoint operators, we obtain

$$|(-\Lambda\mathring{u}, \mathring{v})| \leq (-\Lambda\mathring{u}, \mathring{u})^{1/2}(-\Lambda\mathring{v}, \mathring{v})^{1/2},$$

which is what we were required to prove.

5.2.4 Lower bounds for certain difference operators. In lemma 12 we actually found the energy equivalence constants for the identity operator E and the operator A which corresponds to the difference operator $-\Lambda y = -y_{\bar{x}x}$ for functions which reduce to zero on the ends of the grid $\bar{\omega}$, i.e. we found γ_1 and γ_2 satisfying the inequalities $\gamma_1 E \leq A \leq \gamma_2 E$.

We now find an inequality which relates the operators A and D where $Dy_i = \rho_i y_i$, $1 \leq i \leq N-1$ and $\rho_i \geq 0$. To do this, we must determine Green's difference function for the operator Λ.

Suppose that it is necessary to find the solution of the difference problem

$$\begin{aligned} &\Lambda v_i = v_{\bar{x}x,i} = -f_i, \qquad 1 \leq i \leq N-1, \\ &v_0 = v_N = 0, \end{aligned} \tag{25}$$

on the grid $\bar{\omega}$ defined above. The grid function G_{ik}, which satisfies the conditions

$$\begin{aligned} &\Lambda G_{ik} = G_{\bar{x}x,ik} = -\frac{1}{h}\delta_{ik}, \qquad 1 \leq i \leq N-1, \\ &G_{0k} = G_{Nk} = 0, \end{aligned}$$

for fixed $k = 1, 2, \ldots, N-1$, and where δ_{ik} is the Kronecker delta:

$$\delta_{ik} = \begin{cases} 1, & i = k, \\ 0, & i \neq k, \end{cases}$$

is called *Green's function for the difference operator* Λ.

We list here the basic properties of Green's functions:

1) The Green function is symmetric, $G_{ik} = G_{ki}$, and in addition G_{ik} considered as a function of k for fixed i satisfies the conditions

$$\Lambda G_{ik} = G_{\bar{x}x,ik} = -\frac{1}{h}\delta_{ik}, \qquad 1 \le k \le N-1,$$
$$G_{i0} = G_{iN} = 0.$$

2) the Green function is positive, $G_{ik} > 0$ for $i, k \neq 0, N$.

3) for any grid function y_i satisfying the condition $y_0 = y_N = 0$, we have the representation

$$y_i = -\sum_{k=1}^{N-1} G_{ik}\,\Lambda y_k h, \tag{26}$$

and the solution of problem (25) can be represented in the form

$$v_i = \sum_{k=1}^{N-1} G_{ik} f_k h, \qquad 0 \le i \le N.$$

This assertion is proved using the second Green difference formula (13) and property 1).

Lemma 13. *Suppose $\rho_i \ge 0$ is a grid function defined on ω and not identically zero. For any grid function y_i defined on $\bar{\omega}$ and satisfying the conditions $y_0 = y_N = 0$ we have the estimate*

$$\gamma_1(\rho y, y) \le (y_{\bar{x}}^2, 1)_{\omega^+}, \tag{27}$$

where $1/\gamma_1 = \max_{1 \le i \le N-1} v_i$, and v_i is the solution of the boundary-value problem

$$\begin{aligned} \Lambda v_i &= v_{\bar{x}x,i} = -\rho_i, \qquad 1 \le i \le N-1, \\ v_0 &= v_N = 0. \end{aligned} \tag{28}$$

Proof. Assume that $y_0 = y_N = 0$. Using (26) we obtain

$$\begin{aligned}(\rho y, y) &= \sum_{i=1}^{N-1} \rho_i y_i^2 h = -\sum_{i=1}^{N-1} \rho_i y_i h \left(\sum_{k=1}^{N-1} G_{ik}\Lambda y_k h\right) \\ &= -\sum_{k=1}^{N-1} h\Lambda y_k \left(\sum_{i=1}^{N-1} \rho_i y_i G_{ik} h\right) = -(\Lambda y, w),\end{aligned}$$

where we have denoted

$$w_k = \sum_{i=1}^{N-1} \rho_i y_i G_{ik} h, \qquad 0 \le k \le N.$$

Applying the inequality (24), from this we find

$$(\rho y, y) \le (-\Lambda y, y)^{1/2} (-\Lambda w, w)^{1/2}$$

or by (21)

$$(\rho y, y)^2 \le (y_{\bar{x}}^2, 1)_{\omega^+} (-\Lambda w, w). \tag{29}$$

We now use property 1) of the Green function G_{ik}. We obtain

$$-\Lambda w_k = -\sum_{i=1}^{N-1} h \rho_i y_i \Lambda G_{ik} = \sum_{i=1}^{N-1} \rho_i y_i \delta_{ik} = \rho_k y_k$$

and consequently

$$(-\Lambda w, w) = \sum_{k=1}^{N-1} h \rho_k y_k \left(\sum_{i=1}^{N-1} h \rho_i y_i G_{ik} \right) = \sum_{i=1}^{N-1} \sum_{k=1}^{N-1} a_{ik} y_i y_k,$$

where we have denoted $a_{ik} = h^2 \rho_i \rho_k G_{ik}$, $1 \le i,\ k \le N-1$. Using the inequality $2 y_i y_k \le y_i^2 + y_k^2$, and also the symmetry and positivity of the Green function G_{ik}, from this we find

$$\begin{aligned}(-\Lambda w, w) &\le \sum_{i=1}^{N-1} 0.5\, y_i^2 \sum_{k=1}^{N-1} a_{ik} + \sum_{k=1}^{N-1} 0.5\, y_k^2 \sum_{i=1}^{N-1} a_{ki} \\ &= \sum_{i=1}^{N-1} y_i^2 \sum_{k=1}^{N-1} a_{ik} = \sum_{i=1}^{N-1} \rho_i y_i^2 h \left(\sum_{k=1}^{N-1} \rho_k G_{ik} h \right).\end{aligned}$$

By property 3), the solution of problem (28) can be written in the form

$$v_i = \sum_{k=1}^{N-1} \rho_k G_{ik} h > 0, \qquad 1 \le i \le N-1.$$

Consequently

$$(-\Lambda w, w) = \sum_{i=1}^{N-1} \rho_i y_i^2 v_i h \le \max_{1 \le i \le N-1} v_i (\rho y, y) = \frac{1}{\gamma_1} (\rho y, y).$$

The estimate in the lemma then follows from (29). □

Remark 1. It is possible to show that the function $v_i = 0.5x_i(l - x_i)$, where $x_i = ih \in [0, l]$, is the solution of problem (28) for $\rho_i \equiv 1$. From this follows the estimate

$$\gamma_1(y, y) \le (y_{\bar{x}}^2, 1)_{\omega^+}, \quad \gamma_1 = 8/l^2, \quad y_0 = y_N = 0. \tag{30}$$

Remark 2. Lemma 13 generalizes to the case where y_i only reduces to zero on one end of the grid $\bar{\omega}$. For example, if $y_0 = 0$, then in (27) we have $1/\gamma_1 = \max\limits_{1 \le i \le N} v_i$, where v_i is the solution of the problem $\Lambda v_i = -\rho_i$, $1 \le i \le N$, $v_0 = 0$ with the difference operator Λ defined in (23).

Lemma 14. *Suppose $\rho_i \ge 0$, $d_i \ge 0$ are defined on ω, and the function $a_i \ge c_1 > 0$ is defined on ω^+. For any function y_i defined on $\bar{\omega}$ which satisfies the conditions $y_0 = y_N = 0$, we have the estimate*

$$\gamma_1(\rho y, y) \le (a y_{\bar{x}}^2, 1)_{\omega^+} + (dy, y), \quad 1/\gamma_1 = \max_{1 \le i \le N-1} v_i,$$

where v_i is the solution of the boundary-value problem

$$\Lambda v_i = (a v_{\bar{x}})_{x,i} - d_i v_i = -\rho_i, \qquad 1 \le i \le N-1, \qquad v_0 = v_N = 0. \square$$

Remark 1. If y_i only reduces to zero on one end of the grid $\bar{\omega}$, it is possible to obtain the estimate

$$\gamma_1(\rho y, y) \le (a y_{\bar{x}}^2, 1)_{\omega^+} + (dy, y) + \kappa_0 y_0^2, \tag{31}$$

where $1/\gamma_i = \max\limits_{0 \le i \le N-1} v_i$, and the function v_i is the solution of the problem

$$\Lambda v_i = -\rho_i, \quad 0 \le i \le N-1, \quad v_N = 0,$$

$$\Lambda y_i = \begin{cases} \dfrac{2}{h}(a_1 y_{x,0} - \kappa_0 y_0) - d_0 y_0, & i = 0, \\ (a y_{\bar{x}})_{x,i} - d_i y_i, \quad 1 \le i \le N-1, & \kappa_0 \ge 0. \end{cases} \tag{32}$$

Remark 2. For an arbitrary grid function y_i defined on $\bar{\omega}$, it is possible to obtain the estimate

$$\gamma_1(\rho y, y) \leq (a y_{\bar{x}}^2, 1)_{\omega^+} + (dy, y) + \kappa_0 y_0^2 + \kappa_1 y_N^2, \tag{33}$$

where $\kappa_0 \geq 0$, $\kappa_1 \geq 0$, $\kappa_0 + \kappa_1 + (d, 1) > 0$, and the grid functions $\rho_i \geq 0$, $d_i \geq 0$ are defined on $\bar{\omega}$. Here $1/\gamma_1 = \max\limits_{0 \leq i \leq N} v_i$, where v_i is the solution of the boundary-value problem

$$\Lambda v_i = -\rho_i, \qquad 0 \leq i \leq N,$$

$$\Lambda y_i = \begin{cases} \dfrac{2}{h}(a_1 y_{x,0} - \kappa_0 y_0) - d_0 y_0, & i = 0, \\ (a y_{\bar{x}})_{x,i} - d_i y_i, & 1 \leq i \leq N-1, \\ \dfrac{-2}{h}(a_N y_{\bar{x},N} + \kappa_1 y_N) - d_N y_N, & i = N. \end{cases} \tag{34}$$

The proof of lemma 14 and remarks 1 and 2 is carried out in the same way as for lemma 13. Here we use the Green function for the indicated difference operators Λ which satisfies the properties 1)–3) enumerated above.

Lemma 15. *For a grid function y_i which reduces to zero for $i = N$, we have the estimate*

$$y_0^2 \leq \tanh(\epsilon l)\left[\epsilon(y, y) + \frac{1}{\epsilon}(y_{\bar{x}}^2, 1)_{\omega^+}\right], \quad \epsilon \geq 0. \tag{35}$$

Analogously, the estimate

$$y_N^2 \leq \tanh(\epsilon l)\left[\epsilon(y, y) + \frac{1}{\epsilon}(y_{\bar{x}}^2, 1)_{\omega^+}\right], \quad \epsilon \geq 0,$$

is true for the case where $y_0 = 0$. For an arbitrary grid function y_i defined on the grid $\bar{\omega}$, we have the estimate

$$y_0^2 + y_N^2 \leq \frac{8 + \epsilon^2 l^2}{\epsilon l \sqrt{16 + \epsilon^2 l^2}}\left[\epsilon(y, y) + \frac{1}{\epsilon}(y_{\bar{x}}^2, 1)_{\omega^+}\right], \quad \epsilon > 0. \tag{36}$$

Proof. We shall first prove the validity of the estimate (35). To do this we use remark 1 to lemma 14. We substitute in (32) $a_i \equiv 1/\epsilon$, $d_i \equiv \epsilon$, $\kappa_0 = 0$ and $\rho_0 = 2/h$, $\rho_i = 0$, $1 \leq i \leq N-1$. Then from (31) we obtain the estimate

$$y_0^2 \leq \max_{0 \leq i \leq N-1} v_i \left[\epsilon(y, y) + \frac{1}{\epsilon}(y_{\bar{x}}^2, 1)_{\omega^+}\right],$$

where v_i is the solution of the following auxiliary problem:

$$\begin{aligned} \Lambda v_i &= \frac{1}{\epsilon} v_{\bar{x}x,i} - \epsilon v_i = 0, \quad 1 \le i \le N-1, \\ \Lambda v_0 &= \frac{2}{\epsilon h} v_{x,0} - \epsilon v_0 = -\frac{2}{h}, \quad v_N = 0. \end{aligned} \tag{37}$$

We write out (37) at a point

$$\begin{aligned} v_{i-1} - 2\alpha v_i + v_{i+1} &= 0, \qquad 1 \le i \le N-1, \\ v_1 - \alpha v_0 &= -\epsilon h, \qquad v_N = 0, \end{aligned} \tag{38}$$

where $\alpha = 1 + 0.5\epsilon^2 h^2 \ge 1$.

We have obtained a boundary-value problem for a second-order difference equation with constant coefficients.

Using the general theory developed in Section 1.4.1, and also the properties of the Chebyshev polynomials (see part 2 of the same section), we find that the function

$$v_i = \frac{\epsilon h U_{N-i-1}(\alpha)}{T_N(\alpha)}, \qquad 0 \le i \le N,$$

is the solution of problem (38). Here

$$T_n(\alpha) = \cosh(n \cosh^{-1}\alpha), \quad U_n(\alpha) = \frac{\sinh((n+1)\cosh^{-1}\alpha)}{\sinh(\cosh^{-1}\alpha)}, \quad |\alpha| \ge 1,$$

are the Chebyshev polynomials of degree n of the first and second kinds.

Since $\alpha \ge 1$,

$$\max_{0 \le i \le N-1} v_i = v_0 = \frac{\epsilon h U_{N-1}(\alpha)}{T_N(\alpha)}.$$

Thus, we have obtained the estimate

$$y_0^2 \le v_0 \left[\epsilon(y,y) + \frac{1}{\epsilon}(y_{\bar{x}}^2, 1)_{\omega+}\right]$$

for a grid function y_i satisfying the condition $y_N = 0$. This estimate is precise in the sense that it is an equality if y_i is taken to be the function v_i.

We now estimate v_0 from above for any h. If we denote $\cosh 2z = \alpha$, then $z \geq 0$ and

$$\begin{aligned} \epsilon h &= 2\sinh z, \quad N = l/h = \epsilon l/(2\sinh z), \\ T_N(\alpha) &= \cosh 2Nz = \cosh \omega(z), \\ U_{N-1}(\alpha) &= \frac{\sinh 2Nz}{\sinh 2z} = \frac{\sinh \omega(z)}{2\sinh z \cosh z}, \quad \omega(z) = \frac{\epsilon\, lz}{\sinh z}. \end{aligned} \tag{39}$$

Therefore

$$v_0 = \frac{\sinh \omega(z)}{\cosh z \cosh \omega(z)}$$

Since for fixed ϵ

$$\frac{d\omega}{dz} = \frac{\epsilon l(\sinh z - z\cosh z)}{\sinh^2 z} \leq 0,$$

we have

$$\frac{dv_0}{dz} = \frac{\cosh z \frac{d\omega}{dz} - \sinh z \sinh \omega \cosh \omega}{\cosh^2 z \cosh^2 \omega} \leq 0.$$

Consequently, v_0 is maximal for $z = 0$. This gives the estimate $v_0 = \tanh(\epsilon l)$. The inequality (35) is proved.

Suppose now that y_i is an arbitrary grid function. From remark 2 to lemma 14 for $\alpha_i \equiv 1/\epsilon$, $d_i \equiv \epsilon$, $\kappa_0 = \kappa_1 = 0$, $\rho_0 = \rho_N = 2/h$, $\rho_i = 0$ for $1 \leq i \leq N-1$, we obtain the estimate

$$y_0^2 + y_N^2 \leq \max_{0 \leq i \leq N} \quad v_i \left[\epsilon(y,y) + \frac{1}{\epsilon}(y_{\bar{x}}^2, 1)_{\omega^+}\right],$$

where v_i is the solution of the boundary-value problem

$$\begin{aligned} &\frac{1}{\epsilon} v_{\bar{x}x,i} - \epsilon v_i = 0, \qquad 1 \leq i \leq N-1, \\ &\frac{2}{\epsilon h} v_{x,0} - \epsilon v_0 = -\frac{2}{h}, \quad -\frac{2}{\epsilon h} v_{\bar{x},N} - \epsilon v_N = -\frac{2}{h}, \end{aligned} \tag{40}$$

The solution of problem (40) is the function

$$v_i = \frac{\epsilon h[T_{N-i}(\alpha) + T_i(\alpha)]}{(\alpha^2 - 1)U_{N-1}(\alpha)}, \qquad 0 \leq i \leq N,$$

where α is defined above.

From this we find that

$$\max_{0\le i\le N} v_i = v_0 = v_N = \frac{\epsilon h(1+T_N(\alpha))}{(\alpha^2-1)\,U_{N-1}(\alpha)}. \tag{41}$$

We shall bound this expression from above for any h. Using (39) we obtain

$$v_0 = \frac{1+\cosh\omega(z)}{\cosh z\sinh\omega(z)} = \frac{\cosh\frac{1}{2}\omega(z)}{\cosh z\sinh\frac{1}{2}\omega(z)} \le \frac{\cosh\frac{1}{2}\omega(z)}{\sinh\frac{1}{2}\,\omega(z)} = \varphi(z).$$

Since

$$\frac{d\varphi}{dz} = -\frac{1}{\sinh^2 0.5\omega}\frac{\partial\omega}{\partial z} > 0,$$

the function $\varphi(z)$ is maximal for the maximal $z = z_0$, which is found from the relation ch $2z_0 = 1+\epsilon^2 l^2/8$ ($h \le l/2$). From (39) we obtain that $w(z_0) = 4z_0$. Consequently

$$\varphi(z_0) = \frac{\cosh 2z_0}{\sinh 2z_0} = \frac{1+\epsilon^2 l^2/8}{\sqrt{\epsilon^2 l^2/8+\epsilon^4 l^4/64}} = \frac{8+\epsilon^2 l^2}{\epsilon l\sqrt{16+\epsilon^2 l^2}}.$$

We have obtained the estimate (36). □

Lemmas 13 and 14 can be generalized without any difficulty to the case of an arbitrary non-uniform grid $\bar{\omega}$. In this case the inner product is defined by (4), (6), and the difference operators Λ are changed to the corresponding operators on the non-uniform grid.

Lemma 16. *Suppose that $\rho_i \ge 0$, $d_i \ge 0$ are defined on the arbitrary non-uniform grid $\bar{\omega}$, $\rho_i \not\equiv 0$ and $a_i \ge c_1 > 0$ is defined on ω^+. Let $\kappa_0 \ge 0$, $\kappa_1 \ge 0$ be arbitrary scalars which satisfy the condition $\kappa_0+\kappa_1+(d,1) > 0$. For any grid function y_i defined on $\bar{\omega}$, inequality (33) is valid, where $1/\gamma_1 = \max\limits_{0\le i\le N} v_i$, and v_i is the solution of the problem $\Lambda v_i = -\rho_i$, $0 \le i \le N$. Here the operator Λ is defined by the formulas*

$$\Lambda y_i = \begin{cases} \dfrac{1}{\hbar_0}(a_1 y_{\bar{x},0} - \kappa_0 y_0) - d_0 y_0, & i = 0,\\ (a y_{\bar{x}})_{\hat{x},i} - d_i y_i, & 1 \le i \le N-1,\\ -\dfrac{1}{\hbar_N}(a_N y_{\bar{x},N} + \kappa_1 y_N) - d_N y_N, & i = N. \end{cases} \tag{42}$$

Proof. Lemma 16 is proved in the same way as the preceding lemmas. □

Remark 1. If $a_1 \equiv 1$, $d_1 \equiv 0$, $\rho_1 \equiv 1$, then the inequality (33) takes the form

$$\gamma_1(y,y) \le (y_{\bar{x}}^2, 1)_{\omega^+} + \kappa_0 y_0^2 + \kappa_1 y_N^2, \tag{43}$$

where

$$\gamma_1 = \frac{8(\kappa_0 + \kappa_1 + l\kappa_0\kappa_1)^2}{l(2+l\kappa_0)(2+l\kappa_1)(2\kappa_0 + 2\kappa_1 + l\kappa_0\kappa_1)}.$$

If in addition, $y_0 = y_N = 0$, then the inequality (43) is transformed into inequality (30). If y_i only reduces to zero at one end, for example for $i = N$, then, substituting $y_N = 0$ in (43) and taking the limit as $\kappa_1 \to \infty$, we obtain the estimate

$$\gamma_1(y,y) \le (y_{\bar{x}}^2, 1)_{\omega^+} + \kappa_0 y_0^2, \quad \gamma_1 = \frac{8(1+l\kappa_0)^2}{l^2(2+l\kappa_0)^2}.$$

Remark 2. From the definition (42) of the difference operator Λ and from the first Green difference formula, it follows that

$$(-\Lambda y, y) = (a y_{\bar{x}}^2, 1)_{\omega^+} + (dy, y) + \kappa_0 y_0^2 + \kappa_1 y_N^2.$$

Therefore the inequality (33) of lemma 16 can be written in the form

$$\gamma_1(\rho y, y) \le -(\Lambda y, y).$$

We move on now to derive the estimate (43). We shall find the solution of the problem $\Lambda v_i = -\rho_i$, $0 \le i \le N$ under the assumptions stated in remark 1. We have the boundary-value difference problem

$$v_{\bar{x}x,i} = -1, \qquad 1 \le i \le N-1, \tag{44}$$

$$v_{x,0} = \kappa_0 v_0 - \hbar_0, \qquad i = 0, \tag{45}$$

$$-v_{\bar{x},N} = \kappa_1 v_N - \hbar_N, \qquad i = N. \tag{46}$$

We multiply (44) by $\hbar_i$, sum over i from j to $N-1$, and take into account the boundary condition (46). We obtain

$$\sum_{i=j}^{N-1} v_{\bar{x}x,i}\hbar_i = \sum_{i=j}^{N-1} (v_{\bar{x},i+1} - v_{\bar{x},i}) = v_{\bar{x},N} - v_{\bar{x},j}$$

$$= -\kappa_1 v_N + \hbar_N - v_{\bar{x},j} = -\sum_{i=j}^{N-1} \hbar_i = x_j - 0.5h_j - l + \hbar_N.$$

From this it follows that

$$v_{\bar{x},j} = l - \kappa_1 v_N + 0.5h_j - x_j, \qquad 1 \le j \le N. \tag{47}$$

Substituting $j = 1$ in (47) and taking into account that $\hbar_0 = 0.5h_1$, $v_{\bar{x},1} = v_{x,0} = \kappa_0 v_0 - \hbar_0$, we obtain a relation for v_0 and v_N

$$\kappa_0 v_0 + \kappa_1 v_N = l. \tag{48}$$

Multiplying (47) by h_j and summing over j from 1 to i, we find

$$\sum_{j=1}^{i} v_{\bar{x},j} h_j = v_i - v_0 = (l - \kappa_1 v_N) \sum_{j=1}^{i} h_j - \sum_{j=1}^{i} (x_j - 0.5h_j) h_j.$$

Since $h_j = x_j - x_{j-1}$, $x_j - 0.5h_j = 0.5(x_j + x_{j-1})$, we have

$$\sum_{j=1}^{i} h_j = x_i, \quad \sum_{j=1}^{i} (x_j - 0.5h_j) h_j = 0.5 \sum_{j=1}^{i} (x_j^2 - x_{j-1}^2) = 0.5 x_i^2.$$

Thus

$$\begin{aligned} v_i &= v_0 + x_i(l - \kappa_1 v_N) - 0.5x_i^2 \\ &= v_0 + 0.5(l - \kappa_1 v_N)^2 - 0.5(x_i - l + \kappa_1 v_N)^2, \quad 0 \le i \le N. \end{aligned} \tag{49}$$

Setting $i = N$, we find a second relation for v_0 and v_N

$$v_N = v_0 + l(l - \kappa_1 v_N) - 0.5l^2. \tag{50}$$

From (48), (50) we obtain

$$v_0 = \frac{l(2 + l\kappa_1)}{2(\kappa_0 + \kappa_1 + \kappa_0 \kappa_1 l)}, \qquad v_N = \frac{l(2 + l\kappa_0)}{2(\kappa_0 + \kappa_1 + \kappa_0 \kappa_1 l)}. \tag{51}$$

Since $0 \le l - \kappa_0 v_N < l$, from (49), (51) we find that

$$\max_{0 \le i \le N} v_i \le v_0 + 0.5(l - \kappa_1 v_N)^2 = \frac{l(2 + l\kappa_0)(2 + l\kappa_1)(2\kappa_0 + 2\kappa_1 + l\kappa_0\kappa_1)}{8(\kappa_0 + \kappa_1 + l\kappa_0\kappa_1)^2}.$$

From this and from lemma 16 we obtain the estimate (43). If $y_0 = y_N = 0$, then substituting in (33) $a_i \equiv 1$, $d_i \equiv 0$, $\rho_i \equiv 1$ and taking the limit in (43) as $\kappa_0 \to \infty$ and $\kappa_1 \to \infty$, we obtain the bound (30) with $\gamma_1 = 8/l^2$.

5.2.5 Upper bounds for difference operators. We now obtain upper bounds for certain difference operators.

Lemma 17. *For an arbitrary grid function y_i defined on the non-uniform grid $\bar{\omega}$ the estimate*

$$(ay_{\bar{x}}^2, 1)_{\omega^+} \le \gamma_2(y, y), \tag{52}$$

is valid, where

$$\gamma_2 = \max\left[\frac{4a_1}{h_1^2}, \ \frac{4a_N}{h_N^2}, \ \max_{1\le i\le N-1} \ \frac{2}{\hbar_i}\left(\frac{a_i}{h_i} + \frac{a_{i+1}}{h_{i+1}}\right)\right].$$

If the grid is uniform, then

$$\gamma_2 = \frac{4}{h^2} \max\left[a_1, a_N, \max_{1\le i\le N-1}\left(\frac{a_i + a_{i+1}}{2}\right)\right].$$

If $y_0 = y_N = 0$ then

$$\gamma_2 = \max_{1\le i\le N-1} \frac{2}{\hbar_i}\left(\frac{a_i}{h_i} + \frac{a_{i+1}}{h_{i+1}}\right).$$

Proof. We have

$$\begin{aligned}(ay_{\bar{x}}^2, 1)_{\omega^+} &= \sum_{i=1}^{N} \frac{a_i(y_i - y_{i-1})^2}{h_i} \\ &= \sum_{i=1}^{N} \frac{a_i}{h_i} y_i^2 + \sum_{i=0}^{N-1} \frac{a_{i+1}}{h_{i+1}} y_i^2 - 2\sum_{i=1}^{N} \frac{a_i}{h_i} y_i y_{i-1}.\end{aligned}$$

Using the inequality $2y_iy_{i-1} \le y_i^2 + y_{i-1}^2$, we obtain for $a_i > 0$ that

$$\begin{aligned}(ay_{\bar{x}}^2, 1)_{\omega^+} &\le \sum_{i=1}^{N} \frac{2a_i}{h_i} y_i^2 + \sum_{i=0}^{N-1} \frac{2a_{i+1}}{h_{i+1}} y_i^2 \\ &= \frac{2a_1}{h_1\hbar_0} y_0^2\hbar_0 + \frac{2a_N}{h_N\hbar_N} y_N^2\hbar_N + \sum_{i=1}^{N-1} \frac{2}{\hbar_i}\left(\frac{a_i}{h_i} + \frac{a_{i+1}}{h_{i+1}}\right) y_i^2\hbar_i.\end{aligned}$$

Since $\hbar_0 = 0.5h_1$, $\hbar_N = 0.5h_N$ and $(y, y) = \Sigma_{i=0}^{N} \hbar_i y_i^2$, we obtain the estimate (52) with the indicated value for γ_2. Lemma 17 is proved. □

Lemma 18. *Suppose that $a_i > 0$, $b_i \geq 0$, and σ_0 and σ_1 are non-negative, where $(b,1) + \sigma_0 + \sigma_1 \neq 0$. For an arbitrary grid function y_i defined on the non-uniform grid $\bar{\omega}$ we have the bound*

$$(ay_{\bar{x}}^2, 1)_{\omega^+} + (by, y) + \sigma_0 y_0^2 + \sigma_1 y_N^2 \leq \bar{\gamma}_2(y, y), \tag{53}$$

where $\bar{\gamma}_2 = \gamma_2 + (1 + \gamma_2) \max\limits_{0 \leq i \leq N} v_i$, γ_2 is defined in lemma 17, and v is the solution of the boundary-value problem

$$\begin{aligned}
(av_{\bar{x}})_{\hat{x},i} - v_i &= -b_i, \qquad 1 \leq i \leq N-1, \\
\frac{a_1}{\hbar_0} v_{x,0} - v_0 &= -b_0 - \frac{\sigma_0}{\hbar_0}, \qquad i = 0, \\
-\frac{a_N}{\hbar_N} v_{\bar{x},N} - v_N &= -b_N - \frac{\sigma_1}{\hbar_N}, \qquad i = N.
\end{aligned} \tag{54}$$

Proof. From lemma 16 with $\rho_i = b_i$ for $1 \leq i \leq N-1$, $\rho_0 = b_0 + \sigma_0/\hbar_0$, $\rho_N = b_N + \sigma_1/\hbar_N$ and $\kappa_0 = \kappa_1 = 0$, $d_i \equiv 1$ we obtain the bound

$$(by, y) + \sigma_0 y_0^2 + \sigma_1 y_N^2 = (\rho y, y) \leq \max_{0 \leq i \leq N} v_i [(ay_{\bar{x}}^2, 1)_{\omega^+} + (y, y)],$$

where v_i is the solution of the auxiliary problem (54). Using lemma 17, we have

$$\begin{aligned}
(ay_{\bar{x}}^2, 1)_{\omega^+} + (by, y) + \sigma_0 y_0^2 + \sigma_1 y_N^2 &\leq (1 + c)(ay_{\bar{x}}^2, 1)_{\omega^+} \\
+ c(y, y) &\leq [\gamma_2 + (1 + \gamma_2)c](y, y), \qquad c = \max_{0 \leq i \leq N} v_i.
\end{aligned}$$

Lemma 18 is proved. □

5.2.6 Difference schemes as operator equations in abstract spaces. After having changed the derivatives in the differential equation and boundary conditions into difference derivatives on a certain grid $\bar{\omega}$, we obtain a difference scheme. The difference equations involving the sought-for values of the grid function at the nodes of $\bar{\omega}$ form a system of algebraic equations. This system is linear if the original problem was linear.

The difference scheme is defined by the difference operator which gives the structure of the difference equations at the nodes of the grid, where we are searching for the solution, and by the boundary conditions at the boundary nodes. The difference operator acts in the space of grid functions defined on $\bar{\omega}$.

We shall look at an example. Suppose that we must find a solution of the problem

$$\begin{gathered} u'' = -\varphi(x), \qquad 0 < x < l, \\ u'(0) = \kappa_0 u(0) - \mu_1, \quad u(l) = \mu_2, \quad \kappa_0 \geq 0. \end{gathered} \tag{55}$$

on the interval $0 \leq x \leq l$. On the uniform grid $\bar{\omega} = \{x_i = ih,\ i = 0, 1, \ldots, N, hN = l\}$, we put problem (55) in correspondence with the difference scheme

$$\begin{aligned} &\Lambda y_i = y_{\bar{x}x,i} = -\varphi_i, \qquad 1 \leq i \leq N-1, \\ &\Lambda y_0 = \frac{2}{h}(y_{x,0} - \kappa_0 y_0) = -(\varphi_0 + \frac{2}{h}\mu_1), \\ &y_N = \mu_2. \end{aligned} \tag{56}$$

The difference operator Λ is defined on the $(N+1)$-dimensional set of grid functions on $\bar{\omega}$, and maps it onto the N-dimensional set of functions defined on $\omega^- = \{x_i \in \bar{\omega},\ i = 0, 1, \ldots, N-1\}$. It is clear that the domain and range of the operator Λ are not the same.

We look now at the space $H(\omega^-)$ of grid functions defined on ω^-. The inner product in $H(\omega^-)$ is defined as in example 1 from Section 5.2.1:

$$(u, v) = \sum_{i=1}^{N-1} u_i v_i h + 0.5 h u_0 v_0, \qquad u, v \in H(\omega^-).$$

We define now the linear operator A in the following fashion: $Ay_i = -\Lambda \mathring{y}_i$, $0 \leq i \leq N-1$, where $y \in H(\omega^-)$, $\mathring{y}_i = y_i$ for $0 \leq i \leq N-1$ and $\mathring{y}_N = 0$. Using this definition, we give a detailed description of the operator A:

$$Ay_i = \begin{cases} -\dfrac{2}{h}(y_{x,0} - \kappa_0 y_0), & i = 0, \\ -y_{\bar{x}x,i}, & 1 \leq i \leq N-2, \\ \dfrac{1}{h^2}(2y_{N-1} - y_{N-2}), & i = N-1. \end{cases} \tag{57}$$

The operator A maps $H(\omega^-)$ onto $H(\omega^-)$ and is linear.

We now transform the difference scheme (56). Taking into account the condition $y_N = \mu_2$, we write (56) in the form

$$\begin{aligned} &-\frac{2}{h}(y_{x,0} - \kappa_0 y_0) = f_0 = (\varphi_0 + \frac{2}{h}\mu_1), \\ &-y_{\bar{x}x,i} = f_i = \varphi_i, \qquad 1 \leq i \leq N-2, \\ &\frac{1}{h^2}(2y_{N-1} - y_{N-2}) = f_{N-1} = (\varphi_{N-1} + \frac{1}{h^2}\mu_2). \end{aligned} \tag{58}$$

Comparing (57) and (58) we find that the difference scheme (56) can be written in the form of a first-kind operator equation

$$Ay = f, \tag{59}$$

where y is unknown, f is a given element of the space $H(\omega^-)$, and A is the operator acting in $H(\omega^-)$ defined above.

We list now the basic properties of the operator A.

The operator A is self-adjoint in $H(\omega^-)$, i.e.

$$(Au, v) = (u, Av), \qquad u, v \in H(\omega^-).$$

In fact, $(Au, v) = -(\Lambda \mathring{u}, \mathring{v})$, where $\mathring{u}_N = \mathring{v}_N = 0$. Using Green's second difference formula we obtain

$$\begin{aligned}
(\Lambda \mathring{u}, \mathring{v}) &= \sum_{i=1}^{N-1} \mathring{u}_{\bar{x}x,i} \mathring{v}_i h + (\mathring{u}_{x,0} - \kappa_0 \mathring{u}_0)\mathring{v}_0 \\
&= \sum_{i=1}^{N-1} \mathring{u}_i \mathring{v}_{\bar{x}x,i} h + (\mathring{u}_{\bar{x}} \mathring{v} - \mathring{v}_{\bar{x}} \mathring{u})_N - (\mathring{u}_x \mathring{v} - \mathring{v}_x \mathring{u})_0 \\
&+ (\mathring{u}_x \mathring{v} - \kappa_0 \mathring{u}\mathring{v})_0 = \sum_{i=1}^{N-1} \mathring{u}_i \mathring{v}_{\bar{x}x,i} h + (\mathring{v}_x \mathring{u} - \kappa_0 \mathring{v}\mathring{u})_0 = (\mathring{u}, \Lambda \mathring{v}).
\end{aligned}$$

The assertion has been proved.

The operator A is positive definite, i.e.

$$(Au, u) \geq \gamma_1 (u, u), \qquad u \in H(\omega^-),$$

where

$$\gamma_1 = \frac{8(1 + l\kappa_0)^2}{l^2 (2 + l\kappa_0)^2} \geq \frac{2}{l^2} > 0.$$

This assertion follows from remarks 1 and 2 to lemma 16. By lemma 10, the operator A has a bounded inverse A^{-1}. Therefore, the solution of equation (59) exists and is unique.

For the operator A we have the following upper bound

$$(Au, u) \leq \gamma_2 (u, u), \qquad u \in H(\omega^-),$$

where

$$\gamma_2 = \frac{4}{h^2}\left(1 + \kappa_0 \frac{h}{2}\right),$$

since $y_N = 0$ and

$$(Ay, y) = (y_{\bar{x}}^2, 1)_{\omega^+} + \kappa_0 y_0^2,$$
$$y_0^2 \le \frac{2}{h}(y, y), \qquad (y_{\bar{x}}^2, 1)_{\omega^+} \le \frac{4}{h^2}.$$

The final inequality follows from lemma 17.

As a second example we will look at the difference scheme

$$\begin{aligned}
\Lambda y_i &= (a y_{\bar{x}})_{\hat{x},i} - d_i y_i = -\varphi_i, \qquad 1 \le i \le N-1,\\
\Lambda y_0 &= \frac{1}{\hbar_0}(a_1 y_{x,0} - \kappa_0 y_0) - d_0 y_0 = -\left(\varphi_0 + \frac{1}{\hbar_0}\mu_1\right), \quad i = 0,\\
\Lambda y_N &= -\frac{1}{\hbar_N}(a_N y_{\bar{x},N} + \kappa_1 y_N) - d_N y_N = -\left(\varphi_N + \frac{1}{\hbar_N}\mu_2\right), \quad i = N,
\end{aligned} \tag{60}$$

on the non-uniform grid $\bar{\omega} = \{x_i \in [0, l],\ x_i = x_{i-1} + h_i,\ 1 \le i \le N,\ x_0 = 0,\ x_N = l\}$. The scheme (60) approximates a boundary-value problem of the third kind for an equation with variable coefficients

$$\begin{aligned}
(ku')' - qu &= -\varphi(x), \qquad 0 < x < l,\\
ku' &= \kappa_0 u - \mu_1, \quad x = 0,\\
-ku' &= \kappa_1 u - \mu_2, \quad x = l
\end{aligned}$$

for a corresponding choice of coefficients a_i and d_i, for example for $a_i = k(x_i - 0.5h_i)$ and $d_i = q(x_i)$.

If in the space $H(\bar{\omega})$ of grid functions defined on $\bar{\omega}$ with inner product

$$(u, v) = \sum_{i=0}^{N} u_i v_i \hbar_i, \qquad \hbar_0 = 0.5h, \qquad \hbar_N = 0.5h_N,$$

we define the operator $A = -\Lambda$ and the grid function $f_i = \varphi_i$, $1 \le i \le N-1$, $f_0 = \varphi_0 + \mu_1/\hbar_0$, $f_N = \varphi_N + \mu_2/\hbar_N$, then the difference scheme (60) can be written in the form of the operator equation (59).

The self-adjointness of the operator A mapping $H(\bar{\omega})$ onto $H(\bar{\omega})$ follows from Green's second difference formula.

If $a_i \ge c_1 > 0$, $d_i \ge 0$, $\kappa_0 \ge 0$, $\kappa_1 \ge 0$, $\kappa_0 + \kappa_1 + (d, 1) > 0$, then the operator A is positive definite in $H(\bar{\omega})$, and

$$(Au, u) \ge \gamma_1(u, u), \qquad 1/\gamma_1 = \max_{0 \le i \le N} v_i,$$

where v_i is the solution of the problem $\Lambda v_i = -1$, $0 \le i \le N$. Notice that the positivity of v_i follows from the maximum principle, which is valid for the operator Λ under the indicated conditions.

If $d_i \equiv 0$ then a crude estimate for γ_1 can be obtained as follows. From Green's first difference formula we obtain

$$(Ay, y) = (-\Lambda y, y) = (a y_{\bar{x}}^2, 1)_{\omega^+} + \kappa_0 y_0^2 + \kappa_1 y_1^2.$$

Using the conditions $a_i \geq c_1 > 0$, $1 \leq i \leq N$ we obtain

$$(Ay, y) \geq c_1[(y_{\bar{x}}^2, 1)_{\omega^+} + \bar{\kappa}_0 y_0^2 + \bar{\kappa}_1 y_1^2],$$

where $c_1 \bar{\kappa}_0 = \kappa_0$, $c_1 \bar{\kappa}_1 = \kappa_1$. Since $\kappa_0 + \kappa_1 > 0$, from remark 1 to lemma 16 we obtain the estimate

$$(y_{\bar{x}}^2, 1)_{\omega^+} + \bar{\kappa}_0 y_0^2 + \bar{\kappa}_1 y_1^2 \geq \bar{\gamma}_1(y, y),$$

where

$$\bar{\gamma}_1 = \frac{8(\bar{\kappa}_0 + \bar{\kappa}_1 + l\bar{\kappa}_0\bar{\kappa}_1)^2}{l(2 + l\bar{\kappa}_0)(2 + l\bar{\kappa}_1)(2\bar{\kappa}_0 + 2\bar{\kappa}_1 + l\bar{\kappa}_0\bar{\kappa}_1)}.$$

Substituting here $\bar{\kappa}_0$ and $\bar{\kappa}_1$ we find that $(Au, u) \geq \gamma_1(u, u)$, where

$$\gamma_1 = c_1\bar{\gamma}_1 = \frac{8c_1(\bar{\kappa}_0 + \bar{\kappa}_1 + l\bar{\kappa}_0\bar{\kappa}_1)^2}{l(2 + l\bar{\kappa}_0)(2 + l\bar{\kappa}_1)(2\bar{\kappa}_0 + 2\bar{\kappa}_1 + l\bar{\kappa}_0\bar{\kappa}_1)}.$$

For the operator A we have the following upper bound $(Au, u) \leq \gamma_2(u, u)$ where γ_2 is defined in lemma 18, since

$$(Ay, y) = (a y_{\bar{x}}^2, 1)_{\omega^+} + (dy^2, 1) + \kappa_0 y_0^2 + \kappa_1 y_N^2.$$

In this example the operator A and the difference operator Λ are defined in the same space of grid functions $H(\bar{\omega})$ and differ only in sign. Unlike the first example, the right-hand sides of the difference scheme (60) and the operator equation (56) are the same.

We limited ourselves here to the simplest examples. In the following sections, difference schemes approximating elliptic boundary-value problems in multi-dimensional spaces will be analogously reduced to operator equations in the corresponding finite-dimensional Hilbert spaces of grid functions. We will also study the basic properties of these operators.

From these examples it is clear that the difference schemes can be considered as operator equations with operators in a finite-dimensional normed linear space. These operators characteristically map the whole space into itself.

5.2.7 Difference schemes for elliptic equations with constant coefficients. Let $\bar{G} = \{0 \le x_\alpha \le l_\alpha,\ \alpha = 1,2\}$ be a rectangle, $\bar{\omega} = \{x_{ij} = (ih_1, jh_2) \in \bar{G},\ 0 \le i \le N_1,\ 0 \le j \le N_2,\ h_\alpha N_\alpha = l_\alpha,\ \alpha = 1,2\}$ be a grid in $\bar{G}$, γ be the set of boundary nodes of the grid $\bar{\omega}$. The grid is uniform in each direction x_α with step h_α. We denote by ω the set of interior nodes of the grid. We introduce the space of grid functions $H = H(\omega)$ defined on ω. We define in H the inner product

$$(u, v) = \sum_{i=1}^{N_1-1} \sum_{j=1}^{N_2-1} u(i,j)v(i,j)h_1h_2.$$

We shall look at a Dirichlet difference problem for Poisson's equation on the grid $\bar{\omega}$

$$\begin{aligned} \Lambda y &= \sum_{\alpha=1}^{2} \Lambda_\alpha y = -\varphi(x), \qquad x \in \omega, \\ y(x) &= g(x), \qquad x \in \gamma, \end{aligned} \tag{61}$$

where $\Lambda_\alpha y = y_{\bar{x}_\alpha x_\alpha}$, $\alpha = 1, 2$.

The difference scheme (61) can be written in the form of the operator equation (59). To do this we define the operator A using the formula $Ay = -\Lambda \mathring{y}$, $x \in \omega$, where $y \in H$, $\mathring{y} \in \mathring{H}$ and $y(x) = \mathring{y}(x)$ for $x \in \omega$. Here $\mathring{H}$ is the set of grid functions defined on $\bar{\omega}$ and reducing to zero on γ. The right-hand side f in equation (56) differs from the right-hand side φ in the difference scheme (61) only at the near-boundary nodes

$$f = \varphi + \varphi_1/h_1^2 + \varphi_2/h_2^2,$$

where

$$\varphi_1(x) = \begin{cases} g(0, x_2), & x_1 = h_1 \\ 0, & 2h_1 \le x_1 \le l_1 - 2h_1, \\ g(l_1, x_2), & x_1 = l_1 - h_1, \end{cases}$$

$$\varphi_2(x) = \begin{cases} g(x_1, 0), & x_2 = h_2, \\ 0, & 2h_2 \le x_2 \le l_2 - 2h_2, \\ g(x_1, l_2), & x_2 = l_2 - h_2. \end{cases}$$

We shall investigate the properties of the operator A mapping from $H(\omega)$ into $H(\omega)$.

1. The operator A is self-adjoint:

$$(Au, v) = (u, Av), \qquad u, v \in H(\omega), \tag{62}$$

For the proof we take into account that

$$(A_1 u, v) = (-\Lambda_1 \mathring{u}, \mathring{v}) = -\sum_{j=1}^{N_2-1} h_2 \sum_{i=1}^{N_1-1} h_1\, (\mathring{v}\Lambda_1\mathring{u})_{ij}$$
$$= -\sum_{j=1}^{N_2-1} h_2 \sum_{i=1}^{N_1-1} h_1\, (\mathring{u}\Lambda_1\mathring{v})_{ij} = -(\mathring{u}, \Lambda_1\mathring{v}) = (u, A_1 v),$$

since, by Green's second difference formula, the difference operator Λ_1 on the grid $\bar{\omega}_1 = \{x_1(i) = ih_1,\ 0 \le i \le N_1,\ h_1 N_1 = l_1\}$ satisfies the relation

$$\sum_{i=1}^{N_1-1} h_1\, (\mathring{v}\Lambda_1\mathring{u})_{ij} = \sum_{i=1}^{N_1-1} h_1\, (\mathring{u}_1\Lambda_1\mathring{v})_{ij}$$

and, in addition, it is possible to change the order of the summations in i and j.

Analogously we find that $(A_2 u, v) = (u, A_2 v)$. From this follows (62).

2. The operator A is positive definite and satisfies the inequalities

$$\delta E \le A \le \Delta E, \qquad \delta > 0, \tag{63}$$

where

$$\delta = \sum_{\alpha=1}^{2} \frac{4}{h_\alpha^2} \sin^2 \frac{\pi}{2N_\alpha} \ge \sum_{\alpha=1}^{2} \frac{8}{l_\alpha^2}, \quad \Delta = \sum_{\alpha=1}^{2} \frac{4}{h_\alpha^2} \cos^2 \frac{\pi}{2N_\alpha} < \sum_{\alpha=1}^{2} \frac{4}{h_\alpha^2}. \tag{64}$$

Notice that δ and Δ are the smallest and largest eigenvalues of the Laplace difference operator Λ (see Section 4.2.1).

This assertion is proved in the same way as lemma 12. Thus we have shown that in $H = H(\omega)$

$$A = A^*, \qquad \delta E \le A \le \Delta E, \qquad \delta > 0.$$

If a first-kind boundary condition $y(x) = g(x)$, $x \in \gamma_0$ is given on a part γ_0 of the grid boundary γ, and second- or third-kind boundary conditions are given on the remaining part, then the operator A can be defined as above, where $\overset{\circ}{H}$ is the set of functions which only reduce to zero on γ_0, and $H = H(\omega)$ is the space of grid functions defined on $\omega_0 = \omega \cup (\gamma \backslash \gamma_0)$. For example, let $\gamma_0 = \{x_{ij} \in \omega,\ i = 0,\ 0 \le j \le N_2\}$, and assume that second-kind boundary conditions are given on $\gamma \backslash \gamma_0$. Then the difference scheme is described in the form

$$\begin{aligned} \Lambda y = (\Lambda_1 + \Lambda_2) y = -\varphi(x), \qquad & x \in \omega, \\ y(x) = g(x), \qquad & x \in \gamma_0, \end{aligned}$$

Here

$$\Lambda_2 y = \begin{cases} \dfrac{2}{h_2}\, y_{x_2}, & x_2 = 0, \\ y_{\bar{x}_2 x_2}, & h_2 \le x_2 \le l_2 - h_2, \\ -\dfrac{2}{h_2} y_{\bar{x}_2}, & x_2 = l_2,\ h_1 \le x_1 \le l_1, \end{cases}$$

and the operator Λ_1 is defined by the formulas

$$\Lambda_1 y = \begin{cases} y_{\bar{x}_1 x_1}, & h_1 \le x_1 \le l_1 - h_1, \\ -\dfrac{2}{h_1} y_{\bar{x}_1}, & x_1 = l_1, \quad 0 \le x_2 \le l_2. \end{cases}$$

The inner product in the space $H = H(\omega_0)$ is defined by the formula

$$(u, v) = \sum_{i=1}^{N_1} \sum_{j=0}^{N_2} u(i,j) v(i,j) \hbar_1(i) \hbar_2(j),$$

where

$$\hbar_1(i) = \begin{cases} h_1, & 1 \le i \le N_1 - 1, \\ 0.5 h_1, & i = N_1, \end{cases}$$

$$\hbar_2(j) = \begin{cases} h_2, & 1 \le i \le N_2 - 1, \\ 0.5 h_2, & j = 0, N_2. \end{cases}$$

It is possible to show that the operator $A = A_1 + A_2$ corresponding to the difference operator Λ is self-adjoint in H, and that the estimate (63) is valid with $\delta = \delta_1 + \delta_2$, $\Delta = \Delta_1 + \Delta_2$,

$$\delta_1 = \frac{4}{h_1^2} \sin^2 \frac{\pi}{4N_1}, \quad \Delta_1 = \frac{4}{h_1^2} \cos^2 \frac{\pi}{4N_1}, \quad \delta_2 = 0, \quad \Delta_2 = \frac{4}{h_2^2}.$$

Here δ_α and Δ_α are the smallest and largest eigenvalues of the difference operator Λ_α, $\alpha = 1, 2$.

Notice that the operators A_1 and A_2 commute both for first-kind and second-kind boundary-value problems. Therefore, by the general theory (see Section 5.1.5), the eigenvalues of the operator A are the sum of the eigenvalues of the operators A_1 and A_2: $\lambda(A_1) + \lambda(A_2)$.

5.2.8 Equations with variable coefficients and with mixed derivatives. We look now at a Dirichlet problem for an elliptic equation with variable coefficients in the rectangle $\bar{G} = \{0 \le x_\alpha \le l_\alpha,\ \alpha = 1, 2\}$:

$$\begin{aligned} Lu &= \sum_{\alpha=1}^{2} \frac{\partial}{\partial x_\alpha}\left(k_\alpha(x)\frac{\partial u}{\partial x_\alpha}\right) - q(x)u = -\varphi(x), \quad x \in G, \\ u(x) &= g(x), \qquad x \in \Gamma, \end{aligned} \tag{65}$$

where $k_\alpha(x)$ and $q(x)$ are sufficiently smooth functions satisfying the conditions $0 < c_1 \le k_\alpha(x) \le c_2$, $0 \le d_1 \le q(x) \le d_2$. We denote by $\bar{\omega} = \omega + \gamma$ the grid with steps h_1 and h_2 introduced in Section 5.2.7.

We shall put the problem (65) in correspondence with a Dirichlet difference problem on the grid $\bar{\omega}$:

$$\begin{aligned} \Lambda y &= (\Lambda_1 + \Lambda_2)y - dy = -\varphi(x), \quad x \in \omega, \\ y(x) &= g(x), \qquad\qquad x \in \gamma, \end{aligned} \tag{66}$$

where $\Lambda_\alpha y = (a_\alpha y_{\bar{x}_\alpha})_{x_\alpha}$, $\alpha = 1, 2$, and $a_\alpha(x)$ and $d(x)$ are chosen, for example, as:

$$\begin{aligned} a_1(x_1, x_2) &= k_1(x_1 - 0.5h_1, x_2), \\ a_2(x_1, x_2) &= k_2(x_1, x_2 - 0.5h_2), \qquad d(x) = q(x). \end{aligned}$$

Then the coefficients of the difference scheme satisfy the conditions

$$0 < c_1 \le a_\alpha(x) \le c_2, \quad \alpha = 1, 2, \quad 0 \le d_1 \le d \le d_2. \tag{67}$$

We denote by $H = H(\omega)$ the space of grid functions introduced in the preceding subsection, and by $\mathring{H}$ the set of grid functions which reduce to zero on γ.

We will write out the difference scheme (66) in the form of the operator equation (59), where the operator A is defined in the usual way: $Ay = -\Lambda\mathring{y}$, where $y \in H$, $\mathring{y} \in \mathring{H}$, and $y(x) = \mathring{y}(x)$ for $x \in \omega$.

We denote by $\mathcal{R} = \mathcal{R}_1 + \mathcal{R}_2$, where $\mathcal{R}_\alpha y = y_{\bar{x}_\alpha x_\alpha}$, $\alpha = 1, 2$, the Laplace difference operator and define the corresponding operator in the space H: $Ry = -\mathcal{R}\mathring{y}$, $y \in H$, $\mathring{y} \in \mathring{H}$ and $y(x) = \mathring{y}(x)$ for $x \in \omega$.

Lemma 19. *The operator A is self-adjoint in H, and it satisfies the following bounds*

$$(c_1 + d_1/\Delta)(Ru, u) \leq (Au, u) \leq (c_2 + d_2/\delta)(Ru, u), \tag{68}$$
$$(c_1\delta + d_1)(u, u) \leq (Au, u) \leq (c_2\Delta + d_2)(u, u), \tag{69}$$

where δ and Δ are defined in (64).

Proof. From the conditions (67) and the estimates obtained in the preceding subsection

$$\delta E \leq R \leq \Delta E, \tag{70}$$

it follows that for any $u \in H$ we have the inequalities

$$\frac{d_1}{\Delta}(Ru, u) \leq d_1(u, u) \leq (du, u) \leq d_2(u, u) \leq \frac{d_2}{\delta}(Ru, u). \tag{71}$$

Further, Green's first difference formula gives

$$(A_1 u, u) = -(\Lambda_1 \mathring{u}, \mathring{u}) = \sum_{j=1}^{N_2-1} \sum_{i=1}^{N_1} (a_1 \mathring{u}_{\bar{x}_1}^2)_{ij} h_1 h_2,$$
$$(R_1 u, u) = -(\mathcal{R}_1 \mathring{u}, \mathring{u}) = \sum_{j=1}^{N_2-1} \sum_{i=1}^{N_1} (\mathring{u}_{\bar{x}_1}^2)_{ij} h_1 h_2.$$

By (67), we then obtain the inequality

$$c_1(R_1 u, u) \leq (A_1 u, u) \leq c_2(R_1 u, u).$$

Analogously we find that

$$c_1(R_2 u, u) \leq (A_2 u, u) \leq c_2(R_2 u, u).$$

From this and from (70) we obtain

$$c_1\delta(u, u) \leq c_1(Ru, u) \leq ((A_1 + A_2)u, u) \leq c_2(Ru, u) \leq c_2\Delta(u, u),$$

which together with (71) gives us (68) and (69).

The self-adjointness of the operator A is proven by analogy with the preceding subsection. □

Notice that (68) gives the energy equivalence constants for the operators R and A, where $d_1 \geq 0$ and $\delta \geq 8/l_1^2 + 8/l_2^2$, so that the operators are equivalent with constants which do not depend on the number of nodes in the grid.

We now consider a *Dirichlet problem for an elliptic equation containing mixed derivatives*

$$Lu = \sum_{\alpha,\beta=1}^{2} \frac{\partial}{\partial x_\alpha}\left(k_{\alpha\beta}(x)\frac{\partial u}{\partial x_\beta}\right) = -\varphi(x), \qquad x \in G,$$
$$u(x) = g(x), \qquad x \in \Gamma. \tag{72}$$

We shall assume that the ellipticity conditions are satisfied

$$c_1 \sum_{\alpha=1}^{2} \xi_\alpha^2 \leq \sum_{\alpha,\beta=1}^{2} k_{\alpha\beta}(x)\xi_\alpha\xi_\beta \leq c_2 \sum_{\alpha=1}^{2} \xi_\alpha^2, \qquad x \in \bar{G}, \tag{73}$$

where $c_2 \geq c_1 > 0$, and $\xi = (\xi_1, \xi_2)^T$ is an arbitrary vector.

On the rectangular grid $\bar{\omega}$ the problem (72) can be put in correspondence with the difference scheme

$$\Lambda y = 0.5 \sum_{\alpha,\beta=1}^{2} [(k_{\alpha\beta} y_{\bar{x}_\beta})_{x_\alpha} + (k_{\alpha\beta} y_{x_\beta})_{\bar{x}_\alpha}] = -\varphi(x), \quad x \in \omega,$$
$$y(x) = g(x), \qquad x \in \gamma. \tag{74}$$

We write (74) in the form of the operator equation (59), defining the operator A in the usual way: $Ay = -\Lambda \mathring{y}$, where $y \in H(\omega)$, $\mathring{y} \in \mathring{H}$ and $y(x) = \mathring{y}(x)$ for $x \in \omega$. Here the right-hand side f only differs from the right-hand side φ in equation (74) at the near-boundary nodes. To find an explicit formula for f we write out the difference equation at a boundary node, use the boundary condition and move the known values of $y(x)$ on γ to the right-hand side of the equation.

We shall now show that, if the symmetry condition $k_{12}(x) = k_{21}(x)$ is satisfied, then the operator A is self-adjoint in the space $H = H(\omega)$ defined above. To do this, we write the operator Λ in the form of a sum $\Lambda = (\Lambda_1 + \Lambda_2)/2$, where

$$\Lambda_\alpha y = (k_{\alpha\alpha} y_{\bar{x}_\alpha} + k_{\alpha\beta} y_{\bar{x}_\beta})_{x_\alpha} + (k_{\alpha\alpha} y_{x_\alpha} + k_{\alpha\beta} y_{x_\beta})_{\bar{x}_\alpha},$$
$$\beta = 3 - \alpha, \qquad \alpha = 1, 2.$$

Using the summation by parts formulas (7′) and (9′), we obtain that for any $\mathring{u}, \mathring{v} \in \mathring{H}$

$$(\Lambda_1 \mathring{u}, \mathring{v}) = -\sum_{j=1}^{N_2-1} \sum_{i=1}^{N_1} [(k_{11}\mathring{u}_{\bar{x}_1} + k_{12}\mathring{u}_{\bar{x}_2})\mathring{v}_{\bar{x}_1}]_{ij} h_1 h_2$$
$$-\sum_{j=1}^{N_2-1} \sum_{i=0}^{N_1-1} [(k_{11}\mathring{u}_{x_1} + k_{12}\mathring{u}_{x_2})\mathring{v}_{x_1}]_{ij} h_1 h_2.$$

Taking into account that $\mathring{v}_{\bar{x}_1}$ and $\mathring{v}_{x_1}$ are equal to zero for $j = N_2$ and $j = 0$, respectively, the resulting equality can be written in the form

$$(\Lambda_1 \mathring{u}, \mathring{v}) = -\sum_{j=1}^{N_2} \sum_{i=1}^{N_1} [(k_{11}\mathring{u}_{\bar{x}_1} + k_{12}\mathring{u}_{\bar{x}_2})\mathring{v}_{\bar{x}_1}]_{ij} h_1 h_2$$
$$-\sum_{j=0}^{N_2-1} \sum_{i=0}^{N_1-1} [(k_{11}\mathring{u}_{x_1} + k_{12}\mathring{u}_{x_2})\mathring{v}_{x_1}]_{ij} h_1 h_2. \tag{75}$$

Analogously we find

$$(\Lambda_2 \mathring{u}, \mathring{v}) = -\sum_{i=1}^{N_1} \sum_{j=1}^{N_2} [(k_{22}\mathring{u}_{\bar{x}_2} + k_{21}\mathring{u}_{\bar{x}_1})\mathring{v}_{\bar{x}_2}]_{ij} h_1 h_2$$
$$-\sum_{i=0}^{N_1-1} \sum_{j=0}^{N_2-1} [(k_{22}\mathring{u}_{x_2} + k_{21}\mathring{u}_{x_1})\mathring{v}_{x_2}]_{ij} h_1 h_2. \tag{76}$$

Combining (75) and (76) we obtain

$$(\Lambda \mathring{u}, \mathring{v}) = -0.5 \sum_{i=1}^{N_1} \sum_{j=1}^{N_2} h_1 h_2 \left(\sum_{\alpha,\beta=1}^{2} k_{\alpha\beta} \mathring{u}_{\bar{x}_\alpha} \mathring{v}_{\bar{x}_\beta} \right)_{ij}$$
$$-0.5 \sum_{i=0}^{N_1-1} \sum_{j=0}^{N_2-1} h_1 h_2 \left(\sum_{\alpha,\beta=1}^{2} k_{\alpha\beta} \mathring{u}_{x_\alpha} \mathring{v}_{x_\beta} \right)_{ij}. \tag{77}$$

From this it follows that, if $k_{12} = k_{21}$,

$$(\Lambda \mathring{u}, \mathring{v}) = (\mathring{u}, \Lambda \mathring{v}).$$

Since $(Au, v) = -(\Lambda u, v)$, the operator A is self-adjoint in H.

We now find bounds for the operator A. We substitute the grid function $\mathring{u}$ in place of $\mathring{v}$ in (77), take into account the ellipticity condition (73) and the condition $\mathring{u}(x) = 0$ for $x \in \gamma$. We obtain

$$
\begin{aligned}
-(\Lambda\mathring{u},\mathring{u}) \geq 0.5c_1 \Bigg\{ &\sum_{j=1}^{N_2-1} h_2 \left[\sum_{i=1}^{N_1} (\mathring{u}_{\bar{x}_1})^2_{ij} h_1 + \sum_{i=0}^{N_1-1} (\mathring{u}_{x_1})^2_{ij} h_1 \right] \\
&+ \sum_{i=1}^{N_1-1} h_1 \left[\sum_{j=1}^{N_2} (\mathring{u}_{\bar{x}_2})^2_{ij} h_2 + \sum_{j=0}^{N_2-1} (\mathring{u}_{x_2})^2_{ij} h_2 \right] \Bigg\} \\
= c_1 &\left[\sum_{j=1}^{N_2-1} \sum_{i=1}^{N_1} (\mathring{u}_{\bar{x}_1})^2_{ij} h_1 h_2 + \sum_{i=1}^{N_1-1} \sum_{j=1}^{N_2} (\mathring{u}_{\bar{x}_2})^2_{ij} h_1 h_2 \right] = c_1(-\mathcal{R}\mathring{u},\mathring{u}),
\end{aligned}
$$

where $\mathcal{R}$ is the Laplace difference operator. Analogously we find

$$-(\Lambda\mathring{u},\mathring{u}) \leq c_2(-\mathcal{R}\mathring{u},\mathring{u}).$$

Taking into account the estimate (70), we obtain the following inequalities for the operator A:

$$
\begin{aligned}
c_1(Ru,u) \leq (Au,u) \leq c_2(Ru,u), \\
c_1\delta(u,u) \leq (Au,u) \leq c_2\Delta(u,u),
\end{aligned}
\tag{78}
$$

where δ and Δ are defined in (64). Consequently, the operator A corresponding to the elliptic difference operator with mixed derivatives, and the operator R corresponding to the Laplace difference operator, are energy equivalent with constants c_1 and c_2 which do not depend on the number of nodes in the grid. The operator A has bounds $c_1\delta = 0(1)$ and $c_2\Delta = 0(1/h^2)$ ($h^2 = h_1^2 + h_2^2$), and if the number of nodes in the grid is large, then the operator A is badly conditioned.

Notice that the inequalities (78) remain valid in the case when the differential operator L is approximated by the difference operators

$$\Lambda y = \frac{1}{2}\sum_{\alpha=1}^{2}[(k_{\alpha\alpha}y_{\bar{x}_\alpha})_{x_\alpha} + (k_{\alpha\alpha}y_{x_\alpha})_{\bar{x}_\alpha}] + \frac{1}{2}\sum_{\substack{\alpha,\beta=1 \\ \alpha\neq\beta}}^{2}[(k_{\alpha\beta}y_{x_\beta})_{x_\alpha} + (k_{\alpha\beta}y_{\bar{x}_\beta})_{\bar{x}_\alpha}]$$

or

$$
\begin{aligned}
\Lambda y = &\frac{1}{2}\sum_{\alpha=1}^{2}[(k_{\alpha\alpha}y_{\bar{x}_\alpha})_{x_\alpha} + (k_{\alpha\alpha}y_{x_\alpha})_{\bar{x}_\alpha}] \\
&+ \frac{1}{4}\sum_{\substack{\alpha,\beta=1 \\ \alpha\neq\beta}}^{2}[(k_{\alpha\beta}y_{\bar{x}_\beta})_{x_\alpha} + (k_{\alpha\beta}y_{x_\beta})_{\bar{x}_\alpha} + (k_{\alpha\beta}y_{x_\beta})_{x_\alpha} + (k_{\alpha\beta}y_{\bar{x}_\beta})_{\bar{x}_\alpha}].
\end{aligned}
$$

5.3 Basic concepts from the theory of iterative methods

5.3.1 The steady state method. Above it was shown that difference schemes for elliptic equations can be written in a natural manner in the form of a first-kind operator equation

$$Au = f \tag{1}$$

with an operator A acting in a finite-dimensional Hilbert space H. Linear elliptic equations correspond to linear operators A, and quasi-linear operators correspond to non-linear operators A.

The theory of iterative methods for the operator equation (1) can be set out as one area of the general stability theory for difference schemes. Iterative schemes can be interpreted as steady state methods for the corresponding non-stationary equation. We will clarify this with an example of an equation with a self-adjoint positive-definite and bounded operator A, $A = A^* \geq \delta E$, $\delta > 0$.

Suppose $v = v(t)$ is an abstract function of t with values in H, i.e. $v(t)$ is an element of the space H for each fixed t. We shall look at the abstract Cauchy problem:

$$\frac{dv}{dt} + Av = f, \qquad t > 0, \qquad v(0) = v_0 \in H. \tag{2}$$

We shall show that $\lim_{t\to\infty} \| v(t) - u \| = 0$, where u is the solution of equation (1), i.e. as t increases the solution $v(t)$ of the non-stationary equation (2) converges to the solution u of the stationary (not depending on t) equation (1) (we have "steady state" or "convergence into the stationary regime"). For the error $z(t) = v(t) - u$, we have the homogeneous equation

$$\frac{dz}{dt} + Az = 0, \qquad t > 0, \qquad z(0) = v(0) - u.$$

Forming the inner product of this equation with z:

$$\left(\frac{dz}{dt}, z\right) + (Az, z) = 0$$

and taking into account that

$$\left(\frac{dz}{dt}, z\right) = \frac{1}{2}\frac{d}{dt}(z, z) = \frac{1}{2}\frac{d}{dt} \| z \|^2, \qquad (Az, z) \geq \delta \| z \|^2,$$

we obtain

$$\frac{d}{dt} \| z(t) \|^2 + 2\delta \| z(t) \|^2 \leq 0.$$

After multiplying this inequality by $e^{2\delta t} > 0$ we have

$$\frac{d}{dt} e^{2\delta t} \parallel z(t) \parallel^2 \leq 0,$$

from which it follows that $e^{2\delta t} \parallel z(t) \parallel^2 \leq \parallel z(0) \parallel^2$ or

$$\parallel v(t) - u \parallel \leq e^{-\delta t} \parallel v(0) - u \parallel \rightarrow 0 \quad \text{as} \quad t \rightarrow \infty.$$

Thus, solving equation (2) with any $v_0 \in H$ for a sufficiently large t, we obtain an approximate solution to the original equation (1) to any desired accuracy. The resulting method is called the *steady state method.* Difference analogs of equation (2) also possess the analogous property of damping of the initial data.

5.3.2 Iterative schemes. We look now at the general characteristics of an iterative scheme. Suppose that it is necessary to find the solution of equation (1). We will first assume that A is a linear operator defined in H.

In any iterative method the solution of equation (1) is found from some initial approximation $y_0 \in H$, and a sequence of approximate solutions $y_1, y_2, \ldots, y_k, y_{k+1}, \ldots$ is defined where k is the iteration number. The approximation y_{k+1} is expressed in terms of the already known preceding approximations using a recurrence formula

$$y_{k+1} = F_k(y_0, y_1, \ldots, y_k),$$

where F_k is some function which depends, in general, on the operator A, the right-hand side f, and the iteration number k.

It is said that an iterative method has order m if each successive approximation depends only on the m preceding approximations, i.e.

$$y_{k+1} = F_k(y_{k-m+1}, y_{k-m+2}, \ldots, y_k).$$

High-order iterative schemes require a large amount of storage for intermediate information and therefore in practise we usually limit ourselves to the values $m = 1$ and $m = 2$.

The structure of the iterative scheme depends on the choice of the function F_k. If this function is linear, then the iterative scheme is also called linear. If F_k does not depend on the iteration number k, then the iterative method is called stationary.

We shall look at the general form of a first-order linear iterative scheme. Any such scheme, in correspondence with the definition, can be written in the form

$$y_{k+1} = S_{k+1} y_k + \tau_{k+1} \varphi_{k+1}, \qquad k = 0, 1, \ldots, \tag{3}$$

where S_k are linear operators defined on H, τ_k are some scalar parameters.

Usually an iterative scheme has the following natural requirement: the solution $u = A^{-1}f \in H$ of equation (1) must be, for any f, a fixed point of the iterative approximation process (3), i.e.

$$A^{-1}f = S_{k+1}A^{-1}f + \tau_{k+1}\varphi_{k+1}. \tag{4}$$

It then follows that if we set

$$S_{k+1} = E - \tau_{k+1}B_{k+1}^{-1}A, \quad \varphi_{k+1} = B_{k+1}^{-1}f, \tag{5}$$

where B_{k+1} is an invertible linear operator mapping in H, then condition (4) will be satisfied. Substituting (5) in (3), we obtain as a result of simple transformation

$$B_{k+1}\frac{y_{k+1} - y_k}{\tau_{k+1}} + Ay_k = f, \qquad k = 0, 1, \ldots, \quad y_0 \in H. \tag{6}$$

Preserving the terminology of the theory of difference schemes (see A.A. Samarskii, "The theory of difference schemes", 1977, chapter V), we will call (6) the canonical form of a two-level iterative scheme. Thus, any first-order linear process can be written in the form (6). If $B_{k+1} \equiv E$, then the iterative scheme is called *explicit* since in this case the approximation y_{k+1} is found from the explicit formula

$$y_{k+1} = y_k - \tau_{k+1}(Ay_k - f), \qquad k = 0, 1, \ldots$$

If B_k is different from the identity operator for some k, then the scheme is called *implicit.* If τ_{k+1} depends on the iterative approximation y_k, then the iterative process will be *non-linear.* Obviously, in a stationary iterative process, the operators B_k and the parameters τ_k (more precisely, B_k/τ_k) do not depend on the iteration number k.

Notice that the scheme (6) can be considered as an implicit two-level scheme for the non-stationary equation

$$B(t)\frac{dv}{dt} + Av = f, \qquad t > 0, v(0) = y_0,$$

which is more general than equation (2) considered above. Here the parameter τ_{k+1} can be seen as a step in fictitious time.

The difference between iterative schemes and schemes for non-stationary problems of the form (2) can be described as follows:

1) for any B_{k+1} and τ_{k+1} the solution u of the original equation (1) satisfies (6);

2) the choice of the parameters τ_{k+1} and the operators B_{k+1} is subject only to the requirements that the iteration converge and that the number of arithmetic operations needed to obtain the solution of equation (1) to a desired accuracy be minimized (for non-stationary problems the step length must first satisfy an approximation requirement).

It was assumed above that the operator A was linear. Obviously the scheme (6) can be used to find an approximate solution to equation (1) in the case where the operator A is non-linear. Here the operator B_{k+1} is usually chosen to be linear.

The two-level iterative schemes (6) are very useful. However, three-level schemes which describe second-order iterative processes are also used to solve equation (1). Three-level schemes of the "standard" type have been most investigated. They are written in the form

$$B_{k+1}y_{k+1} = \alpha_{k+1}(B_{k+1} - \tau_{k+1}A)y_k + (1-\alpha_{k+1})B_{k+1}y_{k-1} + \alpha_{k+1}\tau_{k+1}f \quad (7)$$

for $k = 1, 2, \ldots$. Here two sequences of iterative parameters $\{\tau_k\}$ and $\{\alpha_k\}$ are used. In order to realize the scheme (7) it is necessary to supply, in addition to the initial approximation y_0, an additional approximation y_1. Usually this is obtained from y_0 using a two-level scheme (6), i.e.

$$B_1 y_1 = (B_1 - \tau_1 A)y_0 + \tau_1 f, \qquad y_0 \in H. \quad (8)$$

It is possible to show that the solution u of equation (1) is a fixed point for (7), (8).

If $B_k \equiv E$ for all $k = 1, 2, \ldots$, then the scheme (7) is called *explicit*:

$$y_{k+1} = \alpha_{k+1}(E - \tau_{k+1}A)y_k + (1 - \alpha_{k+1})y_{k-1} + \alpha_{k+1}\tau_{k+1}f.$$

Otherwise, the scheme (7) is *implicit.*

5.3.3 Convergence and iteration counts. The basic difference between iterative methods and direct methods is that iterative methods give the exact solution of equation (1) ony as the limit of a sequence of iterative approximations $\{y_k\}$ as $k \to \infty$. Exceptions are the "finite" iterative methods such as the conjugate-direction methods, which obtain the exact solution from any initial approximation after a finite number of operations if A is a linear operator in a finite-dimensional space.

In order to characterize the divergence of an iterative approximation y_k from the exact solution u of problem (1), we introduce the error $z_k = y_k - u$. An iterative process is said to be *convergent in the energy space* H_D

if $\| z_k \|_D \to 0$ as $k \to \infty$. Here H_D is the space generated by a self-adjoint positive-definite operator D in H.

The idea behind the introduction of the energy space H_D is as follows. As we know, a sequence of elements from H which is convergent in one norm also converges in an equivalent norm. Therefore, to investigate a specific iterative scheme it is convenient to choose an energy space H_D in which the operators A and B_k from the iterative scheme possess desirable properties, such as self-adjointness and positive-definiteness.

One of the more important quantitative characteristics of an iterative method is the iteration count. Usually it is necessary to find an approximate solution to equation (1) to some specified accuracy $\epsilon > 0$. If $\| u \|_D = O(1)$, then the following condition must be satisfied

$$\| y_n - u \|_D \le \epsilon \quad \text{for} \quad n \ge n_0(\epsilon). \tag{9}$$

Here $n_0(\epsilon)$ is the minimal number of iterations needed to guarantee that the desired accuracy ϵ has been achieved. This number depends on the initial approximation. The condition (9) can be used to determine the termination point for the iteration if the indicated norm can be effectively computed during the iterative process. For example, if A is non-singular and positive-definite then, choosing as D the operator A^*A, we obtain from (9)

$$\| y_n - u \|_D = \| Ay_n - f \| \le \epsilon,$$

since

$$\begin{aligned}(y_n - u, y_n - u)_D &= (A^*A(y_n - u), y_n - u) \\ &= (Ay_n - Au, Ay_n - Au) \\ &= \| Ay_n - f \|^2 .\end{aligned}$$

To compare the quality of different methods, the iteration count is generally used, determined from the condition

$$\| y_n - u \|_D \le \epsilon \| y_0 - u \|_D \quad \text{for} \quad n \ge n_0(\epsilon). \tag{10}$$

This number indicates how many iterations are sufficient in order that, for any initial approximation y_0, the initial error in H_D will be reduced by a factor of $1/\epsilon$. The condition (10) can also be used as a criterion for terminating the iterative process.

The equation (1) can be put in correspondence with a great number of iterative schemes (6) or (7), (8) with any B_k and τ_k, α_k. Then there arises the problem of choosing a scheme to solve a specific problem. From the point of view of computational mathematics, the most important consideration is constructing those iterative methods which will find the solution of (1) to a given accuracy after a minimal amount of machine time. This economy requirement for a method is natural. For theoretical estimates of the quality of a method, this is often changed to the requirement of minimizing the number of arithmetic operations $Q(\epsilon)$ required to obtain the solution to a given accuracy.

The total volume of computation $Q(\epsilon)$ is equal to

$$Q(\epsilon) = \sum_{k=1}^{n} q_k,$$

where q_k is the number of operations in the computations at iteration k, and n is the number of iterations, $n \geq n_0(\epsilon)$. The problem of constructing an iterative method can be posed as (for the two-level scheme (6)): the operator A is fixed, and the parameters $\{\tau_k, k = 1, 2, \ldots, n\}$ and the operators B_k must be chosen to minimize $Q(\epsilon)$.

In such a general setting this problem scarcely has a solution. Usually the set of operators B_k is given *a priori*, and if the number of operations needed to invert the operator B_k does not depend on k, then $q_k \equiv q$ and $Q(\epsilon) = qn_0(\epsilon)$. In this case, the problem of minimizing $Q(\epsilon)$ leads to the problem of choosing the iterative parameters τ_k from the condition that the number of iterations $n_0(\epsilon)$ be minimized.

In order to set up a hierarchy of methods, it is necessary to compare them using some set of characteristics. Sometimes we use asymptotic estimates for the number of operations or for the number of iterations as the number of unknowns in the difference scheme tends to infinity. However, there exists a real limit on the number of unknowns when we are solving multi-dimensional elliptic equations by a grid method. So, for example, for the three-dimensional Poisson's equation with the number of nodes in each direction being $N \approx 10^2$, we are led to a system of linear algebraic eqations with $M = 10^6$ unknowns. It is hardly appropriate to increase the number of nodes. Therefore it is necessary first of all to compare methods on real grids.

5.3.4 Classification of iterative methods. Iterative methods are characterized by the structure of the iterative scheme, by the energy space H_D in which the convergence of the method is studied, by the termination condition for the iterative process, and also by the algorithm for realizing one iterative step.

We will only look at two-level and three-level iterative schemes, explicit and implicit, for which the termination condition for the iterative process will be the condition

$$\| y_n - u \|_D \leq \epsilon \| y_0 - u \|_D, \qquad \epsilon > 0.$$

In the general theory of iterative methods we look at methods of two types: ones which do and do not use (methods of variational type) *a priori* information about the operators of the iterative scheme. In the first case the iterative parameters τ_k for the scheme (6) and τ_k, α_k for the scheme (7), (8) are chosen so as to minimize the norm of the resolving operator (the operator connecting the initial and final approximations), or the norm of the operator which transforms from one iteration to the next. Here the iterative parameters are chosen so as to ensure the highest convergence rate given the worst initial approximation. In methods of this type, the properties of the initial approximation are not used.

In methods of variational type, the iterative parameters are chosen to minimize certain functionals connected with the original equation. For example, we might choose as a functional the norm of the error at the k-th iteration. In this case, the iterative parameters depend on the preceding iterative approximations and possess properties which take into account the quality of the initial approximation.

In the general theory of iterative methods, we avoid studying the concrete structure of the operators in the iterative scheme — the theory uses the minimum of information of a general functional character concerning the operators. This allows us to achieve our main goal — to indicate the general principles for constructing optimal iterative methods subject to the character and form of *a priori* information about the problem, and also subject to requirements related to the method used to solve the problem. These auxiliary requirements can, for example, consist of the need to construct an optimal method not just for one problem, but for a series of problems with the same operator A but with different right-hand sides.

Undoubtedly, considering the structure of the operator in the problem being solved allows us to construct special iterative methods which possess faster convergence rates than methods from the general theory. This is achieved through a special choice of the operators B_k and the iterative parameters. Special methods have a narrow field of application.

We examine now the role of the operators B_k. For implicit iterative schemes, the choice of the operators B_k must be subordinate to two requirements: guaranteeing a high convergence rate for the method, and the requirement that inverting these operators be simple and economical. These requirements are contradictory. In fact, if we take $B_1 = A$ and $\tau_1 = 1$ in the scheme (6), then for any initial approximation the solution of equation (1) can be obtained after one iteration. In this case, the rate of convergence is maximal, however inverting the operator B_1 is equivalent to solving the original problem.

This indicates, and this will be shown below, that there is no necessity to choose the operator B_k equal to the operator A. It is sufficient that these operators be close in the energy norm. This requirement opens up the possibility of choosing the operators B from a class of easily invertible operators which are close to the operator A in energy.

At the present time, we most commonly use the following approach to construct implicit iterative methods. The operator B_{k+1} is either constructed in explicit form, or the iterative approximation y_{k+1} is found as the result of some auxiliary computational procedure which can be considered as the implicit inversion of the operator B_{k+1}.

In the first case the operator B_{k+1} is usually chosen in the form of the product of some number of easily invertible operators so that the operator B_{k+1} is close to the operator A in some sense. Here the operators entering into the product can themselves depend on parameters which can be looked upon as auxiliary iterative parameters. For example, if $B_k = (E+\omega_k A_1)(E+\omega_k A_2)$, where A_α are operators, then ω_k are scalar parameters. In this case, the variability of the operator B_k is only dependent on the indicated parameters ω_k at iteration k. Then the construction of the operator B_k guarantees a well-defined computational process for finding an approximate solution at each iteration.

We look now at two algorithms for finding a new approximation y_{k+1} in the case when the operator B_{k+1} has a factored form. Suppose that $B_{k+1} = B_{k+1}^1 B_{k+1}^2 \cdots B_{k+1}^p$ and y_{k+1} is found from the two-level iterative scheme (6). In the first algorithm, we solve a sequence of equations

$$B_{k+1}^1 v^1 = F_{k+1}, \quad B_{k+1}^\alpha v^\alpha = v^{\alpha-1}, \quad \alpha = 2,3\ldots,p, \tag{11}$$

where $F_{k+1} = B_{k+1}y_k - \tau_{k+1}(Ay_k - f)$. Clearly, $y_{k+1} = v^p$. Each of the equations (11) must be easy to solve. The algorithm does not require that intermediate information be remembered; once it is obtained, it is immediately used. A deficiency of the algorithm is the necessity of computing the elements $B_{k+1}y_k$, but this can be avoided by a more complex procedure.

The second algorithm has the form of a scheme with a correction:

$$\begin{aligned} y_{k+1} &= y_k - \tau_{k+1} v^p, \\ B_{k+1}^1 v^1 &= Ay_k - f, \\ B_{k+1}^\alpha v^\alpha &= v^{\alpha-1}, \quad \alpha = 2,3\ldots,p. \end{aligned} \tag{12}$$

In this case it is necessary to store the auxiliary values of the preceding iterative approximation y_k until such time as they are no longer needed to find the correction v^p.

In the second version, the construction of the implicit iterative method arises, for example, from the scheme for the correction (12), and the correction v^p is found as an approximate solution of the auxiliary equation

$$R_{k+1}v = r_k, \qquad r_k = Ay_k - f. \tag{13}$$

Suppose that (13) is solved using some two-level iterative scheme. Then the error $z^m = v^m - v$ satisfies the homogeneous equation

$$z^{m+1} = S_{m+1} z^m, \quad m = 0, 1, \ldots, p-1, \quad z^0 = v^0 - v,$$

where S_{m+1} is the operator which transforms from the m-th to the $(m+1)$-st iteration. From this we find

$$z^p = v^p - v = S_p S_{p-1} \cdots S_1 z^0 = T_p(v^0 - v), \quad T_p = \prod_{m=1}^{p} S_m,$$

where T_p is the resolving operator. Substituting here $v = R_{k+1}^{-1} r_k$ and choosing $v_0 = 0$, we obtain

$$v^p = (E - T_p) R_{k+1}^{-1} r_k \quad \text{or} \quad v^p = B_{k+1}^{-1} r_k, \tag{14}$$

where B_{k+1} denotes the operator $R_{k+1}(E - T_p)^{-1}$.

We substitute (14) in (12) and find that y_{k+1} satisfies the two-level scheme (6) with the indicated operator B_{k+1}. If the norm of the operator T_p is small, then the operator B_{k+1} is "close" to the operator R_{k+1}. Therefore it is natural to choose the operator R_{k+1} as some operator close to A.

Chapter 6

Two-Level Iterative Methods

In this chapter we will look at two-level iterative methods for solving the operator equation $Au = f$. The iterative parameters are chosen using ***a priori*** information about the operators from the iterative scheme. In Section 6.1 we pose the problem of choosing the parameters for the two-level scheme. In Section 6.2 and 6.3 this problem is solved in the self-adjoint case. In Section 6.4 we study several ways of choosing the parameters in the non-self-adjoint case based on the volume of ***a priori*** information. In Section 6.5 we look at several sample applications of these methods for solving grid equations.

6.1 Choosing the iterative parameters

6.1.1 The initial family of iterative schemes. In Chapter 5 it was shown that boundary-value difference problems for elliptic equations are special systems of algebraic equations which can be treated as first-kind operator equations

$$Au = f \tag{1}$$

in a real Hilbert space H. In certain special cases such systems can be effectively solved by the direct methods studied in Chapters 1–4. In the general case, an iterative method is one approximate method for solving grid elliptic equations. We will begin our study of iterative methods with the simplest two-level methods — the *Chebyshev method and the simple iteration method.*

In order to find an approximate solution to equation (1) with a non-singular linear operator A defined in H, we consider the *implicit two-level iterative scheme*

$$B\frac{y_{k+1} - y_k}{\tau_{k+1}} + Ay_k = f, \qquad k = 0, 1, \ldots \tag{2}$$

with an arbitrary initial approximation $y_0 \in H$. Here $\{\tau_k\}$ is a sequence of iterative parameters, and B is an arbitrary non-singular linear operator acting in H. The question of the best choice for the operator B will be studied separately; here we will only remark that the operator B must be easy to invert.

The convergence ot the iterative scheme (2) will be studied in the energy space H_D generated by an arbitrary self-adjoint positive-definite operator D in H.

Since the operator B is not fixed, (2) generates a family of iterative schemes which we shall call the *initial family.*

In Chapter 5 it was shown that in order to study the convergence of an iterative method it is necessary to investigate the behavior of the norm of the error $z_k = y_k - u$ in H_D as $k \to \infty$, where y_k is the iterative approximation obtained using scheme (2), and u is the solution of equation (1). The iterative method converges in H_D if the norm of the error z_k in H_D tends to zero as k tends to infinity.

Since the convergence rate depends on the choice of the iterative parameters τ_k, it follows that we should choose them so that the rate of convergence will be maximal.

6.1.2 The problem for the error. We will first investigate the convergence of the two-level iterative schemes (2). For this we obtain an equation for the error z_k.

Substituting $y_k = z_k + u$ in (2) for $k = 0, 1, \ldots$ and using equation (1) we find

$$B\frac{z_{k+1} - z_k}{\tau_{k+1}} + Az_k = 0, \qquad k = 0, 1, \ldots, \; z_0 = y_0 - u,$$

i.e. the error z_k satisfies a homogeneous equation. Solving this equation for z_{k+1}:

$$z_{k+1} = (E - \tau_{k+1}B^{-1}A)z_k$$

and setting $z_k = D^{-1/2}x_k$, we transform to an equation for the equivalent error x_k which will contain one operator. The equation for x_k will have the form

$$x_{k+1} = S_{k+1}x_k, \quad S_{k+1} = E - \tau_{k+1}C, \quad k = 0, 1, \ldots, \tag{3}$$

where $C = D^{1/2}B^{-1}AD^{-1/2}$. Using this change of variable we have

$$\| x_k \| = \| D^{1/2}z_k \| = \| z_k \|_D,$$

therefore the problem of investigating the convergence of the iterative method (2) in H_D leads to the study of the scalar sequence $\| x_k \|$, $k = 1, 2, \ldots$, where x_k is defined in (3).

We now find the solution of equation (3). From (3) we obtain

$$x_k = T_{k,0}x_0, \qquad T_{k,0} = \prod_{i=1}^{k} S_i = S_n S_{n-1} \dots S_1.$$

This leads to the following estimate for the norm of the error z_k in H_D:

$$\| z_k \|_D = \| x_k \| \le \| T_{k,0} \| \, \| x_0 \| = \| T_{k,0} \| \, \| z_0 \|_D . \tag{4}$$

The operator $T_{k,0}$ is called the ***resolving operator*** for the k-th iteration, and S_k is the ***transformation operator*** from the $(k-1)$-st iteration to the k-th iteration.

From the estimate (4) it follows that the iterative method (2) converges in H_D if the norm of the resolving operator $T_{k,0}$ tends to zero as k tends to infinity.

Thus, the problem of investigating the convergence of the iterative scheme (2) in H_D leads to the study of the behavior of the norm of the resolving operator $T_{k,0}$ in the space H as the iteration number k varies.

The resolving operator $T_{k,0}$ is defined by the operator C and by the iterative parameters $\tau_1, \tau_2, \dots, \tau_k$.

Considering the operator C fixed, we pose the problem of choosing the parameters $\{\tau_k\}$ so that the iterative method will converge. Among the convergent iterative methods, the ***optimal*** method will clearly be the one for which the parameters τ_k guarantee the attainment of a given accuracy $\epsilon > 0$ after the minimal number of iterations. Using the estimate (4), this requirement can be phrased in the following equivalent form: for a given n construct a set of iterative parameters $\tau_1, \tau_2, \dots, \tau_n$ for which the norm of the operator $T_{n,0}$ will be minimal.

6.1.3 The self-adjoint case. We now rigorously pose the problem of the best choice for the iterative parameters for the two-level scheme (2). This problem will have a solution under certain assumptions about the operators A, B and D. We now formulate these assumptions.

1) We will assume that the operators A, B, and D are such that the operator $DB^{-1}A$ is self-adjoint in H. If this assumption is satisfied, then we will say that we are considering the self-adjoint case.

2) We will also assume that we are given γ_1 and γ_2 — the constants of energy equivalence for the operator D and $DB^{-1}A$, i.e. the constants from the inequalities

$$\gamma_1 D \le DB^{-1}A \le \gamma_2 D, \qquad \gamma_1 > 0, \qquad DB^{-1}A = (DB^{-1}A)^*. \tag{5}$$

The second assumption defines the type of *a priori* information about the operators for the iterative scheme; this information is used to construct formulas for the iterative parameters in the self-adjoint case. The simplest example, for which the assumption about the self-adjointness of the operator $DB^{-1}A$ is satisfied, is as follows: $A = A^*$, $D = B = E$, i.e. an explicit scheme in the original space H for equation (1) with a self-adjoint operator A. In this case, the *a priori* information consists of bounds for the operator A. More complex examples of possible operators D will be considered below.

Thus, suppose that condition (5) is satisfied. From (5) it follows that the operator $C = D^{-1/2}(DB^{-1}A)D^{-1/2}$ is self-adjoint in H, and γ_1 and γ_2 are its bounds, i.e.

$$\gamma_1 E \leq C \leq \gamma_2 E, \quad \gamma_1 > 0, \quad C = C^* = D^{-1/2}(DB^{-1}A)D^{-1/2}. \tag{6}$$

In fact, substituting $x = D^{-1/2}y$ in the inequalities

$$\gamma_1(Dx, x) \leq (DB^{-1}Ax, x) \leq \gamma_2(Dx, x)$$

we obtain (6). Thus, the assumptions made above about the operators A, B and D are equivalent to the conditions (6).

We now formulate the problem of the optimal choice of the iterative parameters for the scheme (2). From the definition of the resolving operator $T_{k,0}$ and the conditions (6) it follows that the operator $T_{k,0} = T_{k,0}(C)$ is self-adjoint in H and the norm of the polynomial operator $T_{n,0}(C)$ can be estimated in the following way:

$$\| T_{n,0} \| \leq \max_{\gamma_1 \leq t \leq \gamma_2} \left| \prod_{k=1}^{n} (1 - \tau_k t) \right|.$$

From the estimate (4) it follows that in the self-adjoint case the iterative parameters $\tau_1, \tau_2, \ldots, \tau_n$ must be chosen so that the maximum modulus of the polynomial

$$P_n(t) = \prod_{k=1}^{n} (1 - \tau_k t),$$

constructed using these parameters will be minimal on the interval $[\gamma_1, \gamma_2]$, i.e. it is necessary to find the parameters from the condition

$$\min_{\{\tau_k\}} \max_{\gamma_1 \leq t \leq \gamma_2} \left| \prod_{k=1}^{n} (1 - \tau_k t) \right| = \max_{\gamma_1 \leq t \leq \gamma_2} |P_n(t)|.$$

Then we will have the estimate $\| z_n \|_D \leq q_n \| z_0 \|_D$ for the error in the method (2), where

$$q_n = \max_{\gamma_1 \leq t \leq \gamma_2} |P_n(t)|.$$

The problem formulated above is the classical minimax problem. In Section 6.2 we will derive the solution of this problem and construct a set of iterative parameters $\tau_1, \tau_2, \ldots, \tau_n$. The iterative method with this set of parameters is called the *Chebyshev method.* In the literature, this method is also called the *Richardson method.*

6.2 The Chebyshev two-level method

6.2.1 Construction of the set of iterative parameters. In Section 6.1 it was shown that the construction of the optimal set of iterative parameters τ_1, τ_2, $\ldots$, τ_n leads to the finding of a polynomial $P_n(t)$ of the form

$$P_n(t) = \prod_{k=1}^{n} (1 - \tau_k t),$$

whose maximum modulus on the interval $[\gamma_1, \gamma_2]$ is minimal.

We now solve this problem. Since the form of the polynomial is defined by the normalization condition $P_n(0) = 1$, this problem is formulated as follows: among all polynomials of degree n which take on the value 1 for $t = 0$, find the polynomial which deviates least from zero on the interval $[\gamma_1, \gamma_2]$ not containing the point 0.

The solution of this problem was obtained by the Russian mathematician V.A. Markov in 1892 and was mentioned in an appendix. The desired polynomial $P_n(t)$ has the form

$$P_n(t) = q_n \, T_n\left(\frac{1 - \tau_0 t}{\rho_0}\right), \qquad q_n = \frac{1}{T_n\left(\frac{1}{\rho_0}\right)}, \tag{1}$$

where $T_n(x)$ is the n-th degree Chebyshev polynomial of the first type,

$$T_n(x) = \begin{cases} \cos(n \arccos x), & |x| \leq 1, \\ \cosh(n \cosh^{-1} x) & |x| \geq 1, \end{cases}$$

$$q_n = \frac{2\rho_1^n}{1 + \rho_1^{2n}}, \quad \tau_0 = \frac{2}{\gamma_1 + \gamma_2}, \quad \rho_0 = \frac{1 - \xi}{1 + \xi}, \quad \rho_1 = \frac{1 - \sqrt{\xi}}{1 + \sqrt{\xi}}, \quad \xi = \frac{\gamma_1}{\gamma_2}. \tag{2}$$

Here

$$\max_{\gamma_1 \le t \le \gamma_2} |P_n(t)| = q_n.$$

From this we obtain an estimate for the norm of the error z_n in H_D:

$$\| z_n \|_D \le q_n \| z_0 \|_D, \tag{3}$$

where q_n is defined in (2).

We now obtain formulas for the iterative parameters. Since the polynomials on the left- and right-hand sides in (1) both take on the value 1 for $t = 0$, the identity in (1) will only be valid in the case where the roots of the polynomials $P_n(t)$ and $T_n\left(\frac{1-\tau_0 t}{\rho_0}\right)$ are the same. The polynomial $P_n(t)$ has roots $1/\tau_k$, $k = 1, 2, \ldots, n$, and the polynomial $T_n(x)$ has roots equal to

$$-\cos\left(\frac{2i-1}{2n}\pi\right), \qquad i = 1, 2, \ldots, n.$$

If we denote by $\mathcal{M}_n$ the set of roots of the Chebyshev polynomial $T_n(x)$:

$$\mathcal{M}_n = \left\{-\cos\frac{2i-1}{2n}\pi, \quad i = 1, 2, \ldots, n\right\}, \tag{4}$$

then we obtain the following formula for the iterative parameters:

$$\tau_k = \tau_0/(1 + \rho_0 \mu_k), \qquad \mu_k \in \mathcal{M}_n, \qquad k = 1, 2, \ldots, n. \tag{5}$$

Here $\mu_k \in \mathcal{M}_n$ indicates that μ_k runs sequentially through all elements of the set $\mathcal{M}_n$.

From the resulting formula for the parameters τ_k it is clear that the iteration count n must be known in order to compute the iterative parameters. Therefore we now estimate the iteration count. Usually the termination criterion for the iterative process is taken to be the inequality

$$\| z_n \|_D \le \epsilon \| z_0 \|_D$$

and the *iteration count* is the smallest integer n for which this inequality is satisfied.

From (3) it follows that for this method the iteration count is found from the inequality $q_n \le \epsilon$. Using (2), we find the solution to this inequality. We obtain

$$n \ge n_0(\epsilon), \quad n_0(\epsilon) = \ln\left(\frac{1}{\epsilon} + \sqrt{\frac{1}{\epsilon^2} - 1}\right) \Big/ \ln\frac{1}{\rho_1}.$$

Usually a simpler formula for $n_0(\epsilon)$ is used

$$n \geq n_0(\epsilon), \qquad n_0(\epsilon) = \ln\frac{2}{\epsilon}\Big/\ln\frac{1}{\rho_1} \tag{6}$$

After the required iteration count n has been found, the formulas (5) can be used to construct the set of iterative parameters.

Thus, for the implicit two-level scheme

$$B\frac{y_{k+1} - y_k}{\tau_{k+1}} + Ay_k = f, \qquad k = 0, 1, \dots, \qquad y_0 \in H, \tag{7}$$

we have

Theorem 1. *Suppose that*

$$\gamma_1 D \leq DB^{-1}A \leq \gamma_2 D, \quad \gamma_1 > 0, \quad DB^{-1}A = (DB^{-1}A)^*, \quad D = D^* > 0. \tag{8}$$

Then the Chebyshev iterative process (7), (4), (5), (2) *converges in* H_D, *and we have the estimate* (3) *for the error. The iteration count satisfies the estimate* (6). □

From the obtained estimates it follows that, in the self-adjoint case, the convergence rate for the Chebyshev method depends on the ratio $\xi = \gamma_1/\gamma_2$, where the convergence rate increases as ξ grows.

6.2.2 On the optimality of the *a priori* estimate. We now show that, on the class of arbitrary initial approximations y_0, the estimate obtained in theorem 1 for the error of the Chebyshev method is optimal in the case of a finite-dimensional space H. It is sufficient to display an initial approximation y_0 for which the norm of the equivalent error x_k satisfies the equation $\| x_n \| = q_n \| x_0 \|$. We now find the initial error x_0 which guarantees this, and the initial approximation y_0 (using the connection between the errors z_k and x_k, $z_k = D^{-1/2}x_k$) is defined by the formula $y_0 = u + D^{-1/2}x_0$.

We now find the desired x_0. Assume that H is a finite-dimensional space ($H = H_N$). Since the operator C is self-adjoint in H, there exists a complete system of eigenfunctions $v_1, v_2, \dots, v_N$ for the operator C. We denote by λ_k the eigenvalue of the operator C corresponding to the eigenfunction v_k. Assume that the eigenvalues are ordered $\lambda_1 \leq \lambda_2 \leq \dots \leq \lambda_N$. Then the bounds for the operator C can be taken as $\gamma_1 = \lambda_1$ and $\gamma_2 = \lambda_N$.

As the initial error x_0 we choose the eigenfunction v_1. From the equation for the error x_k:

$$x_{k+1} = (E - \tau_{k+1}C)x_k, \quad k = 0, 1, \ldots, \quad x_0 = v_1.$$

and the equation $Cv_k = \lambda_k v_k$ we sequentially obtain

$$x_1 = (E - \tau_1 C)x_0 = (1 - \tau_1\gamma_1)v_1 = (1 - \tau_1\gamma_1)x_0,$$
$$x_2 = (E - \tau_2 C)x_1 = (1 - \tau_1\gamma_1)(E - \tau_2 C)x_0 = (1 - \tau_1\gamma_1)(1 - \tau_2\gamma_1)x_0,$$
$$\cdots\cdots\cdots\cdots\cdots\cdots\cdots\cdots\cdots\cdots\cdots\cdots\cdots\cdots\cdots\cdots$$
$$x_n = \prod_{k=1}^{n}(1 - \tau_k\gamma_1)x_0 = P_n(\gamma_1)(x_0).$$

Substituting $t = \gamma_1$ in (1) and taking into account the equation $1-\tau_0\gamma_1 = \rho_0$, we compute $P_n(\gamma_1) = q_n T_n(1) = q_n$, and consequently,

$$x_n = q_n x_0, \quad \| x_n \| = q_n \| x_0 \|,$$

which is what we were required to prove.

Thus, it has been shown that the *a priori* estimate obtained in theorem 1 is optimal in the class of arbitrary initial approximations.

6.2.3 Sample choices for the operator D. We introduce now some sample choices for the operator D. Recall that the Chebyshev method is being looked at under the assumption that the operator $DB^{-1}A$ is self-adjoint. Below we will indicate the requirements on the operators A and B for which this assumption will be satisfied for the given choice of D. For each concrete choice of the operator D, we will give inequalities which supply the *a priori* information about the operators for the iterative scheme. This information is used to construct the set of iterative parameters in the Chebyshev method.

Example 1. Suppose that A and B are self-adjoint and positive-definite in H. Then it is possible to choose the operator D as one of the following operators: either A or B. If, in addition, the operator B is bounded in H, then it is possible to take $D = AB^{-1}A$. Here the *a priori* information consists of the constants of energy equivalence for the operators A and B:

$$\gamma_1 B \le A \le \gamma_2 B, \qquad \gamma_1 > 0, \qquad B > 0. \tag{9}$$

In fact, it is necessary to show that the following conditions are satisfied: the chosen operator D is self-adjoint and positive-definite in H, the operator $DB^{-1}A$ is self-adjoint in H, and the inequalities (8) and (9) are equivalent.

The self-adjointness of the operators D and $DB^{-1}A$ for all considered cases follows from the self-adjointness of the operators A and B. In the case where $D = A$ or $D = B$, the positive-definiteness of D follows from the positive-definiteness of the operators A and B. We show now that the operator $D = AB^{-1}A$ is also positive definite in H.

Suppose that the conditions on the operators A and B formulated above are satisfied: $A = A^* \geq \alpha E$, $B = B^* \geq \beta E$, $\| Bx \| \leq M \| x \|$, $\alpha, \beta > 0$, $M < \infty$. From these conditions and lemmas 6 and 8 from Section 5.1 we obtain that $B^{-1} \geq \frac{1}{M}E$ and $(Ax, Ax) \geq \alpha(Ax, x) \geq \alpha^2(x, x)$. From this we find that the energy of the operator D is bounded from below

$$\begin{aligned}(Dx, x) &= (AB^{-1}Ax, x) = (B^{-1}Ax, Ax)\\ &\geq \frac{1}{M}(Ax, Ax) \geq \frac{\alpha^2}{M}(x, x), \quad \text{i.e.} \quad D \geq \frac{\alpha^2}{M}E.\end{aligned}$$

Consequently, the positive-definiteness of the operator $D = AB^{-1}A$ has been proved.

We now show that the inequalities (8) and (9) are equivalent for these examples. Suppose that (9) is satisfied:

$$\gamma_1(Bx, x) \leq (Ax, x) \leq \gamma_2(Bx, x), \qquad \gamma_1 > 0, \tag{10}$$

If $D = B$, then $DB^{-1}A = A$ and, consequently, inequalities (10) and (8) are the same. Suppose now that $D = AB^{-1}A$. In this case $DB^{-1}A = AB^{-1}AB^{-1}A$ and, substituting $x = B^{-1}Ay$ in (10) we obtain

$$\gamma_1(AB^{-1}Ay, y) \leq (AB^{-1}Ay, B^{-1}Ay) \leq \gamma_2(AB^{-1}Ay, y)$$

or

$$\gamma_1(Dy, y) \leq (DB^{-1}Ay, y) \leq \gamma_2(Dy, y),$$

i.e. we obtain inequality (8). The reverse path from (8) to (10) is obvious.

If $D = A$ then $DB^{-1}A = AB^{-1}A$. From lemma 9 Section 5.1 it follows that, for self-adjoint and positive-definite operators A and B, inequality (10) and

$$\gamma_1(A^{-1}x, x) \leq (B^{-1}x, x) \leq \gamma_2(A^{-1}x, x), \qquad \gamma_1 > 0$$

are equivalent. Setting here $x = Ay$, we obtain (8). The reverse path is obvious.

This inequality allows us to prove at once the positive-definiteness of D:

$$(Dx, x) \geq \alpha\gamma_1(x, x).$$

In fact,

$$(Dx, x) = (B^{-1}Ax, Ax) \geq \gamma_1(A^{-1}Ax, Ax) = \gamma_1(Ax, x) \geq \gamma_1\alpha(x, x).$$

Example 2. Suppose that the operators A and B are self-adjoint, positive-definite in H, and commutative: $A = A^* > 0$, $B = B^* > 0$, $AB = BA$. If D is taken to be the operator A^2, then the ***a priori*** information can be given in the form of the inequalities (9).

The self-adjointness and positive-definiteness of the operator D follows from the self-adjointness and non-singularity of the operator A. Further, $DB^{-1}A = A(AB^{-1})A$, and since the operators A and B commute, the operators A and B^{-1} also commute. From this and from the self-adjointness of the operators A and B, the self-adjointness of the operator $DB^{-1}A$ follows.

In this case, the inequality (8) has the form

$$\gamma_1(Ax, Ax) \leq (AB^{-1}Ax, Ax) \leq \gamma_2(Ax, Ax), \qquad \gamma_1 > 0.$$

Substituting here $x = A^{-1}B^{1/2}y$ and using the commutativity of the roots of the operator B with the operator A, we find

$$\gamma_1(By, y) \leq (Ay, y) \leq \gamma_2(By, y),$$

i.e. we obtain inequality (9). The path from (9) to (8) is obvious.

We now consider one further example. Suppose that A and B are arbitrary non-singular operators satisfying the conditon

$$B^*A = A^*B. \tag{11}$$

If we choose D to be the operator A^*A then the ***a priori*** information can be given in the form of the inequalities

$$\gamma_1(Bx, Bx) \leq (Ax, Bx) \leq \gamma_2(Bx, Bx), \qquad \gamma_1 > 0. \tag{12}$$

The self-adjointness of the operator D is obvious, and the positive-definiteness follows from the non-singularity of the operator A. Since the operator B

is non-singular, condition (11) can be written in the form of the condition $AB^{-1} = (B^*)^{-1}A^*$, which expresses the self-adjointness of the operator AB^{-1}. From this we obtain that the operator $DB^{-1}A = A^*AB^{-1}A$ is self-adjoint in H. Further, substituting $x = B^{-1}Ay$ in (12), we obtain

$$\gamma_1(Ay, Ay) \leq (AB^{-1}Ay, Ay) \leq \gamma_2(Ay, Ay)$$

or

$$\gamma_1(Dy, y) \leq (DB^{-1}Ay, y) \leq \gamma_2(Dy, y).$$

Thus, from the inequalities (12), (8) follows. The reverse path from (8) to (12) is obvious.

In conclusion we remark that, for the case of self-adjoint, positive-definite, and bounded operators A and B in H, the Chebyshev iterative method converges in H_D, where $D = A, B$, or $AB^{-1}A$ (and if, in addition, A and B commute, then we can also choose $D = A^2$), with an identical rate, defined by the ratio of the constants γ_1 and γ_2 from the inequalities (9).

We particularly comment on the cases $D = AB^{-1}A$ and $D = A^*A$. With this choice of the operator D, the norm of the error in H_D can be computed during the iterative process. In fact, for $D = AB^{-1}A$ we obtain

$$\| z_n \|_D^2 = (Dz_n, z_n) = (B^{-1}Az_n, Az_n) = (B^{-1}r_n, r_n) = (w_n, r_n),$$

and for $D = A^*A$:

$$\| z_n \|_D^2 = (Az_n, Az_n) = (r_n, r_n),$$

where $r_n = Az_n = Ay_n - Au = Ay_n - f$ is the residual at the n-th iteration, and $w_n = B^{-1}r_n$ is the correction. These quantities can be found during the iterative process.

6.2.4 On the computational stability of the method. When studying the convergence of the Chebyshev method it was assumed that the computational process is ideal, i.e. the computations are carried out in infinite precision. In a real computational process, all the computations are performed with a finite number of digits, and at each stage rounding errors appear. Rounding errors in the results of arithmetic operations give rise to computational errors in the method.

In iterative methods, the computational error of the method is formed from the errors arising at each iteration. If the number of iterations is sufficiently great, and the iterative method possesses the property that rounding errors are accumulated at each iterative step, then the computational error of such a method can be so great that it produces a complete loss of accuracy and the iterative approximation y_n will be significantly different from the desired solution. Therefore, for iterative methods it is important to study the mechanism whereby errors arise, and to find those steps of the algorithm which produce growth in the computational error of the method. In a series of cases, certain changes in the computational process allow us to diminish the growth in the computational error and to make the method suitable for practical computation.

Another peculiarity of a real computational process is connected with the presence on a computer of a "machine zero" and a "machine infinity". These concepts characterize the size of numbers which can be represented on a computer. For example, on the BESM-6 computer in single precision it is possible to represent real numbers whose absolute values lie in the range 10^{-19} to 10^{19}. These are also bounds for "machine zero" and "machine infinity". If the result of a computation on the computer is a number which does not lie in this interval, then the computation is stopped and leads to a so-called "machine exception". Therefore the requirement of an "exceptionless" iterative process is natural.

Thus the iterative method must be "exceptionless" and stable with respect to the rounding errors.

In Section 6.2.1 we constructed the Chebyshev two-level method. In theorem 1 it was shown that, if we perform n iterations with the parameters $\tau_k = \tau_0/(1+\rho_0\mu_k)$, $\mu_k \in \mathcal{M}_n$, $k = 1, 2, \dots, n$, then the error z_n will satisfy $\| z_n \|_D \leq q_n \| z_0 \|_D$. For the μ_k we sequentially choose all the elements of the set $\mathcal{M}_n$ in any order.

We now study the computational stability of the Chebyshev method. For definiteness we will say that μ_k is the k-th element of the set $\mathcal{M}_n$. Then a different ordering of the set $\mathcal{M}_n$ will give rise to a different sequence $\{\mu_k\}$ and, consequently, to a different sequence of iterative parameters $\{\tau_k\}$.

From the point of view of an ideal computational process, all the sequences of Chebyshev iterative parameters are equivalent, i.e. each sequence must guarantee that we obtain the same approximation y_n and, consequently, the same accuracy after the completion of n iterations. Different rounding errors in a real computational process lead to non-equivalent sequences of iterative parameters.

We illustrate this assertion with an example. Suppose that on the grid $\bar\omega = \{x_i = ih,\ 0 \leq i \leq N,\ h = 1/N\}$ we are required to find the solution to the following difference problem:

$$\Lambda y = y_{\bar{x}x} - dy = -\varphi(x), \qquad x \in \omega,$$
$$y(0) = 0, \qquad y(1) = 1, \qquad d = \text{constant} > 0.$$

In Section 5.2 it was shown that the difference scheme can be reduced to the operator equation

$$Ay = f, \tag{13}$$

where the operator A is defined as follows: $Ay = -\Lambda\mathring{y}$, where $\mathring{y} \in \mathring{H}$, $\mathring{y}(x) = y(x)$ for $x \in \omega$. Here $\mathring{H}$ is the set of grid functions defined on $\bar{\omega}$ and reducing to zero for $x = 0$ and $x = 1$, and H is the space of grid functions defined in ω with inner product

$$(u, v) = \sum_{x\in\omega} u(x)v(x)h.$$

The right-hand side f of equation (13) differs from the right-hand side φ of the difference scheme only in the near-boundary nodes of the grid: $f(x) = \varphi(x)$, $h \le x \le 1 - 2h$, $f(1 - h) = \varphi(1 - h) + 1/h^2$.

To approximately solve equation (13), we look at the explicit Chebyshev method

$$\frac{y_{k+1} - y_k}{\tau_{k+1}} + Ay_k = f, \qquad k = 0, 1, \ldots, \quad y \in H. \tag{14}$$

Since the operators A and $B = E$ are self-adjoint in H, it follows from the examples considered in Section 6.2.3 that it is sufficient to use the bounds on the operator A: $\gamma_1 E \le A \le \gamma_2 E$, $\gamma_1 > 0$, as the *a priori* information for the Chebyshev method (14) if the operator D is taken to be $B = E$. It is obvious that γ_1 and γ_2 are the same as the smallest and largest eigenvalues of the difference operator Λ, i.e.

$$\gamma_1 = \frac{4}{h^2}\sin^2\frac{\pi h}{2} + d, \qquad \gamma_2 = \frac{4}{h^2}\cos^2\frac{\pi h}{2} + d.$$

The iterative parameters τ_k are computed from the formulas

$$\begin{aligned} \tau_k = \tau_0/(1 + \rho_0\mu_k), \qquad & \mu_k \in \mathcal{M}_n, \qquad k = 1, 2, \ldots, n, \\ \tau_0 = 2/(\gamma_1 + \gamma_2), \qquad & \rho_0 = (\gamma_2 - \gamma_1)/(\gamma_2 + \gamma_1). \end{aligned} \tag{15}$$

We considered three sequences of iterative parameters, defined by the following orderings for $\mathcal{M}_n$:

1) the "forward" sequence

$$\mathcal{M}_n = \mathcal{M}_n^{(1)} = \{\sigma_1, \sigma_2, \ldots, \sigma_n\}, \qquad \text{i.e.} \quad \mu_k = \sigma_k, \quad k = 1, 2, \ldots, n;$$

2) the "reverse" sequence

$$\mathcal{M}_n = \mathcal{M}_n^{(2)} = \{\sigma_n, \sigma_{n-1}, \ldots, \sigma_1\}, \qquad \text{i.e.} \quad \mu_k = \sigma_{n-k+1}, \quad k = 1, 2, \ldots, n;$$

3) the "alternating" sequence

$$\mathcal{M}_n = \mathcal{M}_n^{(3)} = \{\sigma_1, \sigma_n, \sigma_2, \sigma_{n-1} \ldots\},$$
$$\text{i.e. } \mu_{2k-1} = \sigma_k,\ \mu_{2k} = \sigma_{n-k+1},\ k = 1, 2, \ldots, n/2.$$

Here we have denoted

$$\sigma_k = -\cos\frac{2k-1}{2n}\pi.$$

The computation was carried out in the following way: select the number of iterations n and use the scheme (14), (15) for each sequence of iterative parameters to carry out n iterations. The actual accuracy which was achieved after n iterations was defined by the formula

$$\epsilon_{\text{actual}} = \frac{\| y_n - u \|}{\| y_0 - u \|}.$$

For comparison, we computed the value of q_n, where

$$q_n = \frac{2\rho_1^n}{1+\rho_1^{2n}}, \qquad \rho_1 = \frac{1-\sqrt{\xi}}{1+\sqrt{\xi}}, \qquad \xi = \frac{\gamma_1}{\gamma_2},$$

which defines the theoretical accuracy of the method, when the number of iterations is equal to n. In all cases the initial approximation y_0 was set equal to zero on ω. The exact solution of the difference problem $y(x) = x$ corresponds to the right-hand side $\varphi(x) = dx$. The coefficient d was chosen so that γ_1 was equal to 0.1:

$$\gamma_1 = 0.1, \qquad \gamma_2 = 0.1 + \frac{4}{h^2}\cos\pi h, \qquad \frac{1}{\xi} = \frac{40}{h^2}\cos\pi h + 1.$$

The results of the computations for $N = 10$ are given in table 5. In this table, in addition to the indicated sequence of parameters, we give the results for the optimal ordering of the set $\mathcal{M}_n^*$, which will be described below.

These results show that for a real computational process the sequences of iterative parameters are not equivalent. The computations demonstrated two characteristic peculiarities of a real computational process: the possibility of "overflow", arising from growth in the intermediate iterative solutions, and the possibility of loss of accuracy in an exceptionless situation, resulting from the accumulation of rounding errors.

The cause of such computational instability in the method for certain sequences of iterative parameters is the fact that the norm of the transformation operator from iteration to iteration $S_k = E - \tau_k C$ is greater than one for some values of k.

Table 5

n	q_n	ϵ_{actual}			
		$\mathcal{M}_n^{(1)}$	$\mathcal{M}_n^{(2)}$	$\mathcal{M}_n^{(3)}$	$\mathcal{M}_n^{*}$
16	$8.79 \cdot 10^{-1}$	$8.14 \cdot 10^{-1}$	$8.14 \cdot 10^{-1}$	$8.14 \cdot 10^{-1}$	$8.14 \cdot 10^{-1}$
24	$7.58 \cdot 10^{-1}$	$9.62 \cdot 10^{-1}$	$7.11 \cdot 10^{-1}$	$7.11 \cdot 10^{-1}$	$7.11 \cdot 10^{-1}$
32	$6.30 \cdot 10^{-1}$	$3.38 \cdot 10^{3}$	$3.55 \cdot 10^{2}$	$5.63 \cdot 10^{-1}$	$5.63 \cdot 10^{-1}$
40	$5.09 \cdot 10^{-1}$	$3.07 \cdot 10^{7}$	$2.44 \cdot 10^{6}$	$5.03 \cdot 10^{-1}$	$4.85 \cdot 10^{-1}$
48	$4.04 \cdot 10^{-1}$	overflow	$3.46 \cdot 10^{10}$	$2.47 \cdot 10^{0}$	$3.64 \cdot 10^{-1}$
56	$3.17 \cdot 10^{-1}$	–	$1.02 \cdot 10^{15}$	$2.29 \cdot 10^{2}$	$3.10 \cdot 10^{-1}$
64	$2.47 \cdot 10^{-1}$	–	overflow	$1.87 \cdot 10^{4}$	$2.23 \cdot 10^{-1}$
72	$1.92 \cdot 10^{-1}$	–	–	$1.73 \cdot 10^{6}$	$1.72 \cdot 10^{-1}$
80	$1.49 \cdot 10^{-1}$	–	–	overflow	$1.44 \cdot 10^{-1}$
...	...	...	...	...	...
256	$4.97 \cdot 10^{-4}$	–	–	–	$4.80 \cdot 10^{-4}$
...	...	...	...	...	...
512	$1.23 \cdot 10^{-7}$	–	–	–	$1.15 \cdot 10^{-7}$

In fact, since S_k is a self-adjoint operator in H, $\| S_k \| = \sup\limits_{\|x\|=1} |(S_k x, x)|$. Using the bounds γ_1, γ_2 for the operator C

$$\gamma_1 E \leq C \leq \gamma_2 E, \qquad \gamma_1 > 0,$$

we find

$$(1 - \tau_k \gamma_2) E \leq E - \tau_k C \leq (1 - \tau_k \gamma_1) E.$$

We substitute here τ_k from (15) and use the equations $1 - \rho_0 = \tau_0 \gamma_1$, $1 + \rho_0 = \tau_0 \gamma_2$. We obtain

$$-\frac{\rho_0(1-\mu_k)}{1+\rho_0\mu_k} E \leq S_k \leq \frac{\rho_0(1+\mu_k)}{1+\rho_0\mu_k} E$$

and, consequently,

$$\| S_k \| = \begin{cases} \dfrac{\rho_0(1+\mu_k)}{1+\rho_0\mu_k} < 1, & \mu_k \geq 0, \\ \dfrac{\rho_0(1-\mu_k)}{1+\rho_0\mu_k}, & \mu_k < 0. \end{cases}$$

From this it follows that $\| S_k \| > 1$ for $\mu_k < -(1-\rho_0)/(2\rho_0)$. Since $\mu_k \in \mathcal{M}_n$,

$$-\cos\frac{\pi}{2n} \leq \mu_k \leq -\cos\frac{2n-1}{2n}\pi = \cos\frac{\pi}{2n}, \qquad k = 1, 2, \ldots, n,$$

and, consequently, for k large the norm $\| S_k \| > 1$ (this is for k approximately equal to $n/2$). Therefore, if we sequentially use enough parameters τ_k for which the norm of the operator S_k is bigger than one, then it is possible to produce an accumulation of rounding errors and a growth of the iterative approximations which leads to computational instability in the method.

Theorem 1 actually expresses the stability of the iterative scheme in terms of the initial data. In the case of an actual computational process, it is also necessary to investigate the stability of the iterative scheme in terms of the right-hand side, since rounding errors can be interpreted as perturbations in the right-hand side of the iterative scheme at each iteration.

If rounding error is considered then, in place of a homogeneous equation for the equivalent error x_k, we obtain the non-homogeneous equation

$$x_{k+1} = S_{k+1}x_k + \tau_{k+1}\varphi_{k+1}, \qquad k = 0, 1, \ldots. \tag{16}$$

Here $x_k = D^{1/2}(\bar{y}_k - u)$, where $\bar{y}_k$ is the actual iterative approximation.

Solving equation (16), we find

$$x_n = T_{n,0}x_0 + \sum_{j=1}^{n} \tau_j T_{n,j}\varphi_j,$$

where

$$T_{n,j} = \prod_{i=j+1}^{n} S_i, \quad T_{n,n} = E.$$

From this we obtain the following estimate:

$$\| x_n \| \leq \| T_{n,0} \| \, \| x_0 \| + \sum_{j=1}^{n} \tau_j \| T_{n,j} \| \max_{1\leq j\leq n} \| \varphi_j \| . \tag{17}$$

The estimate of the norm of the operator $T_{n,0}$ does not depend on the ordering of the set $\mathcal{M}_n$, and for any sequence of Chebyshev parameters τ_k we have $\| T_{n,0} \| \leq q_n$. The estimate for

$$\sum_{j=1}^{n} \tau_j \| T_{n,j} \|$$

depends on the ordering of the set $\mathcal{M}_n$. From (17) it follows that the set $\mathcal{M}_n$ must be ordered so that this summation takes on its minimal value.

The following lemma indicates the smallest possible value of this summation.

Lemma 1. *If γ_1 and γ_2 are precise bounds on the operator C, then for any ordering of the set $\mathcal{M}_n$ we have the estimate.*

$$\sum_{j=1}^{n} \tau_j \| T_{n,j} \| \geq \frac{1-q_n}{\gamma_1}.$$

Proof. From the definition of the operator $T_{n,j}$ we obtain

$$\tau_j T_{n,j} = (T_{n,j} - T_{n,j-1})C^{-1}, \quad \sum_{j=1}^{n} \tau_j T_{n,j} = (E - T_{n,0})C^{-1}.$$

Since

$$\| (E - T_{n,0})C^{-1} \| = \left\| \sum_{j=1}^{n} \tau_j T_{n,j} \right\| \leq \sum_{j=1}^{n} \tau_j \| T_{n,j} \|,$$

then it is sufficient to estimate the norm of the operator $(E - T_{n,0})C^{-1}$. This operator is self-adjoint in H, and if γ_1 and γ_2 are bounds for the operator C, then

$$\| (E - T_{n,0})C^{-1} \| \leq \max_{\gamma_1 \leq t \leq \gamma_2} \left| \frac{1 - q_n T_n\left(\frac{1-\tau_0 t}{\rho_0}\right)}{t} \right|$$

$$= \frac{1 - q_n T_n\left(\frac{1-\tau_0\gamma_1}{\rho_0}\right)}{\gamma_1} = \frac{1-q_n}{\gamma_1}.$$

Thus, it has been shown that for any $x \in H$ we have the estimate

$$\| (E - T_{n,0})C^{-1}x \| \leq \frac{1 - q_n}{\gamma_1} \| x \| . \tag{18}$$

Since γ_1 is a precise bound for the self-adjoint operator C, γ_1 is smallest eigenvalue of the operator C. Substituting the eigenfunction corresponding to the smallest eigenvalue of the operator C in place of x in (18), we find that (18) becomes an equality. Consequently, we have obtained the estimate $\| (E - T_{n,0})C^{-1} \| = (1 - q_n)/\gamma_1$. The lemma is proved. □

6.2.5 Construction of the optimal sequence of iterative parameters*

6.2.5.1 *The case* $n = 2^p$. The order in which we use the iterative parameters τ_k in the Chebyshev method significantly influences the convergence of the method. Therefore there arises the problem of constructing the best sequence of iterative parameters, guaranteeing the minimal influence of rounding error on the method. Since the sequence of parameters is determined by the ordering of the set $\mathcal{M}_n$, it is necessary to construct the optimal ordering of the set $\mathcal{M}_n$.

We now derive the solution to this problem. Assume initially that the number of iterations is a power of 2: $n = 2^p$. We denote by θ_m the set consisting of the m integers:

$$\theta_m = \{\theta_1^{(m)}, \theta_2^{(m)}, \ldots, \theta_m^{(m)}\}.$$

Starting with the set $\theta_1 = \{1\}$, we construct the set θ_{2^p} according to the following rule. Suppose that the set θ_m has been constructed. Then the set θ_{2m} is defined by the formulas

$$\theta_{2m} = \{\theta_{2i}^{(2m)} = 4m - \theta_i^{(m)}, \quad \theta_{2i-1}^{(2m)} = \theta_i^{(m)}, \quad i = 1, 2, \ldots, m\},$$
$$m = 1, 2, 4, \ldots, 2^{p-1}. \tag{19}$$

It is not difficult to verify that the set θ_{2^k} consists of the odd integers between 1 and $2^{k+1} - 1$.

* The method of ordering the iterative parameters has been published elsewhere: see E.S. Nikolayev, A.A. Samarskii (Journal of Computational and Mathematical Physics, 12, No 4, 1972) for the case of arbitrary n and [8] for the case $n=2^p$.

Using the set θ_{2^p}, we order the set $\mathcal{M}_{2^p}$ as follows:

$$\mathcal{M}_n^* = \left\{-\cos\beta_i, \quad \beta_i = \frac{\pi}{2n}\theta_i^{(n)}, \quad i = 1,2,\ldots,n\right\}, \quad n = 2^p. \tag{20}$$

This is the desired ordering of the set $\mathcal{M}_n$ in the case when $n = 2^p$. For the sequence of iterative parameters corresponding to this ordering we have the estimate

$$\sum_{j=1}^{n} \tau_j \parallel T_{n,j} \parallel \leq \frac{1-q_n}{\gamma_1}.$$

Comparing this estimate with the estimate in lemma 1, we ascertain that this ordering of the set $\mathcal{M}_n^*$ in fact guarantees the minimal influence of rounding error on the convergence of the Chebyshev method.

We now give several examples of construction of the set θ_n.

1) $n = 8$

$$\theta_1 = \{1\}, \quad \theta_2 = \{1,3\}, \quad \theta_4 = \{1,7,3,5\},$$
$$\theta_8 = \{1,15,7,9,3,13,5,11\}.$$

The set θ_8 has been constructed. The set $\mathcal{M}_8^*$ is ordered using the formula (20).

2) $n = 16$. Using the set θ_8 found above, we construct the set θ_{16} using formula (19):

$$\theta_{16} = \{1,31,15,17,7,25,9,23,3,29,13,19,5,27,11,21\}.$$

3) $n = 32$.

$$\theta_{32} = \{1,63,31,33,15,49,17,47,7,57,25,39,9,55,23,41,3,61,$$
$$29,35,13,51,19,45,5,59,27,37,11,53,21,43\}.$$

From the formulas (19) there follows a simple rule for transforming from the set θ_m to the set θ_{2m}: $\theta_{2i-1}^{(2m)} = \theta_i^{(m)}$ and the sum of two neighboring numbers is equal to $4m$:

$$\theta_{2i-1}^{(2m)} + \theta_{2i}^{(2m)} = 4m, \qquad i = 1,2,\ldots,m.$$

An analogous transformation rule applies in the general case, which we now consider.

6.2.5.2 *The general case.* Suppose that the number of iterations n is any integer. We now describe the process for constructing the set θ_n. The elementary stages of this process are the transformations from the set θ_m to the set θ_{2m} and from the set θ_{2m} to the set θ_{2m+1}, where m is an arbitrary integer.

We now formulate the rules for transforming from set to set.

1) Transformation from θ_{2m} to θ_{2m+1} consists of adding the odd number $2m+1$ to the set θ_{2m}.

2) Transformation from θ_m to θ_{2m} is achieved as follows. If the next step is the transformation from θ_{2m} to θ_{4m} or if the transformation from θ_m to θ_{2m} is the last step in the process of constructing θ_n, then we use the formulas introduced above:

$$\theta_{2i-1}^{(2m)} = \theta_i^{(m)}, \quad \theta_{2i-1}^{(2m)} + \theta_{2i}^{(2m)} = 4m, \qquad i = 1, 2, \dots, m. \tag{21}$$

If the next step is the transformation from θ_{2m} to θ_{2m+1}, then we use the formulas

$$\theta_{2i-1}^{(2m)} = \theta_i^{(m)}, \quad \theta_{2i-1}^{(2m)} + \theta_{2i}^{(2m)} = 4m+2, \qquad i = 1, 2, \dots, m. \tag{22}$$

Using these rules and alternating as necessary between transformations from a set with an even number of elements to a set with an odd number of elements and from a set with m elements to a set with $2m$ elements, it is possible, starting from $\theta_1 = \{1\}$ to construct the set θ_n for any n.

We introduce several examples.

1) $n = 15$. In this case the transformation from θ_1 to θ_n is performed according to the following chain:

$$\theta_1 \to \theta_2 \to \theta_3 \to \theta_6 \to \theta_7 \to \theta_{14} \to \theta_{15}.$$

In accordance with the rules laid out above, the transformations from θ_1 to θ_2, from θ_3 to θ_6, and from θ_7 to θ_{14} are achieved using formulas (22), and for the trnasformations from θ_2 to θ_3, from θ_6 to θ_7, and from θ_{14} to θ_{15} it is necessary to add the corresponding odd numbers to the original sets. This gives

$$\begin{aligned} \theta_1 &= \{1\}, \quad \theta_2 = \{1, 5\}, \quad \theta_3 = \{1, 5, 3\}, \\ \theta_6 &= \{1, 13, 5, 9, 3, 11\}, \quad \theta_7 = \{1, 13, 5, 9, 3, 11, 7\}, \\ \theta_{14} &= \{1, 29, 13, 17, 5, 25, 9, 21, 3, 27, 11, 19, 7, 23\}, \\ \theta_{15} &= \{1, 29, 13, 17, 5, 25, 9, 21, 3, 27, 11, 19, 7, 23, 15\}. \end{aligned}$$

The set $\mathcal{M}_{15}^*$ is ordered using formula (20).

2) $n = 25$. To this case corresponds the chain

$$\theta_1 \to \theta_2 \to \theta_3 \to \theta_6 \to \theta_{12} \to \theta_{24} \to \theta_{25},$$

and the transformations from θ_1 to θ_2 and from θ_{12} to θ_{24} are obtained from formulas (22), the transformations from θ_3 to θ_6 and from θ_6 to θ_{12} using formulas (21), and the transformations from θ_2 to θ_3 and from θ_{24} to θ_{25} by adding an odd number. We obtain

$$\begin{aligned}
&\theta_1 = \{1\}, \quad \theta_2 = \{1,5\}, \quad \theta_3 = \{1,5,3\}, \quad \theta_6 = \{1,11,5,7,3,9\},\\
&\theta_{12} = \{1,23,11,13,5,19,7,17,3,21,9,15\},\\
&\theta_{24} = \{1,49,23,27,11,39,13,37,5,45,19,31,7,43,17,33,\\
&\qquad 3,47,21,29,9,41,15,35\},\\
&\theta_{25} = \{1,49,23,27,11,39,13,37,5,45,19,31,7,43,17,33,\\
&\qquad 3,47,21,29,9,41,15,35,25\}.
\end{aligned}$$

The procedure laid out above for constructing the set θ_n for arbitrary n can be formalized. To do this we represent n in the form of an expansion in powers of 2 with integer exponents k_j:

$$n = 2^{k_1} + 2^{k_2} \cdots + 2^{k_s}, \quad k_j \le k_{j-1} - 1, \quad j = 2,3,\ldots,s.$$

We form the following quantities:

$$n_j = \sum_{i=1}^{j} 2^{k_i - k_j}, \qquad j = 1,2,\ldots,s,$$

and set $n_{s+1} = 2n + 1$. Using the formulas (23) we construct the set θ_{n_j}:

$$\theta_{n_j} = \{\theta_i^{(n_j)} = \theta_i^{(n_j-1)}, \quad \theta_{n_j}^{(n_j)} = n_j, \quad i = 1,2,\ldots,n_j - 1\}, \qquad (23)$$

for $j = 1$ we choose $\theta_1 = \{1\}$. Then using formula (24) we construct the sets

$$\theta_{2m} = \{\theta_{2i}^{(2m)} = 4m - \theta_i^{(m)}, \quad \theta_{2i-1}^{(2m)} = \theta_i^{(m)}, \quad i = 1,2,\ldots,m\} \qquad (24)$$

for $m = n_j, 2n_j, 4n_j, \ldots, [(n_{j+1} - 1)/4]$, where $[a]$ is the integer part of a. If $[(n_{j+1} - 1)/4] < n_j$, then the computation in formula (24) is not carried out, the transformation to the next stage is completed. If $j = s$ then the necessary set θ_n has already been constructed. Otherwise set $m = (n_{j+1} - 1)/2$ and construct the set

$$\theta_{2m} = \{\theta_{2i}^{(2m)} = 4m + 2 - \theta_i^{(m)}, \quad \theta_{2i-1}^{(2m)} = \theta_i^{(m)}, \quad i = 1,2,\ldots,m\}. \quad (25)$$

Then j is increased by 1 and the process is repeated, starting with formula (23). As a result we have constructed the set θ_n. The set $\mathcal{M}_n^*$ is ordered according to formula (20).

For the case $n = 2^p$ algorithm (23)–(25) simplifies to the algorithm described by formula (19). In fact, for $n = 2^p$ we obtain $s = 1$, $k_1 = p$, $n_1 = 1$, $n_{s+1} = 2^{p+1} - 1$. Consequently, in algorithm (23)–(25) j only takes on the value one, and the computations proceeds according to formula (24) for $m = 1, 2, 4, \ldots, 2^{p-1}$.

We illustrate a property of the ordering constructed here for the set $\mathcal{M}_n^*$ on the example considered in Section 6.2.1. The given number of iterations was varied from 16 to 512 in increments of 8. For each n the actual precision attained after the completion of n iterations did not surpass the theoretical precision q_n ($q_{512} = 1.23 \cdot 10^{-7}$), and the process was "exceptionless" (see table 5).

6.3 The simple iteration method

6.3.1 The choice of the iterative parameter. In Section 6.2 we solved the problem of constructing the optimal set of iterative parameters τ_k for the two-level scheme

$$B\frac{y_{k+1} - y_k}{\tau_{k+1}} + Ay_k = f, \qquad k = 0, 1, \ldots, \qquad y_0 \in H$$

under the assumption that the operator $DB^{-1}A$ is self-adjoint in H and that we are given γ_1 and γ_2 — the constants of energy equivalence for the operators D and $DB^{-1}A$:

$$\gamma_1 D \le DB^{-1}A \le \gamma_2 D, \qquad \gamma_1 > 0. \tag{1}$$

We obtain now the solution of this problem with the additional limitation $\tau_k \equiv \tau$, i.e. under the assumption that the iterative parameters do not depend on the iteration number k. This problem arises when finding the iterative parameter τ for the stationary two-level scheme

$$B\frac{y_{k+1} - y_k}{\tau} + Ay_k = f, \qquad k = 0, 1, \ldots. \tag{2}$$

We recall now the formulation for the problem indicated above: among all polynomials of degree n of the form

$$Q_n(t) = \prod_{j=1}^{n}(1 - \tau_j t),$$

find the polynomial which deviates least from zero on the interval $[\gamma_1, \gamma_2]$. Because of the above limitation, the polynomial $P_n(t)$ has the form

$$P_n(t) = (1 - \tau t)^n.$$

Therefore the problem stated above is equivalent to the following: among all polynomials of first order which are equal to one for $t = 0$, find the polynomial which deviates least from zero on the interval $[\gamma_1, \gamma_2]$.

This problem is a special case of the problem considered in Section 6.2. In this case $n = 1$, and from the results of Section 6.2.1 it follows that the desired polynomial has the form

$$Q_1(t) = q_1 T_1\left(\frac{1-\tau_0 t}{\rho_0}\right), \quad \tau_0 = \frac{2}{\gamma_1+\gamma_2}, \quad \rho_0 = \frac{1-\xi}{1+\xi}, \quad \xi = \frac{\gamma_1}{\gamma_2},$$

where

$$q_1 = \frac{2\rho_1}{1+\rho_1^2} = \rho_0, \qquad \rho_1 = \frac{1-\sqrt{\xi}}{1+\sqrt{\xi}}.$$

Here $T_1(x)$ is a Chebyshev polynomial of the first kind. Since $T_1(x) = x$, the polynomial $Q_1(t)$ has the form

$$Q_1(t) = 1 - \tau_0 t, \qquad \max_{\gamma_1 \le t \le \gamma_2} |Q_1(t)| = q_1 = \rho_0,$$

therefore

$$P_n(t) = (1 - \tau_0 t)^n.$$

Thus, the optimal value of the parameter τ for the scheme (2) has been found:

$$\tau = \tau_0 = 2/(\gamma_1 + \gamma_2). \tag{3}$$

Since the norm of the resolving operator $T_{n,0}$ for the scheme (2) (see Section 6.1.3) is bounded as follows:

$$\| T_{n,0} \| \le \max_{\gamma_1 \le t \le \gamma_2} |P_n(t)|,$$

for $\tau = \tau_0$ we obtain the estimate $\| T_{n,0} \| \le \rho_0^n$. From this follows the estimate for the error z_n in H_D:

$$\| z_n \|_D \le \rho_0^n \| z_0 \|_D . \tag{4}$$

The iterative method (2), (3) is called the *simple iteration method.*

Thus, we have proved

Theorem 2. *Suppose that the self-adjoint operator $DB^{-1}A$ satisfies the conditions (1). The simple iteration method (2), (3) converges in H_D, and the error satisfies the estimate (4). For the iteration count we have the estimate $n \ge n_0(\epsilon)$, where $n_0(\epsilon) = \ln \epsilon / \ln \rho_0$.* □

Remark. As for the Chebyshev method the *a priori* estimate of the error for the simple iteration method is optimal in the case of a finite-dimensional space.

We now compare the iteration counts for the Chebyshev method and the simple iteration method. From theorem 1 for the case of small ξ we have the following estimate for the iteration count for the Chebyshev method:

$$n \geq n_0(\epsilon), \quad n_0(\epsilon) = \frac{\ln 0.5\epsilon}{\ln \rho_1} \approx \frac{1}{2\sqrt{\xi}} \ln \frac{2}{\epsilon}.$$

From theorem 2 we obtain an estimate for the iteration count for the simple iteration method

$$n \geq n_0(\epsilon), \quad n_0(\epsilon) = \frac{\ln \epsilon}{\ln \rho} \approx \frac{1}{2\xi} \ln \frac{1}{\epsilon}.$$

From these estimates it follows that for $\xi \ll 1$ the iteration count for the Chebyshev method is significantly less than the iteration count for the simple iteration method. For example, for $\xi = 0.01$ the iteration count for the simple iteration method is approximately 10 times larger than for the Chebyshev method.

6.3.2 An estimate for the norm of the transformation operator. In Section 3.1 we investigated te convergence rate for the simple iteration method. There the simple iteration method was considered as a special case of the Chebyshev method. For methodological reasons it will be useful to study the convergence of the simple iteration method independently from the Chebyshev method.

Thus, suppose that we are finding the approximate solution of the equation

$$Au = f$$

using the two-level scheme (2)

$$B\frac{y_{k+1} - y_k}{\tau} + Ay_k = f, \qquad k = 0, 1, \ldots, \qquad y_0 \in H. \tag{5}$$

To study the convergence of the scheme (5) we transform to the problem for the equivalent error $x_k = D^{1/2} z_k$:

$$x_{k+1} = Sx_k, \qquad k = 0, 1, \ldots, \qquad S = E - \tau C, \tag{6}$$

where $C = D^{1/2} B^{-1} A D^{-1/2}$. Using (6), we find an explicit expression for x_n in terms of x_0: $x_n = S^n x_0$, from which follows an estimate for the norm of the error z_n in H_D

$$\| z_n \|_D = \| x_n \| \leq \| S^n \| \, \| x_0 \| = \| S^n \| \, \| z_0 \|_D . \tag{7}$$

We will assume that the operator $DB^{-1}A$ is self-adjoint in H and that the constants γ_1 and γ_2 in the inequalities (1) are given. Under these assumptions the operator C, and in addition the operator S, are self-adjoint in H and γ_1 and γ_2 are bounds on the operator C:

$$\gamma_1 E \leq C \leq \gamma_2 E, \qquad \gamma_1 > 0, \qquad C = C^*. \tag{8}$$

By the self-adjointness of the operator S, we have the equation $\| S^n \| = \| S \|^n$. Therefore from the estimate (7) it follows that the iterative parameter τ must be chosen from the condition that the norm of the transformation operator $S = E - \tau C$ be a minimum.

We have

Lemma 2. *Suppose $S = E - \tau C$ and that the conditions* (8) *are satisfied. The norm of the operator S is minimal for $\tau = \tau_0 = 2/(\gamma_1 + \gamma_2)$, and we have the estimate*

$$\| S \| = \| E - \tau_0 C \| = \rho_0, \quad \rho_0 = (1-\xi)/(1+\xi), \qquad \xi = \gamma_1/\gamma_2.$$

Proof. Since S is self-adjoint in H, from the definition of the norm we obtain

$$\| S \| = \sup_{x \neq 0} \frac{|(Sx, x)|}{(x, x)} = \sup_{x \neq 0} \left| 1 - \tau \frac{(Cx, x)}{(x, x)} \right| = \max_{\gamma_1 \leq t \leq \gamma_2} |1 - \tau t|.$$

Since $\varphi(t) = 1 - \tau t$ is a linear function, the maximum value in modulus of $\varphi(t)$ on the interval $[\gamma_1, \gamma_2]$ can be achieved only at an endpoint of the interval. A direct computation gives

$$\| S \| = \max(|1 - \tau\gamma_1|, |1 - \tau\gamma_2|) = \begin{cases} \varphi_1(\tau) = 1 - \tau\gamma_1, & 0 \leq \tau \leq \tau_0, \\ \varphi_2(\tau) = \tau\gamma_2 - 1, & \tau_0 \leq \tau, \end{cases}$$

where τ_0 is indicated in the lemma. Since the function $\varphi_1(\tau)$ decreases on the interval $[0, \tau_0]$, and $\varphi_2(\tau)$ increases for $\tau \geq \tau_0$, the minimum of the norm of the operator S is achieved for $\tau = \tau_0$ and is equal to $\rho_0 = 1 - \tau_0\gamma_1 = \tau_0\gamma_2 - 1 = (1-\xi)/(1+\xi)$, $\xi = \gamma_1/\gamma_2$. The lemma is proved. □

From lemma 2 and the estimate (7) it follows that for $\tau = \tau_0$ the error for the iterative scheme (5) satisfies the estimate

$$\| z_n \|_D \leq \rho_0^n \| z_0 \|_D \, .$$

Thus, we have obtained another proof of theorem 2 formulated above concerning the convergence of the simple iteration method. Sample choices of the operator D, for which the self-adjointness condition on the operator $DB^{-1}A$ is satisfied, were considered in Section 6.2.3.

6.4 The non-self-adjoint case. The simple iteration method

6.4.1 Statement of the problem. In Sections 6.2, 6.3 we constructed two-level iterative methods for approximating the solution of the linear operator equation

$$Au = f \tag{1}$$

with a non-singular operator A, defined in the real Hilbert space H. It was assumed that the operators A, B, and D were such that the operator $DB^{-1}A$ is self-adjoint in H, and that the constants of energy equivalence γ_1 and γ_2 were given for the operators D and $DB^{-1}A$, where $\gamma_1 > 0$.

Under these assumptions, the problem of the optimal choice of the iterative parameters was solved and they were constructed for the Chebyshev method and for the simple iteration method. In Section 6.2.3 we looked at several sample choices for the operator D and found the conditions for the self-adjointness of the operator $DB^{-1}A$ for each concrete choice of the operator D.

Clearly, if the operators A and B are given, then it is not always possible to pick an operator D for which the operator $DB^{-1}A$ will be self-adjoint in H. Consequently, it is necessary to study iterative methods in the non-self-adjoint case also.

In this subsection we study the simple iteration method for the non-self-adjoint case. We will consider several ways of choosing the iterative parameter based on the volume of *a priori* information about the operators of the iterative scheme.

Thus, suppose that the operator $DB^{-1}A$ is *non-self-adjoint in H*. In order to approximately solve equation (1) we will consider the implicit two-level iterative scheme

$$B\frac{y_{k+1} - y_k}{\tau} + Ay_k = f, \qquad k = 0, 1, \ldots, \qquad y_0 \in H. \tag{2}$$

To investigate the convergence of the scheme (2), we as usual transform to the problem for the equivalent error $x_k = D^{1/2}z_k$

$$x_{k+1} = Sx_k, \qquad k = 0, 1, \ldots, \qquad S = E - \tau C, \tag{3}$$

where $C = D^{1/2}B^{-1}AD^{-1/2}$. Under the assumptions made above, the operator C is non-self-adjoint in H. From these changes and from equation (3) we obtain

$$x_n = S^n x_0, \quad \| x_n \| = \| z_n \|_D \leq \| S^n \| \| x_0 \| = \| S^n \| \| z_0 \|_D . \tag{4}$$

Consequently, the iterative parameter τ must be selected from the condition that the transformation operator S^n have minimal norm.

6.4.2 Minimizing the norm of the transformation operator

6.4.2.1 *The first case.* We now obtain an estimate for the norm of the operator S^n. Since for any operator we have the estimate $\| S^n \| \leq \| S \|^n$, the first way to choose the parameter τ consists of finding the parameter τ from the condition that the norm of the transformation operator S be a minimum. We obtain two types of estimates for the norm of the operator S depending on the volume of *a priori* information about the operator C.

In the first case it is assumed that the *a priori* information consists of the constants γ_1 and γ_2 from the inequalities

$$\gamma_1(x,x) \leq (Cx,x), \qquad (Cx,Cx) \leq \gamma_2(Cx,x), \qquad \gamma_1 > 0. \tag{5}$$

If $C = C^*$, then γ_1 and γ_2 are bounds for the operator C.

Lemma 3. *Suppose that γ_1 and γ_2 are given in the inequalities* (5); *then we have the following estimates for the norm of the operator $S = E - \tau C$ where $\tau = 1/\gamma_2$*

$$\| S \| \leq \rho, \quad \rho = \sqrt{1-\xi}, \quad \xi = \gamma_1/\gamma_2.$$

Proof. Using (5) we obtain

$$\begin{aligned} \| Sx \|^2 &= \| x - \tau Cx \|^2 = (x,x) - 2\tau(Cx,x) + \tau^2(Cx,Cx) \\ &\leq (x,x) - 2\tau(Cx,x) + \tau^2\gamma_2(Cx,x) = \| x \|^2 - \tau(2-\tau\gamma_2)(Cx,x). \end{aligned}$$

From this it follows that if the condition $\tau(2-\tau\gamma_2) > 0$ is satisfied, i.e. $0 < \tau < 2/\gamma_2$ then the norm of the operator S will be less than one. Suppose that this condition is satisfied, then, using (5), we obtain

$$\| Sx \|^2 \leq [1 - \tau\gamma_1(2-\tau\gamma_2)] \| x \|^2$$

and, consequently,

$$\| S \|^2 = \sup_{x\neq 0} \frac{\| Sx \|^2}{\| x \|^2} \leq 1 - \tau\gamma_1(2-\tau\gamma_2).$$

The function $\varphi(\tau) = 1 - \tau\gamma_1(2 - \tau\gamma_2)$ has a minimum at the point $\tau = 1/\gamma_2$, equal to $\varphi(1/\gamma_2) = 1 - \xi$, where $\xi = \gamma_1/\gamma_2$. Therefore, for the indicated value of the parameter τ, the norm of the operator S satisfies the estimate $\| S \| \leq \sqrt{1-\xi}$. The lemma is proved. □

Substituting in (5) the operator $C = D^{-1/2}(DB^{-1}A)D^{-1/2}$, we obtain that the inequality (5) is equivalent to the following inequality:

$$\begin{aligned} &\gamma_1(Dx, x) \leq (DB^{-1}Ax, x), \\ &(DB^{-1}Ax, B^{-1}Ax) \leq \gamma_2(DB^{-1}Ax, x), \quad \gamma_1 > 0. \end{aligned} \tag{6}$$

Substituting in (4) the estimate for the norm of the operator S obtained in lemma 3, we find

$$\| z_n \|_D \leq \rho^n \| z_0 \|_D, \qquad \rho = \sqrt{1-\xi}. \tag{7}$$

Theorem 3. *Suppose γ_1 and γ_2 are the constants from the inequalities* (6). *The simple iteration method* (2) *with the iteration parameter $\tau = 1/\gamma_2$ converges in H_D, and the error z_n satisfies the bound* (7). *The iteration count satisfies the estimate $n \geq n_0(\epsilon)$, where $n_0(\epsilon) = \ln\epsilon/\ln\rho$, $\rho = \sqrt{1-\xi}$, $\xi = \gamma_1/\gamma_2$.* □

We give now sample choices for the operator D and concrete forms of the inequalities (6). In table 6 are laid out: the assumptions on the operators A and B, the operator D, and the form of the inequalities (6). To obtain the concrete form of the inequalities (6) we start from the inequalities (6), or from the equivalent inequalities

$$\gamma_1(DA^{-1}Bx, A^{-1}Bx) \leq (DA^{-1}Bx, x), \quad (Dx, x) \leq \gamma_2(DA^{-1}Bx, x), \tag{8}$$

obtained from (6) by making the change $x = A^{-1}By$.

Notice the inequalities

$$\gamma_1(Bx, Bx) \leq (Ax, Bx), \quad (Ax, Ax) \leq \gamma_2(Ax, Bx), \qquad \gamma_1 > 0.$$

If these conditions are satisfied, then for the special cases examined in table 6 it is possible to choose the operator D as either the operator A^2, if $A = A^*$, or the operator A^*A. For this choice of the operator D, the norm of the operator z_n in H_D can be computed during the iteration process

$$\| z_n \|_D^2 = (Dz_n, z_n) = (Az_n, Az_n) = \| r_n \|^2, \quad r_n = Ay_n - f.$$

Table 6

A and B	D	inequalities
1) $A = A^* > 0$,	A or $B^*A^{-1}B$	$\gamma_1(Bx, A^{-1}Bx) \leq (Bx, x)$,
		$(Ax, x) \leq \gamma_2(Bx, x)$
$B \neq B^* > 0$	A^2 or B^*B	$\gamma_1(Bx, Bx) \leq (Ax, Bx)$,
		$(Ax, Ax) \leq \gamma_2(Ax, Bx)$
2) $A \neq A^* > 0$,	B or $A^*B^{-1}A$	$\gamma_1(Bx, x) \leq (Ax, x)$,
		$(Ax, B^{-1}Ax) \leq \gamma_2(Ax, x)$
$B = B^* > 0$	B^2 or A^*A	$\gamma_1(Bx, Bx) \leq (Ax, Bx)$,
		$(Ax, Ax) \leq \gamma_2(Ax, Bx)$
3) $A \neq A^* > 0$,	A^*A or B^*B	$\gamma_1(Bx, Bx) \leq (Ax, Bx)$,
$B \neq B^* > 0$		$(Ax, Ax) \leq \gamma_2(Ax, Bx)$
4) $A = A^*$, $B = B^*$,	A^2 or B^2	$\gamma_1(Bx, Bx) \leq (Ax, Bx)$,
$AB \neq BA$		$(Ax, Ax) \leq \gamma_2(Ax, Bx)$

We turn now to an estimate for the norm of the operator S. If the operator C is self-adjoint in H, then by (5) it is positive-definite, and consequently, the square root of the operator C exists. Setting $x = C^{-1/2}y$ in the second of the inequalities (5), we obtain that the inequalities (5) are equivalent to the inequalities

$$\gamma_1 E \leq C \leq \gamma_2 E, \qquad \gamma_1 > 0.$$

From lemma 2 under these assumptions we obtain the following estimate for the norm of the operator S: $\| S \| \leq \rho_0$, $\rho_0 = (1-\xi)/(1+\xi)$, $\xi = \gamma_1/\gamma_2$.

Comparing this estimate with the one obtained in lemma 3, we ascertain that the estimate of lemma 3 is cruder and does not lead to the estimate in lemma 2 when the operator C is self-adjoint in H.

6.4.2.2 *The second case.* We obtain now another estimate for the norm of the transformation operator S, one which reduces to the estimate in lemma 2 when C is a self-adjoint operator in H. To do this, we increase the volume of *a priori* information about the operator C, assuming that we are given three scalars γ_1, γ_2, and γ_3:

$$\gamma_1 E \le C \le \gamma_2 E, \qquad \| C_1 \| \le \gamma_3, \qquad \gamma_1 > 0, \qquad \gamma_3 \ge 0, \tag{9}$$

where $C_1 = 0.5(C - C^*)$ is the skew-symmetric part of the operator C.

We have

Lemma 4. *Suppose that the constants γ_1, γ_2, and γ_3 are given in the inequalities* (9). *Then we have the following estimate for the norm of the operator $S = E - \tau C$ for $\tau = \bar{\tau}_0 = \tau_0(1 - \kappa\bar{\rho}_0)$*

$$\| S \| \le \bar{\rho}_0, \qquad \bar{\rho}_0 = (1 - \bar{\xi})/(1 + \bar{\xi}) \tag{9'}$$

where

$$\tau_0 = \frac{2}{\gamma_1 + \gamma_2}, \qquad \kappa = \frac{\gamma_3}{\sqrt{\gamma_1\gamma_2 + \gamma_3^2}}, \qquad \bar{\xi} = \frac{1 - \kappa}{1 + \kappa} \cdot \frac{\gamma_1}{\gamma_2}.$$

Proof. We now derive the proof of lemma 4. Let θ be an arbitrary scalar in the interval $(0, 1)$. We represent the operator S in the following form:

$$S = E - \tau C = [\theta E - \tau C_0] + [(1 - \theta)E - \tau C_1],$$

where $C = 0.5(C + C^*)$ is the self-adjoint part of the operator C. Using the triangle inequality, we obtain an estimate for the norm of the operator S:

$$\| S \| \le \| \theta E - \tau C_0 \| + \| (1 - \theta)E - \tau C_1 \| . \tag{10}$$

We estimate separately the norm of each operator. From (9) and from the equality $(C_0 x, x) = 0.5(Cx, x) + 0.5(C^* x, x) = (Cx, x)$ we obtain that γ_1 and γ_2 are bounds for the self-adjoint operator C_0:

$$\gamma_1 E \le C_0 \le \gamma_2 E, \qquad \gamma_1 > 0.$$

By analogy with lemma 2 we obtain the following estimate for the norm of the operator $\theta E - \tau C_0$:

$$\| \theta E - \tau C_0 \| \le \max\left(|\theta - \tau\gamma_1|, |\theta - \tau\gamma_2|\right) = \begin{cases} \theta - \tau\gamma_1, & 0 \le \tau \le \theta\tau_0, \\ \tau\gamma_2 - \theta, & \tau \ge \theta\tau_0. \end{cases}$$

We now estimate the norm of the operator $(1-\theta)E-\tau C_1$. Since $(C_1x,x)=0$, for all $x\in H$ we obtain

$$\begin{aligned}\| ((1-\theta)E-\tau C_1)x \|^2 &= (1-\theta)^2 \| x \|^2 + \tau^2 \| C_1 x \|^2 \\ &\le ((1-\theta)^2+\tau^2 \| C_1 \|^2) \| x \|^2 .\end{aligned}$$

From this and from (9) follows the estimate $\| (1-\theta)E-\tau C_1 \| \le [(1-\theta)^2+\tau_2\gamma_3^2]^{1/2}$. Substituting this estimate in (10) we have

$$\| S \| \le \begin{cases} \varphi_1(\theta,\tau)=\theta-\tau\gamma_1+\sqrt{(1-\theta)^2+\tau^2\gamma_3^2}, & 0\le\tau\le\tau_0\theta, \\ \varphi_2(\theta,\tau)=\tau\gamma_2-\theta+\sqrt{(1-\theta)^2+\tau^2\gamma_3^2}, & \tau\ge\tau_0\theta. \end{cases}$$

We now choose the parameters τ and θ from the condition that the estimate of the norm of the operator S be a minimum. Notice that the function $\varphi_2(\theta,\tau)$ is monotonic for increasing τ. Therefore, to minimize the norm of the operator S it is sufficient to consider the region $0\le\tau\le\tau_0\theta$, $0<\theta<1$. In this region $\| S \| \le \varphi_1(\theta,\tau)$.

We now investigate the function $\varphi_1(\theta,\tau)$. This function is monotonic for increasing θ, consequently, the minimum is achieved for $\tau=\tau_0\theta$. For this value of the parameter τ we will have

$$\begin{aligned}\| S \| &\le \varphi(\theta)=\varphi_1(\theta,\tau_0\theta)= \\ &= \theta(1-\tau_0\gamma_1)+\sqrt{(1-\theta)^2+\tau_0^2\gamma_3^2\theta^2}=\theta\rho_0+\sqrt{(1-\theta)^2+\tau_0^2\gamma_3^2\theta^2}.\end{aligned}$$

Thus, it is necessary to show that $\min\limits_{0<\theta<1}\varphi(\theta)=\bar{\rho}_0$. We will find the minimum of the function $\varphi(\theta)$. We make a change of variable, setting

$$\theta=(1-x)/(1+a^2),\quad x\in(-a^2,1),\quad a^2=\tau_0^2\gamma_3^2.$$

The function $\varphi(\theta)$ can be written in the form

$$\varphi(\theta)=\bar{\varphi}(x)=\frac{1}{\sqrt{1+a^2}}\left(\sqrt{x^2+a^2}-\frac{\rho_0}{\sqrt{1+a^2}}x\right)+\frac{\rho_0}{1+a^2}. \tag{11}$$

From this it is clear that it is sufficient to find the minimum of the function $v(x)$

$$v(x)=\sqrt{x^2+a^2}-\rho_0 x\Big/\sqrt{1+a^2}$$

in the region $-a^2 < x < 1$. Computing the derivative of the function $v(x)$

$$v'(x) = \frac{x}{\sqrt{x^2+a^2}} - \frac{\rho_0}{\sqrt{1+a^2}},$$
$$v''(x) = \frac{a^2}{(x^2+a^2)^{3/2}} > 0,$$

we find that the equation $v'(x) = 0$ gives the point which minimizes the function $v(x)$. Solving the equation

$$\sqrt{\frac{x^2+a^2}{1+a^2}} = \frac{x}{\rho_0}, \tag{12}$$

we find the desired minimizing point for the function $v(x)$:

$$x_0 = a\rho_0 \Big/ \sqrt{1+a^2-\rho_0^2} \in (0,1), \quad \theta_0 = (1-x_0)/(1+a^2).$$

Substituting (12) in (11), we find the minimal value of the function $\varphi(\theta)$:

$$\varphi(\theta_0) = \sqrt{\frac{x_0^2+a^2}{1+a^2}} + \rho_0\frac{1-x_0}{1+a^2} = \frac{x_0}{\rho_0} + \theta_0\rho_0. \tag{13}$$

It remains to express x_0 and θ_0 in terms of known quantities. Using the notation in lemma 4, we obtain

$$1-\rho_0^2 = \tau_0^2\gamma_1\gamma_2, \quad x_0 = \tau_0\gamma_3\rho_0 \Big/ \sqrt{1-\rho_0^2+\tau_0^2\gamma_3^2} = \kappa\rho_0. \tag{14}$$

From (12) we find

$$a^2 = x_0^2(1-\rho_0^2)\big/(\rho_0^2-x_0^2), \quad 1+a^2 = \rho_0^2(1-x_0^2)\big/\left(\rho_0^2-x_0^2\right).$$

Therefore

$$\theta_0\rho_0 = \frac{1-x_0}{1+a^2}\rho_0 = \frac{\rho_0^2-x_0^2}{\rho_0(1+x_0)} = \frac{\rho_0(1-\kappa^2)}{1+\kappa\rho_0}. \tag{15}$$

We substitute (14) and (15) in (13)

$$\varphi(\theta_0) = \kappa + \frac{\rho_0(1-\kappa^2)}{1+\rho_0\kappa} = \frac{\kappa+\rho_0}{1+\rho_0\kappa} = \frac{(1+\kappa)-\xi(1-\kappa)}{(1+\kappa)+\xi(1-\kappa)} = \frac{1-\bar{\xi}}{1+\bar{\xi}} = \bar{\rho}_0. \tag{16}$$

We now find an expression for the parameter $\tau = \tau_0\theta_0$. Comparing (15) and (16), we obtain

$$\theta_0\rho_0 = \bar{\rho}_0 - \kappa. \tag{17}$$

On the other hand, from (16) it is possible to express ρ_0 in terms of $\bar{\rho}_0$ and κ:

$$\rho_0 = (\bar{\rho}_0 - \kappa)/(1 - \kappa\bar{\rho}).$$

Substituting ρ_0 in (17), we find

$$\theta_0 = 1 - \kappa\bar{\rho}, \qquad \tau = \tau_0(1 - \kappa\bar{\rho}_0).$$

The lemma has been proved. □

The inequality (9) can be written in the following form:

$$\gamma_1(x,x) \le (Cx,x) \le \gamma_2(x,x), \quad (C_1x, C_1x) \le \gamma_3^2(x,x), \quad \gamma_1 > 0.$$

Substituting here $C = D^{-1/2}(DB^{-1}A)D^{-1/2}$ and $C_1 = 0.5D^{-1/2}(DB^{-1}A - (DB^{-1}A)^*)D^{-1/2}$, we obtain the inequalities

$$\begin{gathered} \gamma_1 D \le DB^{-1}A \le \gamma_2 D, \qquad \gamma_1 > 0, \\ \left(D^{-1}\frac{DB^{-1}A - (DB^{-1}A)^*}{2}x, \frac{DB^{-1}A - (DB^{-1}A)^*}{2}x\right) \le \gamma_3^2(Dx,x). \end{gathered} \tag{18}$$

Substituting the estimate (9′) for the norm of the operator S in (4), we find

$$\| z_n \|_D \le \bar{\rho}_0^n \| z_0 \|_D \,. \tag{19}$$

Theorem 4. *Suppose that γ_1, γ_2, and γ_3 are the constants in the inequalities* (18). *The simple iteration method* (2) *with the iterative parameter $\tau = \bar{\tau}_0 = \tau_0(1 - \kappa\bar{\rho}_0)$ converges in H_D, and the error z_n can be estimated by* (19). *For the iteration count we have the estimate $n \ge n_0(\epsilon)$, where $n_0(\epsilon) = \ln\epsilon/\ln\bar{\rho}_0$,*

$$\tau_0 = \frac{2}{\gamma_1 + \gamma_2}, \quad \bar{\rho}_0 = \frac{1 - \bar{\xi}}{1 + \bar{\xi}}, \quad \bar{\xi} = \frac{1 - \kappa}{1 + \kappa} \cdot \frac{\gamma_1}{\gamma_2}, \quad \kappa = \frac{\gamma_3}{\sqrt{\gamma_1\gamma_2 + \gamma_3^2}}. \; \square$$

Remark. Since the iteration count is determined by the quantity $\bar{\xi}$, which can be written in the form

$$\bar{\xi} = \left(\sqrt{\gamma_1/\gamma_2 + (\gamma_3/\gamma_2)^2} - \gamma_3/\gamma_2\right)^2,$$

the operator B should be chosen so that the ratio $\xi = \gamma_1/\gamma_2$ is maximal, and γ_3/γ_2 minimal.

We give here some sample choices for the operator D. If we take D to be the operator A^*A or B^*B, then the inequality (18) can be written in the form

$$\gamma_1(Bx, Bx) \leq (Ax, Bx) \leq \gamma_2(Bx, Bx),$$
$$\| 0.5(AB^{-1} - (B^*)^{-1}A^*) \| \leq \gamma_3. \tag{20}$$

For the case $D = B^*B$ this assertion is obvious, and if $D = A^*A$, then it is necessary to make the change $x = A^{-1}By$ in (18) to obtain the inequalities (20).

If the operator B is self-adjoint, positive-definite, and bounded in H, then it is possible to take D to be the operator B or $A^*B^{-1}A$. In this case, the inequalities (18) are equivalent to the following inequalities:

$$\gamma_1 B \leq A \leq \gamma_2 B, \qquad \gamma_1 > 0,$$
$$(B^{-1}A_1x, A_1x) \leq \gamma_3^2(Bx, x), \quad A_1 = 0.5(A - A^*). \tag{21}$$

For $D = B$ the inequalities (18) and (21) are the same, and for $D = A^*B^{-1}A$, the inequalities (21) follow from the inequalities (18) after making the change $x = A^{-1}By$ in (18).

6.4.3 Minimizing the norm of the resolving operator

6.4.3.1 *The first case.* In Section 6.4.2 an estimate for the norm of the operator S^n was obtained, based on the inequality $\| S^n \| \leq \| S \|^n$. We look now at another way of obtaining an estimate for $\| S^n \|$. This method is based on an estimate for the spectral radius of an operator.

Recall (see Section 5.1) that the ***spectral radius of an operator*** T acting in a complex Hilbert space $\tilde{H}$ is the quantity

$$\rho(T) = \sup_{\|z\|=1} |(Tz, z)|, \qquad z \in \tilde{H}.$$

For a linear bounded operator T the spectral radius satisfies the inequalities

$$\mu(T) \; \| T \| \leq \rho(T) \leq \| T \|, \qquad \rho(T^n) \leq [\rho(T)]^n, \tag{22}$$

where n is a natural number, and $\mu(T) \geq 1/2$.

Using the concept of the spectral radius of an operator, we obtain two estimates for the norm of the operator S^n depending on the type of ***a priori*** information concerning the operator C.

We look now at the case where the *a priori* information is given in the form of constants γ_1, γ_2, and γ_3:

$$\gamma_1 E \le C \le \gamma_2 E, \quad \| C_1 x \| \le \gamma_3 \| x \|, \qquad \gamma_1 > 0, \quad x \in H. \qquad (23)$$

The complex space $\tilde{H}$ is defined in the following fashion: it consists of elements of the form $z = x + iy$, where $x, y \in H$. The inner product in $\tilde{H}$ is defined by the formula

$$(z, w) = (x, u) + i(y, u) - i(x, v) + (y, v),$$
$$z = x + iy, \quad w = u + iv.$$

A linear operator C defined in H is defined in $\tilde{H}$ as follows: $Cz = Cx + iCy$.

From the properties (22), for any integer n we have the estimate

$$\| S^n \| \le \frac{1}{\mu(S^n)} \rho(S^n) \le 2[\rho(S)]^n,$$

therefore it is sufficient to estimate the numerical radius of the operator S.

We have

Lemma 5. *Suppose that γ_1, γ_2, and γ_3 in the inequalities (23) are given. Then we have the estimate*

$$\| S^n \| \le 2\rho^n,$$

of the norm of the operator $S = E - \tau C$ in H for $\tau = \min(\tau_0, \kappa\tau_0)$, where

$$\rho^2 = \begin{cases} 1 - \kappa(1 - \rho_0), & 0 < \kappa < 1, \\ 1 - (2 - 1/\kappa)(1 - \rho_0), & \kappa \ge 1, \end{cases} \qquad \kappa = \frac{\gamma_1(\gamma_1 + \gamma_2)}{2(\gamma_1^2 + \gamma_3^2)},$$
$$\tau_0 = 2/(\gamma_1 + \gamma_2), \quad \rho_0 = (1 - \xi)/(1 + \xi), \qquad \xi = \gamma_1/\gamma_2.$$

Proof. In order to prove the lemma we represent the operator C in the form of the sum $C = C_0 + C_1$, $C_0 = 0.5(C + C^*)$, $C_1 = 0.5(C - C^*)$. We estimate the numerical radius of the operator $S = E - \tau C$. For any $z \in \tilde{H}$ we obtain

$$(Sz, z) = (z, z) - \tau(C_0 z, z) - \tau(C_1 z, z).$$

By the self-adjointness of the operator C_0, the inner product $(C_0 z, z)$ is a real number. Since $C_1 = -C_1^*$, $(C_1 z, z)$ is an imaginary number. Therefore

$$|(Sz,z)|^2 = [(z,z) - \tau(C_0 z,z)]^2 + \tau^2 |(C_1 z,z)|^2. \tag{24}$$

From the inequalities (23) we obtain

$$\gamma_1(z,z) \le (C_0 z,z) \le \gamma_2(z,z), \quad \| C_1 z \| \le \gamma_3 \| z \| . \tag{25}$$

Suppose $\| z \| = 1$. From (25) we find

$$|(z,z) - \tau(C_0 z,z)| \le \max(|1-\tau\gamma_1|, |1-\tau\gamma_2|) = \begin{cases} 1-\tau\gamma_1, & 0 \le \tau \le \tau_0, \\ \tau\gamma_2 - 1, & \tau \ge \tau_0, \end{cases}$$
$$|(C_1 z,z)| \le \| C_1 z \| \, \| z \| \le \gamma_3.$$

Substituting these estimates in (24), we obtain

$$\rho^2(S) = \sup_{\|z\|=1} |(Sz,z)|^2 \le \begin{cases} \varphi_1(\tau) = (1-\tau\gamma_1)^2 + \tau^2\gamma_3^2, & 0 \le \tau \le \tau_0, \\ \varphi_2(\tau) = (1-\tau\gamma_2)^2 + \tau^2\gamma_3^2, & \tau \ge \tau_0. \end{cases}$$

We choose the parameter τ from the condition that the estimate for the numerical radius of the operator S be a minimum. Since the function $\varphi_2(\tau)$ increases with τ for $\tau \ge \tau_0$:

$$\varphi_2'(\tau) = 2[\tau(\gamma_2^2 + \gamma_3^2) - \gamma_2] \ge 2\frac{\gamma_2(\gamma_2 - \gamma_1) + 2\gamma_3^2}{\gamma_1 + \gamma_2} > 0,$$

the minimum of $\rho(S)$ over τ can be found in the region $\tau \le \tau_0$, where the estimate $\rho^2(S) \le \varphi_1(\tau)$ is used for $\rho(S)$.

We now investigate the function $\varphi_1(\tau)$. Since

$$\varphi_1''(\tau) = 2(\gamma_1^2 + \gamma_3^2) > 0,$$

setting the derivative

$$\varphi_1'(\tau) = 2[\tau(\gamma_1^2 + \gamma_3^2) - \gamma_1]$$

to zero, we find the extremal point of the function $\varphi_1(\tau)$

$$\tau = \tau_1 = \frac{\gamma_1}{\gamma_1^2 + \gamma_3^2} = \tau_0 \kappa.$$

For $\tau \le \tau_1$ the function $\varphi_1(\tau)$ decreases, and for $\tau \ge \tau_1$ it increases. Therefore the minimal value of $\varphi_1(\tau)$ is achieved at the point $\tau = \tau_1$ if $\tau_1 \le \tau_0$, and at the point $\tau = \tau_0$ if $\tau_1 \geq \tau_0$. Thus, the optimal value of the parameter τ is $\tau = \min(\tau_0, \tau_0\kappa)$. Then

$$\min_\tau \rho^2(S) \leq \begin{cases} \varphi_1(\tau_0), & \kappa \geq 1, \\ \varphi_1(\tau_1), & 0 \leq \kappa \leq 1. \end{cases}$$

We compute $\varphi_1(\tau_0)$ and $\varphi_1(\tau_1)$. From the definition of κ and from the equality $1 - \tau_0\gamma_1 = \rho_0$ we obtain

$$\kappa = \frac{\tau_0\gamma_1}{\tau_0^2\gamma_1^2 + \tau_0^2\gamma_3^2}, \quad \tau_0^2\gamma_3^2 = \frac{\tau_0\gamma_1}{\kappa} - \tau_0^2\gamma_1^2 = \frac{1-\rho_0}{\kappa} - (1-\rho_0)^2.$$

Further,

$$\begin{aligned}\varphi_1(\tau_0) &= (1-\tau_0\gamma_1)^2 + \tau_0^2\gamma_3^2 = \rho_0^2 + (1-\rho_0)/\kappa - (1-\rho_0)^2 \\ &= 1 - (2 - 1/\kappa)(1-\rho_0), \\ \varphi_1(\tau_1) &= (1-\tau_1\gamma_1)^2 + \tau_1^2\gamma_3^2 = 1 - 2\tau_1\gamma_1 + \tau_1^2(\gamma_1^2 + \gamma_3^2) \\ &= 1 - \tau_1\gamma_1 = 1 - \kappa\tau_0\gamma_1 = 1 - \kappa(1-\rho_0).\end{aligned}$$

Thus, the numerical radius has been estimated. The estimate of the lemma follows from the inequality $\| S^n \| \leq 2[\rho(S)]^n$. The lemma has been proved. □

Using lemma 5, we obtain an estimate for the norm of the error z_n:

$$\| z_n \|_D \leq 2\rho^n \| z_0 \|_D . \tag{26}$$

Theorem 5. *Suppose that γ_1, γ_2, and γ_3 are the constants in the inequalities* (18). *The simple iteration method* (2) *with the iteration parameter $\tau = \min(\tau_0, \kappa\tau_0)$ converges in H_D, and the error z_n is estimated by* (26). *The iteration count is estimated by $n \geq n_0(\epsilon)$, where $n_0(\epsilon = \ln(0.5\epsilon)/\ln(\rho)$, and κ and ρ are defined in lemma* 5. □

Sample choices for the operator D and a concrete form of the inequalities (18) are given in Section 6.2.2.

6.4.3.2 *The second case.* Using the concept of the numerical radius of an operator, we obtain still another estimate for the norm of the operator S^n. We will assume that the *a priori* information is given in the form of the constants γ_1, γ_2, and γ_3, in the inequalities

$$\gamma_1 E \leq C \leq \gamma_2 E, \quad (C_1x, C_1x) \leq \gamma_3(Cx, x), \qquad \gamma_1 > 0. \tag{27}$$

We have

Lemma 6. *Suppose γ_1, γ_2 and γ_3 are given in the inequalities* (27). *Then we have the estimate*

$$\| S^n \| \leq 2\rho^n,$$

of the norm of the operator $S = E - \tau C$ in H for $\tau = \min(\tau_0^*, \kappa\tau_0^*)$ where

$$\rho^2 = \begin{cases} 1 - (2\kappa - 1)\dfrac{1-\rho_0}{1+\rho_0}, & \frac{1}{2} \leq \kappa \leq 1, \\ 1 - \left(2 - \dfrac{1}{\kappa}\right)^2 \dfrac{1-\rho_0}{1+\rho_0}, & \kappa \geq 1, \end{cases} \qquad \kappa = \frac{\gamma_1 + \gamma_2 + \gamma_3}{2(\gamma_1 + \gamma_3)},$$

$$\tau_0^* = 2/(\gamma_1 + \gamma_2 + \gamma_3), \quad \rho_0 = (1-\xi)/(1+\xi), \quad \xi = \gamma_1/\gamma_2.$$

Proof. Representing the operator C in the form $C = C_0 + C_1$, where $C_0 = 0.5(C + C^*)$ and $C_1 = 0.5(C - C^*)$, we obtain

$$|(Sz, z)|^2 = [(z, z) - \tau(C_0 z, z)]^2 + \tau^2 |(C_1 z, z)|^2.$$

From the Cauchy-Schwartz-Bunyakovskij inequality and from the conditions of the lemma we find

$$|(C_1 z, z)|^2 \leq (C_1 z, C_1 z)(z, z) \leq \gamma_3 (C_0 z, z)(z, z).$$

Since for any $z \in \tilde{H}$ we have the inequalities

$$\gamma_1(z, z) \leq (C_0 z, z) \leq \gamma_2(z, z), \qquad \gamma_1 > 0,$$

from the three preceding relations we obtain the following estimate for the numerical radius of the operator S:

$$\rho^2(S) \leq \max_{\gamma_1 \leq t \leq \gamma_2} \varphi(t), \quad \text{where} \quad \varphi(t) = (1 - \tau t)^2 + \tau^2 \gamma_3 t.$$

We now investigate the function $\varphi(t)$. This function can take on its maximal value only at the ends of the interval $[\gamma_1, \gamma_2]$. Therefore

$$\rho^2(S) \leq \begin{cases} \varphi_1(\tau) = (1 - \tau\gamma_1)^2 + \tau^2 \gamma_1 \gamma_3, & 0 \leq \tau \leq \tau_0^*, \\ \varphi_2(\tau) = (1 - \tau\gamma_2)^2 + \tau^2 \gamma_2 \gamma_3, & \tau \geq \tau_0^*. \end{cases}$$

We choose the parameter τ from the condition that the estimate for $\rho(S)$ be minimal. Since the function $\varphi_2(\tau)$ increases with τ for $\tau \geq \tau_0^*$:

$$\varphi_2'(\tau) = 2\gamma_2[\tau(\gamma_2 + \gamma_3) - 1] \geq 2\gamma_2 \frac{\gamma_2 - \gamma_1 + \gamma_3}{\gamma_2 + \gamma_1 + \gamma_3} > 0,$$

the minimum of $\rho(S)$ is found in the region $\tau \leq \tau_0^*$, where the estimate $\rho^2(S) \leq \varphi_1(\tau)$ is used for $\rho(S)$.

The function $\varphi_1(\tau)$ achieves its minimal value at $\tau = \tau_1 = 1/(\gamma_1 + \gamma_3) = \kappa\tau_0^*$, since the function $\varphi_1(\tau)$ decreases for $\tau \leq \tau_1$, and increases for $\tau \geq \tau_1$. Therefore the minimal value of $\varphi_1(\tau)$ on the interval $[0, \tau_0^*]$ is achieved at the point $\tau = \tau_1$ if $\tau_1 \leq \tau_0^*$, and at the point $\tau = \tau_0^*$ if $\tau_1 \geq \tau_0^*$.

Thus, the optimal value of the parameter τ has been found:

$$\tau = \min(\tau_0^*, \kappa\tau_0^*).$$

And thus

$$\min_\tau \rho^2(S) \leq \begin{cases} \varphi_1(\tau_0^*), & \kappa \geq 1, \\ \varphi_1(\tau_1), & \kappa \leq 1. \end{cases}$$

We now compute $\varphi_1(\tau_0^*)$ and $\varphi_1(\tau_1)$. A simple computation gives $\tau_0^* = (2 - 1/\kappa)/\gamma_2$, $\gamma_3 = 1/(\kappa\tau_0^*) - \gamma_1$. Using these relations, we obtain

$$\begin{aligned}\varphi_1(\tau_0^*) &= (1 - \tau_0^*\gamma_1)^2 + (\tau_0^*)^2\gamma_1\gamma_3 = 1 - 2\tau_0^*\gamma_1 + \tau_0^*\gamma_1/\kappa \\ &= 1 - (2 - 1/\kappa)^2\gamma_1/\gamma_2 = 1 - (2 - 1/\kappa)^2(1 - \rho_0)/(1 + \rho_0).\end{aligned}$$

Further

$$\begin{aligned}\varphi_1(\tau_1) &= (1 - \tau_0^*\kappa\gamma_1)^2 + (\tau_0^*)^2\kappa^2\gamma_1\gamma_3 = 1 - \tau_0^*\kappa\gamma_1 \\ &= 1 - (2\kappa - 1)\gamma_1/\gamma_2 = 1 - (2\kappa - 1)(1 - \rho_0)/(1 + \rho_0).\end{aligned}$$

Thus, the numerical radius has been estimated. The estimate in the lemma follows from the inequality $\| S^n \| \leq 2[\rho(S)]^n$. The lemma has been proved. □

Substituting $C = D^{-1/2}(DB^{-1}A)D^{-1/2}$ and $C_1 = 0.5^{-1/2}(DB^{-1}A - (DB^{-1}A)^*)D^{-1/2}$ in the inequality (27), we obtain the inequalities

$$\gamma_1 D \leq DB^{-1}A \leq \gamma_2 D, \qquad \gamma_1 > 0,$$

$$\left(D^{-1}\frac{DB^{-1}A - (DB^{-1}A)^*}{2}x, \frac{DB^{-1}A - (DB^{-1}A)^*}{2}x\right) \leq \gamma_3(DB^{-1}Ax, x). \tag{28}$$

Theorem 6. *Suppose γ_1, γ_2, and γ_3 are the constants in the inequalities (28). The simple iteration method (2) with the iteration parameter $\tau = \min(\tau_0^*, \kappa\tau_0^*)$ converges in H_D, and the error can be estimated by (26). The iteration count is estimated by $n \geq n_0(\epsilon)$, where*

$$n_0(\epsilon) = \ln 0.5\epsilon / \ln \rho,$$

and κ, ρ, and τ_0^* are defined in lemma 6.

We now derive the form of the inequalities (28) for certain sample choices of the operator D. If the operator D is taken to be the operator A^*A or B^*B, then the inequalities (28) can be written in the following form:

$$\begin{aligned} &\gamma_1(Bx, Bx) \leq (Ax, Bx) \leq \gamma_2(Bx, Bx), \qquad \gamma_1 > 0, \\ &\| 0.5(AB^{-1} - (B^*)^{-1}A^*)x \|^2 \leq \gamma_3(A^*x, B^{-1}x). \end{aligned} \tag{29}$$

In fact, the inequalities (29) follow directly from (28) after making the change $D = B^*B$ in (28). For the case $D = A^*A$, it is sufficient to change $x = A^{-1}By$ in (28).

If the operator B is self-adjoint, positive-definite, and bounded in H, then the operator D can be taken to be the operators B or $A^*B^{-1}A$. In this case, the inequalities (28) will have the form

$$\begin{aligned} &\gamma_1 B \leq A \leq \gamma_2 B, \qquad \gamma_1 > 0, \\ &(B^{-1}A_1x, A_1x) \leq \gamma_3(Ax, x), \quad A_1 = 0.5(A - A^*). \end{aligned} \tag{30}$$

Notice that in the case $D = A^*B^{-1}A$ the inequalities (30) follow from (28) after making the change indicated above.

6.4.4 The symmetrization method. To solve the equation $Au = f$ with a non-self-adjoint operator A, the well-known *symmetrization method* can be used. In place of the original equation, we look at the symmetrized equation

$$\tilde{A}u = \tilde{f}, \qquad \tilde{A} = A^*A, \qquad \tilde{f} = A^*f, \tag{31}$$

which is obtained from the original equation by multiplying on the left by the operator conjugate to A. In algebra, this transformation of the equation is called the *first Gauss transformation*.

To approximately solve equation (31), we consider the implicit two-level scheme

$$\tilde{B}\frac{y_{k+1}-y_k}{\tau_{k+1}}+\tilde{A}y_k=\tilde{f}, \qquad k=0,1,\ldots, \qquad y_0\in H, \tag{32}$$

with the self-adjoint positive-definite operator $\tilde{B}$. For the operator D we choose the operator $\tilde{B}$ or $\tilde{A}=A^*A$. In this case the operator $D\tilde{B}^{-1}\tilde{A}$ is self-adjoint in H, therefore the iteration parameters τ_k can be choosen from the formulas of the Chebyshev method investigated in Section 6.2. The *a priori* information for this method with the indicated operator D has the form of the constants of energy equivalence for the operators $\tilde{B}$ and $\tilde{A}=A^*A$

$$\gamma_1\tilde{B}\le A^*A\le\gamma_2\tilde{B}, \qquad \gamma_1>0.$$

The estimate of the convergence rate for the Chebyshev method (32) and the formulas for the iterative parameters are given in theorem 1.

6.5 Sample applications of the iterative methods

6.5.1 A Dirichlet difference problem for Poisson's equation in a rectangle. To illustrate the application of the two-level iterative methods constructed in this chapter we look at the solution of a Dirichlet difference problem for a second-order linear elliptic equation. The difference problem will be treated as the operator equation

$$Au=f \tag{1}$$

in the finite-dimensional space of grid functions. The explicit and implicit Chebyshev methods will be looked at, as well as the simple iteration method.

We will begin our look at examples with a *Dirichlet problem for Poisson's equation in a rectangle.* Suppose that in the rectangle $\bar{G}=\{0\le x_\alpha\le l_\alpha,\ \alpha=1,2\}$ with boundary Γ we must find the solution of Poisson's equation

$$Lu=\frac{\partial^2u}{\partial x_1^2}+\frac{\partial^2u}{\partial x_2^2}=-f(x), \quad x=(x_1,x_2)\in G, \tag{2}$$

which takes on the following values on the boundary Γ

$$u(x)=g(x), \qquad x\in\Gamma. \tag{3}$$

The *Dirichlet difference problem* corresponding to (2), (3) on the rectangular grid

$$\bar{\omega} = \{x_{ij} = (ih_1, jh_2) \in \bar{G},\ 0 \le i \le N_1,\ 0 \le j \le N_2,\ h_\alpha = l_\alpha/N_\alpha,\ \alpha = 1,2\}$$

has the form

$$\Lambda y = \sum_{\alpha=1}^{2} y_{\bar{x}_\alpha x_\alpha} = -\varphi(x), \quad x \in \omega, \quad y(x) = g(x), \quad x \in \gamma, \tag{4}$$

where $\gamma = \{x_{ij} \in \Gamma\}$ is the boundary of the grid $\bar{\omega}$, and

$$\begin{aligned} y_{\bar{x}_1 x_1} &= \frac{1}{h_1^2}(y(i+1,j) - 2y(i,j) + y(i-1,j)), \\ y_{\bar{x}_2 x_2} &= \frac{1}{h_2^2}(y(i,j+1) - 2y(i,j) + y(i,j-1)), \\ y(i,j) &= y(x_{ij}). \end{aligned}$$

In Section 5.2 it was shown that the difference problem (4) can be reduced to the operator equation (1), where the operator A is defined as follows: $Ay = -\Lambda \mathring{y}$, where $y \in H$, $\mathring{y} \in \mathring{H}$ and $\mathring{y}(x) = y(x)$ for $x \in \omega$. Here $\mathring{H}$ is the set of grid functions defined on $\bar{\omega}$ which reduce to zero on γ, and H is the space of grid functions defined on ω with the inner product

$$(u, v) = \sum_{x \in \omega} u(x) v(x) h_1 h_2.$$

The right-hand side of equation (1) only differs from the right-hand side φ of the difference equation (4) at the near-boundary nodes:

$$f(x) = \varphi(x) + \varphi_1(x)/h_1^2 + \varphi_2(x)/h_2^2,$$

$$\varphi_1(x) = \begin{cases} g(0, x_2), & x_1 = h_1, \\ 0, & 2h_1 \le x_1 \le l_1 - 2h_1, \\ g(l_1, x_2), & x_1 = l_1 - h_1, \end{cases}$$

$$\varphi_2(x) = \begin{cases} g(x_1, 0), & x_2 = h_2, \\ 0, & 2h_2 \le x_2 \le l_2 - 2h_2, \\ g(x_1, l_2), & x_2 = l_2 - h_2. \end{cases}$$

Thus, the boundary-value difference problem (4) reduces to the operator equation (1) in the finite-dimensional Hilbert space H.

To approximately solve equation (1) we look at the explicit Chebyshev method ($B = E$):

$$B\frac{y_{k+1}-y_k}{\tau_{k+1}} + Ay_k = f, \quad k = 0,1,\ldots, \quad y_0 \in H, \tag{5}$$

$$\tau_k = \frac{\tau_0}{1+\rho_0\mu_k}, \quad \mu_k \in \mathcal{M}_n^* = \left\{-\cos\frac{(2i-1)\pi}{2n},\ i=1,2,\ldots,n\right\} \tag{6}$$
$$k = 1,2,\ldots,n,$$

$$n \geq n_0(\epsilon), \qquad n_0(\epsilon) = \ln(0.5\epsilon)/\ln\rho_1. \tag{7}$$

In Section 5.2 it was shown that the operator A defined here is self-adjoint in H, and its bounds γ_1 and γ_2 are the smallest and largest eigenvalues of the difference operator Λ, i.e.

$$\gamma_1 E \leq A \leq \gamma_2 E, \qquad \gamma_1 > 0, \tag{8}$$

where

$$\gamma_1 = \sum_{\alpha=1}^{2} \frac{4}{h_\alpha^2}\sin^2\frac{\pi h_\alpha}{2l_\alpha}, \quad \gamma_2 = \sum_{\alpha=1}^{2}\frac{4}{h_\alpha^2}\cos^2\frac{\pi h_\alpha}{2l_\alpha}. \tag{9}$$

The operators A and $B = E$ are self-adjoint and positive-definite in H. Therefore it follows from the examples considered in Section 6.2.3 that γ_1, γ_2 from (8) are the constants for the Chebyshev method (5)–(7), if we choose D to be one of the operators E, A or A^2. Then in the formulas (6), (7)

$$\tau_0 = \frac{2}{\gamma_1+\gamma_2}, \quad \rho_0 = \frac{1-\xi}{1+\xi}, \quad \rho_1 = \frac{1-\sqrt{\xi}}{1+\sqrt{\xi}}, \quad \xi = \frac{\gamma_1}{\gamma_2},$$

where γ_1 and γ_2 are defined in (9).

Since $\gamma_1 = O(1)$ and $\gamma_2 = O(1/h_1^2+1/h_2^2)$, $\xi = O(|h|^2)$, where $|h|^2 = h_1^2 + h_2^2$. Consequently, for this example the asymptotic estimate for the iteration count $n_0(\epsilon)$ has the form

$$n_0(\epsilon) = O\left(\frac{1}{|h|}\ln\frac{2}{\epsilon}\right).$$

In particular, when $\bar{G}$ is a square with side l ($l_1 = l_2 = l$) and the grid $\bar{\omega}$ is square ($h_1 = h_2 = h = l/N$), we have

$$\gamma_1 = \frac{8}{h^2}\sin^2\frac{\pi h}{2l}, \quad \gamma_2 = \frac{8}{h^2}\cos^2\frac{\pi h}{2l}, \quad \xi = \tan^2\frac{\pi h}{2l},$$
$$\tau_0 = \frac{h^2}{4}, \qquad \rho_0 = \cos\frac{\pi h}{l}, \qquad \rho_1 = \frac{1-\sin\frac{\pi h}{l}}{\cos\frac{\pi h}{l}}, \tag{10}$$
$$n_0(\epsilon) \approx \frac{1}{\pi h}\ln\frac{2}{\epsilon} \approx 0.32\ln\frac{2}{\epsilon}.$$

Thus, the iteration count n is proportional to the number of nodes N in one direction. Notice that the number of unknowns in problem (4) is equal to $M = (N-1)^2$, i.e. the iteration count is proportional to the square root of the number of unknowns.

The iterative operator scheme (5) for $B = E$ can be written, using the definition of the operator A and the right-hand side f, in the form of the following difference scheme:

$$y_{k+1} = y_k + \tau_{k+1}(\Lambda y_k + \varphi), \quad x \in \omega, \quad y_k|_\gamma = g, \quad k = 0, 1, \ldots .$$

Substituting here (4), we obtain the computational formulas

$$y_{k+1}(i,j) = \left(1 - \frac{\tau_{k+1}}{\tau_0}\right) y_k(i,j) + \tau_{k+1}\left[\frac{y_k(i+1,j) + y_k(i-1,j)}{h_1^2}\right.$$
$$\left. + \frac{y_k(i,j+1) + y_k(i,j-1)}{h_2^2} + \varphi(i,j)\right],$$
$$1 \le i \le N_1 - 1, \qquad 1 \le j \le N_2 - 1.$$

The initial approximation y_0 is an arbitrary grid function on ω, which takes on the given values $y_0(x) = g(x)$ for $x \in \gamma$, the boundary of the grid.

We now estimate the number of arithmetic operations $Q(\epsilon)$ which are required in order to obtain an approximate solution of the difference problem (4) to a accuracy ϵ using the Chebyshev method (5)–(7).

Considering as given the iterative parameters τ_k, we find that computing y_{k+1} at one node of the grid ω requires nine arithmetic operations. Since the number of interior nodes of the grid $\bar{\omega}$ is equal to $M(N_1-1)(N_2-1)$, realizing one iterative step requires $Q \approx 9N_1N_2$ operations. Therefore $Q_0(\epsilon) = nQ_0 \approx 9nN_1N_2$, where n is the iteration count.

For the special case looked at above, the iteration count n is defined in (10), and, consequently, for this example we obtain

$$Q(\epsilon) \approx 2.9N^3\ln(2/\epsilon).$$

To solve equation (1) we now consider the simple iteration method. The iterative scheme for the simple iteration method has the form (5), and the iterative parameters τ_k and the iteration count n are defined using the formulas in theorem 2:

$$\tau_k \equiv \tau_0 = \frac{2}{\gamma_1+\gamma_2}, \; n \ge n_0(\epsilon) = \frac{\ln\epsilon}{\ln\rho_0}, \; \rho_0 = \frac{1-\xi}{1+\xi}, \; \xi = \frac{\gamma_1}{\gamma_2}, \tag{11}$$

where γ_1 and γ_2 are given in (9). From (9) and (11) we obtain an asymptotic estimate involving h for the iteration count for the simple iteration method

$$n_0(\epsilon) = O\left(\frac{1}{|h|^2}\ln\frac{1}{\epsilon}\right).$$

For the special case considered above we find

$$n_0(\epsilon) \approx \frac{2l^2}{\pi^2 h^2}\ln\frac{1}{\epsilon} \approx 0.2N^2\ln\frac{1}{\epsilon}, \tag{12}$$

i.e. the iteration count for the simple iteration method is proportional to the square of the number of nodes N in one direction (or proportional to the number of unknowns in the equation).

Comparing the estimate for the number of iterations for the Chebyshev method (10) and for the simple iteration method (12), we obtain that the simple iteration method requires significantly more iterations than the Chebyshev method. In order to compare these methods on real grids, for this special case we give the precise iteration counts for various values of the number of nodes N in one direction with $\epsilon = 10^{-4}$ (the first number is the iteration count for the Chebyshev method):

$N = 32$	$n = 101$	$n = 1909$
$N = 64$	$n = 202$	$n = 7642$
$N = 128$	$n = 404$	$n = 30577.$

We now give computational formulas for the simple iteration method applied to this special case:

$$y_{k+1}(i,j) = \frac{1}{4}[y_k(i+1,j) + y_k(i-1,j)$$
$$+ y_k(i,j+1) + y_k(i,j-1)] + \frac{h^2}{4}\varphi(i,j).$$

Because the iteration count for the simple iteration method depends strongly on the number of nodes of the grid N, this method is not used at the present time for solving grid elliptic equations.

6.5.2 A Dirichlet difference problem for Poisson's equation in an arbitrary region. Suppose that, in the arbitrary bounded region G with boundary Γ, we are required to find the solution to Poisson's equation

$$Lu = \frac{\partial^2 u}{\partial x_1^2} + \frac{\partial^2 u}{\partial x_2^2} = -f(x), \qquad x = (x_1, x_2) \in G, \tag{13}$$

which takes on the given values

$$u(x) = g(x), \qquad x \in \Gamma, \tag{14}$$

on the boundary Γ.

For simplicity we consider the case where the intersection of the region G with a straight line passing through the point $x \in G$ and parallel to a coordinate axis consists of one interval.

We cover the surface with a lattice formed by the intersection of straight lines parallel to the coordinate axes and spaced a uniform distance h from each other.

The points of the lattice x_{ij} belonging to G are called the *nodes* of the grid $\omega = \{x_{ij} \in G\}$. We denote by $\Delta_\alpha(x_{ij})$ the interval which is formed by the intersection of G with the straight line drawn through the point $x_{ij} \in \omega$ parallel to the coordinate axis Ox_α, $\alpha = 1, 2$. The endpoints of this interval are called the *boundary nodes* in the direction x_α. The set of all boundary nodes in the direction x_α is denoted by γ_α, and we denote the boundary of the grid region by $\gamma = \gamma_1 \cup \gamma_2$. The sets of interior and boundary nodes form the grid $\bar{\omega} = \omega \cup \gamma$ in the region $\bar{G}$.

We look now at one of the intervals Δ_α. The set of nodes $x_{ij} \in \omega$ lying on this interval is denoted by $\omega_\alpha(x_\beta)$, $\beta = 3 - \alpha$, $\alpha = 1, 2$. We use $\omega_\alpha^+(x_\beta)$ to denote the set consisting of the nodes of $\omega_\alpha(x_\beta)$ and the right endpoint of the interval Δ_α. We define by $\bar{\omega}_\alpha(x_\beta)$ the set consisting of the nodes of $\omega_\alpha(x_\beta)$ and the endpoints of the interval Δ_α. We denote by $x_{ij}^{(+1_\alpha)}$ and $x_{ij}^{(-1_\alpha)}$ the nodes which are closest to the point $x_{ij} \in \omega_\alpha(x_\beta)$ on the right and left, respectively, and which belong to $\bar{\omega}_\alpha(x_\beta)$.

The distances between the nodes x_{ij} and $x_{ij}^{(\pm 1_\alpha)} \in \bar{\omega}$ are called the *steps* $h_\alpha^\pm(x_{ij})$ of the grid ω at the point $x_{ij} \in \omega$. Notice that if all four neighbors $x_{ij}^{(\pm 1_\alpha)}$ to x_{ij} belong to ω, then the steps $h_\alpha^\pm$ are equal to the step-size h of the lattice. At the near-boundary nodes $h_\alpha^\pm \le h$. The steps h_α^+ and h_α^- satisfy the relation

$$h_\alpha^+(x_{ij}) = h_\alpha^-(x_{ij}^{(+1_\alpha)}).$$

We shall put the problem (13), (14) on the grid $\bar{\omega}$ in correspondence with the boundary-value difference problem

$$\Lambda y = \sum_{\alpha=1}^{2} y_{\bar{x}_\alpha \hat{x}_\alpha} = -\varphi(x), \quad x \in \omega, \quad y(x) = g(x), \quad x \in \gamma, \tag{15}$$

where

$$\begin{aligned} y_{\bar{x}_\alpha} &= \frac{1}{h_\alpha^-}(y - y^{-1_\alpha}), \qquad y_{x_\alpha} = \frac{1}{h_\alpha^+}(y^{+1_\alpha} - y), \\ y_{\hat{x}_\alpha} &= \frac{1}{\hbar}(y^{+1_\alpha} - y), \qquad y_{\bar{x}_\alpha \hat{x}_\alpha} = \frac{1}{\hbar}\left(\frac{y^{+1_\alpha} - y}{h_\alpha^+} - \frac{y - y^{-1_\alpha}}{h_\alpha^-}\right), \\ y^{\pm 1_\alpha} &= y(x^{(\pm 1_\alpha)}), \qquad \alpha = 1, 2. \end{aligned}$$

The difference problem (15) reduces to the operator equation (1) and the operator A is defined in the same way as in Section 6.5.1. The inner product in H is computed as follows:

$$(u, v) = \sum_{x \in \omega} u(x) v(x) \hbar^2.$$

We introduce now some notation which will be useful later. For grid functions defined on $\bar{\omega}$ we define the inner products by the formulas

$$\begin{aligned} (u, v)_{\omega_\alpha} &= \sum_{x_\alpha \in \omega_\alpha(x_\beta)} u(x) v(x) \hbar, \\ (u, v)_{\omega_\alpha^+} &= \sum_{x_\alpha \in \omega_\alpha^+(x_\beta)} u(x) v(x) h_\alpha^-(x), \\ (u, v)_\alpha &= ((u, v)_{\omega_\alpha^+}, 1)_{\omega_\beta}, \quad \beta = 3 - \alpha, \quad \alpha = 1, 2. \end{aligned}$$

Using this notation, the inner product in H can be written in the form

$$(u, v) = ((u, v)_{\omega_1}, 1)_{\omega_2} = ((u, v)_{\omega_2}, 1)_{\omega_1}. \tag{16}$$

From the definition of the operator A we obtain that

$$\begin{aligned} (Au, v) &= -(\Lambda \mathring{u}, \mathring{v}) \\ &= -((\mathring{u}_{\bar{x}_1 \hat{x}_1}, \mathring{v})_{\omega_1}, 1)_{\omega_2} - ((\mathring{u}_{\bar{x}_2 \hat{x}_2}, \mathring{v})_{\omega_2}, 1)_{\omega_1}, \end{aligned}$$

and since by Green's difference formula we then have the equation (see Section 5.2.3)

$$-(\mathring{u}_{\bar{x}_\alpha \hat{x}_\alpha}, \mathring{v})_{\omega_\alpha} = (\mathring{u}_{\bar{x}_\alpha}, \mathring{v}_{\bar{x}_\alpha})_{\omega_\alpha^+} = -(\mathring{u}, \mathring{v}_{\bar{x}_\alpha \hat{x}_\alpha})_{\omega_\alpha},$$

from this we obtain

$$(Au,v) = (u,Av),$$
$$(Au,u) = \sum_{\alpha=1}^{2} (\mathring{u}^2_{\bar{x}_\alpha},1)_\alpha, \qquad u \in H,\ \mathring{u} \in \mathring{H}. \tag{17}$$

The first of these equations shows the self-adjointness of the operator A in H.

In order to approximately solve equation (1), we look at the implicit Chebyshev method

$$B\frac{y_{k+1}-y_k}{\tau_{k+1}} + Ay_k = f, \qquad k=0,1,\ldots,\ y_0 \in H,$$

where the operator B is taken to be an easily invertible diagonal operator

$$By = (b_1+b_2)y,\ b_\alpha(x) = \frac{1}{h}\left(\frac{1}{h^+_\alpha(x)} + \frac{1}{h^-_\alpha(x)}\right), \qquad \alpha = 1,2. \tag{18}$$

We now explain this choice of the operator B. If equation (1) is considered as a system of linear algebraic equations with a matrix $\mathcal{A}$ corresponding to the operator A, then the matrix $\mathcal{B}$ corresponding to the operator B is the diagonal part of the matrix $\mathcal{A}$.

Since the operators A and B are self-adjoint and positive-definite in H, the constants γ_1 and γ_2 in the conditions (6), (7) are the constants of energy equivalence for the operators A and B:

$$\gamma_1 B \le A \le \gamma_2 B, \qquad \gamma_1 > 0,$$

if D is taken to be one of the operators A, B, or $AB^{-1}A$.

We now find estimates for γ_1 and γ_2. First of all we show that

$$\gamma_1 + \gamma_2 = 2. \tag{19}$$

Let $u(x)$ be an arbitrary grid function from H. We consider the function $v(x)$ which is defined as follows:

$$v(x_{ij}) = (-1)^{i+j}u(x_{ij}), \qquad x_{ij} \in \omega.$$

We compute the value of the difference operator $\Lambda\mathring{v}$ at the point x_{ij}. We obtain

$$\Lambda\mathring{v}(i,j) = \sum_{\alpha=1}^{2} \frac{1}{h}\left(\frac{\mathring{v}^{+1_\alpha} - \mathring{v}}{h_\alpha^+} - \frac{\mathring{v} - \mathring{v}^{-1_\alpha}}{h_\alpha^-}\right)_{x=x_{ij}}$$
$$= -(-1)^{i+j} \sum_{\alpha=1}^{2} \frac{1}{h}\left(\frac{\mathring{u}^{+1_\alpha} - \mathring{u}}{h_\alpha^+} - \frac{\mathring{u} - \mathring{u}^{-1_\alpha}}{h_\alpha^-}\right)_{x=x_{ij}}$$
$$- 2(-1)^{i+j}(b_1(x_{ij}) + b_2(x_{ij}))\mathring{u}(x_{ij}).$$

Consequently,

$$Av(i,j) = -\Lambda\mathring{v}(i,j) = (-1)^{i+j}(2B - A)u(i,j).$$

Further, since

$$\gamma_1 = \min_{u\neq 0} \frac{(Au,u)}{(Bu,u)}, \quad (Av,v) = 2(Bu,u) - (Au,u),\ (Bv,v) = (Bu,u),$$

then

$$\gamma_2 = \max_{v\neq 0} \frac{(Av,v)}{(Bv,v)} = 2 - \min_{u\neq 0} \frac{(Au,u)}{(Bu,u)} = 2 - \gamma_1.$$

The assertion has been proved.

Using (19) we obtain that, in the formulas (6),

$$\tau_0 = 2/(\gamma_1 + \gamma_2) = 1, \quad \rho_0 = (\gamma_2 - \gamma_1)/(\gamma_2 + \gamma_1) = 1 - \gamma_1.$$

Consequently, to compute the iterative parameters τ_k, it is sufficient to find an estimate for γ_1. From lemma 13 Section 5.2 we obtain that, for any grid function $y \in H$, we have the inequality

$$(b_\alpha \mathring{y}, \mathring{y})_{\omega_\alpha} \le \kappa_\alpha(\mathring{y}^2_{\bar{x}_\alpha}, 1)_{\omega_\alpha^+}, \qquad \alpha = 1,2, \tag{20}$$

where $\kappa_\alpha = \kappa_\alpha(x_\beta) = \max\limits_{x_\alpha \in \omega_\alpha(x_\beta)} v^\alpha(x)$, and $v^\alpha(x)$ is the solution of the following three-point boundary-value problem:

$$\begin{aligned} &v^\alpha_{\bar{x}_\alpha \hat{x}_\alpha} = -b_\alpha(x), && x_\alpha \in \omega_\alpha(x_\beta), \\ &v^\alpha(x) = 0, && x_\alpha \in \gamma_\alpha. \end{aligned} \tag{21}$$

Dividing (20) by κ_α and summing over ω_β, we obtain

$$\left(\frac{b_\alpha}{\kappa_\alpha}\mathring{y}, \mathring{y}\right) \le \left((\mathring{y}^2_{\bar{x}_\alpha}, 1)_{\omega_\alpha^+}, 1\right)_{\omega_\beta} = (\mathring{y}^2_{\bar{x}_\alpha}, 1)_\alpha, \quad \alpha = 1,2.$$

Combining these inequalities, we find

$$\left(\sum_{\alpha=1}^{2} \frac{b_\alpha}{\kappa_\alpha}\mathring{y}, \mathring{y}\right) \le \sum_{\alpha=1}^{2} \left(\mathring{y}^2_{\bar{x}_\alpha}, 1\right)_\alpha . \tag{22}$$

From (17), (18), and (22) it follows that it is possible to take γ_1 to be

$$\gamma_1 = \min_{x\in\omega} \frac{1}{b_1(x)+b_2(x)} \sum_{\alpha=1}^{2} \frac{b_\alpha(x)}{\kappa_\alpha(x_\beta)}. \tag{23}$$

It remains to compute κ_α. To do this, we find the solution to problem (21).

Suppose that the ends of the interval Δ_α, on which the nodes of the grid $\omega_\alpha(x_\beta)$ are located, are $l_\alpha(x_\beta)$ and $L_\alpha(x_\beta)$. From the construction of the lattice on the surface, the steps $h_\alpha^\pm$ are different from h only at the boundary nodes (see figure 3).

Figure 3.

Therefore, on the grid $\omega_\alpha(x_\beta)$, the difference derivative $v_{\bar{x}_\alpha x_\alpha}$ and the right-hand side of equation (21) can be written in the form

$$v_{\bar{x}_\alpha \hat{x}_\alpha} = \frac{1}{h}\left(\frac{v^{+1_\alpha}-v}{h} - \frac{v-v^{-1_\alpha}}{h_\alpha^-}\right), \quad b_\alpha = \frac{1}{h}\left(\frac{1}{h}+\frac{1}{h_\alpha^-}\right), \quad x_\alpha = l_\alpha + h_\alpha^+,$$

$$v_{\bar{x}_\alpha \hat{x}_\alpha} = \frac{1}{h^2}\left(v^{+1_\alpha} - 2v + v^{-1_\alpha}\right), \qquad b_\alpha = \frac{2}{h^2}, \quad l_\alpha + h_\alpha^- < x_\alpha < L_\alpha - h_\alpha^+,$$

$$v_{\bar{x}_\alpha \hat{x}_\alpha} = \frac{1}{h}\left(\frac{v^{+1_\alpha}-v}{h_\alpha^+} - \frac{v-v^{-1_\alpha}}{h}\right), \quad b_\alpha = \frac{1}{h}\left(\frac{1}{h}+\frac{1}{h_\alpha^+}\right), \quad x_\alpha = L_\alpha - h_\alpha^+.$$

A straightforward verification shows that the grid function

$$v^\alpha(x) = \frac{1}{h^2}\left[(x_\alpha - l_\alpha)\left(L_\alpha - x_\alpha + \frac{(h_\alpha^-)^2 - (h_\alpha^+)^2}{L_\alpha - l_\alpha}\right) + h^2 - (h_\alpha^-)^2\right]$$

is the solution of problem (21) for $x_\alpha \in \omega_\alpha(x_\beta)$. Since

$$v^\alpha(x) \le \frac{1}{h^2}(x_\alpha - l_\alpha)(L_\alpha - x_\alpha) + 1,$$

then

$$\kappa_\alpha = \max_{x_\alpha \in \omega_\alpha(x_\beta)} v^\alpha(x) \le \frac{1}{h^2}\left(\frac{L_\alpha - l_\alpha}{2}\right)^2 + 1. \tag{24}$$

Substituting (18) and (24) in (23), we find the estimate for γ_1.

A crude estimate for γ_1 can be obtained as follows. Suppose that the region $\bar{G}$ is embedded in a square with side l. Then $L_\alpha - l_\alpha \le l$ for any α and, consequently, $\kappa_\alpha \le l^2/(4h^2) + 1$, $\alpha = 1, 2$. Substituting this estimate in (23), we obtain $\gamma_1 \ge 4h^2/(l^2 + 4h^2)$, i.e. $\gamma_1 \approx 4h^2/l^2$. Since $\gamma_2 = 2 - \gamma_1$, then $\xi = \gamma_1/\gamma_2 \approx 2h^2/l^2$. Consequently, from the estimate (7) for the iteration count we obtain

$$n_0(\epsilon) \approx \frac{l^2}{2\sqrt{2}h} \ln \frac{2}{\epsilon} \approx 0.35N \ln \frac{2}{\epsilon}, \tag{25}$$

where N is the maximal number of nodes in each direction.

Thus, for the implicit Chebyshev method considered here, the iteration count depends only on the basic step of the grid h and not on the non-uniform steps $h_\alpha^\pm$ at the near-boundary nodes. Comparing the estimate (25) with the estimate (10) obtained earlier, we find that the iteration count for the case of an arbitrary region $\bar{G}$ embedded in a square with side l is the same as for the case where the region $\bar{G}$ is this square.

We give now the computational formulas of the Chebyshev iterative method for solving the Dirichlet difference problem for Poisson's equation in an arbitrary region G:

$$y_{k+1}(x_{ij}) = (1 - \tau_{k+1}) y_k(x_{ij})$$

$$+ \frac{\tau_{k+1}}{b_1(x_{ij}) + b_2(x_{ij})} \left[\frac{1}{h}\left(\frac{y^{+1_1}}{h_1^+} + \frac{y^{-1_1}}{h_1^-} + \frac{y^{+1_2}}{h_2^+} + \frac{y^{-1_2}}{h_2^-}\right) + \varphi\right]_{x = x_{ij}},$$

$$x_{ij} \in \omega, \quad y_k(x) = g(x), \quad x \in \gamma.$$

Note that in Chapter 10 this problem will be solved by another implicit Chebyshev method (the alternate-triangular iterative method), for which the number of arithmetic operations required to realize one step of the method will be somewhat greater than for the method considered here, but the iteration count will be significantly smaller, thus guaranteeing the effectiveness of that method.

6.5.3 A Dirichlet difference problem for an elliptic equation with variable coefficients

6.5.3.1 *The explicit Chebyshev method.* We look now at a Dirichlet problem for a second-order elliptic equation with variable coefficients in the rectangle $\bar{G} = \{0 \le x_\alpha \le l_\alpha,\ \alpha = 1,2\}$:

$$Lu = \sum_{\alpha=1}^{2} \frac{\partial}{\partial x_\alpha}\left(k_\alpha(x)\frac{\partial u}{\partial x_\alpha}\right) - q(x)u = -f(x), \quad x \in G, \qquad u(x) = g(x), \quad x \in \Gamma. \tag{26}$$

On the rectangluar grid

$$\bar{\omega} = \{x_{ij} = (ih_1, jh_2) \in \bar{G}, \quad 0 \le i \le N_1, \quad 0 \le j \le N_2, \quad h_\alpha = l_\alpha/N_\alpha, \quad \alpha = 1,2\}$$

the differential problem (26) corresponds to a difference problem

$$\Lambda y = \sum_{\alpha=1}^{2} (a_\alpha(x) y_{\bar{x}_\alpha})_{x_\alpha} - d(x)y = -\varphi(x), \quad x \in \omega, \qquad y(x) = g(x), \quad x \in \gamma. \tag{27}$$

If the coefficients $k_\alpha(x)$, $q(x)$ and $f(x)$ are sufficiently smooth functions, the coefficients $a_\alpha(x)$, $d(x)$, and $\varphi(x)$ of the difference scheme (27) can, for example, be defined as follows:

$$a_1(x_{ij}) = k_1((i-0.5)h_1, jh_2), \quad a_2(x_{ij}) = k_2(ih_1, (j-0.5)h_2),$$
$$d(x_{ij}) = q(x_{ij}), \qquad \varphi(x_{ij}) = f(x_{ij}).$$

We will assume that the coefficients of the difference scheme (27) satisfy the conditions

$$0 < c_1 \le a_\alpha(x) \le c_2, \quad x \in \bar{\omega}, \qquad 0 \le d_1 \le d(x) \le d_2, \quad x \in \omega,\ \alpha = 1,2. \tag{28}$$

These conditions guarantee the existence and uniqueness of a solution to problem (27).

The difference scheme (27) reduces to the operator equation (1) in the usual way: $Ay = -\Lambda \mathring{y}$, where $y \in H$, $\mathring{y} \in \mathring{H}$, and H is the space of grid functions defined on ω with the inner product

$$(u, v) = \sum_{x \in \omega} u(x)v(x)h_1 h_2.$$

The right-hand side f of equation (1) only differs from the right-hand side φ of the scheme (27) at the near-boundary nodes.

In order to approximately solve equation (1), we will apply the explicit Chebyshev method (5)–(7) ($B = E$). In Section 5.2 it was shown that the operator A defined here is self-adjoint in H. Therefore the *a priori* information for the Chebyshev method has the form of constants γ_1 and γ_2 from the inequalities $\gamma_1 E \le A \le \gamma_2 E$, $\gamma_1 > 0$, if D is chosen as one of the operators E, A, or A^2. We will now find these constants. To do this, we introduce the operator $\mathring{A}$ corresponding to the difference operator $\mathring{\Lambda}$, where $\mathring{\Lambda} y = y_{\bar{x}_1 x_1} + y_{\bar{x}_2 x_2}$, and define the following inner product for grid functions given on $\bar{\omega}$:

$$(u,v)_{\omega_\alpha} = \sum_{x_\alpha \in \omega_\alpha} u(x)v(x)h_\alpha, \qquad (u,v)_{\omega_\alpha^+} = \sum_{x_\alpha \in \omega_\alpha^+} u(x)v(x)h_\alpha,$$

$$(u,v)_\alpha = ((u,v)_{\omega_\alpha^+}, 1)_{\omega_\beta}, \qquad \beta = 3-\alpha, \quad \alpha = 1,2.$$

Here

$$\omega_\alpha = \{x_{\alpha,i} = ih_\alpha,\ 1 \le i \le N_\alpha - 1\},\ \omega_\alpha^+ = \{x_{\alpha,i} = ih_\alpha,\ 1 \le i \le N_\alpha\}.$$

The inner products introduced here are analogs of the inner products defined in Section 6.5.2.

From the definition of the operators A and $\mathring{A}$ and the Green difference formulas (Section 5.2.2) we obtain

$$(Au,u) = -(\Lambda\mathring{u},\mathring{u}) = -((a_1 \mathring{u}_{\bar{x}_1 x_1}, \mathring{u})_{\omega_1}, 1)_{\omega_2} - ((a_2 \mathring{u}_{\bar{x}_2 x_2}, \mathring{u})_{\omega_2}, 1)_{\omega_1}$$

$$+ (du,u) = \sum_{\alpha=1}^{2} (a_\alpha \mathring{u}_{\bar{x}_\alpha}^2, 1)_\alpha + (du,u),$$

$$(\mathring{A}u,u) = -(\mathring{\Lambda}\mathring{u},\mathring{u}) = \sum_{\alpha=1}^{2} (\mathring{u}_{\bar{x}_\alpha}^2, 1)_\alpha, \quad u \in H, \mathring{u} \in \mathring{H}.$$

Taking into account the inequalities (28), from this we obtain operator inequalities of the form

$$c_1 \mathring{A} + d_1 E \le A \le c_2 \mathring{A} + d_2 E. \tag{30}$$

In Section 6.5.1 it was shown that the operator $\mathring{A}$ has the bounds

$$\mathring{\gamma}_1 = \sum_{\alpha=1}^{2} \frac{4}{h_\alpha^2} \sin^2 \frac{\pi h_\alpha}{2l_\alpha}, \qquad \mathring{\gamma}_2 = \sum_{\alpha=1}^{2} \frac{4}{h_\alpha^2} \cos^2 \frac{\pi h_\alpha}{2l_\alpha},$$

i.e. we have the inequalities

$$\mathring{\gamma}_1 E \le \mathring{A} \le \mathring{\gamma}_2 E. \tag{31}$$

From (30) and (31) we find that the operator A has the bounds $\gamma_1 = c_1\mathring{\gamma}_1 + d_1$, $\gamma_2 = c_2\mathring{\gamma}_2 + d_2$.

Thus, the constants γ_1 and γ_2 have been found. Using them, we compute the iterative parameters τ_k from the formulas (6), and we estimate the iteration count n using the formulas (7).

Since $\mathring{\xi} = \mathring{\gamma}_1/\mathring{\gamma}_2 = O(|h|^2)$, then $\xi = \gamma_1/\gamma_2 = O(|h|^2)$ and we have the following asymptotic estimate of the iteration count for this method:

$$n_0(\epsilon) = O\left(\frac{1}{|h|} \ln \frac{2}{\epsilon}\right),$$

where the constant in the estimate depends on the extrema of the coefficients $a_\alpha(x)$ and $d(x)$, i.e. on c_α and d_α, $\alpha = 1, 2$.

In particular, when the region $\bar{G}$ is a square with side l ($l_1 = l_2 = l$), the grid $\bar{\omega}$ is square ($h_1 = h_2 = h = l/N$), and $d \equiv 0$, we obtain

$$\gamma_1 = \frac{8c_1}{h^2} \sin^2 \frac{\pi h}{2l}, \quad \gamma_2 = \frac{8c_2}{h^2} \cos^2 \frac{\pi h}{2l}, \quad \xi = \frac{c_1}{c_2} \tan^2 \frac{\pi h}{2l}$$

and, consequently,

$$n_0(\epsilon) \approx \sqrt{\frac{c_2}{c_1}} \frac{l}{\pi h} \ln \frac{2}{\epsilon} \approx 0.32\sqrt{\frac{c_2}{c_1}} N \ln \frac{2}{\epsilon}.$$

Comparing the estimate (10) with the estimate obtained here of the iteration count for the explicit Chebyshev method applied to the solution of the difference equation (27) with variable coefficients, we find that for this example the iteration count is $\sqrt{c_2/c_1}$ times larger than the iteration count in the constant coefficient case.

We give now the computational formulas for the explicit Chebyshev method (5)–(7) used to solve the difference equation (27). These formulas have the form

$$\begin{aligned} y_{k+1}(i,j) = \alpha_{k+1}(i,j) y_k(i,j) + \tau_{k+1} \Big\{ &\frac{1}{h_1^2}[a_1(i+1,j) y_k(i+1,j) \\ &+ a_1(i,j) y_k(i-1,j)] + \frac{1}{h_2^2}[a_2(i,j+1) y_k(i,j+1) \\ &+ a_2(i,j) y_k(i,j-1)] + \varphi(i,j)\Big\}, \\ &1 \le i \le N_1 - 1, \qquad 1 \le j \le N_2 - 1, \end{aligned}$$

where we have denoted

$$\begin{aligned} \alpha_{k+1}(i,j) = 1 - \tau_{k+1} \Bigg[&\frac{a_1(i+1,j) + a_1(i,j)}{h_1^2} \\ &+ \frac{a_2(i,j+1) + a_2(i,j)}{h_2^2} + d(i,j) \Bigg], \end{aligned}$$

and the initial approximation y_0 is an arbitrary grid function on ω which takes on the following given values on the boundary γ: $y_0(x) = g(x)$ for $x \in \gamma$.

6.5.3.2 *The implicit Chebyshev method.* In order to approximately solve the equation (1) constructed in the preceding subsection which corresponds to the difference scheme (27), we will now look at the simplest implicit Chebyshev method (5)–(7). As in Section 6.5.2, we again choose B as the diagonal part of the operator A

$$By = by,$$
$$b(i,j) = \frac{1}{h_1^2}[a_1(i+1,j) + a_1(i,j)] + \frac{1}{h_2^2}[a_2(i,j+1) + a_2(i,j)] + d(i,j). \quad (32)$$

Since the operators A and B are self-adjoint and positive-definite in H, the *a priori* information for the implicit Chebyshev method (5)–(7) has the form of the constants of energy equivalence for the operators $\gamma_1 B \le A \le \gamma_2 B$, $\gamma_1 > 0$, if D is taken to be one of the operator A, B, or $AB^{-1}A$.

We now find the constants γ_1 and γ_2. Just as in Section 6.5.2, it is shown that $\gamma_1 + \gamma_2 = 2$. Therefore, in the formulas (6) for the iterative parameters τ_k, we have $\tau_0 = 2/(\gamma_1+\gamma_2) = 1$, $\rho_0 = (\gamma_2 - \gamma_1)/(\gamma_2+\gamma_1) = 1 - \gamma_1$.

We now estimate γ_1. From lemma 14 in Section 5.2 it follows that, for any grid function $\mathring{y} \in \overset{\circ}{H}$, we have the inequality

$$(b\mathring{y},\mathring{y})_{\omega_\alpha} \le \kappa_\alpha \left[(a_\alpha \mathring{y}_{\bar{x}_\alpha}^2, 1)_{\omega_\alpha^+} + \frac{1}{2}(d\mathring{y},\mathring{y})_{\omega_\alpha} \right], \quad \alpha = 1,2, \quad (33)$$

where $\kappa_\alpha = \kappa_\alpha(x_\beta) = \max_{x_\alpha \in \omega_\alpha} v^\alpha(x)$, and $v^\alpha(x)$ is the solution of the following three-point boundary-value problem:

$$(a_\alpha v^\alpha_{\bar{x}_\alpha})_{x_\alpha} - \frac{1}{2} d v^\alpha = -b(x) \quad h_\alpha \le x_\alpha \le l_\alpha - h_\alpha,$$
$$v^\alpha(x) = 0, \ x_\alpha = 0, l_\alpha, \ h_\beta \le y_\beta \le l_\beta - h_\beta, \ \beta = 3 - \alpha, \ \alpha = 1,2. \quad (34)$$

Taking the inequalities (33), dividing by κ_α, and successively summing over ω_β, we obtain

$$\left(\frac{b}{\kappa_\alpha}\mathring{y},\mathring{y}\right) \le (a_\alpha \mathring{y}_{\bar{x}_\alpha}^2, 1) + \frac{1}{2}(d\mathring{y},\mathring{y}), \qquad \alpha = 1,2.$$

Combining these inequalities and using (29) we obtain

$$\left(\sum_{\alpha=1}^{2} \frac{1}{\kappa_\alpha} b\mathring{y},\mathring{y}\right) \le \sum_{\alpha=1}^{2} (a_\alpha \mathring{y}_{\bar{x}_\alpha}^2, 1) + (d\mathring{y},\mathring{y}) = (Ay,y).$$

Consequently, it is possible to take γ_1 as

$$\gamma_1 = \min_{x \in \omega} \sum_{\alpha=1}^{2} \frac{1}{\kappa_\alpha} = \min_{x_2 \in \omega_2} \frac{1}{\kappa_1(x_2)} + \min_{x_1 \in \omega_1} \frac{1}{\kappa_2(x_1)}. \tag{35}$$

Thus, to find γ_1 it is necessary to solve equation (23), find $\kappa_\alpha(x_\beta)$, and compute γ_1 using (35). The constant γ_2 is found from the formula $\gamma_2 = 2 - \gamma_1$.

We now obtain an estimate of the iteration count for this implicit Chebyshev method. From the theory of difference schemes it follows that the difference scheme (34) is stable relative to the right-hand side in the uniform metric, i.e. there exists a constant M not depending on the stepsizes of the grid h_1 and h_2, such that the solution of the equation (34) satisfies

$$\max_{x_\alpha \in \omega_\alpha} v^\alpha(x) \le M(b^2, 1)^{1/2}_{\omega_\alpha}.$$

Since $b(x) = O\left(\frac{1}{h^2}\right)$, $h = \min_\alpha h_\alpha$ for $x \in \omega$, from this we obtain that

$$\kappa_\alpha = \max_{x_\alpha \in \omega_\alpha} v^\alpha(x) = O\left(\frac{1}{h^2}\right)$$

and, consequently, $\gamma_1 = O(h^2)$ and $\gamma_2 = O(1)$. Therefore $\xi = \gamma_1/\gamma_2 = O(h^2)$, and for the iteration count we have the same asymptotic estimate in h as for the explicit method

$$n_0(\epsilon) = O\left(\frac{1}{h} \ln \frac{2}{\epsilon}\right).$$

In what way is the implicit iterative method to be prefered over the explicit Chebyshev method considered above? The answer to this question is given in the following theorem, which we state without proof.

Theorem 7 *. *For the iterative scheme* (5)–(7) *with operator A corresponding to the difference scheme* (27), *the best diagonal operator B (i.e. the operator for which the ratio ξ is maximal) is defined by the formulas* (32). □

If the operator B is chosen as the diagonal of the operator A, it follows from theorem 7 that the ratio $\xi = \gamma_1/\gamma_2$ of the constants of energy equivalence γ_1 and γ_2 for the operators A and B will be maximal and, consequently, the iteration count n will be minimal.

---

* This theorem is a special case of a more general theorem that was proved in: G. Forsythe, E.G. Straus, "On best conditioned matrices", Proc. Amer. Math. Soc. 6 (1955), 340–345.

We illustrate the advantages of the implicit method on the following model problem. Suppose the difference scheme (27) is given on a square grid in the unit square: $h_1 = h_2 = h = 1/N$, $l_1 = l_2 = 1$.

The coefficients $a_1(x)$, $a_2(x)$ and $d(x)$ are chosen as follows:

$$a_1(x) = 1 + c[(x_1 - 0.5)^2 + (x_2 - 0.5)^2],$$
$$a_2(x) = 1 + c[0.5 - (x_1 - 0.5)^2 - (x_2 - 0.5)^2],$$
$$d(x) \equiv 0, \quad c > 0.$$

Then in the inequalities (28) we have $c_1 = 1$, $c_2 = 1 + 0.5c$, $d_1 = d_2 = 0$. By changing the parameter c, we obtain the coefficients for the difference scheme (27) with different extrema.

For the explicit method it was shown that the iteration count depends on the ratio c_2/c_1. For the implicit method, the iteration count does not depend on the maximal and minimal values of the coefficients $a_\alpha(x)$, but on certain integral characteristics of these coefficients.

In table 7 we give the iteration counts for the explicit and implicit methods as a function of the ratio c_2/c_1 and of the number of nodes N in one direction. The computations were carried out for $\epsilon = 10^{-4}$. For the case when the parameter $c = 0$, i.e. $a_\alpha(x) \equiv 1$, and Poisson's equation is considered, the iteration counts for the explicit and implicit methods are identical and were given in Section 1.

From the table it follows that for this example the iteration count for the implicit method is significantly less than for the explicit method. The dependence of the iteration count on the ratio c_2/c_1 is weaker for the implicit method than for the explicit one.

In conclusion we give the computational formulas for the implicit Chebyshev method:

$$\begin{aligned} y_{k+1}(i,j) =& (1 - \tau_{k+1}) y_k(i,j) \\ &+ \frac{\tau_{k+1}}{b(i,j)} \left\{ \frac{1}{h_1^2} [a_1(i+1,j) y_k(i+1,j) + a_1(i,j) y_k(i-1,j)] \right. \\ &+ \left. \frac{1}{h_2^2} [a_2(i,j+1) y_k(i,j+1) + a_2(i,j) y_k(i,j+1)] + \varphi(i,j) \right\}, \\ & 1 \le i \le N_1 - 1, \qquad 1 \le j \le N_2 - 1, \end{aligned}$$

where $b(i,j)$ is defined in (32), and the initial approximation y_0 is an arbitrary grid function on ω which takes on the following given values on the boundary γ: $y_0(x) = g(x)$, $x \in \gamma$.

Table 7

$\frac{c_2}{c_1}$	N=32		N=64		N=128	
	implicit	explicit	implicit	explicit	implicit	explicit
2	123	143	246	286	494	571
8	149	286	305	571	616	1142
32	175	571	365	1142	749	2283
128	192	1141	409	2283	856	4565
512	202	2281	436	4565	926	9130

From a comparison of the computational formulas for the explicit and implicit Chebyshev methods it follows that the number of arithmetic operations required to compute y_{k+1} from a given y_k for both methods are practically identical. Since the iteration count for the implicit method is significantly less than for the explicit method, it follows that the implicit method should be given preference.

6.5.4 A Dirichlet difference problem for an elliptic equation with mixed derivatives. In the rectangle $\bar{G} = \{0 \le x_\alpha \le l_\alpha,\ \alpha = 1,2\}$ with boundary Γ it is necessary to solve a *Dirichlet difference problem for an elliptic equation with mixed derivatives*

$$Lu = \sum_{\alpha,\beta=1}^{2} \frac{\partial}{\partial x_\alpha}\left(k_{\alpha\beta}(x)\frac{\partial u}{\partial x_\beta}\right) = -f(x), \qquad x \in G,$$

$$u(x) = g(x), \qquad x \in \Gamma.$$

It is assumed that the symmetry

$$k_{12}(x) = k_{21}(x), \qquad x \in \bar{G}, \tag{36}$$

and ellipticity

$$c_1 \sum_{\alpha=1}^{2} \xi_\alpha^2 \le \sum_{\alpha,\beta=1}^{2} k_{\alpha\beta}\xi_{\alpha\beta}\xi_\beta \le c_2 \sum_{\alpha=1}^{2} \xi_\alpha^2, \qquad c_1 > 0, \tag{37}$$

conditions are satisfied, where $\xi = (\xi_1, \xi_2)^T$ is an arbitrary vector.

On the rectangular grid

$$\bar{\omega} = \{x_{ij} = (ih_1, jh_2) \in \bar{G}, \quad 0 \le i \le N_1,\ 0 \le j \le N_2, \\ h_\alpha = l_\alpha/N_\alpha, \quad \alpha = 1, 2\}$$

the differential problem corresponds to a Dirichlet problem

$$\Lambda y = 0.5 \sum_{\alpha,\beta=1}^{2} [(k_{\alpha\beta} y_{\bar{x}_\beta})_{x_\alpha} + (k_{\alpha\beta} y_{x_\beta})_{\bar{x}_\alpha}] = -\varphi(x),\ x \in \omega, \tag{38}$$
$$y(x) = g(x), \qquad x \in \gamma,$$

where γ is the boundary of the grid $\bar{\omega}$.

In Section 5.2 it was shown that the difference problem (38) reduces to the operator equation (1) in the usual way: $Ay = -\Lambda \mathring{y}$, where $y \in H$, $\mathring{y} \in \mathring{H}$, and $\mathring{y}(x) = y(x)$ for $x \in \omega$. Here $\mathring{H}$ is the set of grid functions defined on $\bar{\omega}$ and mapping to zero on γ, and H is the space of grid function defined on ω with the inner product

$$(u, v) = \sum_{x \in \omega} u(x) v(x) h_1 h_2.$$

It was also shown there that, if condition (36) is satisfied, then the operator A is self-adjoint in H; and if condition (37) is satisfied we have bounds γ_1 and γ_2 equal to

$$\gamma_1 = c_1 \sum_{\alpha=1}^{2} \frac{4}{h_\alpha^2} \sin^2 \frac{\pi h_\alpha}{2 l_\alpha}, \quad \gamma_2 = c_2 \sum_{\alpha=1}^{2} \frac{4}{h_\alpha^2} \cos^2 \frac{\pi h_\alpha}{2 l_\alpha}, \tag{39}$$

i.e.

$$\gamma_1 E \le A \le \gamma_2 E. \tag{40}$$

In order to approximately solve equation (1) corresponding to the difference scheme (38), we consider the explicit Chebyshev method (5)–(7) ($B = E$). Since the operators A and B are self-adjoint and positive-definite in H, the ***a priori*** information has the form of the constants γ_1 and γ_2 in the inequalities (40), and the method converges in H_D, if $D = A$, B, or $AB^{-1}A$.

From (39) we obtain

$$\gamma_1 = O(c_1),\ \gamma_2 = O\left(\frac{c_2}{h^2}\right),\ \xi = \frac{\gamma_1}{\gamma_2} = O\left(\frac{c_1}{c_2} h^2\right),\ h^2 = h_1^2 + h_2^2.$$

Consequently, for this example the asymptotic bound in h for the iteration count $n_0(\epsilon)$ has the form

$$n_0(\epsilon) = O\left(\sqrt{\frac{c_2}{c_1}}\frac{1}{h}\ln\frac{2}{\epsilon}\right).$$

In particular, when $\bar{G}$ is a square with side l and the grid $\bar{\omega}$ is square ($h_1 = h_2 = h = l/N$), we obtain

$$\gamma_1 = \frac{8c_1}{h^2}\sin^2\frac{\pi h}{2l}, \quad \gamma_2 = \frac{8c_2}{h^2}\cos^2\frac{\pi h}{2l}, \quad \xi = \frac{c_1}{c_2}\tan^2\frac{\pi h}{2l},$$
$$n \geq n_0(\epsilon) \approx \sqrt{\frac{c_2}{c_1}}\frac{l}{\pi h}\ln\frac{2}{\epsilon} = 0.32\sqrt{\frac{c_2}{c_1}}N\ln\frac{2}{\epsilon},$$

i.e. the iteration count is again proportional to the number of nodes N in one direction, as in the case of an equation without mixed derivatives.

With this we conclude our examination of sample applications of two-level iterative methods to the solution of elliptic equations. More complex examples will be studied in Chapter 14.

Chapter 7

Three-Level Iterative Methods

In this chapter we study three-level iterative methods for solving the operator equation $Au = f$. The iterative parameters are chosen using *a priori* information about the operators of the scheme. In Section 7.1, an estimate is given for the convergence rate of three-level schemes of standard type. In Sections 7.2, 7.3 the Chebyshev semi-iterative method and the stationary three-level method are considered. In Section 7.4 we investigate the stability of two-level and three-level methods with regard to perturbations of the *a priori* data.

7.1 An estimate of the convergence rate

7.1.1 The basic family of iterative schemes. In Chapter 6, we constructed two-level iterative methods to find an approximate solution to the linear operator equation

$$Au = f \tag{1}$$

with a non-singular operator A acting in a real Hilbert space H. In these methods, the two-level scheme was linked with two iterative approximations y_{k+1} and y_k.

In this chapter, we will study three-level iterative schemes. A three-level iterative scheme for equation (1) is linked with three iterative approximations y_{k+1}, y_k, and y_{k-1}, so that y_{k+1} is defined in terms of y_k and y_{k-1}. In order to realize a three-level scheme, it is necessary to give two initial approximations y_0 and y_1. Usually, y_0 is arbitrary and y_1 is found from a two-level scheme.

We limit ourselves to the study of three-level schemes of standard type. An implicit *standard three-level iterative scheme* has the form

$$\begin{aligned} By_{k+1} &= \alpha_{k+1}(B - \tau_{k+1}A)y_k + (1 - \alpha_{k+1})By_{k-1} + \alpha_{k+1}\tau_{k+1}f, \\ By_1 &= (B - \tau_1 A)y_0 + \tau_1 f, \quad k = 1, 2, \ldots, \quad y_0 \in H, \end{aligned} \tag{2}$$

where y_0 is an arbitrary initial approximation, B is a non-singular linear operator acting in H, α_k and τ_k are iterative parameters. The formulas (2) determine the basic family of three-level iterative schemes.

In order to find a new approximation y_{k+1}, it is possible to proceed as follows. Let $\bar{y}$ be an intermediate iterative approximation which is found from the implicit two-level scheme

$$B\frac{\bar{y}-y_k}{\tau_{k+1}}+Ay_k=f.$$

Then from (2) it follows that y_{k+1} is a linear combination of the approximations $\bar{y}$ and y_{k-1}

$$y_{k+1}=\alpha_{k+1}\bar{y}+(1-\alpha_{k+1})y_{k-1}.$$

Thus, the approximation y_{k+1} is a linear extrapolation along the approximations $\bar{y}$ and y_{k-1}.

If we set $\alpha_k \equiv 1$ in (2), then the three-level scheme (2) reduces to a two-level scheme whose convergence was studied in Chapter 6. Therefore, the introduction of the iterative parameters α_k allows us to conclude that the convergence of the scheme (2) will be no worse than the convergence of the two-level scheme.

Notice that, unlike two-level schemes, the realization of a three-level scheme requires us to store not one, but two iterative approximations y_k and y_{k-1}.

7.1.2 An estimate for the norm of the error. We are interested now in investigating the convergence of the three-level scheme (2) in the energy space H_D generated by a self-adjoint and positive-definite operator D in the space H. To do this we study the behavior of the norm of the error $z_k=y_k-u$ in H_D as $k\to\infty$.

Substituting $y_k=z_k+u$ for $k=0,1,\dots$ in (2) and taking into account equation (1), we find an equation for the error z_k:

$$\begin{aligned}&Bz_{k+1}=\alpha_{k+1}(B-\tau_{k+1}A)z_k+(1-\alpha_{k+1})Bz_{k-1}, \quad k=1,2,\dots,\\&Bz_1=(B-\tau_1 A)z_0, \quad z_0=y_0-u.\end{aligned}$$

We solve this equation for z_{k+1} and, setting $z_k=D^{-1/2}x_k$, we transform to an equation for the equivalent error x_k. The equation for x_k will have the following form:

$$\begin{aligned}&x_{k+1}=\alpha_{k+1}S_{k+1}x_k+(1-\alpha_{k+1})x_{k-1}, \quad k=1,2,\dots,\\&x_1=S_1x_0, \qquad S_k=E-\tau_kC,\end{aligned} \tag{3}$$

where $C=D^{1/2}B^{-1}AD^{-1/2}$.

Due to the substitution $z_k = D^{-1/2}x_k$, we have $\| x_k \| = \| z_k \|_D$, and, consequently, convergence of the scheme (2) in H_D is equivalent to the condition that $\| x_k \| \to 0$ as $k \to \infty$.

We now study the behavior of the norm x_k in H as $k \to \infty$. To do this, we explicitly find the solution of equation (3). Using (3), we sequentially obtain

$$\begin{aligned} x_1 &= (E - \tau_1 C)x_0 = P_1(C)x_0, \\ x_2 &= \alpha_2(E - \tau_2 C)x_1 + (1 - \alpha_2)x_0 = [\alpha_2(E - \tau_2 C)P_1(C) \\ &\quad + (1 - \alpha_2)E]x_0 = P_2(C)x_0, \\ &\cdots\cdots\cdots\cdots\cdots\cdots\cdots\cdots\cdots\cdots \\ x_{k+1} &= \alpha_{k+1}(E - \tau_{k+1}C)P_k(C)x_0 + (1 - \alpha_{k+1})P_{k-1}(C)x_0 \\ &= P_{k+1}(C)x_0 \end{aligned}$$

and so forth.

Consequently, the solution of equation (3) for any k has the form

$$x_k = P_k(C)x_0, \quad k = 0, 1, \ldots, \tag{4}$$

where $P_k(C)$ is an operator polynomial of degree k in the operator C. Since x_0 is arbitrary, the corresponding algebraic polynomial $P_k(t)$ satisfies the following recurrence relations:

$$\begin{aligned} P_{k+1}(t) &= \alpha_{k+1}(1 - \tau_{k+1}t)P_k(t) + (1 - \alpha_{k+1})P_{k-1}(t), \\ P_1(t) &= 1 - \tau_1 t, \quad P_0(t) \equiv 1, \quad k = 1, 2, \ldots. \end{aligned} \tag{5}$$

From (5) it follows that, for any k, the polynomial $P_k(t)$ satisfies the normalization condition $P_k(0) = 1$.

We now estimate the norm of x_k. From (4) we obtain

$$\| x_k \| = \| P_k(C)x_0 \| \leq \| P_k(C) \| \, \| x_0 \|, \quad k = 0, 1, \ldots$$

or, using the substitution $z_k = D^{-1/2}x_k$,

$$\| z_k \|_D \leq \| P_k(C) \| \, \| z_0 \|_D . \tag{6}$$

Thus, an estimate for the norm of z_k has been obtained. From (6) it follows that the method will have the best convergence rate if the norm of the polynomial $P_k(C)$ tends to zero as quickly as possible as $k \to \infty$. Since the polynomial $P_k(C)$ is a function of the iterative parameters $\tau_1, \tau_2, \ldots, \tau_k$ and $\alpha_2, \alpha_3, \ldots, \alpha_k$, these parameters should be chosen from the condition that the norm of the operator polynomial $P_k(C)$ be a minimum. In other words, it is necessary to construct the polynomial of degree k, normalized by the condition $P_k(0) = E$, which has minimal norm.

When we studied the Chebyshev two-level scheme in Chapter 6, this problem was solved under the assumption that the operator $DB^{-1}A$ was self-adjoint in H and that the constants of energy equivalence γ_1 and γ_2 were given for the self-adjoint operators D and $DB^{-1}A$:

$$\gamma_1 D \le DB^{-1}A \le \gamma_2 D, \quad \gamma_1 > 0, \quad DB^{-1}A = (DB^{-1}A)^*. \tag{7}$$

In order to construct three-level iterative methods, we will only consider the self-adjoint case, i.e., we will assume that condition (7) is satisfied.

Suppose that condition (7) is satisfied. We will now find the optimal operator $P_k(C)$ and obtain an *a priori* estimate for the error z_k.

Since $C = D^{1/2}B^{-1}AD^{-1/2} = D^{-1/2}(DB^{-1}A)D^{-1/2}$, then from (7) it follows that the operator C is self-adjoint in H, and γ_1 and γ_2 are its bounds:

$$\gamma_1 E \le C \le \gamma_2 E, \qquad \gamma_1 > 0,\ C = C^*. \tag{8}$$

Then from (8) we have the following estimate for the norm of the operator $P_k(C)$

$$\| P_k(C) \| \le \max_{\gamma_1 \le t \le \gamma_2} |P_k(t)|.$$

Consequently, the optimal polynomial $P_k(t)$ is selected using the following condition: the maximum modulus of this polynomial on the interval $[\gamma_1, \gamma_2]$ is minimal. From Section 6.2 it follows that, because of the condition $P_k(0) = 1$, the desired polynomial has the form

$$P_k(t) = q_k T_k\left(\frac{1 - \tau_0 t}{\rho_0}\right), \quad q_k = \frac{1}{T_k\left(\frac{1}{\rho_0}\right)}, \tag{9}$$

where $T_k(x)$ is the Chebyshev polynomial of the first kind of degree k:

$$T_k(x) = \begin{cases} \cos(k \arccos x), & |x| \le 1, \\ \cosh(k \cosh^{-1} x), & |x| \ge 1, \end{cases}$$

$$\tau_0 = \frac{2}{\gamma_1 + \gamma_2}, \quad \rho_0 = \frac{1-\xi}{1+\xi}, \quad q_k = \frac{2\rho_1^k}{1+\rho_1^{2k}}, \quad \rho_1 = \frac{1-\sqrt{\xi}}{1+\sqrt{\xi}}, \quad \xi = \frac{\gamma_1}{\gamma_2}.$$

Thus we have the estimate

$$\| P_k(C) \| \max_{\gamma_1 \le t \le \gamma_2} |P_k(t)| = q_k, \quad k = 0, 1, \ldots.$$

Substituting this estimate in (6), we find

$$\| z_k \|_D \leq q_k \| z_0 \|_D .$$

Therefore, the convergence rate of the three-level iterative method (2), whose parameters τ_k and α_k are chosen from the condition that the resolving operator have minimal norm, is equal to the convergence rate for the Chebyshev two-level iterative method.

The formulas (9) solve the problem of constructing the fastest converging three-level iterative method. In Section 7.2 we will obtain formulas for the iterative parameters τ_k and α_k for this method, which we will call the *Chebyshev semi-iterative method.*

7.2 The Chebyshev semi-iterative method

7.2.1 Formulas for the iterative parameters. We now find formulas for the iterative parameters α_k and τ_k in the *Chebyshev semi-iterative method.* In Section 7.1, using the three-level iterative scheme of the method

$$\begin{aligned} By_{k+1} &= \alpha_{k+1}(B-\tau_{k+1}A)y_k + (1-\alpha_{k+1})By_{k-1} + \alpha_{k+1}\tau_{k+1}f, \\ By_1 &= (B-\tau_1 A)y_0 + \tau_1 f, \quad k=1,2,\ldots, \quad y_0 \in H, \end{aligned} \tag{1}$$

we obtained an equation for the equivalent error

$$\begin{aligned} x_{k+1} &= \alpha_{k+1}(E-\tau_{k+1}C)x_k + (1-\alpha_{k+1})x_{k-1}, \quad k=1,2,\ldots, \\ x_1 &= (E-\tau_1 C)x_0. \end{aligned} \tag{2}$$

It was shown that, for any k, the solution of this equation has the form

$$x_k = P_k(C)x_0, \qquad k=0,1,\ldots, \tag{3}$$

and the optimal polynomial $P_k(C)$ is defined by the formulas

$$P_k(t) = q_k T_k\left(\frac{1-\tau_0 t}{\rho_0}\right), \quad q_k = \frac{1}{T_k\left(\frac{1}{\rho_0}\right)} = \frac{2\rho_1^k}{1+\rho_1^{2k}}. \tag{4}$$

In order to obtain formulas for the iterative parameters α_k and τ_k, we will find recurrence relations for the polynomial $P_k(t)$.

It is known that, for any x, the Chebyshev polynomial of the first kind $T_k(x)$ satisfies the following recurrence relations (cf. Section 1.4):

$$T_{k+1}(x) = 2xT_k(x) - T_{k-1}(x), \quad k = 1, 2, \ldots,$$
$$T_1(x) = x, \quad T_0(x) \equiv 1. \tag{5}$$

Using (4) and (5) we obtain

$$\frac{P_{k+1}(t)}{q_{k+1}} = 2\left(\frac{1 - \tau_0 t}{\rho_0}\right)\frac{P_k(t)}{q_k} - \frac{P_{k-1}(t)}{q_{k-1}}, \quad k = 1, 2, \ldots, \tag{6}$$

$$P_1(t)/q_1 = (1 - \tau_0 t)/\rho_0, \quad P_0(t)/q_0 \equiv 1. \tag{7}$$

From the definition (4) and the relations (5) it follows that

$$1/q_{k+1} = 2/(\rho_0 q_k) - 1/q_{k-1}, \quad q_1 = \rho_0, \quad q_0 = 1. \tag{8}$$

From this we find

$$q_{k+1}/q_{k-1} = 2q_{k+1}/(\rho_0 q_k) - 1, \quad k = 1, 2, \ldots. \tag{9}$$

Substituting (8), (9) in (6) and (7), we obtain recurrence formulas for the polynomials $P_k(t)$:

$$P_{k+1}(t) = \frac{2}{\rho_0}\frac{q_{k+1}}{q_k}(1 - \tau_0 t)P_k(t) + \left(1 - \frac{2}{\rho_0}\frac{q_{k+1}}{q_k}\right)P_{k-1}(t),$$
$$P_1(t) = 1 - \tau_0 t, \quad P_0(t) \equiv 1, \quad k = 1, 2, \ldots.$$

From this and from (3) we obtain recurrence relations for x_k

$$x_{k+1} = \frac{2}{\rho_0}\frac{q_{k+1}}{q_k}(E - \tau_0 C)x_k + \left(1 - \frac{2}{\rho_0}\frac{q_{k+1}}{q_k}\right)x_{k-1}, \quad k = 1, 2, \ldots,$$
$$x_1 = (E - \tau_0 C)x_0.$$

Comparing these formulas with (2), we get

$$\alpha_{k+1} = 2q_{k+1}/(\rho_0 q_k), \quad \tau_k \equiv \tau_0 = 2/(\gamma_1 + \gamma_2), \quad k = 1, 2, \ldots. \tag{10}$$

Thus, the formulas for the iterative parameters τ_k and α_k have been obtained. We now transform the formula for the parameters α_k. To do this, we compute, using (8), the expression

$$4\frac{\alpha_{k+1}-1}{\alpha_k\alpha_{k+1}} = \frac{4}{\alpha_k}\left(1-\frac{1}{\alpha_{k+1}}\right) = \frac{4}{\alpha_k}\left(1-\frac{\rho_0}{2}\frac{q_k}{q_{k+1}}\right)$$
$$= \frac{4}{\alpha_k}\left[1-\frac{\rho_0}{2}\left(\frac{2}{\rho_0}-\frac{q_k}{q_{k-1}}\right)\right] = \frac{2\rho_0}{\alpha_k}\frac{q_k}{q_{k-1}} = \rho_0^2.$$

From this we obtain $\alpha_{k+1} = 4/(4-\rho_0^2\alpha_k)$, $k = 1, 2, \ldots$. Substituting $k = 0$ in (10) and using (8), we find that $\alpha_1 = 2$.

Thus, we have proved

Theorem 1. *Suppose that*

$$\gamma_1 D \le DB^{-1}A \le \gamma_2 D, \quad \gamma_1 > 0, \quad DB^{-1}A = (DB^{-1}A)^*.$$

The Chebyshev iterative method (2) with iterative parameters

$$\tau_k \equiv 2/(\gamma_1+\gamma_2), \quad \alpha_{k+1} = 4/(4-\rho_0^2\alpha_k), \quad k = 1, 2, \ldots, \quad \alpha_1 = 2, \tag{11}$$

converges in H_D, and the error z_k is estimated by

$$\| z_k \|_D \le q_k \| z_0 \|_D .$$

The iteration count n is estimated by $n \ge n_0(\epsilon)$, where

$$n_0(\epsilon) = \frac{\ln 0.5\epsilon}{\ln \rho_1}, \; \rho_0 = \frac{1-\xi}{1+\xi}, \; \rho_1 = \frac{1-\sqrt{\xi}}{1+\sqrt{\xi}}, \; q_k = \frac{2\rho_1^k}{1+\rho_1^{2k}}, \; \xi = \frac{\gamma_1}{\gamma_2}. \square$$

Remark. A comparison of the Chebyshev semi-iterative method with the Chebyshev two-level scheme shows that for both methods we have the same estimate $\| z_n \|_D \le q_n \| z_0 \|_D$, if n iterations are performed. However, for the two-level method this estimate is only valid after the completion of all the iterations, but for the three-level method the estimate is also valid at any intermediate iteration. Unlike a two-level method, in a three-level method the norm of the error decreases monotonically at intermediate iterations, and this guarantees the computational stability of the three-level method.

7.2.2 Sample choices for the operator D. We give now some sample choices for the operator D and the requirements that must be imposed on the operators A and B so that the conditions of theorem 1 will be satisfied.

In Section 6.2.3 we considered certain choices of the operator D based on properties of the operators A and B. We give those results here.

1) If the operators A and B are self-adjoint and positive-definite in H, then the operator D can be taken to be one of the following operators: A, B, or $AB^{-1}A$. Here the *a priori* information can be given in the form

$$\gamma_1 B \le A \le \gamma_2 B, \qquad \gamma_1 > 0. \tag{12}$$

2) If the operators A and B are self-adjoint, positive-definite, and they commute $A = A^* > 0$, $B = B^* > 0$, $AB = BA$, then D can be taken as the operator A^*A. In this case, the *a priori* information has the form of the inequalities (12).

3) If A and B are non-singular operators which satisfy the condition $B^*A = A^*B$, then D can be taken as the operator A^*A. In this case, the *a priori* information is given in the form of the inequalities

$$\gamma_1(Bx, Bx) \le (Ax, Bx) \le \gamma_2(Bx, Bx), \quad \gamma_1 > 0.$$

If these assumptions are satisfied for the three-level Chebyshev semi-iterative method, then theorem 1 is valid.

7.2.3 The algorithm of the method. We consider now the question of realizing the three-level scheme (1). The algorithm for the method can be described as follows:

1) Using the value of the parameter α_k and the given approximations y_{k-1} and y_k, we find α_{k+1} and τ_{k+1} from the formulas (11) and we compute

$$\varphi = B(\alpha_{k+1} y_k + (1 - \alpha_{k+1}) y_{k-1}) - \alpha_{k+1}\tau_{k+1}(Ay_k - f).$$

The computed φ can be overwritten on y_{k-1}, which is not needed at later iterations;

2) to find the new approximation y_{k+1} we solve the equation $By_{k+1} = \varphi$. The approximation y_1 is found from the equation $By_1 = \varphi$, where $\varphi = By_0 - \tau_1(Ay_0 - f)$. Such an algorithm can be recommended in the case where it is necessary to economize on computer storage.

If computation of the operator B requires many arithmetic operations, but computer storage is not an issue, then it is possible to use the following algorithm:

1) Given y_k compute the residual $r_k = Ay_k - f$;

2) solve the equation for the correction w_k: $Bw_k = r_k$;

3) given α_k, compute α_{k+1} from formulas (11), and the new approximation is found from the formula

$$y_{k+1} = \alpha_{k+1} y_k + (1 - \alpha_{k+1}) y_{k-1} - \alpha_{k+1} \tau_{k+1} w_k,$$

where τ_{k+1} is determined from the formula (11).

This algorithm does not contain the computation of the operator B, but requires auxiliary memory to store r_k and w_k.

7.3 The stationary three-level method

7.3.1 The choice of the iterative parameters. We return now to the formulas for the iterative parameters α_k and τ_k in the Chebyshev semi-iterative method. In Section 7.1 we obtained the following expressions for α_{k+1} and τ_{k+1}:

$$\alpha_{k+1} = 2q_{k+1}/(\rho_0 q_k), \quad \tau_k \equiv \tau_0 = 2/(\gamma_1 + \gamma_2), \quad k = 1, 2, \ldots, \tag{1}$$

where

$$q_k = \frac{2\rho_1^k}{1 + \rho_1^{2k}}, \quad \rho_1 = \frac{1 - \sqrt{\xi}}{1 + \sqrt{\xi}}, \quad \rho_0 = \frac{1 - \xi}{1 + \xi}, \quad \xi = \frac{\gamma_1}{\gamma_2}. \tag{2}$$

The value of the iterative parameter τ_k does not depend on the iteration number k, while the parameter α_k changes, beginning with $\alpha_1 = 2$. We now find the limit value for α_k as k tends to infinity. From (1), (2) we obtain

$$\alpha_{k+1} = 2\rho_1 \left(1 + \rho_1^{2k}\right) / \left(\rho_0 (1 + \rho_1^{2k+2})\right).$$

Since $\rho_1 < 1$ and $\rho_0 = q_1 = 2\rho_1/(1 + \rho_1^2)$, then $\alpha = \lim\limits_{k \to \infty} \alpha_k = 1 + \rho_1^2$, and for sufficiently large k we have $\alpha_k \approx \alpha$. Therefore it is natural to study the *stationary iterative three-level method*

$$\begin{aligned} By_{k+1} &= \alpha(B - \tau A) y_k + (1 - \alpha) B y_{k-1} + \alpha \tau f, \quad k = 1, 2, \ldots, \\ By_1 &= (B - \tau A) y_0 + \tau f, \quad y_0 \in H \end{aligned} \tag{3}$$

with constant (stationary) parameters

$$\alpha = 1 + \rho_1^2, \quad \tau = \tau_0 = \frac{2}{\gamma_1 + \gamma_2}, \quad \rho_1 = \frac{1 - \sqrt{\xi}}{1 + \sqrt{\xi}}, \quad \xi = \frac{\gamma_1}{\gamma_2}, \tag{4}$$

where γ_1 and γ_2 are the constants of energy equivalence for the self-adjoint operators D and $DB^{-1}A$:

$$\gamma_1 D \leq DB^{-1}A \leq \gamma_2 D, \quad \gamma_1 > 0, \quad DB^{-1}A = (DB^{-1}A)^*. \tag{5}$$

7.3.2 An estimate for the rate of convergence. To obtain an estimate for the convergence rate of the stationary three-level method, we transform from (3) to a scheme for the equivalent error $x_k = D^{1/2} z_k$:

$$x_{k+1} = \alpha(E - \tau C)x_k + (1-\alpha)x_{k-1}, \quad k = 1, 2, \ldots,$$
$$x_1 = (E - \tau C)x_0, \quad C = D^{1/2}B^{-1}AD^{-1/2}.$$

From this it follows that x_k can be expressed for any $k \geq 0$ in terms of x_0 as follows:

$$x_k = P_k(C)x_0, \tag{6}$$

where the algebraic polynomial $P_k(t)$ corresponding to $P_k(C)$ is defined by the recurrence relations

$$\begin{aligned} &P_{k+1}(t) = \alpha(1 - \tau t)P_k(t) + (1-\alpha)P_{k-1}(t), \quad k = 1, 2, \ldots, \\ &P_1(t) = 1 - \tau t, \quad P_0(t) \equiv 1. \end{aligned} \tag{7}$$

From (6) we obtain an estimate for the norm of the error z_k in H_D:

$$\| z_k \|_D = \| x_k \| \leq \| P_k(C) \| \, \| x_0 \| = \| P_k(C) \| \, \| z_0 \|_D . \tag{8}$$

Therefore it is necessary to estimate the norm of the operator polynomial $P_k(C)$ in the case where the parameters α and τ are chosen using the formulas (4). From the conditions (5) it follows that C is a self-adjoint operator in H, and γ_1 and γ_2 are its bounds; consequently,

$$\| P_k(C) \| \leq \max_{\gamma_1 \leq t \leq \gamma_2} |P_k(t)|.$$

We now estimate the maximum modulus of the polynomial $P_k(t)$ on the interval $[\gamma_1, \gamma_2]$. To do this we express the polynomial $P_k(t)$ in terms of Chebyshev polynomials. It is convenient for us to consider the polynomial $P_k(t)$, not on the interval $[\gamma_1, \gamma_2]$, but on the standard interval $[-1, 1]$. Setting

$$t = \frac{1 - \rho_0 x}{\tau_0}, \quad \tau_0 = \frac{2}{\gamma_1 + \gamma_2}, \quad \rho_0 = \frac{1 - \xi}{1 + \xi}, \quad \xi = \frac{\gamma_1}{\gamma_2},$$

we map the interval $[\gamma_1, \gamma_2]$ onto $[-1, 1]$. Then

$$P_k(t) = Q_k(x), \quad x \in [-1, 1],$$
$$\max_{\gamma_1 \le t \le \gamma_2} |P_k(t)| = \max_{|x| \le 1} |Q_k(x)|.$$

Taking into account the choice of the parameters α and τ from (4), we obtain from (7) the following recurrence relations for the polynomials $Q_k(x)$:

$$Q_{k+1}(x) = 2\rho_1 x Q_k(x) - \rho_1^2 Q_{k-1}(x), \quad k = 1, 2, \dots,$$
$$Q_1(x) = \rho_0 x, \quad Q_0(x) \equiv 1.$$

Then, using the substitution

$$Q_k = \rho_1^k R_k(x) \tag{9}$$

it is easy to obtain the standard recurrence relation

$$\begin{aligned} R_{k+1}(x) &= 2x R_k(x) - R_{k-1}(x), \quad k = 1, 2, \dots, \\ R_1(x) &= \rho_0 x / \rho_1, \quad R_0(x) \equiv 1. \end{aligned} \tag{10}$$

This relation is satisfied by the Chebyshev polynomial of the first kind $T_k(x)$ with initial conditions $T_1(x) = x$, $T_0(x) \equiv 1$, and the Chebyshev polynomial of the second kind $U_k(x)$:

$$U_k(x) = \begin{cases} \dfrac{\sin((k+1)\arccos x)}{\sin(\arccos x)}, & |x| \le 1, \\ \dfrac{\sinh((k+1)\cosh^{-1} x)}{\sinh(\cosh^{-1} x)}, & |x| \ge 1, \end{cases}$$

with the initial conditions $U_1(x) = 2x$, $U_0(x) \equiv 1$. Using the indicated properties of the polynomials $T_k(x)$ and $U_k(x)$ and the equation $\rho_0 = q_1 = 2\rho_1/(1 + \rho_1^2)$, from (10) we find an expression for the polynomial $R_k(x)$ in terms of the Chebyshev polynomials

$$R_k(x) = \frac{2\rho_1^2}{1 + \rho_1^2} T_k(x) + \frac{1 - \rho_1^2}{1 + \rho_1^2} U_k(x), \quad k \ge 0.$$

Further, using the known estimates

$$\max_{|x| \le 1} |T_k(x)| = T_k(1) = 1,$$
$$\max_{|x| \le 1} |U_k(x)| = U_k(1) = k + 1,$$

we obtain

$$\max_{|x|\leq 1} R_k(x) = R_k(1) = 1 + k(1-\rho_1^2)/(1+\rho_1^2).$$

From this, taking into account the changes made above, we find the following estimate for the norm of the operator polynomial $P_k(C)$:

$$\| P_k(C) \| \leq \rho_1^k \left(1 + k(1-\rho_1^2)/\left(1+\rho_1^2\right)\right). \tag{11}$$

Substituting (11) in (8), we obtain an estimate for the norm of the error z_k in H_D:

$$\| z_k \|_D \leq \bar{q}_k \| z_0 \|_D, \quad \bar{q}_k = \rho_1^k \left(1 + k(1-\rho_1^2)/\left(1+\rho_1^2\right)\right),$$

where $\bar{q} \to 0$ as $k \to \infty$ and $\bar{q}_{k+1} < \bar{q}_k$. Thus, we have proved

Theorem 2. *The stationary three-level iterative method* (3)–(5) *converges in* H_D, *and the error* z_k *is estimated by*

$$\| z_k \|_D \leq \bar{q}_k \| z_0 \|_D, \quad \bar{q}_k = \rho_1^k(1 + k(1-\rho_1^2)/(1+\rho_1^2)). \square$$

Remark. It is possible to show that $\lim\limits_{k\to\infty} q_k/\bar{q}_k = \lim\limits_{k\to\infty} \bar{q}_k/\rho_0^k = 0$, where q_k is defined in theorem 1. Therefore the stationary three-level method converges faster than the simple iteration method, but slower than the Chebyshev two-level method and the Chebyshev semi-iterative method.

7.4 The stability of two-level and three-level methods relative to *a priori* data

7.4.1 Statement of the problem. In order to approximately solve the operator equation $Au = f$, in Chapter 6 we studied the two-level simple iteration and Chebyshev methods, and in Sections 7.2, 7.3 we constructed the Chebyshev semi-iterative method and the three-level stationary method.

Recall that, in order to compute the iterative parameters in these methods, we used ***a priori*** information about the operators in the iterative scheme. In the case where the operator $DB^{-1}A$ is self-adjoint, this information has the form of the constants of energy equivalence γ_1 and γ_2 for the operators D and $DB^{-1}A$:

$$\gamma_1(Dx,x) \leq (DB^{-1}Ax,x) \leq \gamma_2(Dx,x), \quad \gamma_1 > 0. \tag{1}$$

In a series of cases the constants γ_1 and γ_2 can be found precisely, i.e. there exist elements $x \in H$ for which (1) becomes an equality. In other cases, an auxiliary process is used to find γ_1 and γ_2, and these constants are found approximately.

The use of inexact *a priori* information leads to a reduction in the convergence rate, and in some cases to divergence of the method. The goal of the current subsection is to investigate the influence of inexact *a priori* information on the converegnce rate of the methods listed above.

We limit ourselves to the self-adjoint case, i.e. we assume that the operator $DB^{-1}A$ is self-adjoint in H. Suppose that, instead of exact values of γ_1 and γ_2, we are given some approximate values $\bar{\gamma}_1$ and $\bar{\gamma}_2$ in the inequalities (1). We shall look at two-level and three-level methods whose iterative parameters are chosen using $\bar{\gamma}_1$ and $\bar{\gamma}_2$. We first recall the formulas for the iterative parameters.

For the two-level scheme

$$B\frac{y_{k+1}-y_k}{\tau_{k+1}} + Ay_k = f, \quad k = 0, 1, \ldots, \tag{2}$$

the parameters for the simple iteration method are determined from the formula

$$\tau_k = \bar{\tau}_0 = 2/(\bar{\gamma}_1 + \bar{\gamma}_2), \quad k = 1, 2, \ldots, \tag{3}$$

and the parameters for the Chebyshev method are constructed using the formula

$$\begin{aligned} &\tau_k = \bar{\tau}_0/(1+\bar{\rho}_0\mu_k), \quad \mu_k \in \mathcal{M}_n^*, \quad k = 1, 2, \ldots, n, \\ &\bar{\rho}_0 = (1-\bar{\xi})/(1+\bar{\xi}), \quad \bar{\xi} = \bar{\gamma}_1/\bar{\gamma}_2. \end{aligned} \tag{4}$$

For the three-level iterative scheme

$$\begin{aligned} By_{k+1} &= \alpha_{k+1}(B - \tau_{k+1}A)y_k + (1-\alpha_{k+1})B_y y_{k-1} + \alpha_{k+1}\tau_{k+1}f \\ &\qquad k = 1, 2, \ldots, \\ By_1 &= (B - \tau_1 A)y_0 + \tau_1 f \end{aligned} \tag{5}$$

the parameters for the Chebyshev semi-iterative method are determined using the formulas

$$\tau_k \equiv \bar{\tau}_0, \quad \alpha_{k+1} = 4/(4 - \bar{\rho}_0^2\alpha_k), \quad k = 1, 2, \ldots, \quad \alpha_1 = 2, \tag{6}$$

and the parameters for the stationary three-level method are given by the formulas

$$\tau_k = \bar{\tau}_0, \quad \alpha_k = 1+\bar{\rho}_1^2, \quad k = 1, 2, \ldots, \quad \bar{\rho}_1 = \left(1-\sqrt{\bar{\xi}}\right)\Big/\left(1+\sqrt{\bar{\xi}}\right). \tag{7}$$

From the general theory of iterative methods laid out above, we obtain the following estimates for the error $z_k = y_k - u$ with these methods:

1) for the simple iteration method

$$\| z_n \|_D \leq \left(\max_{\gamma_1 \leq t \leq \gamma_2} |1 - \bar{\tau}_0 t| \right)^n \| z_0 \|_D; \tag{8}$$

2) for the Chebyshev two-level method and the Chebyshev semi-iterative method

$$\| z_n \|_D \leq \bar{q}_n \max_{\gamma_1 \leq t \leq \gamma_2} \left| T_n \left(\frac{1 - \bar{\tau}_0 t}{\bar{\rho}_0} \right) \right| \| z_0 \|_D, \tag{9}$$

where $\bar{q}_n = 2\bar{\rho}_1{}^n / (1 + \bar{\rho}_1^{2n})$;

3) for the stationary three-level method

$$\| z_n \|_D \leq \bar{\rho}_1^n \max_{\gamma_1 \leq t \leq \gamma_2} \left| \frac{2\bar{\rho}_1^2}{1 + \bar{\rho}_1^2} T_n \left(\frac{1 - \bar{\tau}_0 t}{\bar{\rho}_0} \right) + \frac{1 - \bar{\rho}_1^2}{1 + \bar{\rho}_1^2} U_n \left(\frac{1 - \bar{\tau}_0 t}{\bar{\rho}_0} \right) \right| \| z_0 \|_D . \tag{10}$$

Here $T_n(x)$ and $U_n(x)$ are the Chebyshev polynomials of the first and second kinds, γ_1 and γ_2 are the exact values of the constants from (1).

These estimates determine the convergence rates for these methods in the case where the iterative parameters are computed using inexact *a priori* information.

7.4.2 Estimates for the convergence rates of the methods. We now estimate the maximum modulus of the polynomials which enter into the estimates (8)–(10). To do this, we make the following changes in (8)–(10): setting $x = (1 - \bar{\tau}_0 t)/\bar{\rho}_0$, and denoting $a = (1 - \bar{\tau}_0 \gamma_2)/\bar{\rho}_0$, $b = (1 - \bar{\tau}_0 \gamma_1)/\bar{\rho}_0$. Then the estimates (8)–(10) will have the form

$$\begin{aligned}
&\| z_n \|_D \leq \bar{\rho}_0^n \left(\max_{a \leq x \leq b} |x| \right)^n \| z_0 \|_D, \\
&\| z_n \|_D \leq \bar{q}_n \max_{a \leq x \leq b} |T_n(x)| \, \| z_0 \|_D, \\
&\| z_n \|_D \leq \bar{\rho}_1^n \max_{a \leq x \leq b} \left| \frac{2\bar{\rho}_1^2}{1 + \bar{\rho}_1^2} T_n(x) + \frac{1 - \bar{\rho}_1^2}{1 + \bar{\rho}_1^2} U_n(x) \right| \| z_0 \|_D .
\end{aligned} \tag{11}$$

We look first at the case where $\bar{\gamma}_1$ and $\bar{\gamma}_2$ are approximations to γ_1 and γ_2 from above and below respectively, i.e.

$$\bar{\gamma}_1 \leq \gamma_1 \leq \gamma_2 \leq \bar{\gamma}_2. \tag{12}$$

In this case, as is easily verified, we have that $-1 \le a \le b \le 1$. From (11) we obtain that the convergence rate for the simple iteration method will be determined by the quantity $\bar{\rho}_0^n$, for the Chebyshev two-level method and the Chebyshev semi-iterative method by the quantity $\bar{q}_n$, and for the stationary three-level method by the quantity $\bar{\rho}_1(1+n(1-\bar{\rho}_1^2)/(1+\bar{\rho}_1^2))$. The iterative methods will converge, but the convergence rates are decreased.

We consider now an example for which condition (12) is satisfied. Suppose

$$\bar{\gamma}_1 = \gamma_1(1-\alpha), \quad \bar{\gamma}_2 = \gamma_2, \quad 0 \le \alpha < 1.$$

In this case $a = -1$, $b < 1$. Therefore, we obtain from (11) the following estimates for the error:

$$\| z_n \|_D \le \bar{\rho}_0^n \| z_0 \|_D,$$
$$\| z_n \|_D \le \bar{q}_n \| z_0 \|_D,$$
$$\| z_n \|_D \le \bar{\rho}_1^n(1+n(1-\bar{\rho}_1^2)/(1+\bar{\rho}_1^2)) \| z_0 \|_D .$$

From the corresponding formulas for the iteration counts, we obtain that for the simple iteration method with an inexact value for γ_1 the iteration count is increased approximately $1/(1-\alpha)$ times over the value for the exact value of γ_1, while for the Chebyshev two-level method and the Chebyshev semi-iterative method, the iteration count is only increased by a factor of $1/\sqrt{1-\alpha}$.

Suppose now that condition (12) is not satisfied. In this case $\max(|a|,|b|) > 1$. We introduce the following notation

$$\frac{1}{\rho_0^*} = \max(|a|,|b|),$$

$$q_n^* = \frac{1}{T_n\left(\frac{1}{\rho_0^*}\right)} = \frac{2\rho_1^{*n}}{1+\rho_1^{*2n}}, \quad \rho_1^* = \frac{\rho_0^*}{1+\sqrt{1-\rho_0^{*2}}}.$$

Using this notation, and also the relationship between the Chebyshev polynomials of the first and second kinds

$$U_n(x) = \left(T_{n+1}^2(x)-1\right)^{1/2} \Big/ \left(T_1^2(x)-1\right)^{1/2}, \quad |x| \ge 1,$$

we obtain

$$\max_{a \le x \le b} |x| = \frac{1}{\rho_0^*}, \quad \max_{a \le x \le b} |T_n(x)| = T_n\left(\frac{1}{\rho_0^*}\right) = \frac{1}{q_n^*} \le \frac{1}{\rho_1^{*n}},$$

$$\max_{a \le x \le b} |U_n(x)| = U_n\left(\frac{1}{\rho_0^*}\right) = \frac{1-\rho_1^{*2(n+1)}}{\rho_1^{*n}(1-\rho_1^{*2})} \le \frac{n+1}{\rho_1^{*n}}.$$

Substituting these estimates in (11), we find

$$\| z_n \|_D \leq \left(\frac{\bar{\rho}_0}{\rho_0^*}\right)^n \| z_0 \|_D, \tag{13}$$

$$\| z_n \|_D \leq \frac{\bar{q}_n}{q_n^*} \| z_0 \|_D, \tag{14}$$

$$\| z_n \|_D \leq (\bar{\rho}_1/\rho_1^*)^n (1 + n(1-\bar{\rho}_1^2)/(1+\bar{\rho}_1^2)) \| z_0 \|_D \tag{15}$$

Notice that if H is a finite-dimensional, then it is possible to select an initial approximation y_0 for which the estimates (13), (14) will become equalities.

We now find a condition which will guarantee the convergence of an iterative method constructed using inexact *a priori* information. Since the ration $\bar{q}_n/q_n^*$ only tends to zero as $n \to \infty$ if $\rho_1^* > \bar{\rho}_1$, and this condition is equivalent to the requirement $\rho_0^* > \bar{\rho}_0$, then from (13)–(15) it follows that the iterative method will converge if

$$\bar{\rho}_0 < \rho_0^*. \tag{16}$$

Using the definitions of ρ_0^*, a, and b we obtain that (16) will hold if $|1-\bar{\tau}_0\gamma_1| < 1$, $|1-\bar{\tau}_0\gamma_2| > 1$.

Solving these inequalities we find

$$\bar{\gamma}_1 + \bar{\gamma}_2 > \gamma_2. \tag{17}$$

Thus, if condition (17) is satisfied, the iterative method constructed using inexact *a priori* information will converge. From the remarks above it follows that, in the case of a finite-dimensional space H, the condition (17) is also necessary for convergence of the method.

We now estimate the actual number of iterations required to achieve a given accuracy ϵ. As before, we will denote by n the iteration count in the case of exact *a priori* information, by $\bar{n}$ the theoretical iteration count computed using the formulas from the corresponding theorems using inexact *a priori* information, and by n^* the actual iteration count required to achieve the accuracy ϵ. From the formulas (13)–(15) it follows that the actual iteration count n^* should be determined from the conditions:

1) for the simple iteration method from the condition $\bar{\rho}_0^n \leq \epsilon \rho_0^{*n}$;

2) for the Chebyshev two-level method and the Chebyshev semi-iterative method from the condition $\bar{q}_n \leq \epsilon q_n^*$.

It is easy to check that $n^* \geq \bar{n}$, $n^* \geq n$, and that the iteration count $\bar{n}$ can be larger or smaller than n. Since the only quantitative characteristic of the iterative method which can be computed beforehand is the theoretical iteration count $\bar{n}$, when implementing the method it is important to estimate by how much the actual iteration count n^* is larger than $\bar{n}$. For a theoretical comparison of the quality of the iterative methods, it is necessary to estimate the ratio n^*/n.

We obtain now the required estimates for one example. Suppose $\bar{\gamma}_1$ and $\bar{\gamma}_2$ are approximations to γ_1 and γ_2 from above and below, respectively,

$$\bar{\gamma}_1 = (1+\alpha)\gamma_1, \quad \bar{\gamma}_2 = (1-\alpha)\gamma_2, \qquad \alpha \geq 0. \tag{18}$$

From condition (17) and the natural requirement $\bar{\gamma}_1 \leq \bar{\gamma}_2$ we obtain that the methods will converge if

$$\alpha < \min(\xi/(1-\xi),\ (1-\xi)/(1+\xi)), \quad \xi = \gamma_1/\gamma_2.$$

For this example we will have the inequalities $n^* \geq n \geq \bar{n}$. In fact, from (18) we obtain

$$\bar{\xi} = \frac{\bar{\gamma}_1}{\bar{\gamma}_2} = \frac{1+\alpha}{1-\alpha}\xi \geq \xi$$

and, consequently,

$$\bar{\rho}_0 \leq \rho_0 = (1-\xi)/(1+\xi), \quad \bar{\rho}_1 \leq \rho_1 = \left(1-\sqrt{\xi}\right)\Big/\left(1+\sqrt{\xi}\right), \quad \bar{q}_n \leq q_n.$$

From this it follows that $n \geq \bar{n}$. We now estimate the quantities which enter into the inequalities (13)–(14). Since

$$\bar{\tau} = 2/(\bar{\gamma}_1 + \bar{\gamma}_2) = \tau_0/(1-\alpha\rho_0) < \tau_0, \quad \tau_0 = 2/(\gamma_1+\gamma_2),$$

then

$$1/\rho_0^* = \max(|a|, |b|) = |a| = (\bar{\tau}_0\gamma_2 - 1)/\bar{\rho}_0.$$

Omitting the simple calculations, we obtain

$$\bar{\rho}_0 = \frac{1-\bar{\xi}}{1+\bar{\xi}} = \frac{\rho_0 - \alpha}{1-\alpha\rho_0},$$

$$\frac{\bar{\rho}_0}{\rho_0^*} = \bar{\tau}_0\gamma_2 - 1 = 1 - \frac{1-\frac{\alpha}{\xi}(1-\xi)}{1+\alpha}(1-\bar{\rho}_0) = 1 - \frac{1-\frac{\alpha}{\xi}(1-\xi)}{1-\alpha\rho_0}(1-\rho_0),$$

$$\bar{\rho}_1 = \frac{1-\sqrt{\bar{\xi}}}{1+\sqrt{\bar{\xi}}} = \frac{\rho_0 - \alpha}{1-\alpha\rho_0 + \sqrt{(1-\alpha^2)(1+\rho_0^2)}},$$

$$\frac{\bar{\rho}_1}{\rho_1^*} = 1 - \frac{(1+\alpha)\sqrt{\xi} + \sqrt{1-\alpha^2} - \frac{\alpha}{\sqrt{\xi}} - \sqrt{\frac{\alpha}{\xi}[1-(1+\alpha)\xi]}}{(1+\alpha)\sqrt{\xi} + \sqrt{1-\alpha^2}}(1-\bar{\rho}_1)$$

$$= 1 - \frac{\left[(1+\alpha)\sqrt{\xi} + \sqrt{1-\alpha^2} - \frac{\alpha}{\sqrt{\xi}} - \sqrt{\frac{\alpha}{\xi}[1-(1+\alpha)\xi]}\right](1+\sqrt{\xi})}{1-\alpha+(1+\alpha)\xi+2\sqrt{(1-\alpha^2)\xi}}(1-\rho_1).$$

We first consider the simple iteration method. From theorem 2 Section 6.3 and from (13) we obtain the following estimates for the iteration counts n^*, $\bar{n}$ and n:

$$n = \frac{\ln \epsilon}{\ln \rho_0} \approx \frac{\ln(1/\epsilon)}{1-\rho_0}, \quad \bar{n} = \frac{\ln \epsilon}{\ln \bar{\rho}_0} \approx \frac{\ln(1/\epsilon)}{1-\bar{\rho}_0}$$

$$n^* = \frac{\ln \epsilon}{\ln(\bar{\rho}_0/\rho_0^*)} \approx \frac{\ln(1/\epsilon)}{1-\bar{\rho}_0/\rho_0^*}.$$

Substituting here the expressions obtained above, we find

$$\frac{n^*}{\bar{n}} \approx \frac{1+\alpha}{1-\frac{\alpha}{\xi}(1-\xi)}, \quad \frac{n^*}{n} \approx \frac{1-\alpha\rho_0}{1-\frac{\alpha}{\xi}(1-\xi)}.$$

If $\alpha \approx c\xi$, where $c < 1$, then from this we obtain

$$n^* \approx \bar{n}/(1-c), \quad n^* \approx n/(1-c).$$

Thus, if $\alpha \approx c\xi$, then the actual iteration count n^* for the simple iteration method is $1/(1-c)$ times larger than the theoretical iteration count $\bar{n}$ computed using the inexact *a priori* information.

We now consider the Chebyshev two-level method and the Chebyshev semi-iterative method. From the definitions of $\bar{q}_n$ and q_n^* we obtain

$$\frac{\bar{q}_n}{q_n^*} = \frac{\bar{\rho}_1^n}{\rho_1^{*n}} \cdot \frac{1+\rho_1^{*2n}}{1+\bar{\rho}_1^{2n}} \leq \frac{2(\bar{\rho}_1/\rho_1^*)^n}{1+(\bar{\rho}_1/\rho_1^*)^{2n}}.$$

Therefore we have the following estimate for the iteration count

$$n^* = \frac{\ln(0.5\epsilon)}{\ln(\bar{\rho}_1/\rho_1^*)} \approx \frac{\ln(2/\epsilon)}{1-\bar{\rho}_1/\rho_1^*}.$$

Further, from theorem 1 Section 6.2 and theorem 1 Section 7.2 we obtain estimates for n and $\bar{n}$:

$$n = \frac{\ln(0.5\epsilon)}{\ln \rho_1} \approx \frac{\ln(2/\epsilon)}{1-\rho_1}, \quad \bar{n} = \frac{\ln(0.5\epsilon)}{\ln \bar{\rho}_1} \approx \frac{\ln(2/\epsilon)}{1-\bar{\rho}_1}.$$

Substituting here the expressions obtained above for the ratio $\bar{\rho}_1/\rho_1^*$ and assuming that $\alpha \approx c\xi$, we find

$$n^*/\bar{n} \approx 1/(1-\sqrt{c}), \quad n^*/n \approx 1/(1-\sqrt{c}).$$

Thus, if $\alpha \approx c\xi$, where $c < 1$, then the actual iteration count n^* for the Chebyshev two-level method and the Chebyshev semi-iterative method is approximately $1/(1-\sqrt{c})$ times larger than the theoretical iteration count $\bar{n}$ computed using the inexact *a priori* information.

Chapter 8

Iterative Methods of Variational Type

In this chapter we look at two-level and three-level iterative methods of variational type. Implementing these methods requires no *a priori* information about the operators of the scheme. In Sections 8.1, 8.2 two-level gradient methods are studied, and in Sections 8.3, 8.4, the three-level conjugate-direction methods are presented. Acceleration of convergence for two-level methods in the self-adjoint case is examined in Section 8.5.

8.1 Two-level gradient methods

8.1.1 The choice of the iterative parameters. In order to find an approximate solution to the linear operator equation

$$Au = f, \tag{1}$$

with a non-singular operator A acting in a real Hilbert space H, we look at the implicit two-level iterative scheme

$$B\frac{y_{k+1} - y_k}{\tau_{k+1}} + Ay_k = f, \quad k = 0, 1, \ldots, \tag{2}$$

with an arbitrary initial approximation $y_0 \in H$ and a non-singular operator B.

We studied the iterative scheme (2) earlier in Chapter 6, where we constructed sets of iterative parameters $\{\tau_k\}$ and gave estimates of the convergence rate for the corresponding iterative methods (the Chebyshev method and the simple iteration method).

Any two-level iterative method constructed on the basis of the scheme (2) is characterized by the operators A and B, the energy space H_D in which the convergence of the method is studied, and by the set of iterative parameters τ_k. The basic question in the theory of iterative methods is the question of the optimal choice of the parameters τ_k.

In Chapter 6 the iterative methods were constructed and the parameters τ_k were chosen based on the condition that either the norm of the transformation operator from iteration to iteration or the norm of the resolving operator be a minimum in H_D. An unfortunate feature of iterative methods based on that principle is that *a priori* information about the operators of the iterative scheme is used to compute the parameters τ_k.

The form of the necessary *a priori* information is determined by the properties of the operators A, B, and D. So, in this case where the operators $DB^{-1}A$ is self-adjoint in the space H, this information takes the form of the constants of energy equivalence for the operators D and $DB^{-1}A$, i.e. the constants γ_1 and γ_2 from the inequalities

$$\gamma_1 D \leq DB^{-1}A \leq \gamma_2 D, \quad \gamma_1 > 0, \tag{3}$$

or the bounds for the operator $DB^{-1}A$ in H_D.

In the non-self-adjoint case, we use either the two scalars γ_1 and γ_2 from the inequalities

$$\gamma_1 D \leq DB^{-1}A, \ (DB^{-1}Ax, B^{-1}Ax) \leq \gamma_2(DB^{-1}Ax, x), \ \gamma_1 > 0, \tag{4}$$

or the three scalars γ_1, γ_2, and γ_3, where γ_1 and γ_2 are the constants from the inequalities (3), and γ_3 is the constant either from the inequality

$$\| 0.5(DB^{-1}A - A^*(B^*)^{-1}D)x \|_{D^{-1}}^2 \leq \gamma_3^2(Dx, x), \tag{5}$$

or from the inequality

$$\| 0.5(DB^{-1}A - A^*(B^*)^{-1}D)x \|_{D^{-1}}^2 \leq \gamma_3(DB^{-1}Ax, x). \tag{6}$$

In many cases, finding the constants γ_1, γ_2, and γ_3 with sufficient accuracy can be a complex independent problem, requiring the use of special computational methods for its solution. If the *a priori* information can be obtained at little computational expense or if we are required to solve a series of problems (1) with different right-hand sides, then it is expedient to once find the necessary *a priori* information and then to use the iterative methods constructed in Chapter 6. This procedure can be recommended if the auxiliary time spent to obtain the *a priori* information is considerably less than the time to solve the whole series of problems (1).

In those cases where it is only necessary to solve one problem (1), or when a good initial approximation is given and computing the constants γ_1, γ_2, and γ_3 is a laborious process, it is better to use iterative methods of variational type, which we are now about to study.

In a two-level method of variational type, no *a priori* information about the operators of the scheme (2) is required to compute the parameters τ_k (except for conditions of a general form: $A = A^* > 0$, $(DB^{-1}A)^* = DB^{-1}A$, etc.), and these methods are based on the following principle. If an approximation y_k is given, and y_{k+1} is found from the scheme (2), then the iterative parameter τ_{k+1} is chosen from the condition that the norm of the error $z_{k+1} = y_{k+1} - u$ in H_D be a minimum, where u is the solution of the equation (1).

The name of the methods is connected with the fact that the sequence y_k constructed using the formula (2), where the parameters τ_k are chosen from the above condition, is a minimizing sequence for the quadratic functional

$$I(y) = (D(y-u), y-u).$$

If D is a positive-definite operator, this functional is bounded below and achieves its minimum, equal to zero, at the solution of equation (1), i.e. for $y = u$. The choice of the parameter τ_{k+1} from the indicated condition guarantees the local minimization of the functional $I(y)$ as we pass from y_k to y_{k+1}, i.e. after one iterative step. In the case of an explicit scheme ($B = E$), the move from y_k to y_{k+1} is performed using the formula

$$y_{k+1} = y_k - \tau_{k+1} r_k, \quad r_k = Ay_k - f.$$

Notice that, for a self-adjoint positive-definite operator A, the move from y_k to y_{k+1} proceeds along the direction $-r_k$, which coincides with the anti-gradient for the functional $(A(y-u), y-u)$ at the level y_k. It is known that the anti-gradient direction is the steepest-descent direction for the functional. Therefore such methods are sometimes called gradient descent methods or simply gradient methods. We shall also use this name for implicit two-level methods of variational type.

Our first problem is to find the parameter τ_{k+1} from the condition that the norm of the error $z_{k+1} = y_{k+1} - u$ in H_D be a minimum.

8.1.2 A formula for the iterative parameters. We now find a formula for the computation of the iterative parameter τ_{k+1}, assuming that the operator A is not singular. We first write out the equation for the error $z_k = y_k - u$, $k = 0, 1, \ldots$. Substituting $y_k = z_k + u$ in the scheme (2), we obtain

$$z_{k+1} = (E - \tau_{k+1} B^{-1} A) z_k, \quad k = 0, 1, \ldots, \quad z_0 = y_0 - u.$$

The change of variables $z_k = D^{-1/2}x_k$ allows us to pass to an equation containing only one operator

$$x_{k+1} = S_{k+1}x_k, \qquad S_k = E - \tau_k C,$$
$$C = D^{-1/2}(DB^{-1}A)D^{-1/2}. \tag{7}$$

Using the equality $\| z_k \|_D = \| x_k \|$, the problem of choosing the parameter τ_{k+1} posed above can be formulated in the following way: choose the parameter τ_{k+1} from the condition that the norm of x_{k+1} in the space H be a minimum.

We now solve this problem. We compute the norm of x_{k+1}:

$$\begin{aligned} \| x_{k+1} \|^2 &= ((E - \tau_{k+1}C)x_k,\ (E - \tau_{k+1}C)x_k) \\ &= \| x_k \|^2 - 2\tau_{k+1}(Cx_k, x_k) + \tau_{k+1}^2(Cx_k, Cx_k) \\ &= (Cx_k, Cx_k)\left[\tau_{k+1} - \frac{(Cx_k, x_k)}{(Cx_k, Cx_k)}\right]^2 + \| x_k \|^2 - \frac{(Cx_k, x_k)^2}{(Cx_k, Cx_k)}. \end{aligned} \tag{8}$$

Since the operator A is not singular, the operator C is also not singular. Therefore, for any x_k, we have $(Cx_k, Cx_k) > 0$, and the minimum norm of x_{k+1} is attained for

$$\tau_{k+1} = \frac{(Cx_k, x_k)}{(Cx_k, Cx_k)}. \tag{9}$$

Substituting (9) in (8), we obtain

$$\| x_{k+1} \| = \rho_{k+1} \| x_k \|, \tag{10}$$

where

$$\rho_{k+1}^2 = 1 - \frac{(Cx_k, x_k)^2}{(Cx_k, Cx_k)(x_k, x_k)}. \tag{11}$$

Thus, the formula (9) defines the optimal value of the iterative parameter τ_{k+1}. Substituting $x_k = D^{1/2}z_k$ in (9), we obtain

$$\tau_{k+1} = \frac{(DB^{-1}Az_k, z_k)}{(DB^{-1}Az_k, B^{-1}Az_k)}, \quad k = 0, 1, \ldots.$$

Taking into account that $Az_k = Ay_k - Au = Ay_k - f = r_k$, the residual, and $B^{-1}r_k = w_k$, the correction, the formula for the parameter τ_{k+1} can be written in the following form:

$$\tau_{k+1} = \frac{(Dw_k, z_k)}{(Dw_k, w_k)}, \quad k = 0, 1, \ldots, \tag{12}$$

and the iterative scheme (2) in the form of an explicit formula for computing y_{k+1}:

$$y_{k+1} = y_k - \tau_{k+1} w_k, \quad k = 0, 1, \dots . \tag{13}$$

The algorithm for realizing this method can be described in the following way:

1) for a given y_k, compute the residual $r_k = Ay_k - f$,
2) solve the equation for the correction $Bw_k = r_k$,
3) compute the parameter τ_{k+1} using the formula (12),
3) find the new approximation y_{k+1} using the formula (13).

The formula (12) is still not suitable for computation, since in addition to the iterative quantities r_k and w_k which are known during the iterative process, it contains the unknown error z_k. In Section 8.2, by choosing a concrete operator D, we obtain formulas for the parameters τ_k which only contain known quantities. But now, we move on to obtain an estimate for the convergence rate of this iterative method.

8.1.3 An estimate of the convergence rate. We now estimate the convergence rate for the two-level gradient methods. Since the iterative parameter τ_{k+1} is chosen from the condition that the norm of the error z_{k+1} in H_D be a minimum, which is equivalent to the condition that the norm of x_{k+1} in H be a minimum, from (7) we obtain

$$\begin{aligned} \| x_{k+1} \| &= \min_{\tau_{k+1}} \| S_{k+1} x_k \| \leq \min_{\tau_{k+1}} \| S_{k+1} \| \, \| x_k \| \\ &= \min_{\tau} \| E - \tau C \| \, \| x_k \| = \rho \| x_k \|,, \quad \rho = \min_{\tau} \| E - \tau C \| . \end{aligned}$$

Comparing this estimate with the equation (10), we find

$$\rho_k \leq \rho \leq 1, \qquad k = 1, 2, \dots . \tag{14}$$

From (10), (14) follows the estimate $\| x_{k+1} \| \leq \rho \| x_k \|$, and using the change of variables $x_k = D^{1/2} z_k$, from this follows an estimate for the norm of the error z_n in the energy space H_D:

$$\| z_n \|_D \leq \rho^n \| z_0 \|_D, \quad \rho = \min_{\tau} \| E - \tau C \| . \tag{15}$$

If the condition $\rho < 1$ is valid, then the two-level gradient method converges in H_D. From the estimate (15) it follows that, in order to decrease the norm of the initial error in H_D by a factor of $1/\epsilon$, it is sufficient to perform $n \geq n_0(\epsilon)$ iterations, where

$$n_0(\epsilon) = \ln \epsilon / \ln \rho. \tag{16}$$

Thus, the convergence rate for the two-level gradient method is determined by the quantity ρ. Recall that in Chapter 6, when studying the simple iteration method under different assumptions about the operator C, estimates for ρ were obtained. The quantity ρ determines the convergence rate for the simple iteration method. Therefore, from the estimate (15) obtained here, it follows that any two-level gradient method converges no slower than the corresponding simple iteration method.

We now give the estimates for ρ obtained in Sections 6.3, 6.4 under various assumptions about the operators A, B, and D.

1. If the operator $DB^{-1}A$ is self-adjoint in H, and γ_1 and γ_2 are the constants from the inequalities (3), then

$$\rho = (1 - \xi)/(1 + \xi), \quad \xi = \gamma_1/\gamma_2. \tag{17}$$

2. Suppose that the operator $DB^{-1}A$ is non-self-adjoint in H;

a) if condition (4) is satisfied, then

$$\rho = \sqrt{1 - \xi}, \quad \xi = \gamma_1/\gamma_2; \tag{18}$$

b) if conditions (3), (5) are satisfied, then

$$\rho = \frac{1 - \xi}{1 + \xi}, \quad \xi = \frac{1 - \kappa\, \gamma_1}{1 + \kappa\, \gamma_2}, \quad \kappa = \frac{\gamma_3}{\sqrt{\gamma_1\gamma_2 + \gamma_3^2}}. \tag{19}$$

Theorem 1. *If the simple iteration method converges for the scheme* (2), *then the two-level gradient method* (2), (12) *also converges. The error* z_n *is estimated by*

$$\| z_n \|_D \leq \rho^n \| z_0 \|_D,$$

where ρ *is defined in* (17) *if the operator* $DB^{-1}A$ *is self-adjoint in* H *and the conditions* (3) *are satisfied,* ρ *is defined in* (18) *if the conditions* (4) *are satisfied for a non-self-adjoint operator* $DB^{-1}A$, *and in* (19) *if the conditions* (3), (15) *are satisfied. An estimate for the iteration count is given in* (16). □

Remark. If equation (1) is considered in a complex Hilbert space, then the iterative parameter τ_{k+1} must be chosen from the formula

$$\tau_{k+1} = \frac{\operatorname{Re}(Dw_k, z_k)}{(Dw_k, w_k)}, \quad k = 0, 1, \ldots.$$

Theorem 1 remains valid, only the conditions (3), (4) should be changed to the inequalities

$$\gamma_1(Dx, x) \le \operatorname{Re}(DB^{-1}Ax, x) \le \gamma_2(Dx, x),$$

$$\gamma_1(Dx, x) \le \operatorname{Re}(DB^{-1}Ax, x),$$

$$(DB^{-1}Ax, B^{-1}Ax) \le \gamma_2 \operatorname{Re}(DB^{-1}Ax, x),$$

where Re z is the real part of the complex number z.

8.1.4 Optimality of the estimate in the self-adjoint case. We shall show that, on the class of arbitrary initial approximations y_0 in the case of a self-adjoint operator $DB^{-1}A$ in a finite-dimensional space H, the *a priori* estimate for the error of the iterative method (2), (12) obtained in theorem 1 is optimal. To do this, it is sufficient to demonstrate some initial approximation x_0 for which the solution of equation (7) satisfies $\| x_{k+1} \| = \rho \| x_k \|$, where ρ is defined in (17).

We now find the desired initial approximation x_0. Assume that H is a finite-dimensional space ($H = H_N$). Since the operator $DB^{-1}A$ is self-adjoint in H, the operator $C = D^{-1/2}(DB^{-1}A)D^{-1/2}$ is also self-adjoint in H. Consequently, there exists a complete system of eigenfunctions $v_1, v_2, \ldots, v_N$ for the operator C. We denote by λ_k the eigenvalue of the operator C corresponding to the eigenfunction v_k, so that $Cv_k = \lambda_k v_k$, $k = 1, 2, \ldots, N$. Suppose $\lambda_1 \le \lambda_2 \le \ldots \le \lambda_N$. Since the inequalities (3) are equivalent to the inequalities

$$\gamma_1 E \le C \le \gamma_2 E, \quad \gamma_1 > 0,$$

then it is possible to choose γ_1 and γ_2 as λ_1 and λ_2 in (3). Then ρ, defined in (17), can be written in the following form: $\rho = (\lambda_N - \lambda_1)/(\lambda_N + \lambda_1)$. We choose the initial approximaton

$$x_0 = \sqrt{\lambda_N} v_1 + \sqrt{\lambda_1} v_N. \tag{20}$$

Then $Cx_0 = \lambda_1 \sqrt{\lambda_n} v_1 + \lambda_N \sqrt{\lambda_1} v_N$. Using the orthonormality of the system of eigenfunctions $v_1, v_2, \ldots, v_N$, we obtain

$$(x_0, x_0) = \lambda_1 + \lambda_N,$$

$$(Cx_0, x_0) = 2\lambda_1 \lambda_N,$$

$$(Cx_0, Cx_0) = \lambda_1 \lambda_N (\lambda_1 + \lambda_N).$$

Substituting these values in (9), (11), we obtain $\tau_1 = 2/(\lambda_1 + \lambda_N)$, $\rho_1 = (\lambda_N - \lambda_1)/(\lambda_N + \lambda_1) = \rho$. From (10) follows the equation $\| x_1 \| = \rho \| x_0 \|$, and from (7) we find x_1:

$$x_1 = \rho \left(\sqrt{\lambda_N} v_1 - \sqrt{\lambda_1} v_N \right).$$

Further computations give

$$Cx_1 = \rho \left(\lambda_1 \sqrt{\lambda_N} v_1 - \lambda_N \sqrt{\lambda_1} v_N \right),$$

$$(x_1, x_1) = \rho^2 (x_0, x_0),$$

$$(Cx_1, x_1) = \rho^2 (Cx_0, x_0),$$

$$(Cx_1, Cx_1) = \rho^2 (Cx_0, Cx_0).$$

Therefore

$$\tau_2 = \frac{(Cx_1, x_1)}{(Cx_1, Cx_1)} = \frac{(Cx_0, x_0)}{(Cx_0, Cx_0)} = \tau_1,$$

$$\rho_2^2 = 1 - \frac{(Cx_1, x_1)^2}{(Cx_1, Cx_1)(x_1, x_1)} = 1 - \frac{(Cx_0, x_0)^2}{(Cx_0, Cx_0)(x_0, x_0)} = \rho_1^2 = \rho^2.$$

Consequently, $\| x_2 \| = \rho \| x_1 \|$. In addition, $x_2 = x_1 - \tau_2 C x_1 = \rho^2 x_0$, i.e. x_2 is proportional to x_0. From this it at once follows that $\tau_3 = \tau_2 = \tau_1$, $\rho_3 = \rho$, and $x_3 = \rho^2 x_1$. Therefore for any k:

$$\tau_k \equiv 2/(\lambda_1 + \lambda_N), \quad \rho_k \equiv \rho = (\lambda_N - \lambda_1)/(\lambda_N + \lambda_1),$$

$$\| x_{k+1} \| = \rho \| x_k \|.$$

The assertion is proved.

Thus we have shown that, if the initial approximation is chosen using the formula (20), then in the two-level gradient method all the parameters τ_k are identical and coincide with the parameter for the simple iteration method (cf. Section 6.3), the errors are proportional, and the convergence rate is just as slow.

Notice that such slow convergence for the method is only true for a special "bad" initial approximation. In the case of a "good" initial approximation, the convergence rate can be significantly better. A more detailed study of the way in which the method's convergence rate changes will be given in the following subsection, but here we will look at one example illustrating the above remark.

We shall show that, if the initial approximation x_0 is taken to be any eigenfunction v_m, then the two-level gradient method converges in one iteration.

Suppose $x_0 = v_m$. Then a simple calculation gives

$$Cx_0 = \lambda_m v_m = \lambda_m x_0, \quad (Cx_0, x_0) = \lambda_m(x_0, x_0),$$
$$(Cx_0, Cx_0) = \lambda_m^2(x_0, x_0), \quad \tau_1 = 1/\lambda_m, \quad \rho_1 = 0,$$

i.e. $x_1 = 0$ or $y_1 = u$.

This is a new property of the two-level gradient methods — the possibility that the convergence rate can increase when a "good" initial approximation is given — which distinguishes them from the two-level methods considered in Chapter 6, which are inflexible with respect to a bad initial approximation.

8.1.5 An asymptotic property of the gradient methods in the self-adjoint case. We look now at an asymptotic property of the two-level gradient methods which they possess in the case of a self-adjoint operator $DB^{-1}A$. This property concerns the fact that the sequence $\{\rho_k\}$, defined in (11), is increasing. Since the quantity ρ_k determines the rate at which the norm of the error decreases from iteration k to $(k+1)$, this property indicates a slower rate of decrease in the norm of the error z_n for large n in comparison with the rate at the beginning of the iterative process. Thus, for sufficiently large n, the convergence rate of the gradient methods becomes practically the same as that for the simple iteration method.

It will be shown that, after a large number of iterations, the errors are almost proportional, if looked at every other iteration. Using this fact, we construct an approximate method for finding the constants γ_1 and γ_2 for the inequalities (3), and in Section 8.5 we construct a process for accelerating the convergence of the two-level iterative methods.

Thus, suppose that the operator $DB^{-1}A$, and in addition the operator C, are self-adjoint in H. We will show that the sequence $\{\rho_k\}$ is increasing. From (10) it follows that

$$\| x_{k+2} \| = \rho_{k+2} \| x_{k+1} \|, \quad \| x_{k+1} \| = \rho_{k+1} \| x_k \| .$$

We now compute the norm of the difference $x_{k+2} - \rho_{k+2}\rho_{k+1}x_k$:

$$\| x_{k+2} - \rho_{k+2}\rho_{k+1}x_k \|^2 = \| x_{k+2} \|^2 - 2\rho_{k+2}\rho_{k+1}(x_{k+2}, x_k)$$
$$+\rho_{k+2}^2\rho_{k+1}^2 \| x_k \|^2 = 2\left(\| x_{k+2} \|^2 - \rho_{k+2}\rho_{k+1}(x_{k+2}, x_k)\right). \quad (21)$$

We compute separately the inner product (x_{k+2}, x_k). From equation (7) we find

$$x_{k+2} = x_{k+1} - \tau_{k+2} C x_{k+1}, \quad x_k = x_{k+1} + \tau_{k+1} C x_k. \tag{22}$$

Taking the inner product of the last equation with Cx_k and using (9), we obtain

$$(Cx_k, x_k) = (x_{k+1}, Cx_k) + \tau_{k+1}(Cx_k, Cx_k) = (x_{k+1}, Cx_k) + (Cx_k, x_k).$$

Consequently, for any k we have the equality

$$(x_{k+1}, Cx_k) = 0, \tag{23}$$

and using the self-adjointness of the operator C — the quantity (Cx_{k+1}, x_k) equals 0.

From (22) and (23) we obtain

$$\begin{aligned}(x_{k+2}, x_k) &= (x_{k+1} - \tau_{k+2} C x_{k+1}, x_k) = (x_{k+1}, x_k) \\ &= (x_{k+1}, x_{k+1} + \tau_{k+1} C x_k) = \| x_{k+1} \|^2 .\end{aligned}$$

Substituting this equation in (21), we find

$$\| x_{k+2} - \rho_{k+2}\rho_{k+1} x_k \|^2 = 2\left(1 - \frac{\rho_{k+1}}{\rho_{k+2}}\right) \| x_{k+2} \|^2 . \tag{24}$$

From (24) it follows that either $\rho_{k+2} > \rho_{k+1}$, or $\rho_{k+1} = \rho_{k+2} = \bar{\rho}$ and $x_{k+2} = \bar{\rho}^2 x_k$. In the latter case, it is clear that for any $n \geq k$ the following equation is satisfied

$$\rho_{n+1} = \bar{\rho}, \qquad x_{n+2} = \bar{\rho}^2 x_n, \tag{25}$$

i.e. the sequence ρ_k tends to a limiting value.

Thus, it has been shown that the sequence $\{\rho_k\}$ is increasing. In Section 8.1.3 it was shown that this sequence is bounded above and, consequently, has a limit. Therefore, after a sufficiently large number of iterations k, we have that $\rho_{k+1} \approx \rho_{k+2}$ and, consequently, $x_{k+2} \approx \rho_{k+2}\rho_{k+1} x_k$, i.e. the errors are almost proportional at every other iteration.

We look now at what happens when the sequence ρ_k converges to its limit value. In this case equation (25) is satisfied, i.e. $x_{n+2} = \bar{\rho}^2 x_n$. Assume that the space H is finite-dimensional, and that $v_1, v_2, \ldots, v_N$ are a system of eigenfunctions for the operator C. We expand x_n in terms of the eigenfunctions

$$x_n = \sum_{k=1}^{N} \alpha_k^{(n)} v_x. \tag{26}$$

From equation (7) we obtain

$$\begin{aligned} x_{n+2} &= (E - \tau_{n+2}C)(E - \tau_{n+1}C)x_n \\ &= \sum_{k=1}^{N}(1 - \tau_{n+2}\lambda_k)(1 - \tau_{n+1}\lambda_k)\alpha_k^{(n)} v_k. \end{aligned}$$

Since $x_{n+2} = \bar{\rho}^2 x_n$, it follows that for all k for which $\alpha_k^{(n)} \neq 0$, we must have that

$$(1 - \tau_{n+2}\lambda_k)(1 - \tau_{n+1}\lambda_k) = \bar{\rho}^2.$$

From this it follows that, in the expansion (26), there are only eigenfunctions corresponding to two different (possibly multiple) eigenvalues. Let them be λ_i and λ_j. Then λ_i and λ_j are the roots of the equation

$$(1 - \tau_{n+2}\lambda)(1 - \tau_{n+1}\lambda) = \bar{\rho}^2. \tag{27}$$

Knowing τ_{n+1}, τ_{n+2} and $\bar{\rho}$, it is possible to find the eigenvalues λ_i and λ_j from this equation.

Ignoring the details, we remark that if, in the expansion of the initial error x_0, the eigenfunctions corresponding to the minimal eigenvalue λ_1 and the maximal eigenvalue λ_N of the operator C are present, then, if the sequence ρ_k converges to its limit value, only these eigenfunctions remain in the expansion (26). Therefore, solving equation (27), we find λ_1 and λ_N.

Having the sequence $\{\rho_k\}$ converge to its limit value for finite n is an exceptional case. In the general case it is only possible to assert that, for sufficiently large n, $\rho_{n+1} \approx \rho_{n+2}$ and $x_{n+2} \approx \rho_{n+2}\rho_{n+1}x_n$.

The existence of this approximate equation leads us to hope that, for sufficiently large n, the roots of the equation

$$(1 - \tau_{n+2}\lambda)(1 - \tau_{n+1}\lambda) = \rho_{n+2}\rho_{n+1} \tag{28}$$

will be good approximations for λ_1 and λ_N and, consequently, also for γ_1 and γ_2 in the inequalities (3).

We describe now a method for finding approximate values for γ_1 and γ_2. Using the iterative scheme (2) with $f = 0$, we perform $n + 2$ iterations with the parameters τ_{k+1} defined in (12). Since, for $f = 0$, the solution of equation (1) is zero ($u = 0$), then $z_k = y_k$ and, consequently, ρ_{k+1} can be found from the formula

$$\rho_{k+1} = \frac{\| z_{k+1} \|_D}{\| z_k \|_D} = \frac{\| y_{k+1} \|_D}{\| y_k \|_D}.$$

Having computed τ_{n+1}, τ_{n+2}, ρ_{n+1} and ρ_{n+2}, we solve equation (28). The roots of this equation are approximations to γ_1 from above and γ_2 from below.

In Section 8.5 we will give an example illustrating this method of finding γ_1 and γ_2.

8.2 Examples of two-level gradient methods

8.2.1 The steepest-descent method. In Section 8.1 we studied the general properties of two-level iterative methods of variational type. These methods can be used to find an approximate solution to the operator equation

$$Au = f \tag{1}$$

with a non-singular operator A. The iterative approximations are computed from a two-level scheme

$$B\frac{y_{k+1} - y_k}{\tau_{k+1}} + Ay_k = f, \quad k = 0, 1, \ldots, \quad y_0 \in H, \tag{2}$$

and the iterative parameters τ_k are found from the formula

$$\tau_{k+1} = \frac{(Dw_k, z_k)}{(Dw_k, w_k)}, \qquad k = 0, 1, \ldots, \tag{3}$$

where $w_k = B^{-1}r_k$ is the correction, $r_k = Ay_k - f$ is the residual, and $z_k = y_k - u$ is the error. The choice of the parameter τ_{k+1} from the formula (3) guarantees that the norm of the error z_{k+1} in H_D will be minimized as we pass from y_k to y_{k+1}.

We look now at special cases of two-level gradient methods. Each specific method is defined by the choice of the operator D and has its own field of applicability. The operator D will be chosen so that the formula (3) for the iterative parameter τ_{k+1} will contain quantities known during the iterative process.

We begin our look at examples with the steepest-descent method. This method can be applied only when A is a self-adjoint and positive-definite operator.

Assume that the operator A is self-adjoint and positive-definite in H. The *steepest-descent method* is characterized by the following operator D: $D = A$. The operator B should be positive-definite in H. Taking into account the equations $Az_k = Ay_k - f = r_k$ and $A = A^*$, from (3) we obtain a formula for the iterative parameter τ_{k+1} in the implicit steepest-descent method

$$\tau_{k+1} = \frac{(r_k, w_k)}{(Aw_k, w_k)}, \quad k = 0, 1, \ldots .$$

In the case of an explicit two-level scheme (2) ($B = E$), we obtain $w_k = B^{-1} r_k = r_k$, and the formula for τ_{k+1} takes the form

$$\tau_{k+1} = \frac{(r_k, r_k)}{(Ar_k, r_k)}, \quad k = 0, 1, \ldots .$$

In the steepest-descent method, the norm of the error z_{k+1} is minimized in the energy space H_A: $\| z_k \|_A = (Az_k, z_k)^{1/2}$. The conditions for the convergence of the method were formulated in theorem 1, from which follows the estimate

$$\| z_n \|_A \leq \rho^n \| z_0 \|_A, \quad n \geq n_0(\epsilon) = \ln \epsilon / \ln \rho.$$

The value of the quantity ρ is determined by the properties of the operators A and B and by the amount of *a priori* information concerning them. Notice that the requirement of self-adjointness for the operator $DB^{-1}A = AB^{-1}A$ in this method is equivalent to the requirement of self-adjointness for the operator B. Therefore

1) if $B = B^*$ and either the conditions (3) of Section 8.1 or the equivalent conditions (cf. Section 6.2.3)

$$\gamma_1 B \leq A \leq \gamma_2 B, \quad \gamma_1 > 0,$$

are satisfied, then

$$\rho = (1 - \xi)/(1 + \xi), \quad \xi = \gamma_1 / \gamma_2;$$

2) if $B \neq B^*$ and either the conditions (4) of Section 8.1 or the equivalent conditions (cf. Section 6.4.2)

$$\gamma_1(Bx, A^{-1}Bx) \leq (Bx, x), \quad (Ax, x) \leq \gamma_2(Bx, x), \quad \gamma_1 > 0,$$

are satisfied, then

$$\rho = \sqrt{1-\xi}, \quad \xi = \gamma_1/\gamma_2.$$

Notice, that if $B = B^*$, then the steepest-descent method possesses the asymptotic property.

8.2.2 The minimal residual method. This method can be applied in the case of any non-self-adjoint non-singular operator A. It is not assumed that each of the operators A and B is positive-definite separately; it is only required that the operator B^*A be positive-definite. The *minimal residual method* is defined by the following choice of the operator D: $D = A^*A$.

The formula (3) for the iterative parameter τ_{k+1} in the minimal residual method has the form

$$\tau_{k+1} = \frac{(Aw_k, r_k)}{(Aw_k, Aw_k)}, \quad k = 0, 1, \ldots.$$

In the case of an explicit scheme (2) ($B = E$), the positive-definiteness of the operator A is required, and the formula for τ_{k+1} has the form

$$\tau_{k+1} = \frac{(Ar_k, r_k)}{(Ar_k, Ar_k)}, \quad k = 0, 1, \ldots.$$

The name of the method is connected with the fact that the norm of the residual is minimized. In fact, for this choice of the operator D we have

$$\| z_k \|_D^2 = (Dz_k, z_k) = (A^*Az_k, z_k) = \| Az_k \|^2 = \| r_k \|^2 .$$

Consequently, for this method the norm of the error in H_D is equal to the norm of the residual, which can be computed during the iterative process and used to control the termination of the iteration.

From theorem 1 follow the estimates

$$\| r_n \| \leq \rho^n \| r_0 \|, \quad n \geq n_0(\epsilon) = \ln \epsilon / \ln \rho.$$

The operator $DB^{-1}A = A^*AB^{-1}A$ will be self-adjoint in H if the operator AB^{-1} is self-adjoint, which is equivalent to the requirement that the operator B^*A be self-adjoint. If this requirement is satisfied, then from the conditions (3) of Section 8.1, which in this case have the form

$$\gamma_1(Ay, Ay) \le (AB^{-1}Ay, Ay) \le \gamma_2(Ay, Ay), \quad \gamma_1 > 0,$$

or after the change of variables $y = A^{-1}Bx$

$$\gamma_1(Bx, Bx) \le (Ax, Bx) \le \gamma_2(Bx, Bx), \quad \gamma_1 > 0, \tag{4}$$

and from theorem 1 it follows that

$$\rho = (1-\xi)/(1+\xi), \quad \xi = \gamma_1/\gamma_2.$$

Notice that the condition $\gamma_1 > 0$ will be satisfied if the operator B^*A is positive-definite. The conditions of self-adjointness and positive-definiteness for the operator B^*A will be satisfied, for example, under the following assumptions: $A = A^* > 0$, $B = B^* > 0$, $AB = BA$.

In this case the inequalities (4) are equivalent to simpler ones. In fact, setting $x = B^{-1/2}y$ in (4) and using the commutativity of the operators A and B, we obtain

$$\gamma_1 B \le A \le \gamma_2 B, \quad \gamma_1 > 0. \tag{5}$$

The conditions of self-adjointness and positive-definiteness for the operator B^*A will also be automatically satisfied in the case where the operator B has the form $B = (A^*)^{-1}B_0$, where B_0 is a self-adjoint and positive-definite operator. In this case, in place of the inequalities (5), it is necessary to use the inequalities

$$\gamma_1 B_0 \le A^*A \le \gamma_2 B_0, \quad \gamma_1 > 0, \tag{6}$$

and, in the formula for the parameter τ_{k+1}, the correction w_k can be found from the equation $B_0 w_k = A^* r_k$.

If the operator B^*A is non-self-adjoint in H, then from the conditions (4) of Section 8.1 or the equivalent conditions

$$\gamma_1(Bx, Bx) \le (Ax, Bx), \quad (Ax, Ax) \le \gamma_2(Ax, Bx), \quad \gamma_1 > 0$$

and from theorem 1, it follows that $\rho = \sqrt{1-\xi}$, $\xi = \gamma_1/\gamma_2$.

8.2.3 The minimal correction method. This method can be applied to solve equation (1) with a non-self-adjoint but positive-definite operator A. It is required that the operator B be a self-adjoint positive-definite and bounded operator. The *minimal correction method* is defined by the following choice of the operator D: $D = A^*B^{-1}A$.

The formula (3) for the iterative parameter τ_{k+1} in the minimal correction method has the form

$$\tau_{k+1} = \frac{(Aw_k, w_k)}{(B^{-1}Aw_k, Aw_k)}, \quad k = 0, 1, \ldots.$$

In the case of an explicit scheme (2) ($B = E$), the minimal correction method and the minimal residual method are the same.

In the minimal correction method, the norm of the correction in H_B is minimized. In fact, for the operator D chosen above, we obtain

$$\| z_k \|_D^2 = (Dz_k, z_k) = (A^*B^{-1}Az_k, z_k) = (w_k, r_k) = (Bw_k, w_k) = \| w_k \|_B^2 .$$

The norm of the correction in H_B can be computed during the iterative process and used to control the termination of the iteration.

From theorem 1 follow the estimates

$$\| w_n \|_B \le \rho^n \| w_0 \|_B, \quad n \ge n_0(\epsilon) = \ln \epsilon / \ln \rho.$$

The operator $DB^{-1}A = A^*B^{-1}AB^{-1}A$ is self-adjoint in H whenever the operator A is. Therefore:

1) if $A = A^*$ and either the conditions (3) of Section 8.1 or the equivalent conditions (cf. Section 6.2.3)

$$\gamma_1 B \le A \le \gamma_2 B, \quad \gamma_1 > 0,$$

are satisfied, then

$$\rho = (1 - \xi)/(1 + \xi), \quad \xi = \gamma_1/\gamma_2;$$

2) if $A \ne A^*$ and either the conditions (4) of Section 8.1 or the equivalent conditions (cf. Section 6.4.2)

$$\gamma_1 B \le A, \quad (Ax, B^{-1}Ax) \le \gamma_2 (Ax, x), \quad \gamma_1 > 0,$$

are satisfied, then

$$\rho = \sqrt{1 - \xi}, \quad \xi = \gamma_1/\gamma_2.$$

Notice that, in comparison with steepest-descent and minimal residual methods, in the minimal correction method it is necessary to invert the operator B not once, but twice, first to compute the correction w_k, and then to compute $B^{-1}Aw_k$.

Notice also that, if $A = A^*$, then the minimal correction method possesses the asymptotic property.

8.2.4 The minimal error method. This method can be applied, just like the minimal residual method, in the case where A is any non-self-adjoint and non-singular operator. The *minimal error method* is defined by the following choice of the operators B and D:

$$B = (A^*)^{-1}B_0, \quad D = B_0,$$

where B_0 is a self-adjoint positive-definite operator in H.

Substituting the chosen operator D in the formula (3) for the iterative parameter τ_{k+1}, and taking into account that $w_k = B^{-1}r_k = B_0^{-1}A^*r_k$, we obtain a formula for τ_{k+1} in the minimal error method

$$\tau_{k+1} = \frac{(r_k, r_k)}{(Aw_k, r_k)}, \quad k = 0, 1, \ldots.$$

The correction w_k is found from the equation $B_0 w_k = A^* r_k$.

In the case of an explicit scheme ($B_0 = E$), the formula for τ_{k+1} has the form

$$\tau_{k+1} = \frac{(r_k, r_k)}{(A^*r_k, A^*r_k)}, \quad k = 0, 1, \ldots.$$

In the minimal error method, the norm of the error in H_{B_0} is minimized. For this method the operator $DB^{-1}A = A^*A$ is self-adjoint in H, and the conditions (3) of Section 8.1 take the form of the inequalities (6). From theorem 1 follow the estimates

$$\| z_n \|_{B_0} \leq \rho^n \| z_0 \|_{B_0}, \quad n \geq n_0(\epsilon) = \ln \epsilon / \ln \rho,$$

where $\rho = (1-\xi)/(1+\xi)$, $\xi = \gamma_1/\gamma_2$, and γ_1 and γ_2 are defined in (6).

The minimal error method always possesses the asymptotic property.

8.2.5 A sample application of two-level methods. To illustrate the application of the constructed two-level gradient methods, we look at the solution of a model problem by the explicit steepest-descent method. As an example, we take a Dirichlet difference problem for Poisson's equation on the square grid $\bar{\omega} = \{x_{ij} = (ih, jh),\ 0 \le i \le N,\ 0 \le j \le N,\ h = 1/N\}$ in the unit square

$$\Lambda u = u_{\bar{x}_1 x_1} + u_{\bar{x}_2 x_2} = -\varphi, \quad x \in \omega, \quad u|_\gamma = g. \tag{7}$$

We introduce the space H consisting of the grid functions defined on ω with inner product $(u, v) = \sum\limits_{x \in \omega} u(x)v(x)h^2$.

The operator A on H is defined in the following way: $Ay = -\Lambda v$, $y \in H$, where $v(x) = y(x)$ for $x \in \omega$ and $v|_\gamma = 0$. The problem (7) can be written in the form of the operator equation

$$Au = f, \tag{8}$$

where f only differs from φ at the near-boundary nodes

$$f = \varphi + \frac{\varphi_1}{h^2} + \frac{\varphi_2}{h^2},$$

$$\varphi_1 = \begin{cases} g(0, x_2), & x_1 = h, \\ 0, & 2h \le x_1 \le 1 - 2h, \\ g(1, x_2), & x_1 = 1 - h, \end{cases} \quad \varphi_2 = \begin{cases} g(x_1, 0), & x_2 = h, \\ 0, & 2h \le x_2 \le 1 - 2h, \\ g(x_1, 1), & x_2 = 1 - h. \end{cases}$$

The operator A is self-adjoint and positive-definite in H. Therefore it is possible to apply the steepest-descent method in order to solve the equation (8). The explicit iterative scheme has the form

$$\frac{y_{k+1} - y_k}{\tau_{k+1}} + Ay_k = f \quad \text{or} \quad y_{k+1} = y_k - \tau_{k+1} r_k, \quad k = 0, 1, \ldots,$$

and the iterative parameters τ_k are found from the formula

$$\tau_{k+1} = \frac{(r_k, r_k)}{(Ar_k, r_k)}, \quad r_k = Ay_k - f, \quad k = 0, 1, \ldots.$$

We give now the computational formulas and count the number of arithmetic operations expended during one iteration.

Taking into account the definition of the operator A and the right-hand side f, the computational formulas can be written in the following form:

$$
\begin{aligned}
1)\quad & r_k(x_{ij}) &&= -(y_k)_{\bar{x}_1 x_1} - (y_k)_{\bar{x}_2 x_2} - \varphi(x_{ij}), \quad 1 \le i,\ j \le N-1,\\
& y_k|_\gamma &&= g;\\
2)\quad & (r_k, r_k) &&= \sum_{i=1}^{N-1}\sum_{j=1}^{N-1} r_k^2(x_{ij})h^2,\\
& (Ar_k, r_k) &&= -\sum_{i=1}^{N-1}\sum_{j=1}^{N-1} r_k(x_{ij})\left[(r_k)_{\bar{x}_1 x_1} + (r_k)_{\bar{x}_2 x_2}\right]h^2, r_k|_\gamma = 0,\\
& \tau_{k+1} &&= \frac{(r_k, r_k)}{(Ar_k, r_k)};\\
3)\quad & y_{k+1}(x_{ij}) = y_k(x_{ij}) - \tau_{k+1} r_k(x_{ij}), && \quad 1 \le i, j \le N-1.
\end{aligned}
$$

The initial approximation y_0 is an arbitrary grid function in ω which takes on the following values on γ: $y_0|_\gamma = g$.

We now count the number of arithmetic operations. If the computation of the difference derivatives is carried out using the formula

$$u_{\bar{x}_1 x_1} + u_{\bar{x}_2 x_2} = \frac{1}{h^2}(u_{i+1,j} + u_{i-1,j} + u_{i,j+1} + u_{i,j-1} - 4u_{ij}),$$

then the computation of r_k requires $6(N-1)^2$ additions and $2(N-1)^2$ multiplications and divisions. Computing (r_k, r_k) requires $(N-1)^2$ additions and $(N-1)^2$ multiplications, (Ar_k, r_k) requires $6(N-1)^2$ additions and $2(N-1)^2$ multiplications, y_{k+1} requires $(N-1)^2$ additions and $(N-1)^2$ multiplications. In total, $14(N-1)^2$ additions and $6(N-1)^2$ multiplications and divisions are needed. Exactly half of this total number of operations is required to compute the inner products, i.e. to compute the iterative parameter τ_{k+1}. Consequently, one step of the steepest-descent method is twice as laborious as one step of the simple iteration method or the Chebyshev method, where the parameter τ_{k+1} is known *a priori*. For implicit methods, this difference will be less, since computing the inner products requires just the same number of operations as in the explicit method, but the total is increased by the additional arithmetic operations expended to invert the operator B.

We compute now the total number of arithmetic operations $Q(\epsilon)$ which are needed to achieve a relative accuracy ϵ. To do this, it is necessary to estimate the iteration count $n_0(\epsilon)$. In Section 8.2.1, the following estimate was obtained:

$$n_0(\epsilon) = \frac{\ln \epsilon}{\ln \rho}, \quad \rho = \frac{1-\xi}{1+\xi}, \quad \xi = \frac{\gamma_1}{\gamma_2},$$

where γ_1 and γ_2 are, in the case of an explicit scheme, the bounds for the operator A: $\gamma_1 E \le A \le \gamma_2 E$.

For this example, γ_1 and γ_2 are the same as the minimal and maximal eigenvalues δ and Δ of the Laplace difference operator Λ. It is known that

$$\delta = \frac{8}{h^2}\sin^2\frac{\pi h}{2}, \quad \Delta = \frac{8}{h^2}\cos^2\frac{\pi h}{2}.$$

Therefore

$$\rho = \frac{1-\xi}{1+\xi} = 1 - 2\sin^2\frac{\pi h}{2}, \quad \xi = \frac{\delta}{\Delta} = \tan^2\frac{\pi h}{2},$$

and, consequently, if $h \ll 1$, then

$$n_0(\epsilon) \approx \frac{2\ln\frac{1}{\epsilon}}{\pi^2 h^2} \approx 0.2N^2 \ln\frac{1}{\epsilon}.$$

If we consider the operations of addition, multiplication, and division to be equivalent, then one iteration requires approximately $20N^2$ operations. Therefore, the total number of arithmetic operations will be estimated by $Q(\epsilon) \approx 4N^4 \ln\frac{1}{\epsilon}$.

8.3 Three-level conjugate-direction methods

8.3.1 The choice of the iterative parameters. An estimate of the convergence rate. In Section 8.1, in order to find an approximate solution to the linear equation

$$Au = f \tag{1}$$

with a non-singular operator A, we looked at two-level iterative methods of variational type. The iterative scheme for these methods has the form

$$B\frac{y_{k+1} - y_k}{\tau_{k+1}} + Ay_k = f, \quad k = 0, 1, \ldots, \quad y_0 \in H, \tag{2}$$

and the iterative parameters τ_{k+1} are chosen from the condition that the norm of the error z_{k+1} in the energy space H_D be a minimum. Recall that sequence y_k constructed according to the formula (2) leads to a step-by-step minimization of the functional $I(y) = (D(y-u), y-u)$, whose minimum is achieved at the solution of equation (1), i.e. for $y = u$.

Such a strategy of local minimization, however, is not optimal, since we are interested in the global minimum of the functional $I(y)$ on a finite grid; if some value of this functional is given, we would like to find the desired minimum after a minimal number of iterations. Local minimization at each iterative step does not lead us to the solution by the shortest path.

It is natural to try and choose the parameters τ_k from the condition that the norm of the error z_n in H_D be a minimum after n steps, i.e. after moving from y_0 to y_n. We met with an analogous situation when we studied the Chebyshev method and the simple iteration method in Chapter 6. It turned out that the method converged faster if the iterative parameters were chosen by minimizing the norm of the resolving operator, and not the transformation operator from iteration to iteration. This property is also valid for iterative methods of variational type. It will be shown that the iterative methods considered in this section converge significantly faster if the iterative parameters τ_k are chosen according to the rule given above. In addition, when H is a finite-dimensional space, these methods are semi-iterative methods for any initial guess, i.e. the exact solution of equation (1) can be obtained after a finite number of iterations.

We move on now to construct the *conjugate-direction method.* We will assume that the operator $DB^{-1}A$ is self-adjoint and positive-definite in H, and we will perform n iterations of the scheme (2). Transforming from the problem for the error $z_k = y_k - u$ to the problem for $x_k = D^{1/2}z_k$ we obtain as before,

$$x_{k+1} = S_{k+1}x_k, \quad k = 0, 1, \ldots, n-1,$$

$$S_k = E - \tau_k C, \quad C = D^{1/2}B^{-1}AD^{-1/2}.$$

From this we find

$$x_n = T_n x_0, \quad T_n = \prod_{j=1}^{n}(E - \tau_j C). \tag{3}$$

The resolving operator T_n is an operator polynomial of degree n in the operator C with coefficients depending on the parameters $\tau_1, \tau_2, \ldots, \tau_n$

$$T_n = P_n(C) = E + \sum_{j=1}^{n} a_j^{(n)} C^j, \quad a_n^{(n)} \neq 0. \tag{4}$$

Using the equality $\| x_n \| = \| z_n \|_D$, the problem presented above concerning the choice of the iterative parameters τ_k can be formulated as follows: among all polynomials of the form (4), choose the one for which the norm of $x_n = P_n(C)x_0$ is minimal; in other words, choose the coefficients $a_1^{(n)}, a_2^{(n)}, \ldots, a_n^{(n)}$ of the polynomial $P_n(C)$ from the condition that the norm of x_n in H be a minimum.

This problem will be solved in the next subsection, but first we obtain an estimate for the convergence rate of the conjugate-direction method constructed on the basis of the above-formulated principle for choosing the parameters. We obtain this estimate using *a priori* information about the operators of the scheme in the form of γ_1 and γ_2 — the constants of energy equivalence for the self-adjoint operators D and $DB^{-1}A$:

$$\gamma_1 D \le DB^{-1}A \le \gamma_2 D, \quad \gamma_1 > 0, \quad DB^{-1}A = (DB^{-1}A)^*. \tag{5}$$

Suppose that $P_n(C)$ is the desired polynomial. Then from (3), (4) follows an estimate for x_n:

$$\| x_n \| = \| P_n(C)x_0 \| = \min_{\{Q_n\}} \| Q_n(C)x_0 \| \le \min_{\{Q_n\}} \| Q_n(C) \| \, \| x_0 \|,$$

where the minimum is taken over the polynomials $Q_n(C)$ normalized, as in (4), by the condition that $Q_n(0) = E$.

We now estimate the minimal norm of the polynomial $Q_n(C)$. From (5) it follows that the operator $C = D^{-1/2}(DB^{-1}A)D^{-1/2}$ is self-adjoint in H, and γ_1 and γ_2 are its bounds: $C = C^*$, $\gamma_1 E \le C \le \gamma_2 E$, $\gamma_1 > 0$. Therefore we have the estimate

$$\min_{\{Q_n\}} \| Q_n(C) \| \le \min_{\{Q_n\}} \max_{\gamma_1 \le t \le \gamma_2} |Q_n(t)|.$$

From the results in Section 6.2 it follows that the problem of constructing the polynomial, normed by the condition $Q_n(0) = 1$, whose maximum modulus on the interval $[\gamma_1, \gamma_2]$ is minimal, is solved by the Chebyshev polynomial of the first kind, for which

$$\max_{\gamma_1 \le t \le \gamma_2} |Q_n(t)| = q_n, \quad q_n = \frac{2\rho_1^n}{1+\rho_1^{2n}}, \quad \rho_1 = \frac{1-\sqrt{\xi}}{1+\sqrt{\xi}}, \quad \xi = \frac{\gamma_1}{\gamma_2}.$$

Consequently, we have the estimate $\| x_n \| \le q_n \| x_0 \|$ for x_n.

Thus, we have proved

Theorem 2. *If the conditions* (5) *are satisfied, then the iterative conjugate-direction method converges in* H_D, *and the error* z_n *is estimated for any* n *by* $\| z_n \|_D \le q_n \| z_0 \|_D$. *Also, the iteration count can be estimated by*

$$n \ge n_0(\epsilon) = \ln(0.5\epsilon)/\ln \rho_1,$$

where $\rho_1 = (1-\sqrt{\xi})/(1+\sqrt{\xi})$, $\xi = \gamma_1/\gamma_2$. □

8.3.2 Formulas for the iterative parameters. The three-level iterative scheme. We move on now to construct the polynomial $P_n(C)$. Using (3) and (4), we compute the norm of x_n:

$$\| x_n \|^2 = (P_n(C)x_0, P_n(C)x_0)$$
$$= \| x_0 \|^2 + 2\sum_{j=1}^{n} a_j^{(n)}(C^j x_0, x_0) + \sum_{j=1}^{n}\sum_{i=1}^{n} a_j^{(n)} a_i^{(n)}(C^j x_0, C^i x_0).$$

The norm of x_n is a function of the parameters $a_1^{(n)}, a_2^{(n)}, \ldots, a_n^{(n)}$. Setting the partial derivatives of $\| x_n \|^2$ with respect to $a_j^{(n)}$

$$\frac{\partial \| x_n \|^2}{\partial a_j^{(n)}} = 2\sum_{i=1}^{n} a_i^{(n)}(C^j x_0, C^i x_0) + 2(C^j x_0, x_0), \quad j = 1, 2, \ldots, n,$$

equal to zero, we obtain a system of linear algebraic equations

$$\sum_{i=1}^{n} a_i^{(n)}(C^j x_0, C^i x_0) + (C^j x_0, x_0) = 0, \quad j = 1, 2, \ldots, n. \tag{6}$$

For a self-adjoint and positive-definite operator C in H, the system (6) gives the conditions for the norm of x_n in H to be a minimum.

Thus, the problem of constructing the optimal polynomial $P_n(C)$ has in principle been solved. The coefficients of the polynomial $a_1^{(n)}, a_2^{(n)}, \ldots, a_n^{(n)}$ are found by solving the system (6). But first, we construct the formulas for computing the iterative approximation y_n. One way of doing this is to use the iterative scheme (2). However, this requires us to find the roots of the polynomial $P_n(t)$ and then to choose the τ_k as the inverse of these roots. Such a scheme is not economical.

A second method is to use the coefficients of the polynomial to compute y_n. From (3), (4), and the change of variable $x_k = D^{1/2}z_k$, where $z_k = y_k - u$, we obtain

$$y_n - u = D^{-1/2}P_n(C)D^{1/2}(y_0 - u). \tag{7}$$

Using (4) and the equation $D^{-1/2}C^j D^{1/2} = (B^{-1}A)^j$, we find

$$D^{-1/2}P_n(C)D^{1/2} = E + \sum_{j=1}^{n} a_j^{(n)}(B^{-1}A)^j.$$

Substituting this equation in (7), we obtain

$$y_n = y_0 + \sum_{j=1}^{n} a_j^{(n)}(B^{-1}A)^j(y_0 - u) = y_0 + \sum_{j=1}^{n} a_j^{(n)}(B^{-1}A)^{j-1}w_0, \tag{8}$$

where w_0 is the correction, $w_0 = B^{-1}A(y_0 - u) = B^{-1}r_0$, $r_0 = Ay_0 - f$.

This method is also not optimal. For each new n we must again solve the system (6).

We shall now show that the sequence $y_1, y_2, \ldots, y_k, \ldots$, constructed according to (6), (8) for $n = 1, 2, \ldots$, can be found using the following three-level scheme

$$\begin{gathered} By_{k+1} = \alpha_{k+1}(B - \tau_{k+1}A)y_k + (1 - \alpha_{k+1})By_{k-1} + \alpha_{k+1}\tau_{k+1}f, \\ k = 1, 2, \ldots, \\ By_1 = (B - \tau_1 A)y_0 + \tau_1 f, \quad y_0 \in H. \end{gathered} \tag{9}$$

To do this it is necessary to display the set of parameters $\{\tau_k\}$ and $\{\alpha_k\}$ for which the norm of the equivalent error x_k will be minimal for any k. From the equation for the error x_k in the case of the scheme (9)

$$\begin{aligned} x_{k+1} &= \alpha_{k+1}(E - \tau_{k+1}C)x_k + (1 - \alpha_{k+1})x_{k-1}, \quad k = 1, 2, \ldots, \\ x_1 &= (E - \tau_1 C)x_0, \end{aligned} \tag{10}$$

we obtain that $x_k = P_k(C)x_0$, where the polynomial $P_k(C)$ has the form (4) ($n = k$). Therefore, if the parameters $\{\tau_k\}$ and $\{\alpha_k\}$ are chosen so that for any $n = 1, 2, \ldots$ the conditions (6) are satisfied, then the iterative approximations y_n constructed according to (9) will be the same as the approximations constructed using the formulas (6), (8) for any n.

We now construct the desired set of parameters $\{\tau_k\}$ and $\{\alpha_k\}$. To do this we require

Lemma. *The conditions*

$$(Cx_j, x_n) = 0, \quad j = 1, 2, \ldots, n, \tag{11}$$

are necessary and sufficient for the norm of x_n in H to be a minimum for any $n \geq 1$.

Proof. From (4), (6) it follows that the conditions (6) — the conditions for the norm of x_n to be a minimum — are equivalent to:

$$(C^j x_0, x_n) = 0, \quad j = 1, 2, \ldots, n, \tag{12}$$

for any $n = 1, 2, \ldots$. From this we obtain, for $j \leq n - 1$,

$$(Cx_0, x_n) + \sum_{i=2}^{j+1} a_{i-1}^{(j)}(C^i x_0, x_n) = (Cx_j, x_n) = 0,$$

i.e. the conditions (11) are necessary.

We now prove the sufficiency of the conditions (11). Suppose that the conditions (11) are satisfied. We shall show that then the conditions (12) are also satisfied. From (11) for $j = 0$ we obtain that the equations (12) are valid for $j = 1$. The validity of (12) for $j \geqq 2$ is proved by induction. Suppose that the conditions (12) are satisfied for $j \leq k$, i.e. $(C^j x_0, x_n) = 0$, $j = 1, 2, \dots, k$. We shall show that they are also satisfied for $j = k+1$, if the conditions (11) are satisfied.

From (11) for $j = k$ we obtain

$$0 = (Cx_k, x_n) = (CP_k(C)x_0, x_n)$$
$$= (Cx_0, x_n) + \sum_{j=1}^{k} a_j^{(k)} (C^{j+1} x_0, x_n) = a_k^{(k)} (C^{k+1} x_0, x_n).$$

Consequently, $(C^{k+1} x_0, x_n) = 0$. The lemma is proved. □

We now use the lemma to construct the set of parameters $\{\tau_k\}$ and $\{\alpha_k\}$ for the scheme (9). To simplify the presentation, we will assume that y_1 in the scheme (9) is found from the general formula (9) with $\alpha_1 = 1$.

We look now at the scheme (10). Since x_1 is found from a two-level scheme, it follows from Section 8.1 that we should choose the optimal parameter τ_1 from the formula

$$\tau_1 = \frac{(Cx_0, x_0)}{(Cx_0, Cx_0)}.$$

The construction of the parameters $\tau_2, \tau_3, \dots$ and $\alpha_2, \alpha_3, \dots$ will be carried out sequentially. Suppose that the iterative parameters $\tau_1, \tau_2, \dots, \tau_k$ and $\alpha_1, \alpha_2, \dots, \alpha_k$ have already been choosen in an optimal way. Since these parameters determine the approximations $y_1, y_2, \dots, y_k$, it follows from the lemma that the conditions

$$(Cx_j, x_i) = 0, \quad j = 0, 1, \dots, i-1, \quad i = 1, 2, \dots, k, \tag{13}$$

are satisfied.

We now choose the parameters τ_{k+1} and α_{k+1} which determine the approximation y_{k+1}. From the lemma it follows that the norm of x_{k+1} will be a minimum if the conditions

$$(Cx_j, x_{k+1}) = 0, \quad j = 0, 1, \dots, k, \tag{14}$$

are satisfied.

From these conditions we find the parameters τ_{k+1} and α_{k+1}. We shall first show that (13) implies that the conditions (14) are satisfied for $j \leq k-2$, and then from the remaining two conditions (14) for $j = k-1$ and $j = k$, we obtain the formulas for τ_{k+1} and α_{k+1}.

Thus, suppose $j \leq k-2$. From (10) and (13) we find

$$(x_{k+1}, Cx_j) = \alpha_{k+1}(x_k, Cx_j) - \alpha_{k+1}\tau_{k+1}(Cx_k, Cx_j) \\ + (1-\alpha_{k+1})(x_{k-1}, Cx_j) = -\alpha_{k+1}\tau_{k+1}(Cx_k, Cx_j).$$

We shall show that $(Cx_k, Cx_j) = 0$ for $j \leq k-2$. In fact, from (10) for $k = j$ we obtain

$$Cx_j = \frac{1}{\tau_{j+1}}x_j - \frac{1}{\tau_{j+1}\alpha_{j+1}}[x_{j+1} - (1-\alpha_{j+1})x_{j-1}], \quad j \geq 0. \tag{15}$$

Using the self-adjointness of the operator C and the conditions (13), from this we obtain for $j \leq k-2$

$$(Cx_k, Cx_j) \\ = \frac{1}{\tau_{j+1}}(Cx_j, x_k) - \frac{1}{\tau_{j+1}\alpha_{j+1}}[(Cx_{j+1}, x_k) - (1-\alpha_{j+1})(Cx_{j-1}, x_k)] = 0.$$

Consequently, $(x_{k+1}, Cx_j) = 0$ for $j \leq k-2$.

We now find τ_{k+1} and α_{k+1}. Setting $j = k-1$ and $j = k$ in (14), we obtain from (10) and (13)

$$\begin{aligned} 0 &= (Cx_{k-1}, x_{k+1}) \\ &= -\alpha_{k+1}\tau_{k+1}(Cx_k, Cx_{k-1}) + (1-\alpha_{k+1})(Cx_{k-1}, x_{k-1}), \\ 0 &= (Cx_k, x_{k+1}) = \alpha_{k+1}[(Cx_k, x_k) - \tau_{k+1}(Cx_k, Cx_k)]. \end{aligned} \tag{16}$$

From the second equation, we at once find the parameter τ_{k+1}:

$$\tau_{k+1} = \frac{(Cx_k, x_k)}{(Cx_k, Cx_k)}. \tag{17}$$

From the first equation we eliminate the expression (Cx_k, Cx_{k-1}). To do this, we set $j = k-1$ in (15) and take the inner product with Cx_k on the left- and right-hand sides of (15).

Taking into account the self-adjointness of the operator C, from the condition (13) we obtain

$$(Cx_k, Cx_{k-1}) = \frac{1}{\tau_k}(Cx_{k-1}, x_k) - \frac{1}{\tau_k\alpha_k}(Cx_k, x_k) + \frac{1-\alpha_k}{\tau_k\alpha_k}(Cx_{k-2}, x_k)$$
$$= -\frac{1}{\tau_k\alpha_k}(Cx_k, x_k).$$

Substituting this expression in (16), we obtain

$$\frac{\alpha_{k+1}\tau_{k+1}}{\alpha_k\tau_k}\frac{(Cx_k, x_k)}{(Cx_{k-1}, x_{k-1})} + (1-\alpha_{k+1}) = 0.$$

From this equation we find a recurrence formula for the parameter α_{k+1}:

$$\alpha_{k+1} = \left(1 - \frac{\tau_{k+1}}{\tau_k}\frac{(Cx_k, x_k)}{(Cx_{k-1}, x_{k-1})}\frac{1}{\alpha_k}\right)^{-1}. \tag{18}$$

Thus, assuming that the iterative parameters $\tau_1, \tau_2, \ldots, \tau_k$ and $\alpha_1, \alpha_2, \ldots, \alpha_k$ have already been found, we obtain formulas for the parameters τ_{k+1} and α_{k+1}. Since $\alpha_1 = 1$ and

$$\tau_1 = \frac{(Cx_0, x_0)}{(Cx_0, Cx_0)},$$

the formulas (17), (18) define the parameters τ_{k+1} and α_{k+1} for any k.

Substituting $x_k = D^{1/2}z_k$ in (17) and (18) and taking into account that

$$C = D^{-1/2}(DB^{-1}A)D^{-1/2} \quad \text{and} \quad Az_k = r_k, \quad B^{-1}r_k = w_k,$$

we obtain the following formulas for the iterative parameters τ_{k+1} and α_{k+1}:

$$\tau_{k+1} = \frac{(Dw_k, z_k)}{(Dw_k, w_k)}, \quad k = 0, 1, \ldots, \tag{19}$$

$$\alpha_{k+1} = \left(1 - \frac{\tau_{k+1}}{\tau_k}\frac{(Dw_k, z_k)}{(Dw_{k-1}, z_{k-1})}\frac{1}{\alpha_k}\right)^{-1}, \tag{20}$$
$$k = 1, 2, \ldots, \quad \alpha_1 = 1.$$

Thus, the conjugate-direction method is described by the three-level scheme (9), where the iterative parameters τ_{k+1} and α_{k+1} are chosen using the formulas (19), (20). For this method, theorem 2 proved above is valid.

From the formulas (19), (20) it follows that the iterative parameters τ_{k+1} in the conjugate-direction method and in the two-level gradient methods are chosen using the very same formulas, and that it is not necessary to compute any auxiliary inner products in order to compute the parameters α_{k+1}. Therefore to compute the iterative parameters in the two-level and three-level methods of variational type, practically the same expenditure of arithmetic operations is required. At the same time, it follows from theorems 1 and 2 that conjugate-direction methods converge substantially faster than gradient methods.

We shall show now that, if H is a finite-dimensional space ($H = H_N$), then the conjugate-direction methods converge after a finite number of iterations not exceeding the dimension of the space. In fact, from the lemma it follows that the conjugate-direction methods must satisfy the equations $(Cx_j, x_n) = (x_j, x_n)_C = 0$, $j = 0, 1, \dots, n-1$ for the equivalent errors x_k. Consequently, the system of vectors $x_0, x_1, \dots, x_n$ must be a C-orthogonal system in H_N for any n. And since it is impossible to construct more than N orthogonal vectors in H_N, it follows that $x_N = 0$ and $z_N = y_N - u = 0$. Thus, on the class of arbitrary initial approximations y_0, the conjugate-direction methods converge in N iterations to the exact solution of equation (1).

For special initial approximations y_0, these methods converge in even fewer iterations. In fact, suppose that y_0 is such that the expansion of x_0 in eigenfunctions of the operator C contains $N_0 < N$ functions, i.e. x_0 belongs to an invariant subspace of H_{N_0} relative to the operator C. Then it is obvious that all $x_k \in H_{N_0}$. Therefore in this case the iterative process will converge after N_0 iterations.

From the remarks above it does not follow that the convergence estimate for the method obtained in theorem 2 is very crude, and that equality in $\| z_n \|_D = q_n \| z_0 \|_D$ is never achieved. It is possible to construct an example of equation (1) and show that for any $n < N$ there is an initial approximation y_0 such that this equation is satisfied.

8.3.3 Variants of the computational formulas. We give now several different ways of realizing the three-level conjugate-direction methods. From (9), (19) and (20) we obtain the following algorithm:

1) given y_0, compute the residual $r_0 = Ay_0 - f$;
2) solve the equation for the correction $Bw_0 = r_0$;
3) compute the parameter τ_1 using the formula (19);
4) find the approximation y_1 from the formula $y_1 = y_0 - \tau_1 w_0$.

Then, for $k = 1, 2, \dots$ sequentially perform the following operations:

5) compute the residual $r_k = Ay_k - f$ and solve the equation for the correction $Bw_k = r_k$;

6) using the formulas (19), (20), compute the parameters τ_{k+1} and α_{k+1};
7) find the approximation y_{k+1} from the formula

$$y_{k+1} = \alpha_{k+1}y_k + (1-\alpha_{k+1})y_{k-1} - \alpha_{k+1}\tau_{k+1}w_k.$$

Thus, in this description of the algorithm, it is necessary to store y_{k-1}, y_k, and w_k in order to find y_{k+1}. Below we will indicate the specific form of (19) and (20) for certain concrete choices of the operator D. Here we will limit ourselves to the comment that it is possible to use the residual r_k in place of the correction w_k. To compute the residual, either the equation $r_k = Bw_k$ (if the computation of Bw_k is not a laborious process) or the definition of the residual $r_k = Ay_k - f$ can be used.

In practice, there are still other algorithms for realizing the conjugate-direction method. We give one of these. To do this, we shall treat the scheme (9) as a scheme with a correction. From (9) we obtain

$$y_{k+1} = y_k - a_{k+1}s_k, \quad s_{k+1} = w_{k+1} + b_{k+1}s_k, \quad k = 0,1,\ldots, \quad s_0 = w_0, \tag{21}$$

where $w_k = B^{-1}r_k$, $r_k = Ay_k - f$, and the parameters a_{k+1} and b_k are linked with α_{k+1} and τ_{k+1} by the following formulas:

$$a_{k+1} = \alpha_{k+1}\tau_{k+1}, \quad b_k = (\alpha_{k+1}-1)\alpha_k\tau_k/(\alpha_{k+1}\tau_{k+1}).$$

We now obtain an expression for b_k and a_{k+1}. From (19), (20) we find

$$b_k = (Dw_k, z_k)/(Dw_{k-1}, z_{k-1}), \quad k = 1,2,\ldots. \tag{22}$$

From these formulas, it is easy to obtain the recurrence formulas for a_{k+1}, however it is also possible to find an explicit expression for a_{k+1}:

$$a_{k+1} = \frac{(Cx_k, x_k)}{(Ds_k, s_k)} = \frac{(Dw_k, z_k)}{(Ds_k, s_k)}, \quad k = 0,1,\ldots. \tag{23}$$

The formulas (21), (22), and (23) describe a second algorithm for the conjugate-direction method. Here the computations are carried out in the following order:

1) given y_0, compute the residual $r_0 = Ay_0 - f$, solve the equation $Bw_0 = r_0$ for the correction w_0, and set $s_0 = w_0$;
2) using (23) find the parameter a_1 and compute $y_1 = y_0 - a_1 s_0$. Then for $k = 1,2,\ldots$ sequentially perform the operations:
3) compute the residual $r_k = Ay_k - f$ and solve the equation for the correction $Bw_k = r_k$;

4) using (22), compute the parameter b_k and find s_k from the formula $s_k = w_k + b_k s_{k-1}$;
5) from (23) determine the parameter a_{k+1} and compute the approximation y_{k+1} using

$$y_{k+1} = y_k - a_{k+1} s_k.$$

Notice that in this algorithm it is necessary to store y_k, w_k, and s_k, i.e. the same amount of intermediate information as in the first algorithm.

8.4 Examples of three-level methods

8.4.1 Special cases of the conjugate-direction methods. In Section 8.3, we constructed the three-level iterative conjugate-direction methods, and used them to solve the linear equation

$$Au = f. \tag{1}$$

The iterative approximations are computed from the three-level scheme

$$\begin{gathered} By_{k+1} = \alpha_{k+1}(B - \tau_{k+1}A)y_k + (1 - \alpha_{k+1})By_{k-1} + \alpha_{k+1}\tau_{k+1}f, \\ k = 1, 2, \ldots, \\ By_1 = (B - \tau_1 A)y_0 + \tau_1 f, \quad y_0 \in H, \end{gathered} \tag{2}$$

and the iterative parameters α_{k+1} and τ_{k+1} are found from the formulas

$$\begin{gathered} \tau_{k+1} = \frac{(Dw_k, z_k)}{(Dw_k, w_k)}, \quad k = 0, 1, \ldots, \\ \alpha_{k+1} = \left(1 - \frac{\tau_{k+1}}{\tau_k} \frac{(Dw_k, z_k)}{(Dw_{k-1}, z_{k-1})} \cdot \frac{1}{\alpha_k}\right)^{-1}, \quad k = 1, 2, \ldots, \quad \alpha_1 = 1, \end{gathered} \tag{3}$$

where $w_k = B^{-1}r_k$ is the correction, $r_k = Ay_k - f$ is the residual, and $z_k = y_k - u$ is the error.

The choice of the parameters α_k and τ_k from the formulas (3) guarantees that the norm of the error z_n in H_D after the move from y_0 to y_n will be a minimum for any n, if $DB^{-1}A$ is a self-adjoint and positive-definite operator.

We look now at special cases of the conjugate-direction methods, defined by specific choices of the operator D. In Section 8.2 four examples of two-level methods were considered. Each of these two-level methods corresponds to a particular three-level iterative conjugate-direction method. We shall list these methods along with the appropriate conditions on the operators A and B which guarantee the self-adjointness of the operator $DB^{-1}A$. For these methods, theorem 2 is valid, and the form of the inequalities which define the constants γ_1 and γ_2 will be indicated in the description of the corresponding method.

1) *The conjugate-gradient method.*
The operator D: $D = A$.
The conditions: $\gamma_1 B \le A \le \gamma_2 B$, $\gamma_1 > 0$,
$A = A^* > 0$, $B = B^* > 0$.
The formulas for the iterative parameters:

$$\tau_{k+1} = \frac{(r_k, w_k)}{(Aw_k, w_k)}, \quad \alpha_{k+1} = \left(1 - \frac{\tau_{k+1}}{\tau_k} \frac{(r_k, w_k)}{(r_{k-1}, w_{k-1})} \frac{1}{\alpha_k}\right)^{-1}.$$

2) *The conjugate-residual method.*
The operator D: $D = A^*A$.
The conditions: $\gamma_1(Bx, Bx) \le (Ax, Bx) \le \gamma_2(Bx, Bx)$, $\gamma_1 > 0$,
$B^*A = A^*B$.

If it is also assumed that $A = A^* > 0$, $B = B^* > 0$, $AB = BA$, then the conditions have the form

$$\gamma_1 B \le A \le \gamma_2 B, \quad \gamma_1 > 0.$$

The formulas for the iterative parameters:

$$\tau_{k+1} = \frac{(Aw_k, r_k)}{(Aw_k, Aw_k)}, \quad \alpha_{k+1} = \left(1 - \frac{\tau_{k+1}}{\tau_k} \frac{(Aw_k, r_k)}{(Aw_{k-1}, r_{k-1})} \cdot \frac{1}{\alpha_k}\right)^{-1}.$$

3) *The conjugate-correction method.*
The operator D: $D = AB^{-1}A$.
The conditions: $\gamma_1 B \le A \le \gamma_2 B$, $\gamma_1 > 0$, $B = B^* > 0$.
The formulas for the iterative parameters:

$$\tau_{k+1} = \frac{(Aw_k, w_k)}{(B^{-1}Aw_k, Aw_k)}, \quad \alpha_{k+1} = \left(1 - \frac{\tau_{k+1}}{\tau_k} \frac{(Aw_k, w_k)}{(Aw_{k-1}, w_{k-1})} \cdot \frac{1}{\alpha_k}\right)^{-1}.$$

4) *The conjugate-error method.*
The operator D: $D = B_0$.
The conditions: $B = (A^*)^{-1}B_0$, $\gamma_1 B_0 \le A^*A \le \gamma_2 B_0$, $B_0 = B_0^* > 0$.
The formulas for the iterative parameters:

$$\tau_{k+1} = \frac{(r_k, r_k)}{(Aw_k, r_k)}, \quad \alpha_{k+1} = \left(1 - \frac{\tau_{k+1}}{\tau_k} \frac{(r_k, r_k)}{(r_{k-1}, r_{k-1})} \cdot \frac{1}{\alpha_k}\right)^{-1}.$$

8.4.2 Locally optimal three-level methods. We turn now to the procedure examined in Section 8.3 for constructing the iterative parameters α_{k+1} and τ_{k+1} for the conjugate-direction method. Recall that the parameters α_{k+1} and τ_{k+1} were chosen from the conditions $(Cx_{k-1}, x_{k+1}) = 0$ and $(Cx_k, x_{k+1}) = 0$ under the assumption that the iterative approximations $y_1, y_2, \ldots, y_k$ guaranteed that

$$(Cx_j, x_i) = 0, \quad j = 0, 1, \ldots, i-1, \quad i = 1, 2, \ldots, k. \tag{4}$$

In an ideal computational process, the conditions (4) are satisfied, therefore the choice of the parameters α_{k+1} and τ_{k+1} from the formulas obtained in Section 8.3 guarantees the minimization of the norm of the error z_{k+1} in H_D after the move from y_0 to y_{k+1}. In an actual computational process, where the influence of rounding errors is taken into account, the iterative approximations $y_1, y_2, \ldots, y_k$ will be computed inexactly, and consequently the conditions (4) will not be satisfied. In many cases, this can lead to slower convergence rates for the method, and sometimes even to divergence.

We construct now a modified conjugate-direction method which does not possess this deficiency. In order to approximately solve the equation $Au = f$ we look at the three-level iterative scheme

$$\begin{gathered} By_{k+1} = \alpha_{k+1}(B - \tau_{k+1}A)y_k + (1 - \alpha_{k+1})By_{k-1} + \alpha_{k+1}\tau_{k+1}f, \\ k = 1, 2, \ldots, \end{gathered} \tag{5}$$

with arbitrary approximations y_0 and $y_1 \in H$. Assuming that y_k and y_{k-1} are known, we choose the parameters α_{k+1} and τ_{k+1} from the condition that the norm of the error z_{k+1} in H_D be a minimum, i.e. from a local optimization condition for one step of the three-level scheme.

We solve this problem under the natural assumption that the operator $DB^{-1}A$ is positive-definite. To do this, we transform to the equation for the equivalent error $x_k = D^{-1/2}z_k$:

$$x_{k+1} = \alpha_{k+1}(E - \tau_{k+1}C)x_k + (1 - \alpha_{k+1})x_{k-1}, \quad C = D^{1/2}B^{-1}AD^{-1/2}. \tag{6}$$

To simplify the exposition, we introduce the notation

$$1 - \alpha_{k+1} = a, \quad \tau_{k+1}\alpha_{k+1} = b \tag{7}$$

and rewrite (6) in the following form:

$$x_{k+1} = x_k - a(x_k - x_{k-1}) - bCx_k. \tag{8}$$

The problem is then: choose a and b from the condition that the norm of x_{k+1} in H be a minimum. We now compute the norm of x_{k+1}. From (8) we obtain

$$\begin{aligned} \| x_{k+1} \|^2 &= \| x_k \|^2 + a^2 \| x_k - x_{k-1} \|^2 + b^2 \| Cx_k \|^2 \\ &\quad - 2a(x_k, x_k - x_{k-1}) - 2b(Cx_k, x_k) + 2ab(Cx_k, x_k - x_{k-1}). \end{aligned}$$

Setting the partial derivatives with respect to a and b equal to zero, we obtain a system in the parameters a and b

$$\begin{aligned} \| x_k - x_{k-1} \|^2 a + (Cx_k, x_k - x_{k-1})b &= (x_k, x_k - x_{k-1}), \\ (Cx_k, x_k - x_{k-1})a + \| Cx_k \|^2 b &= (Cx_k, x_k). \end{aligned} \tag{9}$$

The determinant of this system is equal to $\| x_k - x_{k-1} \|^2 \| Cx_k \|^2 - (Cx_k, x_k - x_{k-1})^2$ and, by the Cauchy-Schwartz-Bunyakovskij inequality, is only zero when $x_k - x_{k-1}$ is proportional to Cx_k: $x_k - x_{k-1} = dCx_k$. In this case, the equations of the system are proportional, and they reduce to a single equation

$$(b + ad) \| Cx_k \|^2 = (Cx_k, x_k). \tag{10}$$

Since this implies that (8) has the form $x_{k+1} = x_k - (b + ad)Cx_k$, then, assuming that $a = 0$ in (10), we obtain from (7), (10)

$$\alpha_{k+1} = 1, \quad \tau_{k+1} = \frac{(Cx_k, x_k)}{(Cx_k, Cx_k)}. \tag{11}$$

If the determinant is not zero, then, solving the system (9), we obtain

$$a = \frac{\| Cx_k \|^2 (x_k, x_k - x_{k-1}) - (Cx_k, x_k)(Cx_k, x_k - x_{k-1})}{\| x_k - x_{k-1} \|^2 \| Cx_k \|^2 - (Cx_k, x_k - x_{k-1})^2},$$

$$b = \frac{(Cx_k, x_k)}{(Cx_k, Cx_k)}(1 - a) + \frac{(Cx_k, x_{k-1})}{(Cx_k, Cx_k)}a.$$

Hence, using the notation (7), we find the formulas for the parameters α_{k+1} and τ_{k+1}:

$$\alpha_{k+1} = \frac{(Cx_k, x_k - x_{k-1})(Cx_k, x_{k-1}) - (x_{k-1}, x_k - x_{k-1})(Cx_k, Cx_k)}{(Cx_k, Cx_k)(x_k - x_{k-1}, x_k - x_{k-1}) - (Cx_k, x_k - x_{k-1})^2},$$

$$\tau_{k+1} = \frac{(Cx_k, x_k)}{(Cx_k, Cx_k)} + \frac{1 - \alpha_{k+1}}{\alpha_{k+1}} \frac{(Cx_k, x_{k-1})}{(Cx_k, Cx_k)}, \quad k = 1, 2, \ldots. \tag{12}$$

The formulas (11) obtained earlier can be considered as a special case of the general formulas (12), with $\alpha_{k+1} = 1$ if the denominator in the expression for α_{k+1} vanishes.

The formulas (12) are more complex than the formulas for the parameters α_{k+1} and τ_{k+1} in the conjugate-direction method obtained in Section 8.3. Here it is necessary to compute auxiliary inner products. However, the iterative process (5), (12) constructed in this section is less susceptible to the influence of rounding errors, and errors from previous iterations are damped.

The connection between the locally-optimal three-level methods and the conjugate-direction methods is laid out in

Theorem 3. *If the initial approximation y_1 for the method* (5), (12) *is chosen as follows:*

$$By_1 = (B - \tau_1 A)y_0 + \tau_1 f, \quad \tau_1 = \frac{(Dw_0, z_0)}{(Dw_0, w_0)}, \tag{13}$$

and the operator $DB^{-1}A$ is self-adjoint, then the method (5), (12) *is identical to the conjugate-direction method.*

Proof. The proof is carried out by induction. From the conditions of the theorem it follows that the approximations y_1 obtained here and in the conjugate-direction method are the same. Suppose that the approximations $y_1, y_2, \ldots,$ y_k are the same. We shall prove that the y_{k+1} constructed according the formulas (5), (12) is identical to the approximation y_{k+1} from the conjugate-direction method.

From the above assumptions it follows that the iterative parameters $\tau_1, \tau_2, \ldots, \tau_k$ and $\alpha_2, \alpha_3, \ldots, \alpha_k$ are also identical for both methods. If we can prove that the parameters τ_{k+1} and α_{k+1} are the same in these methods as well, then the assertion of theorem 3 will be proved.

Since $y_1, y_2, \ldots, y_k$ are the iterative approximations for the conjugate-direction method, it follows from the lemma that the conditions

$$(Cx_j, x_i) = 0, \quad j = 0, 1, \ldots, i-1, \quad i = 1, 2, \ldots, k, \tag{14}$$

are satisfied. Substituting (14) for $j = k-1$ and $i = k$ in (12) and using the self-adjointness of the operator C, we obtain

$$\alpha_{k+1} = \frac{(x_{k-1}, x_k - x_{k-1})(Cx_k, Cx_k)}{(Cx_k, x_k)^2 - \| Cx_k \|^2 \| x_k - x_{k-1} \|^2}, \quad \tau_{k+1} = \frac{(Cx_k, x_k)}{(Cx_k, Cx_k)}. \tag{15}$$

Thus the parameters τ_{k+1} from the locally-optimal method and the conjugate-direction method are the same. It remains to prove that the parameters α_{k+1} are the same.

From (6) and (13) we obtain

$$\begin{aligned} x_k - x_{k-1} &= (\alpha_k - 1)(x_{k-1} - x_{k-2}) - \alpha_k \tau_k C x_{k-1}, \quad k = 2, 3, \ldots, \\ x_1 - x_0 &= -\tau_1 C x_0. \end{aligned} \tag{16}$$

From (16) it follows that the difference $x_k - x_{k-1}$ is a linear combination of $Cx_0, Cx_1, \ldots, Cx_{k-1}$ and has the following form

$$\begin{aligned} x_k - x_{k-1} &= -\alpha_k \tau_k C x_{k-1} + \sum_{j=0}^{k-2} \beta_j C x_j, \quad k \geq 2, \\ x_1 - x_0 &= -\tau_1 C x_0, \end{aligned} \tag{17}$$

where the coefficients β_j are expressed in terms of $\tau_1, \tau_2, \ldots, \tau_{k-1}$ and $\alpha_2, \alpha_3, \ldots, \alpha_{k-1}$. Taking the inner product of the left- and right-hand sides of (17) with x_{k-1} and $x_k - x_{k-1}$ and taking into account (14), we obtain

$$\begin{aligned} (x_{k-1}, x_k - x_{k-1}) &= -\alpha_k \tau_k (C x_{k-1}, x_{k-1}), \\ \| x_k - x_{k-1} \|^2 &= \alpha_k \tau_k (C x_{k-1}, x_{k-1}). \end{aligned} \tag{18}$$

Substituting (18) in the expression (15) for α_{k+1} and using the formula for the parameter τ_{k+1}, we obtain

$$\begin{aligned} \alpha_{k+1} &= \frac{\alpha_k \tau_k (C x_{k-1}, x_{k-1})(Cx_k, Cx_k)}{\alpha_k \tau_k (C x_{k-1}, x_{k-1})(Cx_k, Cx_k) - (Cx_k, x_k)^2} \\ &= \left(1 - \frac{\tau_{k+1}(Cx_k, x_k)}{\tau_k (C x_{k-1}, x_{k-1})} \cdot \frac{1}{\alpha_k} \right)^{-1}, \end{aligned}$$

which is the same as the formula for the parameter α_{k+1} in the conjugate-direction method. The theorem has been proved. □

Substituting $x_k = D^{1/2}z_k$ and $C = D^{-1/2}(DB^{-1}A)D^{-1/2}$ in (12), we obtain the following version of the formulas for the parameters α_{k+1} and τ_{k+1}:

$$\begin{aligned}\alpha_{k+1} &= \frac{(Dw_k, z_k - z_{k-1})(Dw_k, z_{k-1}) - (Dz_{k-1}, y_k - y_{k-1})(Dw_k, w_k)}{(Dw_k, w_k)(D(z_k - z_{k-1}), y_k - y_{k-1}) - (Dw_k, z_k - z_{k-1})^2}, \\ \tau_{k+1} &= \frac{(Dw_k, z_k)}{(Dw_k, w_k)} + \frac{1-\alpha_{k+1}}{\alpha_{k+1}}\frac{(Dw_k, z_{k-1})}{(Dw_k, w_k)}.\end{aligned} \tag{19}$$

If we introduce notation for the inner products

$$a_k = (Dw_k, z_k), \quad b_k = (Dw_k, z_{k-1}), \quad c_k = (Dz_k, y_k - y_{k-1}),$$
$$d_k = (Dz_{k-1}, y_k - y_{k-1}), \quad e_k = (Dw_k, w_k),$$

then the formulas (19) can be rewritten in the form

$$\alpha_{k+1} = \frac{(a_k - b_k)b_k - d_k e_k}{(c_k - d_k)e_k - (a_k - b_k)^2}, \quad k = 1, 2, \ldots, \quad \alpha_1 = 1,$$
$$\tau_{k+1} = \frac{a_k}{e_k} + \frac{1-\alpha_{k+1}}{\alpha_{k+1}}\frac{b_k}{e_k}, \quad k = 0, 1, \ldots.$$

We now give expressions for a_k, b_k, c_k, d_k, and e_k for specific choices of the operator D:

1) $D = A$, $A = A^*$.

$$a_k = (w_k, r_k), \quad b_k = (w_k, r_{k-1}), \quad c_k = (r_k, y_k - y_{k-1}),$$
$$d_k = (r_{k-1}, y_k - y_{k-1}), \quad e_k = (Aw_k, w_k).$$

2) $D = A^*A$.

$$a_k = (Aw_k, r_k), \quad b_k = (Aw_k, r_{k-1}), \quad c_k = (r_k, r_k - r_{k-1}),$$
$$d_k = (r_{k-1}, r_k - r_{k-1}), \quad e_k = (Aw_k, Aw_k).$$

3) $D = A^*B^{-1}A$, $B = B^*$.

$$a_k = (Aw_k, w_k), \quad b_k = (Aw_k, w_{k-1}), \quad c_k = (w_k, r_k - r_{k-1}),$$
$$d_k = (w_{k-1}, r_k - r_{k-1}), \quad e_k = (B^{-1}Aw_k, Aw_k).$$

8.5 Accelerating the convergence of two-level methods in the self-adjoint case

8.5.1 An algorithm for the acceleration process. In Section 8.1.5 it was shown that, in the case of a self-adjoint operator $DB^{-1}A$, the two-level gradient methods possess an asymptotic property. It appears that, after a large number of iterations, the convergence rate of the method drops significantly in comparison with the initial iterations. It was also shown that, after a large number of iterations, the errors are nearly proportional if examined at every other step.

Using this property, we now construct an algorithm for accelerating the convergence of two-level gradient methods.

To solve the equation

$$Au = f \tag{1}$$

we look at the two-level gradient iterative method

$$B\frac{y_{k+1} - y_k}{\tau_{k+1}} + Ay_k = f, \quad k = 0, 1, \dots, \quad y_0 \in H, \tag{2}$$

$$\tau_{k+1} = \frac{(Dw_k, z_k)}{(Dw_k, w_k)}, \quad k = 0, 1, \dots. \tag{3}$$

Assume that the operator $DB^{-1}A$ is self-adjoint in H. Then the iterative method possesses the asymptotic property, and after a sufficiently large number of iterations k we have the approximate equation

$$z_{k+2} \approx \rho^2 z_k, \quad z_k = y_k - u. \tag{4}$$

We look first at the case where (4) is actually an equality, i.e. $z_{k+2} = \rho^2 z_k$. Using the already-computed approximations y_k and y_{k+2}, we construct a new approximation using the formula

$$y = \alpha y_{k+2} + (1 - \alpha) y_k, \quad \alpha = 1/(1 - \rho^2). \tag{5}$$

For the error $z = y - u$, we obtain

$$z = \alpha z_{k+2} + (1 - \alpha) z_k = (\alpha\rho^2 + 1 - \alpha) z_k = [1 - \alpha(1 - \rho^2)] z_k = 0.$$

Consequently, in the case where (4) is actually an equality, the linear combination (5) of the approximations y_k and y_{k+2} gives the exact solution to equation (1).

As was noted in Section 8.1, exact equality in (4) is a remarkable case which is only true for special initial approximations. In the general case, (4) is only an approximate equality, and the discussion above only allows us to hope that some linear combination of y_k and y_{k+2} will give a better approximation to the solution of the original problem.

We shall now find the best of these linear combinations. Let y_k, y_{k+1}, and y_{k+2} be the iterative approximations obtained using the formulas (2), (3). We will look for a new approximation y using the formula

$$y = \alpha y_{k+2} + (1-\alpha) y_k. \tag{6}$$

We will choose the parameter α so that the norm of the error $z = y - u$ in H_D will be minimal.

First of all, using the scheme (2), we eliminate y_{k+2} from (6). We obtain

$$By_{k+2} = (B - \tau_{k+2}A)y_{k+1} + \tau_{k+2}f$$

and, after substituting for y_{k+2} in (6), we will have

$$By = \alpha(B - \tau_{k+2}A)y_{k+1} + (1-\alpha)By_k + \alpha\tau_{k+2}f, \tag{7}$$

where y_{k+1} is found from the two-level scheme

$$By_{k+1} = (B - \tau_{k+1}A)y_k + \tau_{k+1}f. \tag{8}$$

If we assume that y_k is some given initial approximation, then the scheme (7), (8) is the same as the scheme for the conjugate-direction method, and the parameters τ_{k+1} and α_{k+1} are the same as the parameters for the conjugate-direction method. From the theory of this method it follows (Section 8.4.1, formula (3)) that the optimal value of the parameter α is given by the formula

$$\alpha = \frac{1}{1 - \dfrac{\tau_{k+2}}{\tau_{k+1}} \dfrac{(Dw_{k+1}, z_{k+1})}{(Dw_k, z_k)}}. \tag{9}$$

Thus, we have solved the problem of choosing the optimal parameter α. The formulas (6), (9) define the acceleration procedure.

Notice that, instead of using (6), it is possible to compute y from the following two-level scheme:

$$\begin{aligned} B\bar{y}_{k+1} &= (B - \bar{\tau}_{k+1}A)y_k + \bar{\tau}_{k+1}f, \\ By &= (B - \bar{\tau}_{k+2}A)\bar{y}_{k+1} + \bar{\tau}_{k+2}f, \end{aligned} \tag{10}$$

where $\bar{\tau}_{k+1}$ and $\bar{\tau}_{k+2}$ are the roots of the equation

$$\tau^2 - \alpha(\tau_{k+1} + \tau_{k+2})\tau + \alpha\tau_{k+1}\tau_{k+2} = 0.$$

$\bar{\tau}_{k+1}$ should be taken to be the smaller root.

Using (10) in place of (6) enables us to avoid the use of auxiliary storage for intermediate information.

8.5.2 An estimate of the effectiveness. We now estimate the effectiveness of the acceleration procedure. First of all, we compute the norm of the error $z = y - u$ in H_D by transforming the expression (9) for α.

Making the change of variables $z_k = D^{-1/2}x_k$ in (9) gives

$$\alpha = \left(1 - \frac{\tau_{k+2}}{\tau_{k+1}} \frac{(Cx_{k+1}, x_{k+1})}{(Cx_k, x_k)}\right)^{-1}, \quad C = D^{1/2}B^{-1}AD^{1/2}. \tag{11}$$

From (10) and (11) in Section 8.1 we have

$$\| x_{k+1} \| = \rho_{k+1} \| x_k \|, \quad \rho_{k+1}^2 = 1 - \frac{(Cx_k, x_k)^2}{(Cx_k, Cx_k) \| x_k \|^2}. \tag{12}$$

From the formulas (9) in Section 8.1 we obtain

$$\tau_{k+1} = \frac{(Cx_k, x_k)}{(Cx_k, Cx_k)}. \tag{13}$$

Using (12) and (13), we find

$$\frac{\tau_{k+2}}{\tau_{k+1}} \frac{(Cx_{k+1}, x_{k+1})}{(Cx_k, x_k)} = \frac{1 - \rho_{k+2}^2}{1 - \rho_{k+1}^2} \rho_{k+1}^2.$$

Substituting this expression in (11) gives

$$\alpha = \frac{1 - \rho_{k+1}^2}{1 - 2\rho_{k+1}^2 + \rho_{k+1}^2 \rho_{k+2}^2}. \tag{14}$$

We now compute the norm of the error $z = y - u$ in H_D. From (6) it follows that

$$z = \alpha z_{k+2} + (1 - \alpha) z_k.$$

Then, for the equivalent errors $z_k = D^{1/2}x_k$ and $x = D^{1/2}z$ we will have $x = \alpha x_{k+2} + (1 - \alpha)x_k$. We now compute the norm of x in H. We obtain

$$\| x \|^2 = \alpha^2 \| x_{k+2} \|^2 + 2\alpha(1 - \alpha)(x_{k+2}, x_k) + (1 - \alpha)^2 \| x_k \|^2 .$$

From the equality $(x_{k+2}, x_k) = \| x_{k+1} \|^2$ proved in Section 8.1.5 it follows that

$$\| x \|^2 = \alpha^2 \| x_{k+2} \|^2 + 2\alpha(1-\alpha) \| x_{k+1} \|^2 + (1-\alpha)^2 \| x_k \|^2 .$$

Substituting here the expression (14) for α and using (12), we obtain

$$\| x \|^2 = \frac{\rho_{k+2}^2 - \rho_{k+1}^2}{\rho_{k+1}^2(1 - 2\rho_{k+1}^2 + \rho_{k+1}^2\rho_{k+2}^2)} \| x_{k+2} \|^2 < \| x_{k+2} \|^2 . \tag{15}$$

Since $\rho_{k+1} \le \rho_{k+2} \le \rho < 1$, then

$$1 - 2\rho_{k+1}^2 + \rho_{k+1}^2\rho_{k+2}^2 \ge (1-\rho^2)^2,$$

consequently, the norm of x is estimated by

$$\| x \|^2 \le \left(\frac{\rho_{k+2}^2}{\rho_{k+1}^2} - 1 \right) \frac{\| x_{k+2} \|^2}{(1-\rho^2)^2}.$$

From the asymptotic property, for sufficiently large k we have $\rho_{k+1} \approx \rho_{k+2}$, therefore the effect of the acceleration procedure will be significant.

Notice that, although the acceleration procedure is effective for large k, this procedure can be used at any iteration. We suggest that from time to time the iterative process using the two-level scheme (2), (3) be interrupted, and a new approximation be computed using the above formulas. The iterative process can be terminated with such an approximation if the computed y_{k+2} satisfies the inequality

$$\frac{\rho_{k+2}^2 - \rho_{k+1}^2}{\rho_{k+1}^2(1 - 2\rho_{k+1}^2 + \rho_{k+1}^2\rho_{k+2}^2)} \| z_{k+2} \|_D^2 \le \epsilon^2 \| z_0 \|_D^2 .$$

In fact, in this case we obtain from (15) that $\| y - u \|_D \le \epsilon \| y_0 - u \|_D$, i.e. the required accuracy ϵ has been achieved.

8.5.3 An example. To illustrate the effectiveness of this acceleration procedure for the two-level gradient methods, we look at the solution of a model problem using an implicit steepest-descent method. As an example, we use a Dirichlet difference problem for Laplace's equation on the square grid $\bar{\omega} = \{x_{ij} = (ih, jh), 0 \le i \le N, 0 \le j \le N, h = 1/N\}$ in the unit square

$$\begin{aligned} &\Lambda u = \Lambda_1 u + \Lambda_2 u = 0, \quad x \in \omega, \quad u|_\gamma = 0, \\ &\Lambda_\alpha u = u_{\bar{x}_\alpha x_\alpha}, \quad \alpha = 1, 2. \end{aligned} \tag{16}$$

We introduce the space H consisting of the grid functions defined on ω with the inner product

$$(u,v)=\sum_{x\in\omega} u(x)v(x)h^2.$$

The operator A is defined as follows: $A=A_1+A_2$, $A_\alpha y=-\Lambda_\alpha v$, $y\in H$, where $v(x)=y(x)$ for $x\in\omega$ and $v|_\gamma=0$.

The problem (16) can be written in the form of the operator equation

$$Au=f, \qquad f=0. \tag{17}$$

We choose B to be the following factored operator: $B=(E+\omega A_1)(E+\omega A_2)$, $\omega>0$, where ω is an iterative parameter.

Since the operators A_1 and A_2 are self-adjoint and commutative in H, the operators A and B are self-adjoint in H. In addition, it is easy to show that the operators A and B are positive-definite in H. Consequently, it is possible to use the implicit steepest-descent method

$$B\frac{y_{k+1}-y_k}{\tau_{k+1}}+Ay_k=f, \quad \tau_{k+1}=\frac{(w_k,r_k)}{(Aw_k,w_k)}, \quad k=0,1,\ldots \tag{18}$$

to solve equation (17).

In this method, $D=A$ and $DB^{-1}A=AB^{-1}A$. Since the operator $DB^{-1}A$ is self-adjoint in H, this method possesses the asymptotic property. From the theory for steepest-descent method (cf. Section 8.2.1) it follows that the convergence rate for the method in this case is determined by the ratio $\xi=\gamma_1/\gamma_2$, where γ_1 and γ_2 are the constants of energy equivalence for the operators A and B: $\gamma_1 B\le A\le\gamma_2 B$, $\gamma_1>0$.

Therefore, the iterative parameter ω is chosen from the condition that ξ be maximal. In Section 11.2 it will be proved that the optimal value of ω is defined by the formula

$$\omega=\frac{1}{\sqrt{\delta\Delta}}, \quad \delta=\frac{4}{h^2}\sin^2\frac{\pi h}{2}, \quad \Delta=\frac{4}{h^2}\cos^2\frac{\pi h}{2},$$

where

$$\gamma_1=\frac{2\delta}{(1+\sqrt{\eta})^2}, \quad \gamma_2=\frac{(\Delta+\delta)\sqrt{\eta}}{(1+\sqrt{\eta})^2}, \quad \xi=\frac{2\sqrt{\eta}}{1+\eta}, \quad \eta=\frac{\delta}{\Delta}.$$

For this example

$$\omega=\frac{h^2}{2\sin\pi h}, \quad \gamma_1=\frac{2}{h^2}\frac{\sin^2\pi h}{1+\sin\pi h}, \quad \gamma_2=\frac{2}{h^2}\frac{\sin\pi h}{1+\sin\pi h}, \quad \xi=\sin\pi h.$$

Table 8

k	$\frac{\Vert z_k \Vert_D}{\Vert z_0 \Vert_D}$	ρ_k	$\gamma_1^{(k)}$	$\gamma_2^{(k)}$	τ_k
1	$3.6 \cdot 10^{-1}$	0.36203	—	—	$5.392 \cdot 10^{-3}$
2	$2.3 \cdot 10^{-1}$	0.63810	77.31858	236.1883	$7.809 \cdot 10^{-3}$
3	$1.8 \cdot 10^{-1}$	0.76998	40.59796	232.1435	$6.911 \cdot 10^{-3}$
4	$1.4 \cdot 10^{-1}$	0.81178	26.87824	233.4976	$8.644 \cdot 10^{-3}$
...	...	...	...	...	...
26	$3.9 \cdot 10^{-3}$	0.85175	18.27141	230.5962	$8.876 \cdot 10^{-3}$
27	$3.4 \cdot 10^{-3}$	0.85178	18.26983	230.6607	$7.338 \cdot 10^{-3}$
28	$2.9 \cdot 10^{-3}$	0.85183	18.27026	230.7191	$8.872 \cdot 10^{-3}$
29	$2.4 \cdot 10^{-3}$	0.85186	18.26895	230.7771	$7.335 \cdot 10^{-3}$
...	...	...	...	...	...
46	$1.6 \cdot 10^{-4}$	0.85226	18.26677	231.4121	$8.845 \cdot 10^{-3}$
47	$1.4 \cdot 10^{-4}$	0.85227	18.26632	231.4375	$7.318 \cdot 10^{-3}$
48	$1.2 \cdot 10^{-4}$	0.85229	18.26664	231.4612	$8.843 \cdot 10^{-3}$
49	$9.9 \cdot 10^{-5}$	0.85230	18.26623	231.4849	$7.317 \cdot 10^{-3}$

We give here the results of the computations with the initial approximation y_0 equal to $y_0(x) = e^{(x_1 - x_2)}$ for $x \in \omega$, $y_0|_\gamma = 0$. The required accuracy ϵ was taken to be 10^4, $N = 40$.

In table 8, for several values of the iteration number k, we list: $\Vert z_k \Vert_D / \Vert z_0 \Vert_D$ — the relative error at the k-th iteration, $\rho_k = \Vert z_k \Vert_D / \Vert z_{k-1} \Vert_D$ — the value which characterizes the decrease in the norm of the error from the $(k-1)$-st to the k-th iteration, $\gamma_1^{(k)}$ and $\gamma_2^{(k)}$ — the approximations to γ_1 and γ_2 found as the roots of the quadratic equation

$$(1 - \tau_k \gamma)(1 - \tau_{k-1}\gamma) = \rho_k \rho_{k-1}, \quad k = 2, 3, \ldots,$$

and the iterative parameters τ_k.

The given accuracy ϵ was achieved after 49 iterations of the scheme (18). For $\epsilon = 10^{-4}$, the theoretical iteration count is equal to 59. The values for ρ_k listed in the table illustrate well the asymptotic property of the method. It is clear that, as the iteration number increases, the convergence rate of the method slows. An accuracy of $4 \cdot 10^{-3}$ was achieved after 26 iterations, and increasing the accuracy by a factor of 40 required an additional 23 iterations.

The quantity ρ_k grows monotonically, and for $k = 26$ we have $\rho_{k+1} - \rho_k \approx 3 \cdot 10^{-5}$. The iterative parameters τ_k and τ_{k+2} become almost equal.

For comparison with the approximate values $\gamma_1^{(k)}$ and $\gamma_2^{(k)}$, here are the exact values of γ_1 and γ_2:

$$\gamma_1 = 18.26556, \quad \gamma_2 = 232.8036.$$

After 49 iterations, γ_1 had been found to an accuracy of 0.004%, and γ_2 to an accuracy of 0.6%.

The procedure described in Section 8.5.1 was used to accelerate the method. A new approximation y was constructed from y_{26} and y_{28} (found from the scheme (18)) using the formulas (6), (9). The given accuracy $\epsilon = 10^{-4}$ was then achieved. An application of the acceleration procedure described in this section to the two-level gradient methods allows us to decrease the number of iterations required for this example by a factor of 1.8.

Chapter 9

Triangular Iterative Methods

In this chapter we study implicit two-level iterative methods whose operators B correspond to triangular matrices. In Section 1 we look at the Gauss-Seidel method and formulate sufficient conditions for its convergence. In Section 2, the successive over-relaxation method is investigated. Here the choice of the iteration parameter is examined, and an estimate is obtained for the spectral radius of the transformation operator. In Section 3 a general matrix iterative scheme is investigated, selection of the iterative parameter is examined, and the method is shown to converge in H_A.

9.1 The Gauss-Seidel method

9.1.1 The iterative scheme for the method. In the preceding chapters we have developed the general theory of two-level and three-level iterative methods and applied them to find an approximate solution to the first-kind equation

$$Au = f. \tag{1}$$

This theory points to choices for the iterative parameters and gives estimates for the iteration counts for these methods; also, the theory uses a minimum of general information about the operators in the iterative scheme. By avoiding a study of the specific structure of the operators in the iterative scheme, we are able to develop the theory from a unified point of view and construct implicit iterative methods which are optimal for a class of operators B.

In the general theory of iterative methods it was proved that the effectiveness of a method depends essentially on the choice of the operator B. This choice depends both on the number of iterations required to achieve a given accuracy ϵ, and on the number of operations needed to complete one iteration. Neither of these quantities on its own can serve as a criterion for the effectiveness of an iterative method. Let us explain. Suppose that the operators A and B are self-adjoint and positive-definite in H. From the theory of iterative methods it follows that, if the operator D is taken to be one of the operators A, B, or $AB^{-1}A$, then the iteration counts for the methods examined in Chapters 6–8 (the Chebyshev and simple iteration methods, methods of variational type, etc.) are determined by the ratio $\xi = \gamma_1/\gamma_2$, where γ_1 and γ_2 are the constants of energy equivalence for the operators A and B: $\gamma_1 B \le A \le \gamma_2 B$.

Therefore, if we choose $B = A$, then we obtain the maximal possible value $\xi = 1$, and the iterative methods will give the exact solution of equation (1) after one iteration for any initial guess. Consequently, this choice of the operator B minimizes the iteration count. However, to realize this single iterative step it is necessary to invert the operator B, i.e. the operator A. It is obvious that the iteration count will be maximal.

On the other hand, for an explicit scheme with $B = E$, the operation count for one iteration is minimal, but for this choice of B the iteration count will become quite large.

Thus there arises the problem of optimally choosing the operator B in order to minimize the total volume of computational work needed to obtain the solution to a given accuracy.

Naturally, this problem cannot be solved when stated in such a general context. At the present time, the development of iterative methods proceeds by constructing easily invertible operators B, and then choosing from among these the operators with the best ratio γ_1/γ_2. Easily-invertible or economical operators are usually considered to be those which can be inverted using a total number of operations that is proportional or nearly proportional to the number of unknowns. Examples of such operators are those which correspond to diagonal, tridiagonal, and triangular matrices, and also their products. A more complex example is the Laplace difference operator in a rectangle which, as was shown in Chapter 4, can be inverted using direct methods in a small number of arithmetic operations.

Notice that taking B to be a diagonal operator allows us to reduce the iteration count in comparison with an explicit scheme. However, asymptotically the order of dependence of the iteration count on the number of unknowns remains the same as for an explicit scheme. The use of triangular operators B is a more promising direction to follow.

In this chapter and in Chapter 10 we will examine universal two-level implicit iterative methods whose operators B correspond to triangular matrices (triangular methods) or products of triangular matrices (alternate-triangular methods).

Our study of these methods begins with the simplest — the ***Gauss-Seidel method.***

We look at the system of linear algebraic equations (1), which when expanded takes the form

$$\sum_{j=1}^{M} a_{ij}u_j = f_i, \quad i = 1, 2, \ldots, M.$$

In this case, we will work with equation (1) in a finite-dimensional space $H = H_M$.

The Gauss-Seidel iterative method, under the assumption that the diagonal elements of the matrix $A = (a_{ij})$ are non-zero ($a_{ii} \neq 0$), can be written in the following form:

$$\sum_{j=1}^{i} a_{ij}y_j^{(k+1)} + \sum_{j=i+1}^{M} a_{ij}y_j^{(k)} = f_i, \quad i = 1, 2, \ldots, M, \tag{2}$$

where $y_j^{(k)}$ is the j-th component of the k-th iterative approximation. The initial guess is taken to be some arbitrary vector.

The definition of the $(k+1)$-st iteration begins with $i = 1$:

$$a_{11}y_1^{(k+1)} = -\sum_{j=2}^{M} a_{1j}y_j^{(k)} + f_1.$$

Since $a_{11} \neq 0$, we can use this equation to find $y_1^{(k+1)}$. For $i = 2$ we obtain

$$a_{22}y_2^{(k+1)} = -a_{21}y_1^{(k+1)} - \sum_{j=3}^{M} a_{2j}y_j^{(k)} + f_2.$$

Suppose that $y_1^{(k+1)}, y_2^{(k+1)}, \ldots, y_{i-1}^{(k+1)}$ have been found. Then $y_i^{(k+1)}$ is found from the equation

$$a_{ii}y_i^{(k+1)} = -\sum_{j=1}^{i-1} a_{ij}y_j^{(k+1)} - \sum_{j=i+1}^{M} a_{ij}y_j^{(k)} + f_i. \tag{3}$$

From the formula (3) it is clear that the algorithm for the Gauss-Seidel method is extremely simple. The value of $y_i^{(k+1)}$ found using (3) is overwritten on $y_i^{(k)}$.

We now estimate the number of operations required to perform one iteration. If all the a_{ij} are non-zero, then the computations in (3) require $M-1$ additions, $M-1$ multiplications, and one division. Therefore, realizing one iterative step involves $2M^2-M$ arithmetic operations.

If each row of the matrix A only contains m non-zero elements, the situation for grid elliptic equations, then performing one iteration requires $2mM-M$ operations, i.e. the number of operations is proportional to the number of unknowns M.

We now write out the Gauss-Seidel iterative method (2) in matrix form. To do this, we represent A as the sum of a diagonal, a lower-triangular, and an upper-triangular matrix

$$A=\mathcal{D}+L+U, \tag{4}$$

where

$$L=\begin{Vmatrix} 0 & & & & 0 \\ a_{21} & 0 & & & \\ a_{31} & a_{32} & 0 & & \\ \vdots & & \ddots & \ddots & \\ a_{M1} & a_{M2} & \dots & a_{M,M-1} & 0 \end{Vmatrix}, \quad U=\begin{Vmatrix} 0 & a_{12} & a_{13} & \dots & a_{1M} \\ & 0 & a_{23} & \dots & a_{2M} \\ & & & \ddots & \vdots \\ & 0 & & 0 & a_{M-1,M} \\ & & & & 0 \end{Vmatrix},$$

$$\mathcal{D}=\begin{Vmatrix} a_{11} & & & \\ & a_{22} & & 0 \\ & 0 & \ddots & \\ & & & a_{MM} \end{Vmatrix}$$

We denote by $y_k=(y_1^{(k)},y_2^{(k)},\dots,y_M^{(k)})$ the k-th iterative approximation:

Using this notation, we write the Gauss-Seidel method in the form

$$(\mathcal{D}+L)y_{k+1}+Uy_k=f, \quad k=0,1,\dots.$$

We can reduce this iterative scheme to the canonical form for two-level schemes

$$(\mathcal{D}+L)(y_{k+1}-y_k)+Ay_k=f, \quad k=0,1,\dots, \quad y_0\in H. \tag{5}$$

Comparing (5) with the canonical form

$$B\frac{y_{k+1}-y_k}{\tau_{k+1}}+Ay_k=f, \quad k=0,1,\dots, \quad y_0\in H,$$

we find that $B = \mathcal{D} + L$, $\tau_k \equiv 1$. The scheme (5) is implicit, the operator B is a triangular matrix and consequently is not self-adjoint in H.

We have examined the so-called point or scalar Gauss-Seidel method assuming that the elements a_{ij} of the matrix A are scalars. A block or vector Gauss-Seidel method can be constructed analogously in the case where the a_{ii} are square matrices, in general of different dimensions, and the a_{ij} are rectangular matrices for $i \neq j$. In this case, y_i and f_i are vectors whose dimensions correspond to the dimensions of the matrices a_{ii}.

Assuming that the matrices a_{ii} are non-singular, the block Gauss-Seidel method can be written in the form (2) or in the canonical form (5).

9.1.2 Sample applications of the method. We look now at the application of the Gauss-Seidel method to find an approximate solution of a Dirichlet difference problem for Poisson's equation and for an elliptic equation with variable coefficients in a rectangle.

Suppose that, on the rectangular grid $\bar{\omega} = \{x_{ij} = (ih_1, jh_2) \in \bar{G}, 0 \le i \le N_1, 0 \le j \le N_2, h_\alpha = l_\alpha/N_\alpha, \alpha = 1,2\}$ in the rectangle $\bar{G} = \{0 \le x_\alpha \le l_\alpha, \alpha = 1,2\}$, it is necessary to find the solution to a Dirichlet difference problem for Poisson's equation

$$\Lambda y = \sum_{\alpha=1}^{2} y_{\bar{x}_\alpha x_\alpha} = -\varphi(x), \quad x \in \omega, \quad y(x) = g(x), \quad x \in \gamma, \tag{6}$$

where $\gamma = \{x_{ij} \in \Gamma\}$ is the boundary of the grid $\bar{\omega}$.

In this example the unknowns are $y(i,j) = y(x_{ij})$ at the interior nodes of the grid. If we order the unknowns using the natural ordering along the rows of the grid ω, starting with the bottom row, then the difference scheme (6) can be written in the form of the following system of algebraic equations:

$$\begin{aligned} -\frac{1}{h_1^2}y(i-1,j) - \frac{1}{h_2^2}y(i,j-1) + \left(\frac{2}{h_1^2} + \frac{2}{h_2^2}\right) y(i,j) \\ -\frac{1}{h_1^2}y(i+1,j) - \frac{1}{h_2^2}y(i,j+1) = \varphi(i,j) \end{aligned}$$

for $i = 1,2,\ldots,N_1-1$, $j = 1,2,\ldots,N_2-1$ and $y(x) = g(x)$ for $x \in \gamma$. Here the unknowns $y(i-1,j)$ and $y(i,j-1)$ precede $y(i,j)$, but $y(i+1,j)$ and $y(i,j+1)$ follow after $y(i,j)$. Since each equation involves no more than five unknowns, each row of the matrix A has no more than five non-zero elements.

For this system, the point Gauss-Seidel method will have the following form:

$$\left(\frac{2}{h_1^2}+\frac{2}{h_2^2}\right)y_{k+1}(i,j)=\frac{1}{h_1^2}y_{k+1}(i-1,j)+\frac{1}{h_2^2}y_{k+1}(i,j-1)$$
$$+\frac{1}{h_1^2}y_k(i+1,j)+\frac{1}{h_2^2}y_k(i,j+1)+\varphi(i,j),$$
$$1\le i\le N_1-1,\quad 1\le j\le N_2-1,$$

where $y_k(x)=g(x)$, $x\in\gamma$ for any $k\ge 0$.

The computations begin at the point $i=1$, $j=1$ and continue either along the rows or along the columns of the grid ω. Finding $y_{k+1}(i,j)$ requires 7 arithmetic operations, and $7M$ operations are required to perform one iteration, where $M=(N_1-1)(N_2-1)$ is the number of unknowns in the problem.

For this example the operator B (in the finite-dimensional space H of grid functions defined on ω with inner product $(u,v)=\Sigma_{x\in\omega}u(x)v(x)h_1h_2$, $u,v\in H$) is defined as follows:

$$By=(\mathcal{D}+L)y=\left(\frac{2}{h_1^2}+\frac{2}{h_2^2}\right)\mathring{y}(i,j)-\frac{1}{h_1^2}\mathring{y}(i-1,j)-\frac{1}{h_2^2}\mathring{y}(i,j-1)$$
$$=\left(\frac{1}{h_1^2}+\frac{1}{h_2^2}\right)\mathring{y}+\sum_{\alpha=1}^{2}\frac{1}{h_\alpha}\mathring{y}_{\bar{x}_\alpha},$$

where $y\in H$, $\mathring{y}\in\mathring{H}$, and $\mathring{y}(x)=y(x)$ for $y\in\omega$. Here $\mathring{H}$ is the set of grid functions defined on $\bar{\omega}$ which vanish on γ.

We look now at the *block Gauss-Seidel method.* If we denote by $Y_j=(y(1,j),y(2,j)\dots,y(N_1-1,j))^T$ the vector which consists of the unknowns in row j of the grid, then, as was shown in Section 1.1, the difference problem (6) can be written in the form of a three-level system of vector equations:

$$\begin{aligned}&-Y_{j-1}+CY_j-Y_{j+1}=F_j,\quad j=1,2,\dots,N_2-1,\\&Y_0=F_0,\quad Y_{N_2}=F_{N_2},\end{aligned}\tag{7}$$

where C is a square tridiagonal matrix of dimension $(N_1-1)\times(N_1-1)$, defined as follows:

$$(CY_j)_i=(2y-h_2^2y_{\bar{x}_1x_1})_{ij},\quad y_{0j}=y_{N_1j}=0.$$

The right-hand sides F_j are defined by the formulas

$$F_j = \left(h_2^2\varphi(1,j) + \frac{h_2^2}{h_1^2} g(0,j), h_2^2\varphi(2,j), \dots, h_2^2\varphi(N_1-2,j), \right.$$
$$\left. h_2^2\varphi(N_1-1,j) + \frac{h_2^2}{h_1^2} g(N_1,j) \right)^T \quad \text{for} \quad j = 1,2,\dots,N_2-1,$$
$$F_j = (g(1,j), g(2,j), \dots, g(N_1-1,j))^T \quad \text{for} \quad j = 0, N_2.$$

The block Gauss-Seidel method for the system (7) has the form

$$\begin{aligned} CY_j^{(k+1)} &= Y_{j-1}^{(k+1)} + Y_{j+1}^{(k)} + F_j, \quad j = 1,2,\dots,N_2-1, \\ Y_0^{(k+1)} &= F_0, \quad Y_{N_2}^{(k)} = F_{N_2}, \quad k = 0,1,\dots, \end{aligned} \tag{8}$$

and finding $Y_j^{(k+1)}$ requires the inversion of the tridiagonal matrix C.

If we write out the scheme (8) at a point of the grid, then we obtain the following formulas:

$$\begin{aligned} &-\frac{1}{h_1^2} y_{k+1}(i-1,j) + \left(\frac{2}{h_1^2} + \frac{2}{h_2^2}\right) y_{k+1}(i,j) - \frac{1}{h_1^2}(i+1,j) \\ &\quad = \frac{1}{h_2^2} y_{k+1}(i,j-1) + \frac{1}{h_2^2} y_k(i,j+1)\varphi(i,j), \\ &1 \le i \le N_1 - 1, \quad 1 \le j \le N_2 - 1, \end{aligned} \tag{9}$$

where $y_k(x) = g(x)$, $x \in \gamma$ for any $k \ge 0$. In order to find y_{k+1} on the j-th row it is necessary to solve the three-point boundary-value problem (9) with a known right-hand side using, for example, the elimination method, and then to overwrite the resulting solution on the j-th row of y_k.

The block Gauss-Seidel method corresponds to the following operator B:

$$By = \frac{1}{h_2^2}\mathring{y} + \frac{1}{h_2}\mathring{y}_{\bar{x}_2} + \mathring{y}_{\bar{x}_1 x_1}, \quad y \in H, \ \mathring{y} \in \mathring{H}.$$

Suppose now that, on the grid $\bar{\omega}$, it is necessary to find the solution of a Dirichlet difference problem for an elliptic equation with variable coefficients

$$\begin{aligned} &\Lambda y = \sum_{\alpha=1}^{2} (a_\alpha(x) y_{\bar{x}_\alpha})_{x_\alpha} - d(x)y = -\varphi(x), \quad x \in \omega, \\ &y(x) = g(x), \quad x \in \gamma, \\ &0 < c_1 \le a_\alpha(x) \le c_2, \quad x \in \bar{\omega}, \ \alpha = 1,2, \ 0 \le d_1 \le d(x) \le d_2, \ x \in \omega. \end{aligned} \tag{10}$$

For this problem, the point Gauss-Seidel method with the unknowns ordered by rows of the grid has the following form:

$$\left(\frac{a_1(i+1,j)+a_1(i,j)}{h_1^2}+\frac{a_2(i,j+1)+a_2(i,j)}{h_2^2}+d(i,j)\right)y_{k+1}(i,j)$$
$$=\frac{a_1(i,j)}{h_1^2}y_{k+1}(i-1,j)+\frac{a_2(i,j)}{h_2^2}y_{k+1}(i,j-1)$$
$$+\frac{a_1(i+1,j)}{h_1^2}y_k(i+1,j)+\frac{a_2(i,j+1)}{h_2^2}y_k(i,j+1)+\varphi(i,j)$$

for $i=1,2,\dots,N_1-1$ and $j=1,2,\dots,N_2-1$, where $y_k(x)=g(x)$ when $x\in\gamma$ for any $k\geq 0$.

The operator B in the canonical form of the iterative scheme for this example is defined as follows:

$$By(x_{ij})=\left(\frac{a_1(i+1,j)}{h_1^2}+\frac{a_2(i,j+1)}{h_2^2}+d(i,j)\right)\mathring{y}(i,j)$$
$$+\frac{a_1(i,j)}{h_1}\mathring{y}_{\bar{x}_1}+\frac{a_2(i,j)}{h_2}\mathring{y}_{\bar{x}_2},\quad y\in H,\ \mathring{y}\in\mathring{H},$$

where the space H and the set $\mathring{H}$ are as defined above.

9.1.3 Sufficient conditions for convergence. We formulate now certain sufficient conditions for the convergence of the Gauss-Seidel method. We require

Theorem 1. *Suppose that the operator A in equation* (1) *is self-adjoint and positive-definite in H. Then the two-level iterative process*

$$B\frac{y_{k+1}-y_k}{\tau}+Ay_k=f,\quad k=0,1,\dots,\quad y_0\in H,\ \tau>0,\tag{11}$$

converges in H_A if the operator $B-0.5\tau A$ is positive-definite in H, i.e. if the condition

$$B>\frac{\tau}{2}A\tag{12}$$

is satisfied.

Proof. From (11) we obtain the following problem for the error $z_k=y_k-u$:

$$B\frac{z_{k+1}-z_k}{\tau}+Az_k=0,\quad k=0,1,\dots,\quad z_0=y_0-u.\tag{13}$$

We will establish the basic energy identity for z_k. We represent z_k in (13) in the form

$$z_k = \frac{1}{2}(z_{k+1} + z_k) - \frac{\tau}{2}\left(\frac{z_{k+1} - z_k}{\tau}\right)$$

and obtain

$$\left(B - \frac{\tau}{2}A\right)\frac{z_{k+1} - z_k}{2} + \frac{1}{2}A(z_{k+1} + z_k) = 0.$$

We take the inner product of the left- and right-hand sides of this equation with $2(z_{k+1} - z_k)$ and take into account that, for a self-adjoint operator A, we have $(A(z_{k+1} + z_k), z_{k+1} - z_k) = (Az_{k+1}, z_{k+1}) - (Az_k, z_k)$. As a result we obtain the basic energy identity

$$2\tau\left(\left(B - \frac{\tau}{2}A\right)\frac{z_{k+1} - z_k}{\tau}, \frac{z_{k+1} - z_k}{\tau}\right) + \| z_{k+1} \|_A^2 - \| z_k \|_A^2 = 0.$$

From this and from the inequality $B - 0.5\tau A > 0$, $\tau > 0$ it follows that $\| z_{k+1} \|_A^2 \leq \| z_k \|_A^2$, i.e. the sequence $\{\| z_k \|_A^2\}$ is non-increasing, is bounded below by zero, and is convergent. Then from the energy identity it follows that

$$\lim_{k\to\infty}\left(\left(B - \frac{\tau}{2}A\right)\frac{z_{k+1} - z_k}{\tau}, \frac{z_{k-1} - z_k}{\tau}\right) = 0. \tag{14}$$

Further, from the inequality $B - 0.5\tau A > 0$ it follows that $\| z_{k+1} - z_k \| \to 0$ as $k \to \infty$. Using (13) to make the change $A^{1/2}z_k = -A^{-1/2}B(z_{k+1} - z_k)/\tau$, we obtain that $\| z_k \|_A^2 \leq \| A^{-1} \| \| B \|^2 \| z_{k+1} - z_k \|^2 / \tau^2 \to 0$ as $k \to \infty$. □

We now formulate a sufficient condition for the convergence of the Gauss-Seidel method.

Theorem 2. *If the operator A is self-adjoint and positive-definite in H, then the Gauss-Seidel method* (4), (5) *converges in H_A.*

Proof. From (5) and from theorem 1 it follows that it is sufficient to verify the inequality $\mathcal{D} + L - 0.5A > 0$. Since $A = A^*$, we then have that $U = L^*$ in (4), and

$$((\mathcal{D} + L - 0.5A)x, x) = 0.5((\mathcal{D} + L - U), x, x) = 0.5(\mathcal{D}x, x).$$

Since A is a positive-definite operator, then for the point Gauss-Seidel method we have $a_{ii} > 0$, $1 \leq i \leq M$, and for the block Gauss-Seidel method, the matrices a_{ii} satisfy $a_{ii} = a_{ii}^* > 0$. Consequently, $\mathcal{D} = \mathcal{D}^* > 0$. Thus, $\mathcal{D} + L - 0.5A > 0$. □

We now give without proof another condition which implies the convergence of the Gauss-Seidel method.

Theorem 3. *If the operator A is self-adjoint and non-singular, and all $a_{ii} > 0$, then the Gauss-Seidel method converges for any initial guess if and only if A is positive-definite.* □

In order to estimate the convergence rate for the Gauss-Seidel method, we use a different type of assumption.

For example, if the condition

$$\sum_{j\neq i} |a_{ij}| \leq q|a_{ii}|, \quad i = 1, 2, \ldots, M, \; q < 1, \tag{15}$$

is satisfied, then the Gauss-Seidel method converges at a geometric rate with multiplier q, and for the error z_n we have the estimate $\| z_n \| \leq q^n \| z_0 \|$, where $\| z_n \| = \max\limits_{1\leq i\leq M} |y_i^{(n)} - u_i|$.

In fact, from (3) we obtain the homogeneous equation

$$a_{ii} z_i^{(k+1)} = -\sum_{j=1}^{i-1} a_{ij} z_j^{(k+1)} - \sum_{j=i+1}^{M} a_{ij} z_j^{(k)}.$$

for the error $z_i^{(k)} = y_i^{(k)} - u_i$. From this we find

$$\begin{aligned} \left| z_i^{(k+1)} \right| &\leq \sum_{j=1}^{i-1} \left| \frac{a_{ij}}{a_{ii}} \right| \left| z_j^{(k+1)} \right| + \sum_{j=i+1}^{M} \left| \frac{a_{ij}}{a_{ii}} \right| \left| z_j^{(k)} \right| \\ &\leq \sum_{j=1}^{i-1} \left| \frac{a_{ij}}{a_{ii}} \right| \| z_{k+1} \| + \sum_{j=i+1}^{M} \left| \frac{a_{ij}}{a_{ii}} \right| \| z_k \| . \end{aligned} \tag{16}$$

From (15) we obtain

$$\sum_{j=i+1}^{M} \left| \frac{a_{ij}}{a_{ii}} \right| \leq q - \sum_{j=1}^{i-1} \left| \frac{a_{ij}}{a_{ii}} \right| \leq q \left(1 - \sum_{j=1}^{i-1} \left| \frac{a_{ij}}{a_{ii}} \right| \right).$$

Substituting this estimate in (16), we obtain the following inequality:

$$\left| z_i^{(k+1)} \right| \leq \sum_{j=1}^{i-1} \left| \frac{a_{ij}}{a_{ii}} \right| \| z_{k+1} \| + q \left(1 - \sum_{j=1}^{i-1} \left| \frac{a_{ij}}{a_{ii}} \right| \right) \| z_k \| . \tag{17}$$

Suppose that $\max_i |z_i^{(k+1)}|$ is achieved for some $i = i_0$, so that $\| z_{k+1} \| = |z_{i_0}^{(k+1)}|$. From (17) for $i = i_0$ we obtain

$$\left(1 - \sum_{j=1}^{i_0-1} \left|\frac{a_{i_0 j}}{a_{i_0 i_0}}\right|\right) \| z_{k+1} \| \le q \left(1 - \sum_{j=1}^{i_0-1} \left|\frac{a_{i_0 j}}{a_{i_0 i_0}}\right|\right) \| z_k \|;$$

from this follows the estimate $\| z_{k+1} \| \le q \| z_k \| \le \ldots \le q^{k+1} \| z_0 \|$. The assertion is proved.

The condition (15) indicates that A is a matrix with diagonal dominance. For the sample applications of the Gauss-Seidel method in Section 9.1.2, condition (15) is not satisfied ($q = 1$). In these examples, the operator A is self-adjoint and positive-definite in H. Therefore, by theorem 2, it is only possible to assert that the method converges in H_A. An estimate for the rate of convergence in H_A will be given below after we examine the general scheme for the triangular iterative methods.

9.2 The successive over-relaxation method

9.2.1 The iterative scheme. Sufficient conditions for convergence. In order to accelerate the convergence of the Gauss-Seidel method, we shall modify it, introducing an iterative parameter ω into the iterative scheme so that

$$(\mathcal{D} + \omega L)\frac{y_{k+1} - y_k}{\omega} + A y_k = f, \quad k = 0, 1, \ldots, \quad y_0 \in H, \tag{1}$$

where, as before, the matrix A is represented in the form of a sum

$$A = \mathcal{D} + L + U. \tag{2}$$

The Gauss-Seidel method corresponds to the value $\omega = 1$.

Comparing (1) with the canonical form of a two-level iterative scheme, we find that

$$B = \mathcal{D} + \omega L, \qquad \tau_k \equiv \omega.$$

In both the Gauss-Seidel method and this method, the operator B corresponds to a lower-triangular matrix, so that introducing the parameter ω does not take us out of the class of triangular iterative methods. All that is new is the question of how the choose the parameter ω.

If we write out the iterative scheme (1) in terms of the components of the vector y_{k+1}, then we obtain the following formulas:

$$a_{ii}y_i^{(k+1)} = (1-\omega)a_{ii}y_i^{(k)} - \omega\sum_{j=1}^{i-1} a_{ij}y_j^{(k+1)} - \omega\sum_{j=i+1}^{M} a_{ij}y_j^{(k)} + \omega f_i \qquad (3)$$

for $i = 1, 2, \ldots, M$ (once it has been found, $y_i^{(k+1)}$ is overwritten on $y_i^{(k)}$); performing one iteration requires approximately the same number of operations as in the Gauss-Seidel method.

For $\omega > 1$, the iterative method (1) is called the ***successive over-relaxation method***, for $\omega = 1$ it is called ***full relaxation***, and for $\omega < 1$ it is called ***under-relaxation***.

In Section 9.1 it was proved that the Gauss-Seidel method converges in H_A in the case where A is a self-adjoint and positive-definite operator. In addition to these requirements, an additional condition on the iterative parameter ω is needed to prove the convergence of a relaxation method. We now formulate additional conditions.

Theorem 4. *If the operator A is self-adjoint and positive-definite in H, and the parameter ω satisfies the condition $0 < \omega < 2$, then the relaxation method (1) converges in H_A.*

Proof. From theorem 1, it follows that it is sufficient to prove that the inequality $\mathcal{D} + \omega L > 0.5\omega A$ is satisfied for $\omega > 0$. Since $A = A^* > 0$, the operator $\mathcal{D}$ is self-adjoint and positive-definite in H, and $U = L^*$. Therefore, using equation (14) of Section 9.1, we obtain

$$\begin{aligned}((\mathcal{D} + \omega L)x, x) &= (1 - 0.5\omega)(\mathcal{D}x, x) + 0.5\omega\,((\mathcal{D} + 2L)x, x)\\ &= (1 - 0.5\omega)(\mathcal{D}x, x) + 0.5\omega(Ax, x).\end{aligned}$$

The assertion of the theorem then follows if $\omega < 2$. □

Remark. Theorem 4 is valid both for the point relaxation method where the a_{ij} in (3) are scalars, and for the block or vector relaxation method where the a_{ij} in (3) are matrices of the appropriate dimensions.

9.2.2 The choice of the iterative parameter. Theorem 4 gives sufficient conditions for the convergence of the relaxation method, leaving open the question of the optimal choice of the parameter ω. A peculiarity of the iterative process (1) is that the iterative parameter ω enters into the operator $B = \mathcal{D} + \omega L$

which is a non-self-adjoint operator in H. We have already dealt with the non-self-adjoint case in Section 6.4, where the iterative parameter for the simple iteration method was chosen under a variety of conditions, for example from the condition that the norm of the transformation operator from iteration to iteration be a minimum. Here it is necessary to take into account the above mentioned peculiarity of this iterative scheme. Choosing the parameter ω by minimizing the norm of the transformation operator from iteration to iteration in H_A will be examined in Section 9.3, where we will look at a general scheme for three-level iterative methods. In this subsection, we will choose the parameter ω for the relaxation method from the condition that the spectral radius of this transformation operator should be a minimum.

Recall the definition of the spectral radius of an operator

$$\rho(S) = \lim_{n\to\infty} \sqrt[n]{\| S^n \|} = \max_k |\lambda_k|, \tag{4}$$

where the λ_k are the eigenvalues of the operator S. The spectral radius possesses the following properties:

$$\rho(S^n) = \rho^n(S), \qquad \rho(S) \le \| S \| \tag{5}$$

and $\rho(S) = \| S \|$ if S is a self-adjoint operator in H. For an arbitrary operator S, from (5) we obtain $\rho^n(S) = \rho(S^n) \le \| S^n \|$. On the other hand, for sufficiently large n we will have from (4) that $\rho^n(S) \approx \| S^n \|$.

We move on now to state the problem of choosing the optimal parameter ω for the iterative scheme (1). We first obtain the problem for the error $z_k = y_k - u$. From (1) we find

$$(\mathcal{D} + \omega L)\frac{z_{k+1} - z_k}{\omega} + Az_k = 0, \quad k = 0, 1, \ldots, \quad z_0 = y_0 - u$$

or

$$z_{k+1} = Sz_k, \quad k = 0, 1, \ldots, \quad S = E - \omega(\mathcal{D} + \omega L)^{-1}A. \tag{6}$$

Using (6), we express z_n in terms of z_0:

$$z_n = S^n z_0, \quad \| z_n \| \le \| S^n \| \| z_0 \| . \tag{7}$$

The operator S is a non-self-adjoint operator in H which depends on the parameter ω. The problem of the optimal choice for the parameter ω is formulated as follows: find ω from the condition that the spectral radius of the operator S be a minimum.

We should note that we are not minimizing the norm of the solution operator S^n, as is implied by (7), but we are minimizing the spectral radius $\rho(S)$ of the transformation operator S, for which we have the estimate $\rho^n(S) \leq \| S^n \|$. However, because of the approximation $\rho^n(S) \approx \| S^n \|$, it can be expected that for sufficiently large n this method of choosing ω should be successful.

The solution of the problem formulated above is a complex issue, however with certain additional assumptions about the operator A, this problem can be successfully solved.

Assumption 1. The operator A is self-adjoint and positive-definite in H ($U = L^*$, and $\mathcal{D} = \mathcal{D}^* > 0$).

Assumption 2. The operator A is such that for any complex $z \neq 0$, the eigenvalues μ of the generalized eigenvalue problem

$$\left(zL + \frac{1}{z} U \right) x - \mu \mathcal{D} x = 0$$

do not depend on z.

Using these assumptions, we can prove the following assertion, which we will require later.

Lemma 1. *If the operator A satisfies assertions 1 and 2, then the eigenvalues of the problem*

$$Ax - \lambda \mathcal{D} x = 0 \tag{8}$$

are real, positive, and, if λ is an eigenvalue, then $2 - \lambda$ is also an eigenvalue.

Proof. In fact, it follows from the self-adjointness and positive-definiteness of the operator A that the eigenvalues λ are real and positive. Further, suppose that λ is an eigenvalue for the problem (8), i.e.

$$Ax - \lambda \mathcal{D} x = (L + U)x - (\lambda - 1)\mathcal{D} x = 0, \quad x \neq 0.$$

By assumption 2 we will have that

$$(-L - U)y - (\lambda - 1)\mathcal{D} y = 0 \quad \text{or} \quad Ay - (2 - \lambda)\mathcal{D} y = 0.$$

From this, the assertion of the lemma follows. □

We move on now to solve the problem of optimally choosing the parameter ω. To do this, it is necessary to estimate the spectral radius of the transformation operator $S = E - \omega(\mathcal{D} + \omega L)^{-1}A$, i.e. to estimate the eigenvalues μ of the operator S:

$$Sx - \mu x = 0. \tag{9}$$

We will assume that assumptions 1 and 2 are satisfied. The following lemma establishes the correspondence between the eigenvalues μ of the problem (9) and the eigenvalues λ of the problem (8).

Lemma 2. *For $\omega \neq 1$ the eigenvalues* (8) *and* (9) *are related by the equation*

$$(\mu + \omega - 1)^2 = \omega^2 \mu (1 - \lambda)^2. \tag{10}$$

Proof. Suppose that μ and λ are eigenvalues for the problems (9) and (8). From the definition of the operator S and the decomposition $A = \mathcal{D} + L + U$ it follows that (9) can be written in the form

$$\frac{1 - \mu - \omega}{\omega}\mathcal{D}x - (\mu L + U)x = 0, \quad x \neq 0. \tag{11}$$

We shall first prove that, if $\omega \neq 1$, then all the μ are non-zero. In fact, suppose that $\mu = 0$. Then (11) takes the form

$$\frac{1 - \omega}{\omega}\mathcal{D}x - Ux = 0.$$

Since U is an upper-triangular matrix, and $\mathcal{D}$ is a diagonal (block-diagonal) positive-definite matrix (positive-definite because of assumption 1), this equation is only valid for $x \neq 0$ when $\omega = 1$. Consequently, having assumed that $\mu = 0$ for $\omega \neq 1$, we have arrived at a contradiction.

Dividing the left- and right-hand sides of (11) by $\sqrt{\mu}$, we obtain

$$\frac{1 - \mu - \omega}{\omega\sqrt{\mu}}\mathcal{D}y - \left(\sqrt{\mu}L + \frac{1}{\sqrt{\mu}}U\right)x = 0.$$

By assumption 2, from this we find

$$\frac{1 - \mu - \omega}{\omega\sqrt{\mu}}\mathcal{D}y - (L + U)y = 0$$

or

$$Ay - \left(1 + \frac{1 - \mu - \omega}{\omega\sqrt{\mu}}\right)\mathcal{D}y = 0.$$

Comparing this equation with (8), we obtain the relation

$$\frac{\mu - \omega - 1}{\omega\sqrt{\mu}} = 1 - \lambda.$$

This concludes the proof of lemma 2. □

Remark. The self-adjointness of the operator A was not used in the proof of lemma 2. Relation (10) is also valid in the case where A is any non-self-adjoint operator as long as the operator $\mathcal{D}$ is non-singular.

From lemma 1 it follows that the eigenvalues λ are distributed along the real axis, are symmetric with respect to the point $\lambda = 1$, and $\lambda \in [\lambda_{\min}, 2 - \lambda_{\min}]$, $\lambda > 0$. Therefore from lemma 2 we obtain that to each $\lambda_i = 1$ there corresponds a $\mu_i = 1 - \omega$, and to each pair λ_i and $2 - \lambda_i$ corresponds a pair of non-zero μ_i, obtained by solving equation (10) with $\lambda = \lambda_i$. Consequently, all the μ_i can be found as the roots of the quadratic equation (10), where λ is taken to be each of the λ_i lying on the interval $[\lambda_{\min}, 1]$.

9.2.3 An estimate of the spectral radius. We now find the optimal value of the parameter ω and estimate the spectral radius of the operator S. To do this, we investigate equation (10):

$$\mu^2 + \left[2(\omega - 1) - \omega^2(1 - \lambda)^2\right] \mu + (\omega - 1)^2 = 0, \tag{12}$$

where $\lambda_{\min} \le \lambda \le 1$ and $0 < \omega < 2$.

Solving equation (12), we find the two roots

$$\begin{aligned} \mu_1(\lambda, \omega) &= \left(\frac{\omega(1 - \lambda) + \sqrt{\omega^2(1 - \lambda)^2 - 4(\omega - 1)}}{2} \right)^2, \\ \mu_2(\lambda, \omega) &= \left(\frac{\omega(1 - \lambda) - \sqrt{\omega^2(1 - \lambda)^2 - 4(\omega - 1)}}{2} \right)^2. \end{aligned} \tag{13}$$

An examination of the discriminant of equation (12) shows that, if $\omega > \omega_0 > 1$ where

$$\omega_0 = \frac{2}{1 + \sqrt{\lambda_{\min}(2 - \lambda_{\min})}} \in (1, 2), \tag{14}$$

then for any $\lambda \in [\lambda_{\min}, 1]$ the roots μ_1 and μ_2 are complex and $|\mu_1| = |\mu_2| = \omega - 1$. Therefore, if $\omega > \omega_0$, then the spectral radius of the operator S is equal to $\rho(S) = \omega - 1$ and increases with ω. If $\omega = \omega_0$, then

$$\mu_1(\lambda_{\min}, \omega_0) = \mu_2(\lambda_{\min}, \omega_0) = \omega_0 - 1,$$

and for $\lambda_{\min} < \lambda \leq 1$ the roots μ_1 and μ_2 again will be complex and $|\mu_1| = |\mu_2| = \omega_0 - 1$. Consequently, in the region $\omega \geq \omega_0$, the optimal value is $\omega = \omega_0$ which corresponds to the value $\rho(S) = \omega_0 - 1$.

Suppose now that $1 < \omega < \omega_0$. We investigate the behavior of the roots μ_1 and μ_2, defined in formula (13), as functions of the variable λ for a fixed ω.

If λ lies in the interval $[\lambda_{\min}, \lambda_0]$,

$$\lambda_{\min} \leq \lambda \leq \lambda_0 = 1 - 2\frac{\sqrt{\omega - 1}}{\omega} < 1,$$

then the discriminant $\omega^2(1-\lambda)^2 - 4(\omega-1)^2$ is non-negative and, consequently, the roots μ_1 and μ_2 are real, and μ_1 is the maximal root.

We shall show that $\mu_1(\lambda, \omega)$ is a decreasing function of λ on the interval $[\lambda_{\min}, \lambda_0]$. Differentiating (12) with respect to λ and using (13), we obtain

$$\frac{\partial \mu_1}{\partial \lambda} = -\frac{2\omega\mu_1}{\sqrt{\omega^2(1-\lambda)^2 - 4(\omega - 1)}} < 0.$$

Consequently, for $1 < \omega < \omega_0$, the root $\mu_1(\lambda, \omega)$ decreases as we vary λ from $\lambda_{\min}$ to λ_0, taking on the following values:

$$\mu_1(\lambda_{\min}, \omega) = \left(\frac{\omega(1 - \lambda_{\min}) + \sqrt{\omega^2(1 - \lambda_{\min})^2 - 4(\omega - 1)}}{2}\right)^2,$$

$$\mu_1(\lambda_0, \omega) = \omega - 1.$$

Further, if λ changes between λ_0 and 1, then the roots μ_1 and μ_2 are complex and equal in modulus: $|\mu_1| = |\mu_2| = \omega - 1$. Consequently, if $1 < \omega < \omega_0$, then

$$\rho(S) = \mu_1(\lambda_{\min}, \omega) = \left(\frac{\omega(1 - \lambda_{\min}) + \sqrt{\omega^2(1 - \lambda_{\min})^2 - 4(\omega - 1)}}{2}\right)^2. \quad (15)$$

If $\omega < 1$, then all the roots of equation (12) are real, μ_1 is the maximal root, and its value decreases as we vary λ between $\lambda_{\min}$ and 1. Consequently, for $\omega < 1$ the spectral radius of the operator S is defined by formula (15). Since for $\omega = 1$ the non-null μ_k satisfy equation (12), we have that (15) is also valid for $\omega = 1$.

Thus, if $0 < \omega < \omega_0$, then the spectral radius of the operator S is defined by formula (15). We shall show that $\mu_1(\lambda_{\min}, \omega)$ decreases for ω in the interval $0 < \omega < \omega_0$.

Since if $\omega < \omega_0$ the root μ_1 decreases as λ varies for $\lambda \le \lambda_0$, and since $\mu_1(0, \omega) = 1$, we have that $\mu_1(\lambda_{\min}, \omega) < 1$.

Further, from (15) we obtain

$$\frac{\partial \mu_1(\lambda_{\min}, \omega)}{\partial \omega} = \sqrt{\mu_1}\left(1 - \lambda_{\min} + \frac{\omega(1-\lambda_{\min})^2 - 2}{\sqrt{\omega^2(1-\lambda_{\min})^2 - 4(\omega-1)}}\right)$$

$$= \frac{\sqrt{\mu_1}}{\omega} \frac{[\omega^2(1-\lambda_{\min})^2 - 2(\omega-1) + (1-\lambda_{\min})\omega\sqrt{\omega^2(1-\lambda_{\min})^2 - 4(\omega-1)} - 2]}{\sqrt{\omega^2(1-\lambda_{\min})^2 - 4(\omega-1)}}.$$

Substituting here (13), in conclusion we find

$$\frac{\partial \mu_1}{\partial \omega} = \frac{2\sqrt{\mu_1}(\mu_1 - 1)}{\omega\sqrt{\omega^2(1-\lambda_{\min})^2 - 4(\omega-1)}} < 0.$$

The assertion has been proved. Consequently, in the region $\omega \le \omega_0$ the optimal value is $\omega = \omega_0$, to which corresponds

$$\rho(S) = \omega_0 - 1 = \frac{1 - \sqrt{\lambda_{\min}(2-\lambda_{\min})}}{1 + \sqrt{\lambda_{\min}(2-\lambda_{\min})}} = \left(\frac{1-\sqrt{\eta}}{1+\sqrt{\eta}}\right)^2, \quad \eta = \frac{\lambda_{\min}}{2-\lambda_{\min}}.$$

Notice that it follows from the above derivation that if δ is a lower bound for $\lambda_{\min}$, i.e. $\delta \le \lambda_{\min}$, and if ω is chosen using formula (14) with $\lambda_{\min}$ replaced by δ, then $\omega_0 \le \omega$ and

$$\rho(S) \le \left(\frac{1-\sqrt{\eta}}{1+\sqrt{\eta}}\right)^2, \quad \eta = \frac{\delta}{2-\delta}.$$

Thus we have proved

Theorem 5. *Suppose that assumptions* 1 *and* 2 *are satisfied and that* δ *is the constant from the inequality*

$$\delta \mathcal{D} \le A, \qquad \delta > 0. \tag{16}$$

Then the spectral radius of the transformation operator S *in the iterative scheme* (1) *with the optimal value of the parameter* ω

$$\omega = \omega_0 = \frac{2}{1 + \sqrt{\delta(2-\delta)}}, \tag{17}$$

can be estimated by

$$\rho(S) \le \left(\frac{1-\sqrt{\eta}}{1+\sqrt{\eta}}\right)^2, \quad \eta = \frac{\delta}{2-\delta}, \tag{18}$$

where, if equality holds in (16), *then equality also holds in* (18). □

The iterative method (1), (17) is the *successive over-relaxation method* since $\omega_0 > 1$.

9.2.4 A Dirichlet difference problem for Poisson's equation in a rectangle. We look now at an application of the successive over-relaxation method for finding an approximate solution to a Dirichlet difference problem for Poisson's equation defined on the rectangular grid $\bar{\omega} = \{x_{ij} = (ih_1, jh_2) \in \bar{G},\ 0 \le i \le N_1,\ 0 \le j \le N_2,\ h_\alpha = l_\alpha/N_\alpha,\ \alpha = 1,2\}$ in the rectangle $\bar{G} = \{0 \le x_\alpha \le l_\alpha,\ \alpha = 1,2\}$:

$$\Lambda y = \sum_{\alpha=1}^{2} y_{\bar{x}_\alpha x_\alpha} = -\varphi(x), \quad x \in \omega, \quad y(x) = g(x), \quad x \in \gamma. \tag{19}$$

The operator A in the space H of grid functions defined on ω with inner product

$$(u, v) = \sum_{x \in \omega} u(x) v(x) h_1 h_2$$

is defined in the usual way: $Ay = -\Lambda \mathring{y}$, $y \in H$, $\mathring{y} \in \mathring{H}$. As we already know, the operator A for problem (19) is self-adjoint and positive-definite in H. Consequently, assumption 1 is satisfied.

We look first at the *point successive over-relaxation method.* If the unknowns are ordered by rows of the grid ω, then the difference scheme (19) can be written in the form of the following system of algebraic equations:

$$-\frac{1}{h_1^2} y(i-1,j) - \frac{1}{h_2^2} y(i,j-1) + \left(\frac{2}{h_1^2} + \frac{2}{h_2^2}\right) y(i,j)$$
$$-\frac{1}{h_1^2} y(i+1,j) - \frac{1}{h_2^2} y(i,j+1) = \varphi(i,j)$$

for $i = 1, 2, \ldots, N_1 - 1$, $j = 1, 2, \ldots, N_2 - 1$ and $y_k(x) = g(x)$, $x \in \gamma$.

This notation for the operator A corresponds to a representation of A in the form of a sum $A = \mathcal{D} + L + U$, where

$$\mathcal{D}y = \left(\frac{2}{h_1^2} + \frac{2}{h_2^2}\right) y,$$

$$Ly(i,j) = -\frac{1}{h_1^2}\mathring{y}(i-1,j) - \frac{1}{h_2^2}\mathring{y}(i,j-1),$$

$$Uy(i,j) = -\frac{1}{h_2^2}\mathring{y}(i+1,j) - \frac{1}{h_2^2}\mathring{y}(i,j+1).$$

For this system, the point successive over-relaxation method corresponding to (3) will have the following form:

$$\left(\frac{2}{h_1^2} + \frac{2}{h_2^2}\right) y_{k+1}(i,j) = (1-\omega)\left(\frac{2}{h_1^2} + \frac{2}{h_2^2}\right) y_k(i,j) + \omega\left[\frac{1}{h_1^2} y_{k+1}(i-1,j)\right.$$
$$\left. + \frac{1}{h_2^2} y_{k+1}(i,j-1) + \frac{1}{h_1^2} y_k(i+1,j) + \frac{1}{h_2^2} y_k(i,j+1) + \varphi(i,j)\right]$$

for $i = 1, 2, \ldots, N_1 - 1$, $j = 1, 2, \ldots, N_2 - 1$, where $y_k(x) = g(x)$ if $x \in \gamma$ for any $k \geq 0$.

The computations, as in the Gauss-Seidel method, begin at the point $i = 1$, $j = 1$ and continue either along the rows or along the columns of the grid ω. Once $y_{k+1}(i,j)$ is found, it is overwritten on $y_k(i,j)$.

We will prove now that for this example assumption 2 is satisfied. To do this, it is necessary to show that, for any complex $z \neq 0$, the eigenvalues μ of the problem

$$z\left(\frac{1}{h_1^2} y(i-1,j) + \frac{1}{h_2^2} y(i,j-1)\right) + \frac{1}{z}\left(\frac{1}{h_1^2} y(i+1,j) + \frac{1}{h_2^2} y(i,j+1)\right)$$
$$+ \mu\left(\frac{2}{h_1^2} + \frac{2}{h_2^2}\right) y(i,j) = 0, \quad 1 \leq i \leq N_1 - 1, \quad 1 \leq j \leq N_2 - 1,$$
$$y(x) = 0, \quad x \in \gamma$$

do not depend on z.

Setting here

$$y(i,j) = z^{i+j} v(i,j), \quad 0 \leq i \leq N_1, \quad 0 \leq j \leq N_2,$$

we obtain

$$\frac{1}{h_1^2}v(i-1,j)+\frac{1}{h_2^2}v(i,j-1)+\frac{1}{h_1^2}v(i+1,j)+\frac{1}{h_2^2}v(i,j+1)$$
$$+\mu\left(\frac{2}{h_1^2}+\frac{2}{h_2^2}\right)v(i,j)=0,$$
$$1\le i\le N_1-1,\quad 1\le j\le N_2-1,\qquad v(x)=0,\quad x\in\gamma.$$

Consequently, μ does not depend on z.

It remains to find the optimal value of the parameter ω. To do this, we must find an estimate from below the minimal eigenvalue of the problem (8), which in this case can be written in the form

$$y_{\bar{x}_1x_1}+y_{\bar{x}_2x_2}+\lambda\left(\frac{2}{h_1^2}+\frac{2}{h_2^2}\right)y=0,\quad x\in\omega,\quad y(x)=0,\quad x\in\gamma.$$

Since the eigenvalues of the Laplace difference operator $\Lambda y=y_{\bar{x}_1x_1}+y_{\bar{x}_2x_2}$ are known

$$\overset{\circ}{\lambda}_k=\frac{4}{h_1^2}\sin^2\frac{k_1\pi h_1}{2l_1}+\frac{4}{h_2^2}\sin^2\frac{k_2\pi h_2}{2l_2},\quad k_\alpha=1,2,\dots,N_\alpha-1,$$

then

$$\lambda_k=\overset{\circ}{\lambda}_k\Big/\left(\frac{2}{h_1^2}+\frac{2}{h_2^2}\right)=\frac{2h_2^2}{h_1^2+h_2^2}\sin^2\frac{k_1\pi h_1}{2l_1}+\frac{2h_1^2}{h_1^2+h_2^2}\sin^2\frac{k_2\pi h_2}{2l_2}.$$

Consequently,

$$\lambda_{\min}=\frac{2h_2^2}{h_1^2+h_2^2}\sin^2\frac{\pi h_1}{2l_1}+\frac{2h_1^2}{h_1^2+h_2^2}\sin^2\frac{\pi h_2}{2l_2},$$

and the parameter ω_0 is found from formula (14). In particular, when $\bar{G}$ is a square with side l ($l_1=l_2=l$) and the grid is square ($N_1=N_2=N$), we have

$$\lambda_{\min}=2\sin^2\frac{\pi}{2N},\quad \omega_0=\frac{2}{1+\sin\frac{\pi}{N}},\quad \eta=\tan^2\frac{\pi}{2N},$$
$$\rho(S)=\frac{1-\sin\frac{\pi}{N}}{1+\sin\frac{\pi}{N}}\approx 1-\frac{2\pi}{N}.$$

Notice that the spectral radius of the transformation operator for the corresponding point Gauss-Seidel method is estimated by formula (15), where we set $\omega=1$. This gives $\rho(S)=(1-\lambda_{\min})^2=\cos^2\frac{\pi}{N}$, which is significantly worse than for the successive over-relaxation method.

We look now at the block successive over-relaxation method. If we put together the unknowns $y(i,j)$ from the j-th row of the grid in one block, then the block notation for the operator A corresponds to the following representation $A = \mathcal{D} + L + U$, where

$$\mathcal{D}y = -\frac{1}{h_1^2}\mathring{y}(i-1,j) + \left(\frac{2}{h_1^2} + \frac{2}{h_2^2}\right)\mathring{y}(i,j) - \frac{1}{h_1^2}\mathring{y}(i+1,j),$$

$$Ly(i,j) = -\frac{1}{h_2^2}\mathring{y}(i,j-1), \quad Uy(i,j) = -\frac{1}{h_2^2}\mathring{y}(i,j+1).$$

The computational formulas for the block successive over-relaxation method have the form

$$\begin{aligned}
&-\frac{1}{h_1^2}y_{k+1}(i-1,j) + \left(\frac{2}{h_1^2} + \frac{2}{h_2^2}\right)y_{k+1}(i,j) - \frac{1}{h_2^2}y_{k+1}(i+1,j)\\
&= (1-\omega)\left(-\frac{1}{h_1^2}y_k(i-1,j) + \left(\frac{2}{h_1^2} + \frac{2}{h_2^2}\right)y_k(i,j) - \frac{1}{h_2^2}y_k(i+1,j)\right)\\
&\quad + \omega\left(\frac{1}{h_2^2}y_{k+1}(i,j-1) + \frac{1}{h_2^2}y_k(i,j+1) + \varphi(i,j)\right),\\
&\quad 1 \le i \le N_1 - 1, \quad 1 \le j \le N_2 - 1,
\end{aligned}$$

where $y_k(x) = g(x)$, $x \in \gamma$ for all $k \ge 0$. In order to find y_{k+1} on the j-th row it is necessary to solve a three-point boundary-value problem using, for example, the elimination method.

We shall show that for this example assumption 2 is satisfied, i.e. the eigenvalues μ for the problem

$$\begin{aligned}
&z\frac{1}{h_2^2}y(i,j-1) + \frac{1}{z}\frac{1}{h_2^2}y(i,j+1) + \mu\left(-\frac{1}{h_1^2}y(i-1,j)\right.\\
&\quad \left. + \left(\frac{2}{h_1^2} + \frac{2}{h_2^2}\right)y(i,j) - \frac{1}{h_2^2}y(i+1,j)\right) = 0,\\
&\quad 1 \le i \le N_1 - 1, \quad 1 \le j \le N_2 - 1, \quad y(x) = 0, \quad x \in \gamma
\end{aligned}$$

do not depend on z. This is easy to establish if we make the change of variables $y(i,j) = z^j v(i,j)$, $0 \le i \le N_1$, $0 \le j \le N_2$.

We find now the optimal value of the parameter ω. The corresponding problem (8) has the form

$$\begin{aligned}
&y_{\bar{x}_1x_1} + y_{\bar{x}_2x_2} + \lambda\left(\frac{2}{h_2^2}y - y_{\bar{x}_1x_1}\right) = 0, \quad x \in \omega\\
&y(x) = 0, \quad x \in \gamma.
\end{aligned} \tag{20}$$

It is not difficult to verify that the eigenfunctions of problem (20) are

$$y_k(x) = \sin\frac{k_1\pi x_1}{l_1}\sin\frac{k_2\pi x_2}{l_2}. \tag{21}$$

Substituting (21) in (20) we find

$$\lambda_k = \frac{\mathring{\lambda}_{k_1} + \mathring{\lambda}_{k_2}}{\frac{2}{h_2^2} + \mathring{\lambda}_{k_1}}, \quad k_\alpha = 1, 2, \ldots, N_\alpha - 1, \quad k = (k_1, k_2),$$

where

$$\mathring{\lambda}_{k_\alpha} = \frac{4}{h_\alpha^2}\sin^2\frac{k_\alpha\pi h_\alpha}{2l_\alpha}, \quad k_\alpha = 1, 2, \ldots, N_\alpha - 1, \quad \alpha = 1, 2.$$

From this we obtain

$$\lambda_{\min} = \frac{2h_2^2\sin^2\frac{\pi h_1}{2l_1} + 2h_1^2\sin^2\frac{\pi h_2}{2l_2}}{2h_2^2\sin^2\frac{\pi h_1}{2l_1} + h_1^2}.$$

For the special case considered above we have

$$\lambda_{\min} = \frac{4\sin^2\frac{\pi}{2N}}{1 + 2\sin^2\frac{\pi}{2N}}, \quad \omega_0 = \frac{2 + 4\sin^2\frac{\pi}{2N}}{(1 + \sqrt{2}\sin\frac{\pi}{2N})^2},$$

$$\eta = 2\sin^2\frac{\pi}{2N}, \quad \rho(S) = \left(\frac{1 - \sqrt{2}\sin\frac{\pi}{2N}}{1 + \sqrt{2}\sin\frac{\pi}{2N}}\right)^2 \approx 1 - 2\sqrt{2}\frac{\pi}{N}.$$

Comparing the estimates of the spectral radius for the block and point successive over-relaxation methods, we find that the block method will converge $\sqrt{2}$ times as fast as the point method. On the other hand, the block method requires more arithmetic operations to perform one iteration than the point method.

In conclusion we give the operation counts for the point successive over-relaxation method as a function of the number of nodes N in one direction for $\epsilon = 10^{-4}$. As a model problem we take the difference scheme (19) on a square grid with $N_1 = N_2 = N$ and $\varphi(x) \equiv 0$, $g(x) \equiv 0$. The initial approximation $y_0(x)$ is chosen as follows: $y_0(x) = 1$, $x \in \omega$, $y_0(x) = 0$, $x \in \gamma$.

The iterative process will be terminated if the condition

$$\| z_n \|_A \le \epsilon \| z_0 \|_A \tag{22}$$

is satisfied.

From the theory for the method it follows that the error z_n can be estimated by $\| z^n \|_A \leq \| S^n \|_A \| z_0 \|_A$, and since the spectral radius of the operator is less than or equal to any norm of the operator, $\rho^n(S) \leq \| S^n \|_A$. Therefore the condition $\rho^n(S) \leq \epsilon$ cannot be used to estimate the required iteration count.

We give now the iteration count n, defined using condition (22), and for comparison the iteration count n^*, which is obtained using the inequality $\rho^n(S) \leq \epsilon$:

$$N = 32 \quad n = 65 \quad n^* = 47$$
$$N = 64 \quad n = 128 \quad n^* = 94$$
$$N = 128 \quad n = 257 \quad n^* = 187$$

A comparison of the iteration counts for the successive over-relaxation method and the explicit Chebyshev method (examined for problem (19) in Section 6.5.1) shows that the successive over-relaxation method requires approximately 1.6 times fewer iterations than the explicit Chebyshev method. The operation counts for one iteration in these methods are practically identical.

9.2.5 A Dirichlet difference problem for an elliptic equation with variable coefficients. We look now at an application of the successive over-relaxation method for finding an approximate solution to a Dirichlet difference problem for an equation with variable coefficients in a rectangle

$$\begin{aligned} \Lambda y = \sum_{\alpha=1}^{2} (a_\alpha(x) y_{\bar{x}_\alpha})_{x_\alpha} - d(x) y = -\varphi(x), \quad x \in \omega, \\ y(x) = g(x), \quad x \in \gamma, \end{aligned} \tag{23}$$

assuming that the following conditions are satisfied:

$$\begin{aligned} 0 < c_1 \leq a_\alpha(x) \leq c_2, \quad x \in \bar{\omega}, \quad \alpha = 1, 2, \\ 0 \leq d_1 \leq d(x) \leq d_2, \quad x \in \omega. \end{aligned} \tag{24}$$

For problem (23), the point successive over-relaxation method with the unknowns ordered according to the rows of the grid ω is described by the formula

$$\begin{aligned} b(i,j) y_{k+1}(i,j) &= (1-\omega) b(i,j) y_k(i,j) \\ &+ \omega \left[\frac{a_1(i,j)}{h_1^2} y_{k+1}(i-1,j) + \frac{a_2(i,j)}{h_2^2} y_{k+1}(i,j-1) \right. \\ &+ \left. \frac{a_1(i+1,j)}{h_1^2} y_k(i+1,j) + \frac{a_2(i,j+1)}{h_2^2} y_k(i,j+1) + \varphi(i,j) \right], \\ &1 \leq i \leq N_1 - 1, \quad 1 \leq j \leq N_2 - 1, \end{aligned} \tag{25}$$

where

$$b(i,j) = \frac{a_1(i,j) + a_1(i+1,j)}{h_1^2} + \frac{a_2(i,j) + a_2(i,j+1)}{h_2^2} + d(i,j)$$

and $y_k(x) = g(x)$, $x \in \gamma$ for any $k \geq 0$.

For this example, the operators $\mathcal{D}$, L and U are defined as follows:

$$\mathcal{D}y = by$$

$$Ly(i,j) = -\frac{a_1(i,j)}{h_1^2}\mathring{y}(i-1,j) - \frac{a_2(i,j)}{h_2^2}\mathring{y}(i,j-1),$$

$$Uy(i,j) = -\frac{a_1(i+1,j)}{h_1^2}\mathring{y}(i+1,j) - \frac{a_2(i,j+1)}{h_2^2}\mathring{y}(i,j+1).$$

Assumptions 1 and 2 are satisfied, as was proved for the example from Section 9.2.4.

In order to find the parameter ω, it is necessary to estimate the constant δ in the inequality $A \geq \delta\mathcal{D}$. This problem was solved earlier in Section 6.5.3, where the simplest implicit Chebyshev method was considered for the difference problem (23). We give here the estimate for δ:

$$\delta = \min_{0<x_2<l_2} \frac{1}{\kappa_1(x_2)} + \min_{0<x_1<l_1} \frac{1}{\kappa_2(x_1)},$$

where $\kappa_\alpha(x_\beta) = \max_{0<x_\alpha<l_\alpha} v^\alpha(x)$, $\beta = 3-\alpha$, $\alpha = 1,2$, and $v^\alpha(x)$ is the solution of the following three-point boundary-value problem:

$$\left(a_\alpha v^\alpha_{\bar{x}_\alpha}\right)_{x_\alpha} - \frac{1}{2}dv^\alpha = -b(x), \quad h_\alpha \leq x_\alpha \leq l_\alpha - h_\alpha,$$

$$v^\alpha(x) = 0, \quad x_\alpha = 0, l_\alpha,$$

$$h_\beta \leq x_\beta \leq l_\beta - x_\beta, \quad \beta = 3-\alpha, \quad \alpha = 1,2.$$

The iterative parameter ω is found using formula (17):

$$\omega = \omega_0 = \frac{2}{1+\sqrt{\delta(2-\delta)}}.$$

In order to compare the successive over-relaxation method with the simplest implicit Chebyshev method examined in Section 6.5.3, we give the iteration counts for the successive over-relaxation method applied to the following model problem. Assume that the difference scheme (23) is given on a square grid with $N_1 = N_2 = N$ and $\varphi(x) = 0$, $g(x) = 0$. The coefficients $a_1(x)$, $a_2(x)$, and $d(x)$ are chosen as follows:

$$a_1(x) = 1 + c\left[(x_1 - 0.5)^2 + (x_2 - 0.5)^2\right]$$

$$a_2(x) = 1 + c\left[0.5 - (x_1 - 0.5)^2 - (x_2 - 0.5)^2\right]$$

$$d(x) \equiv 0, \quad c > 0.$$

Here in the inequalities (24) we have $c_1 = 1$, $c_2 = 1 + 0.5c$, $d_1 = d_2 = 0$. The initial guess for the successive over-relaxation method (25) is chosen as follows: $y_0(x) = 1$, $x \in \omega$, $y_0(x) = 0$, $x \in \gamma$, and the iterative process is terminated when condition (22) is satisfied.

In table 9 we give the iteration counts for the relaxation method as a function of the ratio c_2/c_1 and of the number of nodes N in one direction for $\epsilon = 10^{-4}$. For the case when $a_\alpha(x) \equiv 1$ and $d(x) \equiv 0$, the iteration counts for the successive over-relaxation method were given in Section 9.2.4.

Table 9

c_1/c_2	2	8	32	128	512
$N = 32$	65	81	95	96	98
$N = 64$	129	164	192	193	195

From table 9 it follows that the iteration counts for the successive over-relaxation method applied to the model probelm are approximately half as large as the iteration counts for the simplest implicit Chebyshev method. Since the number of arithmetic operations required to perform one iteration for these methods are identical, the successive over-relaxation method is twice as effective as the simplest implicit Chebyshev method.

9.3 Triangular methods

9.3.1 The iterative scheme. In Sections 9.1, 9.2, two methods were studied — the Gauss-Seidel method and the relaxation method. These methods belong to a class of implicit two-level methods whose operator B corresponds to a triangular or block-triangular matrix. In canonical form, the iterative scheme of the methods has the following form:

$$(\mathcal{D}+\omega L)\frac{y_{k+1}-y_k}{\omega}+Ay_k=f, \quad k=0,1,\ldots, \quad y_0\in H, \tag{1}$$

where $\mathcal{D}$ and L are the operators in the decomposition of A into the sum of diagonal, lower- and upper-triangular matrices

$$A=\mathcal{D}+L+U. \tag{2}$$

The Gauss-Seidel method corresponds to the value of the parameter $\omega=1$.

For the case of a self-adjoint and positive-definite operator A in H, a sufficient condition for the convergence of the iterative method (1) in H_A has the form

$$0<\omega<2. \tag{3}$$

In Section 9.2 we looked at the question of the optimal choice of the iterative parameter ω. Assuming that assumptions 1 and 2 are satisfied and that *a priori* information is given in the form of the constant δ from the inequality

$$\delta\mathcal{D}\le A, \qquad \delta>0, \tag{4}$$

we proved that the optimal value of ω, which minimizes the spectral radius of the transformation operator S of the scheme (1), is defined by the formula

$$\omega=\omega_0=\frac{2}{1+\sqrt{\delta(2-\delta)}}. \tag{6}$$

In Sections 9.2.4, 9.2.5, sample problems were looked at for which assumptions 1 and 2 were satisfied. These assumptions are also satisfied for more complex problems, for example for a five-point difference scheme approximating a Dirichlet difference problem for an elliptic equation with variable coefficients on a non-uniform grid in an arbitrary region.

There exist, however, sample problems for which assumption 2 is not satisfied. To this class belong a Dirichlet difference problem for an elliptic equation with mixed derivatives, a high-order Dirichlet difference problem, and others.

The non-global choice of the iterative parameter ω and the absence of estimates for the convergence rate of the method in any norm are particular deficiencies of the theory developed in Section 9.2.

In this section we will look at a general scheme for triangular iterative methods, in which the iterative parameter ω is chosen in order to minimize the norm of the transformation operator in H_A. Here we will also find an estimate for the convergence rate of the method in H_A under the assumption that A is a self-adjoint and positive-definite operator.

Our examination of the triangular methods begins with a transformation of the iterative scheme (1). We introduce the operators R_1 and R_2 in the following way:

$$R_1 = \frac{1}{2}\mathcal{D} + L, \quad R_2 = \frac{1}{2}\mathcal{D} + U.$$

Then the decomposition (2) will have the form

$$A = R_1 + R_2, \tag{6}$$

and if A is a self-adjoint operator in H, then the operators R_1 and R_2 will be conjugate to each other

$$R_1 = R_2^*. \tag{7}$$

Substituting $L = R_1 - 1/2\mathcal{D}$ in (1) and denoting

$$\tau = 2\omega/(2-\omega), \tag{8}$$

we write the iterative scheme (1) in the equivalent form

$$(\mathcal{D} + \tau R_1)\frac{y_{k+1} - y_k}{\tau} + Ay_k = f, \quad k = 0, 1, \ldots, \quad y_0 \in H, \tag{9}$$

where from (3), (8) we have $\tau > 0$.

The scheme (9) can be considered independently from the scheme (1). Namely, suppose that the operator A is self-adjoint in H and is represented using (6) in the form of a sum of mutually-conjugate operators R_1 and R_2, and that $\mathcal{D}$ is an arbitrary self-adjoint and positive-definite operator defined in H. The iterative scheme (9) will be called the *canonical form of the triangular iterative methods.* We will also use the term triangular methods in the case where the matrices corresponding to the operators R_1 and R_2 are not triangular, and the matrix corresponding to the operator $\mathcal{D}$ is not a diagonal matrix.

From theorem 1 it follows that, for a positive-definite operator A, the iterative method (9) converges in H_A for $\tau > 0$. In fact, here it is sufficient to prove that $\mathcal{D} + \tau R_1 > 0.5\tau A$. From (7) we obtain

$$(Ax, x) = (R_1 x, x) + (R_2 x, x) = 2(R_1 x, x) = 2(R_2 x, x) \tag{10}$$

and, consequently,

$$((\mathcal{D} + \tau R_1)x, x) = (\mathcal{D}x, x) + 0.5\tau(Ax, x) > 0.5\tau(Ax, x),$$

which is what we were required to prove.

In conclusion we remark that the Gauss-Seidel method corresponds to the value $\tau = 2$ in the scheme (9), and the successive over-relaxation method corresponds to the value $\tau = 2/\sqrt{\delta(2-\delta)}$.

9.3.2 An estimate of the convergence rate. We now estimate the convergence rate in H_A of the iterative scheme (9), assuming that A is a self-adjoint positive-definite operator in H.

A transformation to the error $z_k = y_k - u$ in (9) gives a homogeneous scheme for z_k

$$B\frac{z_{k+1} - z_k}{\tau} + Az_k = 0, \quad k = 0, 1, \ldots, \quad z_0 = y_0 - u, \quad B = \mathcal{D} + \tau R_1,$$

from which we obtain

$$\begin{aligned} z_{k+1} &= Sz_k, \quad k = 0, 1, \ldots, \quad S = E - \tau B^{-1} A, \\ \| z_{k+1} \|_A &\le \| S \|_A \| z_k \|_A . \end{aligned} \tag{11}$$

We estimate now the norm of the transformation operator S in H_A. From the definition of the norm of an operator we obtain

$$\begin{aligned} \| S \|_A^2 &= \sup_{x \ne 0} \frac{(ASx, Sx)}{(Ax, x)} \\ &= \sup_{x \ne 0} \left[1 - 2\tau \frac{(B^{-1}Ax, Ax)}{(Ax, x)} + \tau^2 \frac{(AB^{-1}Ax, B^{-1}Ax)}{(Ax, x)} \right]. \end{aligned} \tag{12}$$

We transform the expression in square brackets. Using (10) and the definition of the operator B we obtain

$$(By, y) = (\mathcal{D}y, y) + \tau(R_1 y, y) = (\mathcal{D}y, y) + 0.5\tau(Ay, y).$$

From this we find $\tau^2(Ay,y) = 2\tau(By,y) - 2\tau(\mathcal{D}y,y)$ or, after making the change of variables $y = B^{-1}Ax$,

$$\tau^2(AB^{-1}Ax, B^{-1}Ax) = 2\tau(B^{-1}Ax, Ax) - 2\tau(\mathcal{D}B^{-1}Ax, B^{-1}Ax).$$

Substituting this expression in (12), we will have

$$\| S \|_A^2 = \sup_{x\neq 0}\left[1 - 2\tau\frac{(\mathcal{D}B^{-1}Ax, B^{-1}Ax)}{(Ax,x)}\right].$$

We now make a further transformation. Setting $x = (B^*)^{-1}\mathcal{D}^{1/2}y$, we obtain

$$\frac{(\mathcal{D}B^{-1}Ax, B^{-1}Ax)}{(Ax,x)} = \frac{(Cy,Cy)}{(Cy,y)}, \quad C = \mathcal{D}^{1/2}B^{-1}A(B^*)^{-1}\mathcal{D}^{1/2}.$$

Since the operator C is self-adjoint and positive-definite in H, then, setting $y = C^{-1/2}\mathcal{D}^{-1/2}B^*v$, we find

$$\frac{(\mathcal{D}B^{-1}Ax, B^{-1}Ax)}{(Ax,x)} = \frac{(Av,v)}{(B\mathcal{D}^{-1}B^*v,v)}.$$

Thus, finally we have

$$\| S \|_A^2 = \sup_{v\neq 0}\left[1 - 2\tau\frac{(Av,v)}{(B\mathcal{D}^{-1}B^*v,v)}\right].$$

From this we obtain, if γ_1 is the quantity in the inequality

$$\gamma_1 B\mathcal{D}^{-1}B^* \leq A, \tag{13}$$

then

$$\| S \|_A \leq (1 - 2\tau\gamma_1)^{1/2}. \tag{14}$$

Since γ_1 depends on the parameter τ, the optimal value for τ can be found, if we have obtained an expression for γ_1 under additional assumptions about the operators $\mathcal{D}$, R_1, and R_2.

9.3.3 The choice of the iterative parameter. We now select the iterative parameter τ. We require

Lemma 3. *Suppose that δ and Δ are the constants in the inequalities*

$$\delta\mathcal{D} \le A, \quad R_1\mathcal{D}^{-1}R_2 \le \frac{\Delta}{4}A, \quad \delta > 0. \tag{15}$$

then in (13)

$$\gamma_1 = \delta \Big/ \left(1 + \tau\delta + \tau^2\frac{\delta\Delta}{4}\right) \tag{16}$$

Proof. Since $B^* = \mathcal{D} + \tau R_2$, then

$$\begin{aligned} B\mathcal{D}^{-1}B^* = (\mathcal{D} + \tau R_1)\mathcal{D}^{-1}(\mathcal{D} + \tau R_2) &= \mathcal{D} + \tau(R_1 + R_2) + \tau^2 R_1\mathcal{D}^{-1}R_2 \\ &= \mathcal{D} + \tau A + \tau^2 R_1\mathcal{D}^{-1}R_2. \end{aligned}$$

Using assumption (15), from this we obtain

$$B\mathcal{D}^{-1}B^* \le (1/\delta + \tau + \tau^2\Delta/4)A.$$

The lemma is proved. □

Thus, if the *a priori* information has the form of the constants δ and Δ in the inequalities (15), then γ_1 is estimated in formula (16).

Substituting (16) in (14) we obtain

$$\| S \|_A^2 \le \varphi(\tau) = 1 - 2\tau\delta \Big/ \left(1 + \tau\delta + \tau^2\frac{\delta\Delta}{4}\right).$$

It remains to minimize the function $\varphi(\tau)$. Setting the derivative $\varphi'(\tau)$ equal to zero, we find

$$\varphi'(\tau) = \frac{2\delta\left(\tau^2\frac{\delta\Delta}{4} - 1\right)}{\left(1 + \tau\delta + \tau^2\frac{\delta\Delta}{4}\right)^2} = 0, \quad \tau_0 = \frac{2}{\sqrt{\delta\Delta}}.$$

Since for $\tau < \tau_0$ the derivative $\varphi'(\tau) < 0$, and for $\tau > \tau_0$ the derivative $\varphi'(\tau) > 0$, then for $\tau = \tau_0$ the function $\varphi(\tau)$ achieves its minimum value, equal to $\varphi(\tau_0) = (1 - \sqrt{\eta})/(1 + \sqrt{\eta})$, $\eta = \delta/\Delta$. Thus, we have proved

Theorem 6. *Suppose that A and $\mathcal{D}$ are self-adjoint positive-definite operators in H, and δ and Δ are the constants in* (15). *The triangular iterative method* (9), (6) *for $\tau = \tau_0 = 2/\sqrt{\delta\Delta}$ converges in H_A, and the error z_n is estimated by $\| z_n \|_A \le \rho^n \| z_0 \|_A$. The iteration count n is estimated by $n \ge n_0(\epsilon)$,*

$$n_0(\epsilon) = \ln \epsilon / \ln \rho,$$

where

$$\rho = \left(\frac{1-\sqrt{\eta}}{1+\sqrt{\eta}}\right)^2, \quad \eta = \frac{\delta}{\Delta}. \square$$

9.3.4 An estimate for the convergence rates of the Gauss-Seidel and relaxation methods. Theorem 6 allows us to obtain estimates for the convergence rates in H_A of the Gauss-Seidel and successive over-relaxation methods examined earlier. In Sections 9.1.2 and 9.2.4, these methods were applied to find an approximate solution to a Dirichlet difference problem for Poisson's equation on the rectangular grid

$$\begin{aligned} &\bar{\omega} = \{x_{ij} = (ih_1, jh_2),\ 0 \le i \le N_1,\ 0 \le j \le N_2,\ h_\alpha = l_\alpha/N_\alpha,\ \alpha = 1,2\}, \\ &\qquad \Lambda y = y_{\bar{x}_1 x_1} + y_{\bar{x}_2 x_2} = -\varphi(x), \quad x \in \omega, \\ &\quad y(x) = g(x), \quad x \in \gamma. \end{aligned}$$

The iterative scheme for these methods had the form (1), where

$$\begin{aligned} \mathcal{D}y &= \left(\frac{2}{h_1^2} + \frac{2}{h_2^2}\right) \mathring{y}, \\ Ly(i,j) &= -\frac{1}{h_1^2}\mathring{y}(i-1,j) - \frac{1}{h_2^2}\mathring{y}(i,j-1), \\ Uy(i,j) &= -\frac{1}{h_2^2}\mathring{y}(i+1,j) - \frac{1}{h_2^2}\mathring{y}(i,j+1). \end{aligned}$$

For the Gauss-Seidel method $\omega = 1$, and for the successive over-relaxation method ω was found from formula (5), where δ in inequality (4) was estimated as follows:

$$\delta = \frac{2h_2^2}{h_1^2 + h_2^2} \sin^2 \frac{\pi h_1}{2l_1} + \frac{2h_1^2}{h_1^2 + h_2^2} \sin^2 \frac{\pi h_2}{2l_2}. \tag{17}$$

We now put the scheme (1) for this example in the form (9). To do this, we define the operators R_1 and R_2:

$$R_1 y = \left(\frac{1}{2}\mathcal{D} + L\right) y = \frac{1}{h_1}\mathring{y}_{\bar{x}_1} + \frac{1}{h_2}\mathring{y}_{\bar{x}_2},$$

$$R_2 y = \left(\frac{1}{2}\mathcal{D} + U\right) y = -\frac{1}{h_1}\mathring{y}_{x_1} - \frac{1}{h_2}\mathring{y}_{x_2}.$$

It is obvious that

$$(R_1 + R_2)y = Ay = -\Lambda \mathring{y} = -\mathring{y}_{\bar{x}_1 x_1} - \mathring{y}_{\bar{x}_2 x_2}.$$

It is easy to establish the mutual-conjugacy of the operators R_1 and R_2 using Green's difference formula. As was noted above, the Gauss-Seidel method corresponds to the value $\tau = 2$ in the scheme (9), and the successive over-relaxation method corresponds to the value $\tau = 2/\sqrt{\delta(2-\delta)}$, where δ was defined in (17).

From (11), (14), and lemma 3 it follows that, in order to obtain estimates of the convergence rates in H_A for these methods, it is necessary to find δ and Δ from the inequalities (15). The constant δ has already been found. We now find Δ. From the definition of the operators $\mathcal{D}$, R_1, and R_2, we obtain

$$(R_1 \mathcal{D}^{-1} R_2 y, y) = 0.5\frac{h_1^2 h_2^2}{h_1^2 + h_2^2}(R_2 y, R_2 y). \tag{18}$$

Further

$$(R_2 y, R_2 y) = \frac{1}{h_1^2}(\mathring{y}_{x_1}^2, 1) + \frac{2}{h_1 h_2}(\mathring{y}_{x_1}, \mathring{y}_{x_2}) + \frac{1}{h_2^2}(\mathring{y}_{x_2}^2, 1)$$
$$\leq \left(\frac{1}{h_1^2} + \frac{1}{h_2^2}\right)\left[(\mathring{y}_{x_1}^2, 1) + (\mathring{y}_{x_2}^2, 1)\right] \leq \frac{h_1^2 + h_2^2}{h_1^2 h_2^2}(Ay, y).$$

Substituting this estimate in (18), we obtain

$$(R_1 \mathcal{D}^{-1} R_2 y, y) \leq \frac{1}{2}(Ay, y)$$

and, consequently, $\Delta = 2$ in inequality (15).

We now estimate the convergence rates for the Gauss-Seidel and successive over-relaxation methods.

From (11) we obtain

$$\| z_n \|_A \leq \| S \|_A^n \| z_0 \|_A$$

and, consequently, to achieve an accuracy ϵ it is sufficient to complete $n \geq n_0(\epsilon)$ iterations, where $n_0(\epsilon) = \ln \epsilon / \ln \| S \|_A$. From (14) we find

$$n_0(\epsilon) = 2 \ln \epsilon / \ln \| S \|_A^2 \geq \ln \frac{1}{\epsilon} / (\tau \gamma_1). \tag{19}$$

From (16) for the Gauss-Seidel method we obtain ($\tau = 2$)

$$\tau \gamma_1 = 2\delta / (1 + 4\delta) \tag{20}$$

and in particular, where $N_1 = N_2 = N$, $l_1 = l_2 = l$, we will have from (17), (19), and (20)

$$\delta = 2 \sin^2 \frac{\pi}{2N}, \quad \tau \gamma_1 \approx 4 \sin^2 \frac{\pi}{2N} \approx \frac{\pi^2}{N^2},$$
$$n_0(\epsilon) \approx \frac{N^2}{\pi^2} \ln \frac{1}{\epsilon} \approx 0.1 N^2 \ln \frac{1}{\epsilon}.$$

In Section 6.5.1 for the explicit simple iteration method applied to this special case, we obtained the following estimate for the iteration count: $n_0(\epsilon) \approx 0.2 N^2 \ln \frac{1}{\epsilon}$. Comparing these estimates, we find that the Gauss-Seidel method requires roughly half as many iterations as the simple iteration method. The character of the dependence of the iteration count on the number of nodes N in each direction is identical for these methods — the iteration count is proportion to N^2.

We now look at the successive over-relaxation method. Substituting in (16)

$$\delta = 2 \sin^2 \frac{\pi}{2N}, \quad \Delta = 2 \quad \text{and} \quad \tau = \frac{2}{\sqrt{\delta(2 - \delta)}} = \frac{2}{\sin \frac{\pi}{N}},$$

we obtain

$$\tau \gamma_1 = \frac{2 \tan \frac{\pi}{2N}}{2 + 2 \tan \frac{\pi}{2N} + \tan^2 \frac{\pi}{2N}} \approx \tan \frac{\pi}{2N} \approx \frac{\pi}{2N}.$$

From (19) we find the following estimate of the iteration count for the successive over-relaxation method:

$$n_0(\epsilon) \approx \frac{2N}{\pi} \ln \frac{1}{\epsilon} \approx 0.64 N \ln \frac{1}{\epsilon}, \tag{21}$$

i.e. the iteration count for the successive over-relaxation method is proportional to the number of nodes N in one direction.

In conclusion we give an estimate for the iteration count which follows from theorem 6. For the value of the parameter τ

$$\tau = \tau_0 = \frac{2}{\sqrt{\delta\Delta}} = \frac{1}{\sin\frac{\pi}{N}}$$

the iteration count will be estimated by

$$n \geq n_0(\epsilon) = \ln\frac{1}{\epsilon} \Big/ \sin\frac{\pi}{2N} \approx 0.64 N \ln\frac{1}{\epsilon}.$$

Notice that the estimate (21) is somewhat excessive. To see this, it is necessary to compare the value of $n_0(\epsilon)$ computed from formula (21) with the iteration count derived in Section 9.2.4. In this connection, here the iteration count was estimated from the inequality $\| S \|_A^n \leq \epsilon$, and in Section 9.2.4, the iteration was terminated using the condition $\| S^n \|_A \leq \epsilon$.

Chapter 10

The Alternate-Triangular Method

In this chapter we study the alternate-triangular iterative method* for solving an operator equation with a self-adjoint operator. In Section 10.1 the general theory of the method is laid out, described by the construction of the iterative scheme and by indicating the set of iterative parameters. The method is illustrated on a sample problem — a Dirichlet difference problem for Poisson's equation in a rectangle. In Section 10.2 this method is applied to the solution of elliptic difference equations with variable coefficients and mixed derivatives in a rectangle. A variant of the alternate-triangular method is constructed in Section 10.3 to solve an elliptic equation with variable coefficients on a non-uniform grid in a arbitrary region.

10.1 The general theory of the method

10.1.1 The iterative scheme. In Section 9.3 we studied the triangular iterative method for solving the equation

$$Au = f. \tag{1}$$

The iterative scheme of this method has the form

$$B\frac{y_{k+1} - y_k}{\tau_{k+1}} + Ay_k = f, \quad k = 0, 1, \ldots, \quad y_0 \in H, \tag{2}$$

* The method was presented by A.A Samarskii in 1964 (cf. Journal of Computational Mathematics and Mathematical Physics, 4, No. 3, 1964) and improved in [8].

where $\tau_k \equiv \tau$, and the operator $B = B_1 = \mathcal{D} + \tau R_1$ is defined by the following decomposition of the operator A into a sum of operators:

$$A = R_1 + R_2, \quad R_1 = R_2^*, \quad A = A^* > 0. \tag{3}$$

Concerning the operator $\mathcal{D}$, it is assumed that it is self-adjoint and positive-definite in H, i.e.

$$\mathcal{D} = \mathcal{D}^* > 0. \tag{4}$$

The triangular iterative method is actually a class of methods whose iterative parameters are chosen using *a priori* information about the operators of the iterative scheme. For the triangular method the primary information consists of the constants δ and Δ from the inequalities

$$\delta \mathcal{D} \le A, \quad R_1 \mathcal{D}^{-1} R_2 \le \frac{\Delta}{4} A, \quad \delta > 0. \tag{5}$$

The parameter τ found in Section 9.3.3 allows us to achieve an accuracy ϵ after $n_0 = O(\ln(1/\epsilon)\sqrt{\eta})$ iterations, where $\eta = \delta/\Delta$.

Notice that, because the operator B is not self-adjoint, we are not able to use a set of parameters τ_k in the iterative scheme (2) and thus increase the convergence rate of the method. However, the simple way in which the operator B is constructed and the possibility of an expansion (3) for the operator A have encouraged the study of possible variants of the triangular method. As a result, the alternate-triangular method was constructed, which combines the simplicity of the operator B with the possibility of using a set of parameters τ_k in the scheme (2).

We proceed now to the study of the alternate-triangular method. The iterative scheme of the method has the form (2), where the operator B is defined as follows:

$$B = (\mathcal{D} + \omega R_1)\mathcal{D}^{-1}(\mathcal{D} + \omega R_2), \quad \omega > 0. \tag{6}$$

Here ω is an iterative parameter which has yet to be chosen. We will further assume that, for the scheme (2), (6), the conditions (3), (4) are satisfied and that δ and Δ in the inequalities (5) are given.

We now list several properties of the operator B defined in (6). If the operator $\mathcal{D} + \omega R_1$ corresponds to a triangular matrix, and $\mathcal{D}$ to a diagonal matrix, then B corresponds to the product of two triangular matrices and a diagonal matrix. In this case, inversion of the operator B is not a complex problem.

We shall show that the operator B is self-adjoint in H, and if the operator $\mathcal{D}$ is bounded, then B is positive-definite. In fact, by (3) we have the equation

$$(Au, u) = 2(R_1 u, u) = 2(R_2 u, u) > 0.$$

From this and from (4) it follows that the operators $B_1 = \mathcal{D} + \omega R_1$ and $B_2 = \mathcal{D} + \omega R_2$ are conjugate and positive-definite: $B_1^* = (\mathcal{D} + \omega R_1)^* = \mathcal{D} + \omega R_2 = B_2$, $B_\alpha > \mathcal{D} > 0$, $\alpha = 1, 2$, therefore

$$B^* = (B_1 \mathcal{D}^{-1} B_2)^* = B_2^* \mathcal{D}^{-1} B_1^* = B_1 \mathcal{D}^{-1} B_2 = B.$$

Further, since $\mathcal{D}$ is a self-adjoint, bounded, and positive-definite operator, the inverse operator $\mathcal{D}^{-1}$ will be positive-definite in H. Consequently, using the inequality $(\mathcal{D}^{-1}x, x) \geq d(x, x)$, $d > 0$ which express the positive-definiteness of the operator $\mathcal{D}^{-1}$, we have

$$(Bu, u) = (\mathcal{D}^{-1} B_2 u, B_2 u) \geq d \parallel B_2 u \parallel^2 > 0.$$

From (2), (6) *it is clear that, in order to define* y_{k+1} *given* y_k, *it is* necessary to solve the equation

$$(\mathcal{D} + \omega R_1)\mathcal{D}^{-1}(\mathcal{D} + \omega R_2) y_{k+1} = \varphi_k, \quad k = 0, 1, \ldots,$$

where $\varphi_k = B y_k - \tau_{k+1}(A y_k - f)$. This leads to the solution of two equations

$$(\mathcal{D} + \omega R_1) v = \varphi_k, \quad (\mathcal{D} + \omega R_2) y_{k+1} = \mathcal{D} v.$$

It is possible to obtain a second algorithm for realizing the scheme (2), (6), based on writing it in the form of a scheme with a correction

$$y_{k+1} = y_k - \tau_{k+1} w_k, \quad B w_k = r_k,$$

where $r_k = A y_k - f$ is the residual. The correction w_k is found by solving two equations

$$(\mathcal{D} + \omega R_1) \bar{w}_k = r_k, \quad (\mathcal{D} + \omega R_2) w_k = \mathcal{D} \bar{w}_k.$$

In this algorithm, we avoid having to compute $B y_k$, but it is necessary to simultaneously store both y_k and the intermediate quantities r_k, $\bar{w}_k$, and w_k.

10.1.2 Choice of the iterative parameters. We are interested now in investigating the convergence of the iterative scheme (2), (6). Since the operators A and B are self-adjoint and positive-definite in H, it is possible to study the convergence in H_D, where D is taken to be one of the operators A, B, or $AB^{-1}A$ (in the latter case, B must be a bounded operator). For such an operator D, the operator $DB^{-1}A$ will obviously be self-adjoint in H, and, consequently, according to the classification of Chapter 6, we have an iterative scheme in the self-adjoint case.

Using the results of Section 6.2, we can at once indicate the optimal set of iterative parameters τ_k for the scheme (2), (6). Assume that γ_1 and γ_2 are taken from the inequalities

$$\gamma_1 B \leq A \leq \gamma_2 B, \quad \gamma_1 > 0. \tag{7}$$

Then the Chebyshev set of parameters $\{\tau_k\}$ is defined by the formulas

$$\begin{gathered}\tau_k = \frac{\tau_0}{1+\rho_0\mu_k}, \quad \mu_k \in \mathcal{M}_n^* = \left\{\cos\frac{(2i-1)\pi}{2n},\ 1 \leq i \leq n\right\}, \quad 1 \leq k \leq n, \\ \tau_0 = \frac{2}{\gamma_1+\gamma_2}, \quad \rho_0 = \frac{1-\xi}{1+\xi}, \quad n \geq n_0(\epsilon) = \frac{\ln(0.5\epsilon)}{\ln\rho_1}, \quad \xi = \frac{\gamma_1}{\gamma_2},\end{gathered} \tag{8}$$

and the error $z_n = y_n - u$ of the iterative scheme (2), (6), (8) can be estimated using

$$\| z_n \|_D \leq q_n \| z_0 \|_D, \quad q_n = \frac{2\rho_1^n}{1+\rho_1^{2n}} \leq \epsilon, \quad \rho_1 = \frac{1-\sqrt{\xi}}{1+\sqrt{\xi}}. \tag{9}$$

This result is from the general theory of two-level iterative methods. For the scheme (2), (6) the *a priori* information consists of the constants δ and Δ in the inequalities (5). Therefore, one of the problems is to obtain expressions for γ_1 and γ_2 in terms of δ and Δ. Further, since the operator B depends on the iterative parameter ω, then γ_1 and γ_2 are functions of ω: $\gamma_1 = \gamma_1(\omega)$, $\gamma_2 = \gamma_2(\omega)$. Since it follows from the estimate (9) that the maximal convergence rate will be attained when the ratio $\xi = \gamma_1/\gamma_2$ is maximal, we arrive at the problem of choosing the parameter ω so as to maximize ξ. These problems are solved by

Lemma 1. *Suppose that conditions* (3), (4) *are satisfied, the operator B is defined by formula* (6), *and that the constants δ and Δ are given in the inequalities* (5). *Then in the inequalities* (7) *we have*

$$\gamma_1 = \delta \Big/ \left(1 + \omega\delta + \frac{1}{4}\omega^2\delta\Delta\right), \quad \gamma_2 = 1/(2\omega). \tag{10}$$

The ratio $\xi = \gamma_1/\gamma_2$ is maximal if

$$\omega = \omega_0 = 2/\sqrt{\delta\Delta}, \tag{11}$$

where

$$\gamma_1 = \frac{\delta}{2(1+\sqrt{\eta})}, \quad \gamma_2 = \frac{\delta}{4\sqrt{\eta}}, \quad \xi = \frac{2\sqrt{\eta}}{1+\sqrt{\eta}}, \quad \eta = \frac{\delta}{\Delta}. \tag{12}$$

Proof. We write the operator B in the form

$$B = (\mathcal{D} + \omega R_1)\mathcal{D}^{-1}(\mathcal{D} + \omega R_2) = \mathcal{D} + \omega(R_1 + R_2) + \omega^2 R_1 \mathcal{D}^{-1} R_2. \tag{13}$$

Taking into account that $A = R_1 + R_2 \geq \delta\mathcal{D}$ or $\mathcal{D} \leq \frac{1}{\delta}A$, we obtain an upper bound for B

$$B \leq \left(\frac{1}{\delta} + \omega + \frac{1}{4}\omega^2\Delta\right) A = \frac{1}{\gamma_1} A,$$

i.e. $A \geq \gamma_1 B$, where γ_1 is defined in (10).

We now transform formula (13):

$$\begin{aligned} B &= \mathcal{D} - \omega(R_1 + R_2) + \omega^2 R_1\mathcal{D}^{-1}R_2 + 2\omega(R_1 + R_2) \\ &= (\mathcal{D} - \omega R_1)\mathcal{D}^{-1}(\mathcal{D} - \omega R_2) + 2\omega A. \end{aligned}$$

From this follows

$$(By, y) = 2\omega(Ay, y) + (\mathcal{D}^{-1}(\mathcal{D} - \omega R_2)y, (\mathcal{D} - \omega R_2)y).$$

Using the positive-definiteness of the operator $\mathcal{D}^{-1}$, we obtain $(By, y) \geq 2\omega(Ay, y)$, i.e. $A \leq \gamma_2 B$. Thus, γ_1 and γ_2 have been found. We look further at the relation

$$\xi = \xi(\omega) = \gamma_1/\gamma_2 = 2\omega\delta \Bigg/ \left(1 + \omega\delta + \frac{\omega^2\delta\Delta}{4}\right).$$

Setting the derivative

$$\xi'(\omega) = \frac{2\delta(1 - \omega^2\delta\Delta/4)}{\left(1 + \omega\delta + \frac{\omega^2\delta\Delta}{4}\right)^2}$$

equal to zero, we find $\omega = \omega_0 = 2/\sqrt{\delta\Delta}$. At this point the function $\xi(\omega)$ achieves its maximum since $\xi''(\omega_0) < 0$. Substituting this value of ω in (10), we obtain (12). We now show that $\delta \leq \Delta$, $\eta \leq 1$. In fact, using the equation

$(Ax, x) = 2(R_2x, x)$ and the Cauchy-Schwartz-Bunyakovskij inequality, from (5) we obtain

$$\delta(\mathcal{D}x, x) \le (Ax, x) = \frac{(Ax, x)^2}{(Ax, x)} = 4\frac{(R_2x, x)^2}{(Ax, x)}$$
$$= 4\frac{(\mathcal{D}^{-1/2}R_2x, \mathcal{D}^{1/2}x)^2}{(Ax, x)} \le 4\frac{(\mathcal{D}^{-1/2}R_2x, \mathcal{D}^{-1/2}R_2x)}{(Ax, x)}(\mathcal{D}^{1/2}x, \mathcal{D}^{1/2})$$
$$= 4\frac{(R_1\mathcal{D}^{-1}R_2x, x)}{(Ax, x)}(\mathcal{D}x, x) \le \Delta(\mathcal{D}x, x),$$

which is what we were required to prove. The lemma is proved. □

Theorem 1. *Suppose that the conditions of lemma 1 are satisfied. Then, for the alternate-triangular method (2), (6), (11) with the Chebyshev parameters* τ_k *defined by the formulas (8) and (12), the estimate (9) is valid. In order to achieve* $\| z_n \|_D \le \epsilon \| z_0 \|_D$*, it is sufficient to perform* n *iterations, where* $n \ge n_0(\epsilon)$*,*

$$n_0(\epsilon) = \ln\frac{2}{\epsilon} \Big/ (2\sqrt{2}\sqrt[4]{\eta}), \quad \eta = \delta/\Delta\,.$$

Here $D = A$*,* B*, or* $AB^{-1}A$*.*

Proof. To prove this theorem, just use lemma 1 and the formulas (8) for the iterative parameters and the iteration count. □

We look now at one procedure which is used to construct implicit iterative schemes. Suppose that R is a given self-adjoint and positive-definite operator in H, which is energy equivalent to A with constants c_1 and c_2:

$$c_1R \le A \le c_2R, \quad c_1 > 0, \tag{14}$$

and to an operator B with constants $\mathring{\gamma}_1$ and $\mathring{\gamma}_2$:

$$\mathring{\gamma}_1B \le R \le \mathring{\gamma}_2B, \quad \mathring{\gamma}_1 > 0. \tag{15}$$

We will assume that the operators A and B are self-adjoint. From (14) and (15) we obtain the following inequalities: $\gamma_1B \le A \le \gamma_2B$, $\gamma_1 = c_1\mathring{\gamma}_1$, $\gamma_2 = c_2\mathring{\gamma}_2$. This allows us to construct an operator B which arises, not from the decomposition (3) of the operator A, but from the decomposition of the operator R, which can be chosen from a large class of different operators which may not include A. Here the constants $\mathring{\gamma}_1$ and $\mathring{\gamma}_2$ in (15) can be computed once and for all, and the problem of obtaining the *a priori* information for the method reduces to finding c_1 and c_2 in (14).

Thus, suppose that the operator R is represented in the form of the sum of conjugate operators R_1 and R_2:

$$R = R^* > 0, \quad R = R_1 + R_2, \quad R_1 = R_2^*, \tag{16}$$

and in place of (5) we have the inequalities

$$\delta \mathcal{D} \leq R, \quad R_1 \mathcal{D}^{-1} R_2 \leq \frac{\Delta}{4} R, \quad \delta > 0. \tag{17}$$

The operator B for the scheme (2) is constructed using (6). Then by lemma 1 for $\omega = \omega_0 = 2/\sqrt{\delta\Delta}$ in the inequalities (15) we have

$$\mathring{\gamma}_1 = \frac{\delta}{2(1+\sqrt{\eta})}, \quad \mathring{\gamma}_2 = \frac{\delta}{4\sqrt{\eta}}, \quad \mathring{\xi} = \frac{\mathring{\gamma}_1}{\mathring{\gamma}_2} = \frac{2\sqrt{\eta}}{1+\sqrt{\eta}}, \quad \eta = \frac{\delta}{\Delta}. \tag{18}$$

From this follows

Theorem 2. *Suppose that* $A = A^* > 0$, $\mathcal{D} = \mathcal{D}^* > 0$, *the conditions* (16) *are satisfied, and that* c_1 *and* c_2 *in* (14) *and* δ *and* Δ *in* (17) *are given. Then the estimate* (9) *is valid for the alternate-triangular method* (2), (6), (8), (11) *with the Chebyshev parameters* τ_k, *where* $\gamma_1 = c_1\mathring{\gamma}_1$ *and* $\gamma_2 = c_2\mathring{\gamma}_2$, *and* $\mathring{\gamma}_1$ *and* $\mathring{\gamma}_2$ *are defined in* (18). *In order to achieve* $\| z_n \|_D \leq \epsilon \| z_0 \|_D$ *it is sufficient to perform* n *iterations, where*

$$n \geq n_0(\epsilon), \quad n_0(\epsilon) = \frac{\ln(2/\epsilon)}{2\sqrt{2}\sqrt[4]{\eta}}\sqrt{\frac{c_2}{c_1}}, \quad \eta = \frac{\delta}{\Delta}. \square$$

10.1.3 A method for finding δ and Δ. From theorems 1 and 2 it follows that, in order to apply the alternate-triangular method, it is necessary to find the two scalars δ and Δ in the inequalities (5) or (17). In the examples of grid elliptic equations looked at below, these constants will be found explicitly or an algorithm for computing them will be given. To do this, it is natural to exploit the structure of the operators A, R_1, R_2, and $\mathcal{D}$. For the general theory of iterative methods, which does not consider the concrete structure of the operators, it is necessary to assume that there is some general way of finding the *a priori* information required by the method.

This can be based on the use of the asymptotic property of the iterative methods of variational type (cf. Section 8.1.5). Suppose that the operators A and B are self-adjoint and positive-definite in H. If in the iterative scheme

$$B\frac{v_{k+1} - v_k}{\tau_{k+1}} + Av_k = 0, \quad k = 0, 1, \ldots, \quad v_0 \neq 0 \tag{19}$$

the parameters τ_{k+1} are chosen using the formula for the steepest-descent method

$$\tau_{k+1}=\frac{(w_k,r_k)}{(Aw_k,w_k)},\quad k=0,1,\ldots,\quad r_k=Av_k,\quad Bw_k=r_k, \tag{20}$$

and if, for sufficiently large n, the roots $x_1 \le x_2$ of the equation

$$(1-\tau_n x)(1-\tau_{n-1}x)=\rho_n\rho_{n-1},\quad \rho_n=\frac{\| v_n \|_A}{\| v_{n-1} \|_A}, \tag{21}$$

are found, where $\| \cdot \|_A$ is the norm in H_A, then x_1 and x_2 will be approximations to γ_1 and γ_2 in the inequalities (7) from above and below, respectively.

We shall use this method. We look at the iterative scheme (2), (3), (6). Notice that, by lemma 1, for (7) we have $\gamma_2 = 1/(2\omega)$, and only γ_1 depends on the *a priori* data. We shall attempt to avoid finding δ and Δ separately for (5), but instead to directly find an expression for γ_1 as a function of the iterative parameter ω. We will give this expression in the form (10), and indicate the corresponding values of δ and Δ. Then from lemma 1 we find ω_0 using formula (11) and γ_1 and γ_2 corresponding to ω_0 using formula (12). For the set of parameters τ_k we use (8).

We now find the desired expression for γ_1. We take $\omega = 0$ and use the method (19)–(21) to find x_1. Assuming that, in (19), (20), a sufficient number of iterations have been performed, and taking into account that for $\omega = 0$ the operator $B = \mathcal{D}$, we obtain the approximate inequality

$$x_1\mathcal{D} \le A,\quad x_1 > 0. \tag{22}$$

Further, we take $\omega = \omega_1 > 0$ and use the method (19)–(20) to find $\bar{x}_1$, so that $\bar{x}_1 > 0$ and

$$\bar{x}_1 B \le A \quad \text{or} \quad \bar{x}_1(\mathcal{D}+\omega_1 A+\omega_1^2 R_1\mathcal{D}^{-1}R_2) \le A, \tag{23}$$

where it is clear that $\bar{x}_1\omega_1 < 1$. We write (23) in the form

$$\bar{x}_1\mathcal{D}+\bar{x}_1\omega_1^2 R_1\mathcal{D}^{-1}R_2 \le (1-\bar{x}_1\omega_1)A$$

and add it to (22), after first multiplying by some as yet to be determined coefficient $\alpha > 0$. We obtain

$$(\alpha x_1+\bar{x}_1)\mathcal{D}+\bar{x}_1\omega_1^2 R_1\mathcal{D}^{-1}R_2 \le (1-\bar{x}_1\omega_1+\alpha)A. \tag{24}$$

We divide this inequality by $\alpha x_1 + \bar{x}_1$, add the term ωA to the right- and left-hand sides, and choose α from the condition

$$\bar{x}_1\omega_1^2 = \omega^2(\alpha x_1+\bar{x}_1); \tag{25}$$

then the transformed inequality will have the form

$$\mathcal{D} + \omega A + \omega^2 R_1 \mathcal{D}^{-1} R_2 = B \le \frac{1}{\gamma_1} A,$$

where

$$\frac{1}{\gamma_1} = \frac{1}{\gamma_1(\omega)} = \omega + \frac{1 - \bar{x}_1\omega_1 + \alpha}{\alpha x_1 + \bar{x}_1}. \tag{26}$$

From (25) we find α:

$$\alpha = \bar{x}_1(\omega_1^2 - \omega^2)/(\omega^2 x_1).$$

Since α must be positive, the expression (26) will be valid for $0 < \omega < \omega_1$. Substituting this value of α in (26), we obtain

$$\frac{1}{\gamma_1} = \frac{1}{x_1} + \omega + \frac{x_1 - \bar{x}_1 - x_1\bar{x}_1\omega_1}{x_1\bar{x}_1\omega_1^2}\omega^2.$$

Comparing this expression with (10), we obtain that it is possible to take Δ to be

$$\delta = x_1, \quad \Delta = 4\frac{x_1 - \bar{x}_1 - x_1\bar{x}_1\omega_1}{x_1\bar{x}_1\omega_1^2}.$$

Notice that ω_0, found from these values of δ and Δ using (11), will belong to the interval $(0, \omega_1)$, if the inequality $2\bar{x}_1 \le x_1(1 - \bar{x}_1\omega_1)$ is satisfied. If this inequality is not satisfied, then we must increase ω_1 and carry out these computations again (we recommend taking $\omega_1 = 2/x_1$).

10.1.4 A Dirichlet difference problem for Poisson's equation in a rectangle. We shall illustrate the alternate-triangular method using an example of a Dirichlet difference problem for Poisson's equation in the rectangle $\bar{G} = \{0 \le x_\alpha \le l_\alpha,\ \alpha = 1, 2\}$:

$$\begin{aligned} \Lambda y &= y_{\bar{x}_1 x_1} + y_{\bar{x}_2 x_2} = -\varphi(x), \quad x \in \omega, \\ y(x) &= g(x), \quad x \in \gamma \end{aligned} \tag{27}$$

on the grid $\bar{\omega} = \{x_{ij} = (ih_1, jh_2) \in \bar{G},\ 0 \le i \le N_1,\ 0 \le j \le N_2,\ h_\alpha = l_\alpha/N_\alpha,\ \alpha = 1, 2\}$ with boundary γ.

For this example, H is the space of grid functions defined on ω with inner product

$$(u, v) = \sum_{x \in \omega} u(x)v(x)h_1h_2.$$

The operator A is defined by the equation $Ay = -\Lambda\mathring{y}$, where $y \in H$, $\mathring{y} \in \mathring{H}$ and $y(x) = \mathring{y}(x)$, $x \in \omega$, and $\mathring{y}(x) = 0$ for $x \in \gamma$. The right-hand side f is defined in the usual way: $f(x) = \varphi(x) + \frac{1}{h_1^2}\varphi_1(x) + \frac{1}{h_2^2}\varphi_2(x)$, where

$$\varphi_1(x) = \begin{cases} g(0, x_2), & x_1 = h_1, \\ 0, & 2h_1 \le x_1 \le l_1 - 2h_1, \\ g(l_1, x_2), & x_1 = l_1 - h_1, \end{cases}$$

$$\varphi_2(x) = \begin{cases} g(x, 0), & x_2 = h_2, \\ 0, & 2h_2 \le x_2 \le l_2 - 2h_2, \\ g(x_1, l_2), & x_2 = l_2 - h_2. \end{cases}$$

The problem (27) can then be written in the form of equation (1).

The operator A is self-adjoint and positive-definite in H, since it corresponds to the Laplace difference operator with Dirichlet boundary conditions.

We are now interested in constructing the operator B. We will consider a classical variant of the alternate-triangular method, where in (6) we set

$$\mathcal{D} = E. \tag{28}$$

We define now the difference operators $\mathcal{R}_1$ and $\mathcal{R}_2$, which map onto grid functions on $\bar{\omega}$, in the following way:

$$\mathcal{R}_1 y = -\sum_{\alpha=1}^{2} \frac{1}{h_\alpha} y_{\bar{x}_\alpha}, \quad \mathcal{R}_2 y = \sum_{\alpha=1}^{2} \frac{1}{h_\alpha} y_{x_\alpha}, \qquad x \in \omega.$$

It is obvious that $\mathcal{R}_1 + \mathcal{R}_2 = \Lambda$. Using Green's difference formulas, it is easy to obtain the following equation for grid functions $\mathring{y}(x) \in \mathring{H}$, $\mathring{u}(x) \in \mathring{H}$, i.e. for functions defined on $\bar{\omega}$ and vanishing on γ,

$$(\mathcal{R}_1\mathring{y}, \mathring{u}) = (\mathring{y}, \mathcal{R}_2\mathring{u}). \tag{29}$$

We define the operators R_1 and R_2 on H in the following fashion: $R_\alpha y = -\mathcal{R}_\alpha\mathring{y}$, $\alpha = 1, 2$, where $y \in H$, $\mathring{y} \in \mathring{H}$, and $y(x) = \mathring{y}(x)$, $x \in \omega$. Then, using the definition of the difference operators $\mathcal{R}_\alpha$ and equation (29), we have that the conditions (3) are satisfied, i.e. $A = R_1 + R_2$, $R_1 = R_2^*$. Taking into account (28), we obtain from (6) the following form of the operator B:

$$B = (E + \omega R_1)(E + \omega R_2).$$

We now find the *a priori* information required to implement the alternate-triangular method. In this case, it has the form of the constants δ and Δ in the inequalities $\delta E \leq A$, $R_1 R_2 \leq (\Delta/4)A$. Clearly, it is possible to take δ to be the smallest eigenvalue of the Laplace difference operator

$$\delta = \frac{4}{h_1^2}\sin^2\frac{\pi h_1}{2l_1} + \frac{4}{h_2^2}\sin^2\frac{\pi h_2}{2l_2}.$$

An estimate for Δ was found in Section 9.3.4 (the operators R_1 and R_2 defined here are identical to those in that section). We have $\Delta = 4/h_1^2 + 4/h_2^2$.

Thus, the necessary information about δ and Δ has been found. From lemma 1 we find the optimal value for the parameter ω_0, and also γ_1 and γ_2. The iterative parameters τ_k are computed using the formulas (8). In particular, when $N_1 = N_2 = N$, $l_1 = l_2 = l$, we obtain

$$\delta = \frac{8}{h^2}\sin^2\frac{\pi}{2N}, \quad \Delta = \frac{8}{h^2}, \quad \eta = \frac{\delta}{\Delta} = \sin^2\frac{\pi}{2N},$$
$$\xi = \frac{2\sqrt{\eta}}{1+\sqrt{\eta}} \approx 2\sqrt{\eta} = 2\sin\frac{\pi}{2N} \approx \frac{\pi}{N}, \quad \omega_0 = \frac{h^2}{4\sin\frac{\pi h}{2l}}.$$

From theorem 1, we find that the iteration count n can be bounded using $n \geq n_0(\epsilon)$, where

$$n_0(\epsilon) = \frac{\ln(2/\epsilon)}{2\sqrt{2}\sqrt[4]{\eta}} \approx \frac{\sqrt{N}}{2\sqrt{\pi}}\ln\frac{2}{\epsilon} \approx 0.28\sqrt{N}\ln\frac{2}{\epsilon},$$

i.e. the iteration count is proportional to the fourth root of the number of unknowns in the problem.

In Section 9.3.4, we obtained the following iteration count estimate for the successive over-relaxation method applied to the solution of the difference problem (27):

$$n \geq n_0(\epsilon) \approx 0.64N\ln(1/\epsilon), \quad (N = l/h).$$

A comparison of the relaxation method with the alternate-triangular method shows that the latter method is clearly better. Although performing one iteration of the alternate-triangular method requires twice as many operations as the relaxation method, its iteration count is low enough that it is still an effective method.

For the difference problem (27) we now give the iteration counts for the alternate-triangular method with the Chebyshev parameters as a function of the number of nodes N in one direction of the square grid $\bar{\omega}$ for $\epsilon = 10^{-4}$:

$$N = 32 \quad n = 16$$
$$N = 64 \quad n = 23$$
$$N = 128 \quad n = 32$$

A comparison with the iteration count for the successive over-relaxation method described in Section 9.2.4 shows that the relaxation method requires approximately 3.5–7.5 times as many iterations as the alternate-triangular method.

Remark 1. If the rectangle $\bar{G}$ is replaced by the p-dimensional parallelipiped $\bar{G} = \{0 \le x_\alpha \le l_\alpha,\ \alpha = 1, 2, \dots, p\}$, and we consider a Dirichlet difference problem for Poisson's equation

$$\Lambda y = \sum_{\alpha=1}^{p} y_{\bar{x}_\alpha x_\alpha} = -\varphi(x), \quad x \in \omega,$$
$$y(x) = g(x), \quad x \in \gamma,$$

on the rectangular grid $\bar{\omega} = \{x_i = (i_1 h_1, i_2 h_2, \dots, i_p h_p) \in \bar{G},\ 0 \le i_\alpha \le N_\alpha,\ h_\alpha N_\alpha = l_\alpha,\ \alpha = 1, 2, \dots, p\}$, then the difference operators $\mathcal{R}_1$ and $\mathcal{R}_2$ are defined as follows:

$$\mathcal{R}_1 y = -\sum_{\alpha=1}^{p} \frac{1}{h_\alpha} y_{\bar{x}_\alpha}, \quad \mathcal{R}_2 y = \sum_{\alpha=1}^{p} \frac{1}{h_\alpha} y_{x_\alpha}, \quad x \in \omega.$$

In this case, we set

$$\delta = \sum_{\alpha=1}^{p} \frac{4}{h_\alpha^2} \sin^2 \frac{\pi h_\alpha}{2 l_\alpha}, \quad \Delta = \sum_{\alpha=1}^{p} \frac{4}{h_\alpha^2},$$

in the inequalities (5) for $\mathcal{D} = E$, since

$$\| \mathcal{R}_2 \mathring{y} \|^2 = (\mathcal{R}_1 \mathcal{R}_2 \mathring{y}, \mathring{y}) = \left\| \sum_{\alpha=1}^{p} \frac{1}{h_\alpha} \mathring{y}_{x_\alpha} \right\|^2$$
$$\le \sum_{\alpha=1}^{p} \frac{1}{h_\alpha^2} \sum_{\alpha=1}^{p} \| \mathring{y}_{x_\alpha} \|^2 \le \left(\sum_{\alpha=1}^{p} \frac{1}{h_\alpha^2} \right) (A\mathring{y}, \mathring{y}).$$

Then in the case $N_1 = N_2 = \ldots = N_p = N$, $l_1 = l_2 = \ldots = l_p = l$ we have the following estimate for the iteration count

$$n \geq n_0(\epsilon), n_0(\epsilon) \approx \frac{\sqrt{N}}{2\sqrt{\pi}} \ln \frac{2}{\epsilon} \approx 0.28\sqrt{N} \ln \frac{2}{\epsilon},$$

which does not depend on the dimension p.

We look now at certain questions connected with the implementation of the alternate-triangular method for the problem (27). In Section 10.1.1, two different iterative algorithms were presented for finding y_{k+1} given y_k. We look first at the second algorithm. For $\mathcal{D} = E$ it has the following form:

$$\begin{aligned} &r_k = Ay_k - f, \\ &(E + \omega_0 R_1)\bar{w}_k = r_k, \quad (E + \omega_0 R_2)w_k = \bar{w}_k, \\ &y_{k+1} = y_k - \tau_{k+1} w_k, \quad k = 0, 1, \ldots. \end{aligned} \tag{30}$$

This algorithm is used when the parameters τ_k are chosen using, not the formulas (9), but the formulas for the iterative methods of variational type.

Using the definition of the operators A, R_1, and R_2 in terms of the difference operators Λ, $\mathcal{R}_1$, and $\mathcal{R}_2$, the formulas (30) can be written in the following form:

$$\begin{aligned} r_k(i,j) = &\left(\frac{2}{h_1^2} + \frac{2}{h_2^2}\right) y_k(i,j) - \frac{1}{h_1^2}\left[y_k(i-1,j) + y_k(i+1,j)\right] \\ &- \frac{1}{h_2^2}\left[y_k(i,j-1) + y_k(i,j+1)\right] - \varphi(i,j), \\ &1 \leq i \leq N_1 - 1, \quad 1 \leq j \leq N_2 - 1, \quad y_k|_\gamma = g. \end{aligned} \tag{31}$$

$$\begin{aligned} &\bar{w}_k(i,j) = \alpha \bar{w}_k(i-1,j) + \beta \bar{w}_k(i,j-1) + \kappa r_k(i,j), \\ &\qquad i = 1, 2, \ldots, N_1 - 1, \quad j = 1, 2, \ldots, N_2 - 1, \\ &\bar{w}_k(0,j) = 0, \quad 1 \leq j \leq N_2 - 1, \quad \bar{w}_k(i,0) = 0, \quad 1 \leq i \leq N_1 - 1. \end{aligned} \tag{32}$$

Here the computations are carried out starting at the point $i = 1$, $j = 1$ either along the rows of the grid $\bar{\omega}$, i.e. with i increasing and j fixed, or along the columns with j increasing and i fixed.

$$\begin{aligned} &w_k(i,j) = \alpha w_k(i+1,j) + w_k(i,j+1) + \kappa \bar{w}_k(i,j), \\ &\qquad i = N_1 - 1, N_1 - 2, \ldots, 1, \quad j = N_2 - 1, N_2 - 2, \ldots, 1, \\ &w_k(N_1,j) = 0, \quad 1 \leq j \leq N_2 - 1, \quad w_k(i,N_2) = 0, \quad 1 \leq i \leq N_1 - 1. \end{aligned} \tag{33}$$

Here the computations are carried out starting at the point $i = N_1 - 1$, $j = N_2 - 1$, either along the rows or along the columns of the grid with the corresponding indices i or j decreasing. As a result, y_{k+1} is defined by the formulas

$$\begin{aligned} &y_{k+1}(i,j) = y_k(i,j) - \tau_{k+1} w_k(i,j), \\ &\qquad 1 \le i \le N_1 - 1, \quad 1 \le j \le N_2 - 1, \\ &y_{k+1}|_\gamma = g. \end{aligned} \tag{34}$$

We have used the following notation:

$$\begin{aligned} \alpha &= \frac{\omega_0 h_2^2}{h_1^2 h_2^2 + \omega_0(h_1^2 + h_2^2)}, \quad \beta = \frac{\omega_0 h_1^2}{h_1^2 h_2^2 + \omega_0(h_1^2 + h_2^2)}, \\ \kappa &= \frac{h_1^2 h_2^2}{h_1^2 h_2^2 + \omega_0(h_1^2 + h_2^2)}. \end{aligned} \tag{35}$$

Since α, β, $\kappa > 0$ and $\alpha + \beta + \kappa = 1$, the computations in (32) and (33) are stable. An elementary count of the arithmetic operations in the algorithm (31)–(35) gives $Q_+ = 10(N_1 - 1)(N_2 - 1)$ additions and subtractions and $Q_* = Q_+$ multiplications, or in total, $Q = 20(N_1 - 1)(N_2 - 1)$.

Taking into account the estimate found earlier for the iteration count, we obtain that the compuation of the solution to the difference problem (27) using the algorithm (31)–(35) with accuracy ϵ in the case $N_1 = N_2 = N$, $l_1 = l_2 = l$ requires

$$Q(\epsilon) \approx 5.6 N^2 \sqrt{N} \ln(2/\epsilon)$$

arithmetic operations.

We look now at the first algorithm, which in this case has the form

$$\begin{aligned} &\varphi_k = (E + \omega_0 R_1)(E + \omega_0 R_2) y_k - \tau_{k+1}(A y_k - f), \\ &(E + \omega_0 R_1) v = \varphi_k, \quad (E + \omega_0 R_2) y_{k+1} = v. \end{aligned} \tag{36}$$

In this algorithm, we move to a different notation; it will be convenient for us to work with grid functions defined on $\bar{\omega}$ and vanishing on γ. These functions are the same as y_k, v, and y_{k+1} on ω and, as usual, are denoted $\mathring{y}_k$, $\mathring{v}$, and $\mathring{y}_{k+1}$. In order to obtain an approximation to the solution of problem (27), we proceed as follows: $y_k(x) = \mathring{y}_k(x)$ for $x \in \omega$ and $y_k(x) = g(x)$, $x \in \gamma$.

Here, in order to transform to the (pointwise) difference notation in (36), it is necessary to define the difference operator $\bar{\mathcal{R}}$, which corresponds to the product of the operators $R_1 R_2$. Notice, that from the definition of the operators R_1 and R_2, we can write them in the following form:

$$R_1 y = \begin{cases} \frac{1}{h_1} y_{\bar{x}_1} + \frac{1}{h_2} y_{\bar{x}_2}, & 2 \le i \le N_1 - 1,\ 2 \le j \le N_2 - 1, \\ \frac{1}{h_1^2} y + \frac{1}{h_2} y_{\bar{x}_2}, & i = 1,\ 2 \le j \le N_2 - 1, \\ \frac{1}{h_1} y_{\bar{x}_1} + \frac{1}{h_2^2} y, & 2 \le i \le N_1 - 1,\ j = 1, \\ \left(\frac{1}{h_1^2} + \frac{1}{h_2^2} \right) y, & i = j = 1. \end{cases}$$

$$R_2 y = \begin{cases} -\frac{1}{h_1} y_{x_1} - \frac{1}{h_2} y_{x_2}, & 1 \le i \le N_1 - 2,\ 1 \le j \le N_2 - 2, \\ \frac{1}{h_1^2} y - \frac{1}{h_2} y_{x_2}, & i = N_1 - 1,\ 1 \le j \le N_2 - 2, \\ -\frac{1}{h_1} y_{x_1} + \frac{1}{h_2^2} y, & 1 \le i \le N_1 - 2,\ j = N_2 - 1, \\ \left(\frac{1}{h_1^2} + \frac{1}{h_2^2} \right) y, & i = N_1 - 1,\ j = N_2 - 1. \end{cases}$$

Compuations show that, if the operator $\bar{\mathcal{R}}$ is defined as follows

$$\bar{\mathcal{R}} y = \sum_{\alpha=1}^{2} \frac{1}{h_\alpha^2} y_{\bar{x}_\alpha x_\alpha} + \frac{1}{h_1 h_2} (y_{x_2 \bar{x}_1} + y_{x_1 \bar{x}_2}) + q y, \quad x \in \omega,$$

where

$$q(i,j) = \begin{cases} 0, & 2 \le i \le N_1 - 1,\ 2 \le j \le N_2 - 1, \\ \frac{1}{h_1^4}, & i = 1,\ 2 \le j \le N_2 - 1, \\ \frac{1}{h_2^4}, & 2 \le i \le N_1 - 1,\ j = 1, \\ \frac{1}{h_1^4} + \frac{1}{h_2^4}, & i = 1,\ j = 1, \end{cases}$$

then $R_1 R_2 y = -\bar{\mathcal{R}} \mathring{y}$, where $y \in H$, $\mathring{y} \in \mathring{H}$ and $y(x) = \mathring{y}(x)$ for $x \in \omega$.

Using the definition of the operators A, R_1, and R_2, and also the expression above for $R_1 R_2$, we write the algorithm (36) in the form

$$\begin{aligned} \varphi(i,j) = {} & \left[d_{k+1} - q(i,j) \omega_0^2 \right] \mathring{y}_k(i,j) + a_{k+1} \left[\mathring{y}_k(i+1,j) \right. \\ & \left. + \mathring{y}_k(i-1,j) \right] + b_{k+1} \left[\mathring{y}_k(i,j+1) + \mathring{y}_k(i,j-1) \right] \\ & + c \left[\mathring{y}(i-1,j+1) + \mathring{y}_k(i+1,j-1) \right] + \tau_{k+1} f(i,j), \end{aligned} \tag{37}$$

where we have denoted

$$a_{k+1} = \frac{\tau_{k+1}}{h_1^2} - \frac{\omega_0}{h_1^2}\left(1 + \frac{\omega_0}{h_1^2} + \frac{\omega_0}{h_2^2}\right),$$

$$b_{k+1} = \frac{\tau_{k+1}}{h_2^2} - \frac{\omega_0}{h_2^2}\left(1 + \frac{\omega_0}{h_1^2} + \frac{\omega_0}{h_2^2}\right),$$

$$c = \frac{\omega_0^2}{h_1^2 h_2^2}, \quad d_{k+1} = 1 - 2(a_{k+1} + b_{k+1} + c).$$

Further

$$\begin{aligned} v(i,j) &= \alpha v(i-1,j) + \beta v(i,j-1) + \kappa\varphi_k(i,j), \\ &\quad i = 1,2,\ldots,N_1-1, \quad j = 1,2,\ldots,N_2-1, \\ v(0,j) &= 0, \quad 1 \le j \le N_2 - 1, \quad v(i,0) = 0, \quad 1 \le i \le N_1 - 1, \end{aligned} \tag{38}$$

$$\begin{aligned} \mathring{y}_{k+1}(i,j) &= \alpha\mathring{y}_{k+1}(i+1,j) + \beta\mathring{y}_{k+1}(i,j+1) + \kappa v(i,j), \\ &\quad i = N_1-1, N_1-2,\ldots,1, \quad j = N_2-1, N_2-2,\ldots,1, \\ \mathring{y}_{k+1}|_\gamma &= 0, \end{aligned} \tag{39}$$

where α, β, and κ are defined in (35). A count of the number of arithmetic operations gives $Q_+ = 11N_1N_2 - 10(N_1+N_2) + 10$ additions and subtractions and $Q_* = Q_+$ multiplications; in total, $Q = 22N_1N_2 - 20(N_1 + N_2) + 20$. This is approximately 1.1 times larger than for the algorithm (31)–(34). The advantage of the algorithm (37)–(39) is that it does not require auxiliary memory for the storage of the intermediate quantities $\varphi_k(i,j)$, $v(i,j)$; once again, $\mathring{y}_{k+1}(i,j)$ is overwritten on $\mathring{y}_k(i,j)$.

Remark 2. In the p-dimensional case, the operator $\bar{\mathcal{R}}$ has the form

$$\bar{\mathcal{R}}y = \sum_{\alpha=1}^{p} \frac{1}{h_\alpha^2} y_{\bar{x}_\alpha x_\alpha} + \sum_{\alpha=1}^{p} \sum_{\substack{\beta=1 \\ \beta\neq\alpha}}^{p} \frac{1}{h_\alpha^2 h_\beta^2} y_{x_\beta \bar{x}_\alpha} + qy,$$

where

$$q(i_1, i_2, \ldots, i_p) = \sum_{\alpha=1}^{p} \frac{\delta_{i_\alpha,1}}{h_\alpha^4}, \quad \delta_{i,j} = \begin{cases} 0, & i \neq j, \\ 1, & i = j. \end{cases}$$

Remark 3. The block alternate-triangular method corresponds to the following definition for the difference operators $\mathcal{R}_1$ and $\mathcal{R}_2$:

$$\mathcal{R}_1 y = -\frac{1}{2} y_{\bar{x}_1 x_1} - \frac{1}{h_2} y_{\bar{x}_2}, \quad \mathcal{R}_2 y = -\frac{1}{2} y_{\bar{x}_1 x_1} + \frac{1}{h_2} y_{x_2}.$$

In this case, it is necessary to use the three-point elimination method to invert the operator B. This leads to an increase in the computational work for one iteration which is not compensated for by the small decrease in the iteration count (by a factor of 1.2).

10.2 Boundary-value difference problems for elliptic equations in a rectangle

10.2.1 A Dirichlet problem for an equation with variable coefficients. We look now at an application of the alternate-triangular method in order to find the solution of a Dirichlet difference problem for an elliptic equation without mixed derivatives

$$\begin{aligned} \Lambda y = \sum_{\alpha=1}^{2} (a_\alpha(x) y_{\bar{x}_\alpha})_{x_\alpha} = -\varphi(x), \quad x \in \omega, \\ y(x) = g(x), \quad x \in \gamma, \end{aligned} \tag{1}$$

in a rectangle where $\bar{\omega} = \omega \cup \gamma$ is a rectangular grid with steps h_1 and h_2: $\bar{\omega} = \{x_{ij} = (ih_1, jh_2),\ 0 \le i \le N_1,\ 0 \le j \le N_2,\ h_\alpha N_\alpha = l_\alpha,\ \alpha = 1, 2\}$. We will assume that the coefficients $a_\alpha(x)$ satisfy the conditions

$$0 < c_1 \le a_\alpha(x) \le c_2, \quad \alpha = 1, 2. \tag{2}$$

We will also require, for fixed j, $1 \le j \le N_2 - 1$, that the number of nodes of the grid ω for which $(a_1)_{x_1} = O(h_1^{-1})$ will be finite and not depend on h_1. This indicates that the corresponding coefficient in the differential equation will, for each fixed x_2, have a finite number of points of discontinuity in the direction x_1. An analogous requirement must be satisfied by $(a_2)_{x_2}$.

The difference problem (1) reduces to the operator equation

$$Au = f \tag{3}$$

in the usual way. Here H is the space of grid functions defined on ω with the inner product

$$(u,v)=\sum_{x\in\omega}u(x)v(x)h_1h_2,$$

$$Ay=-\Lambda\mathring{y},\ y\in H,\ \mathring{y}\in\mathring{H}\quad\text{and}\quad y(x)=\mathring{y}(x)\quad\text{for}\quad x\in\omega;$$

$$f(x)=\varphi(x)+\frac{1}{h_1^2}\varphi_1(x)+\frac{1}{h_2^2}\varphi_2(x),$$

where

$$\varphi_1(x)=\begin{cases}a_1(h_1,x_2)g(0,x_2), & x_1=h_1,\\ 0, & 2h_1\le x_1\le l_1-2h_1,\\ a_1(l_1,x_2)g(l_1,x_2), & x_1=l_1-h_1,\end{cases}$$

$$\varphi_2(x)=\begin{cases}a_2(x_1,h_2)g(x_1,0), & x_2=h_2,\\ 0, & 2h_2\le x_2\le l_2-2h_2,\\ a_2(x_1,l_2)g(x_1,l_2), & x_2=l_2-h_2.\end{cases}$$

Using the Green difference formulas, we find that the operator A is self-adjoint in H and satisfies

$$(Ay,y)=-(\Lambda\mathring{y},\mathring{y})=\sum_{\alpha=1}^{2}\left(a_\alpha\mathring{y}_{\bar{x}_\alpha}^2,1\right)_\alpha,\tag{4}$$

where

$$(u,v)_\alpha=\sum_{x_\alpha=h_\alpha}^{l_\alpha}\sum_{x_\beta=h_\beta}^{l_\beta-h_\beta}u(x)v(x)h_\alpha h_\beta,\quad \beta=3-\alpha,\ \alpha=1,2.$$

In order to find an approximate solution to this equation, we look at the alternate-triangular method constructed using a regularizer $R\neq A$:

$$\begin{gathered}B\frac{y_{k+1}-y_k}{\tau_{k+1}}+Ay_k=f,\quad k=0,1,\ldots,\ y_0\in H,\\ B=(E+\omega R_1)(E+\omega R_2),\quad R_1=R_2^*,\ R=R_1+R_2.\end{gathered}\tag{5}$$

The regularizer R is chosen as follows:

$$Ry=-\mathcal{R}\mathring{y},\quad \mathcal{R}y=y_{\bar{x}_1x_1}+y_{\bar{x}_2x_2},\quad \mathring{y}\in\mathring{H},\tag{6}$$

and the operators R_1 and R_2 are defined by the formulas

$$R_\alpha y=-\mathcal{R}_\alpha\mathring{y},\quad \mathcal{R}_1y=-\sum_{\alpha=1}^{2}\frac{1}{h_\alpha}y_{\bar{x}_\alpha},\quad \mathcal{R}_2y=\sum_{\alpha=1}^{2}\frac{1}{h_\alpha}y_{x_\alpha}.\tag{7}$$

In Section 10.1.4 it was shown that, for the operators R_1 and R_2 defined here, we have that

$$\delta E \leq R, \quad R_1 R_2 \leq (\Delta/4)R,$$

$$\delta = \sum_{\alpha=1}^{2} \frac{4}{h_\alpha^2} \sin^2 \frac{\pi h_\alpha}{2l_\alpha}, \quad \Delta = \sum_{\alpha=1}^{2} \frac{4}{h_\alpha^2}.$$

Further, using the Green difference formulas, we obtain

$$(Ry, y) = -(\mathcal{R}\mathring{y}, \mathring{y}) = \sum_{\alpha=1}^{2} \left(\mathring{y}_{\bar{x}_\alpha}^2, 1\right)_\alpha. \tag{8}$$

Consequently, (2), (4), and (8) imply that $c_1 R \leq A \leq c_2 R$, $c_1 > 0$. Since the operator R is self-adjoint and positive-definite in H, theorem 2 is valid for this method with $\mathcal{D} = E$, and this theorem indicates how to choose the iterative parameters ω and $\{\tau_k\}$. From this theorem we also obtain an estimate for the error

$$\| z_n \|_D \leq q_n \| z_0 \|_D, \quad D = A, B \quad \text{or} \quad AB^{-1}A,$$

where

$$q_n = \frac{2\rho_1^n}{1+\rho_1^{2n}}, \quad \rho_1 = \frac{1-\sqrt{\xi}}{1+\sqrt{\xi}}, \quad \xi = \frac{c_1}{c_2} \frac{2\sqrt{\eta}}{1+\sqrt{\eta}}, \quad \eta = \frac{\delta}{\Delta}.$$

For small η we obtain an estimate for the iteration count:

$$n \geq n_0(\epsilon), \quad n_0(\epsilon) = \sqrt{\frac{c_2}{c_1}} \frac{\ln(2/\epsilon)}{2\sqrt{2}\sqrt[4]{\eta}}, \quad \eta = \frac{\pi^2}{4N^2}.$$

From this it follows that the iteration count is proportional to $\sqrt{c_2/c_1}$, and thus it is appropriate to use the method (5), (7) when this ratio is not too large.

10.2.2 A modified alternate-triangular method*. We continue our study of the alternate-triangular method for the difference problem (1) in the case where the coefficients $a_\alpha(x)$ are rapidly varying, i.e. the ration c_2/c_1 is large.

* See A.B. Kucherov and E.S. Nikolaev (Journal of Computational Mathematics and Mathematical Physics, 16, No. 5, 1976; 17, No. 3, 1977).

To solve equation (3), we look now at a modified variant of the alternate-triangular method

$$B\frac{y_{k+1}-y_k}{\tau_{k+1}}+Ay_k=f, \quad k=0,1,\ldots,$$
$$B=(\mathcal{D}+\omega R_1)\mathcal{D}^{-1}(\mathcal{D}+\omega R_2), \quad R_1=R_2^*,\ R_1+R_2=A, \tag{9}$$

where we set $\mathcal{D}y=d(x)y$, $x\in\omega$. Here $d(x)$ is some positive grid function on ω which we will define later. In this case, $\mathcal{D}$ is a self-adjoint and positive-definite operator in H. The grid function $d(x)$ in (9) plays the role of an auxiliary iterative parameter and allows us to take into account the peculiarities of the operator A at each node x of the grid ω.

We define now the operators R_α as follows: $R_\alpha y=-\mathcal{R}_\alpha\mathring{y}$, $y\in H$ and $\mathring{y}\in\mathring{H}$, where

$$\mathcal{R}_1y=-\sum_{\alpha=1}^{2}\left(\frac{a_\alpha}{h_\alpha}y_{\bar{x}_\alpha}+\frac{a_{\alpha x_\alpha}}{2h_\alpha}y\right),$$
$$\mathcal{R}_2y=\sum_{\alpha=1}^{2}\left(\frac{a_\alpha^{+1}}{h_\alpha}y_{x_\alpha}+\frac{a_{\alpha x_\alpha}}{2h_\alpha}y\right), \quad x\in\omega, \tag{10}$$

and $a_1^{\pm1}(x)=a_1(x_1\pm h_1,x_2)$, $a_2^{\pm1}(x)=a_2(x_1,x_2\pm h_2)$.

We shall show that the operators R_1 and R_2 are conjugate in H. To do this, it is sufficient to prove that $(\mathcal{R}_1\mathring{y},\mathring{v})=(\mathring{y},\mathcal{R}_2\mathring{v})$, $\mathring{y}\in\mathring{H}$, $\mathring{v}\in\mathring{H}$. From the Green difference formulas for functions which vanish on γ, and from the formula for the difference derivative of the product of two grid functions $(yv)_{x_\alpha}=y^{+1}v_{x_\alpha}+y_{x_\alpha}v$, it follows that

$$\begin{aligned}(\mathcal{R}_1\mathring{y},\mathring{v})&=-\sum_{\alpha=1}^{2}\frac{1}{h_\alpha}(a_\alpha\mathring{y}_{\bar{x}_\alpha},\mathring{v})-\sum_{\alpha=1}^{2}\frac{1}{2h_\alpha}(a_{\alpha x_\alpha}\mathring{y},\mathring{v})\\&=\sum_{\alpha=1}^{2}\left[\frac{1}{h_\alpha}(\mathring{y},(a_\alpha\mathring{v})_{x_\alpha})-\frac{1}{2h_\alpha}(a_{\alpha x_\alpha}\mathring{y},\mathring{v})\right]\\&=\sum_{\alpha=1}^{2}\left[\frac{1}{h_\alpha}(\mathring{y},a_\alpha^{+1}\mathring{v}_{x_\alpha})+\frac{1}{2h_\alpha}(\mathring{y},a_{\alpha x_\alpha}\mathring{v})\right]=(\mathring{y},\mathcal{R}_2\mathring{v}).\end{aligned}$$

The assertion is proved.

Since $R_1+R_2=A$, by theorem 1 the *a priori* information for the alternate-triangular method (9) has the form of the constants δ and Δ in the inequalities

$$\delta\mathcal{D} \le A, \quad R_1\mathcal{D}^{-1}R_2 \le \frac{\Delta}{4}A, \quad \delta > 0. \tag{11}$$

Since the ratio $\eta = \delta/\Delta$ determines the iteration count, the grid function $d(x)$ should be chosen in order to maximize this ratio.

We examine now the choice of the function $d(x)$ and the estimates for δ and Δ. We shall first prove one inequality.

Lemma 2. *Suppose that $p_\alpha(x)$, $q_\alpha(x)$, $u_\alpha(x)$, and $v_\alpha(x)$, $\alpha = 1, 2$ are grid functions defined on ω. Then for any $x \in \omega$ we have that*

$$\begin{aligned}\left[\sum_{\alpha=1}^{2}(p_\alpha u_\alpha + q_\alpha v_\alpha)\right]^2 &\le (1+\epsilon)(|p_1| + \kappa_1|q_1|)\left(|p_1|u_1^2 + \frac{|q_1|}{\kappa_1}v_1^2\right)\\ &\quad + \frac{1+\epsilon}{\epsilon}(|p_2| + \kappa_2|q_2|)\left(|p_2|u_2^2 + \frac{|q_2|}{\kappa_2}v_2^2\right),\end{aligned} \tag{12}$$

where $\epsilon(x)$, $\kappa_1(x)$ and $\kappa_2(x)$ are arbitrary positive grid functions on ω.

Proof. Using the inequality $2ab \le \epsilon a^2 + b^2/\epsilon$, $\epsilon > 0$, we obtain

$$\begin{aligned}&\left[\sum_{\alpha=1}^{2}(p_\alpha u_\alpha + q_\alpha v_\alpha)\right]^2\\ &\quad = (p_1u_1 + q_1v_1)^2 + 2(p_1u_1 + q_1v_1)(p_2u_2 + q_2v_2) + (p_2u_2 + q_2v_2)^2\\ &\quad \le (1+\epsilon)(p_1u_1 + q_1v_1)^2 + \frac{1+\epsilon}{\epsilon}(p_2u_2 + q_2v_2)^2.\end{aligned} \tag{13}$$

Using this inequality again, we find

$$\begin{aligned}(p_\alpha u_\alpha + q_\alpha v_\alpha)^2 &= p_\alpha^2 u_\alpha^2 + 2p_\alpha q_\alpha u_\alpha v_\alpha + q_\alpha^2 v_\alpha^2\\ &\le p_\alpha^2 u_\alpha^2 + |p_\alpha|\,|q_\alpha|\left(\kappa_\alpha u_\alpha^2 + \frac{1}{\kappa_\alpha}v_\alpha^2\right) + q_\alpha^2 v_\alpha^2\\ &= (|p_\alpha| + \kappa_\alpha|q_\alpha|)\left(|p_\alpha|u_\alpha^2 + \frac{|q_\alpha|}{\kappa_\alpha}v_\alpha^2\right), \quad \kappa_\alpha > 0,\ \alpha = 1, 2.\end{aligned}$$

Substituting this result in (13), we obtain (12). The lemma is proved. □

Exploiting the inequality (12), and also the definition of the operators R_1 and R_2, we find that

$$(R_1\mathcal{D}^{-1}R_2y,y) = (\mathcal{D}^{-1}R_2\mathring{y}, R_2\mathring{y})$$
$$= \left(\frac{1}{d}\left[\sum_{\alpha=1}^{2}\left(\frac{a_\alpha^{+1}}{h_\alpha}\mathring{y}_{x_\alpha} + \frac{a_{\alpha x_\alpha}}{2h_\alpha}\mathring{y}\right)\right]^2, 1\right)$$
$$\leq \left(\frac{(1+\epsilon)}{dh_1^2}(a_1^{+1} + 0.5h_1\kappa_1|a_{1x_1}|)\left(a_1^{+1}\mathring{y}_{x_1}^2 + \frac{0.5h_1|a_{1x_1}|}{\kappa_1 h_1^2}\mathring{y}^2\right), 1\right)$$
$$+ \left(\frac{(1+\epsilon)}{d\epsilon h_2^2}(a_2^{+1} + 0.5h_2\kappa_2|a_{2x_2}|)\left(a_2^{+1}\mathring{y}_{x_2}^2 + \frac{0.5h_2|a_{2x_2}|}{\kappa_2 h_2^2}\mathring{y}^2\right), 1\right).$$

Notice that in (12) we have replaced p_α, q_α, u_α and v_α by

$$p_\alpha = \frac{a_\alpha^{+1}}{h_\alpha},\ q_\alpha = 0.5a_{\alpha x_\alpha},\ u_\alpha = \mathring{y}_{x_\alpha},\ v_\alpha = \frac{1}{h_\alpha}\mathring{y},\ \alpha = 1,2.$$

We will require in this result that κ_1 be only a function of x_2, and κ_2 only of x_1, i.e. we set

$$\kappa_\alpha = \kappa_\alpha(x_\beta), \quad \beta = 3-\alpha, \quad \alpha = 1,2. \tag{14}$$

We denote

$$\epsilon = \epsilon(x) = \frac{a_2^{+1} + 0.5h_2\kappa_2|a_{2x_2}|}{a_1^{+1} + 0.5h_1\kappa_1|a_{1x_1}|}\cdot\frac{h_1^2\theta_2(x_1)}{h_2^2\theta_1(x_2)} \tag{15}$$

and define $d(x)$ as follows:

$$d(x) = \sum_{\alpha=1}^{2}(a_\alpha^{+1} + 0.5h_\alpha\kappa_\alpha|a_{\alpha x_\alpha}|)\frac{\theta_\alpha}{h_\alpha^2}, \tag{16}$$

where $\theta_\alpha = \theta_\alpha(x_\beta)$, $\beta = 3-\alpha$, $\alpha = 1,2$ is a positive grid function on ω, not yet defined.

Substituting (15) and (16) in the inequality obtained above, we will have

$$(R_1\mathcal{D}^{-1}R_2y,y) \leq \sum_{\alpha=1}^{2}\left(\frac{a_\alpha^{+1}}{\theta_\alpha}\mathring{y}_{x_\alpha}^2, 1\right) + \sum_{\alpha=1}^{2}\left(\frac{|a_{\alpha x_\alpha}|}{2h_\alpha\theta_\alpha\kappa_\alpha}\mathring{y}^2, 1\right).$$

Since θ_α does not depend on x_α, then, using the inner product $(,)_\alpha$ introduced earlier, we obtain that

$$\left(\frac{a_\alpha^{+1}}{\theta_\alpha}\mathring{y}_{x_\alpha}^2, 1\right) \leq \left(\frac{a_\alpha}{\theta_\alpha}\mathring{y}_{\bar{x}_\alpha}^2, 1\right)_\alpha, \quad \alpha = 1,2.$$

Consequently,

$$(R_1\mathcal{D}^{-1}R_2y,y) \le \sum_{\alpha=1}^{2}\left(\frac{a_\alpha}{\theta_\alpha}\mathring{y}_{\bar{x}_\alpha}^2,1\right)_\alpha + \sum_{\alpha=1}^{2}\left(\frac{|a_{\alpha x_\alpha}|}{2h_\alpha\theta_\alpha\kappa_\alpha}\mathring{y}^2,1\right). \tag{17}$$

We will now choose θ_α and κ_α. We denote

$$\omega_1 = \{x_1 = ih_1, \quad 1 \le i \le N_1 - 1, \quad h_1N_1 = l_1\},$$
$$\omega_1^+ = \{x_1 = ih_1, \quad 1 \le i \le N_1, \quad h_1N_1 = l_1\}$$

and define

$$(u,v)_{\omega_1} = \sum_{x_1\in\omega_1} u(x)v(x)h_1,$$
$$(u,v)_{\omega_1^+} = \sum_{x_1\in\omega_1^+} u(x)v(x)h_1.$$

We analogously introduce ω_2 and ω_2^+, and also $(u,v)_{\omega_2}$ and $(u,v)_{\omega_2^+}$. Then it is easy to see that

$$\begin{aligned}(u,v) &= ((u,v)_{\omega_1},1)_{\omega_2} = ((u,v)_{\omega_2},1)_{\omega_1},\\ (u,v)_\alpha &= ((u,v)_{\omega_\alpha^+},1)_{\omega_\beta} \quad \beta = 3-\alpha,\ \alpha = 1,2.\end{aligned} \tag{18}$$

Suppose now that $b_\alpha(x_\beta) = \max\limits_{x_\alpha\in\omega_\alpha} v^\alpha(x)$, $\alpha = 1,2$, $x_\beta \in \omega_\beta$, where $v^\alpha(x)$ is, for fixed x_β, the solution of the following three-point boundary-value problem:

$$\begin{aligned}\left(a_\alpha v_{\bar{x}_\alpha}^\alpha\right)_{x_\alpha} &= -\frac{a_\alpha^{+1}}{h_\alpha^2}, \quad h_\alpha \le x_\alpha \le l_\alpha - h_\alpha,\\ v^\alpha(x) &= 0, \quad x_\alpha = 0, l_\alpha, \quad x_\beta \in \omega_\beta.\end{aligned} \tag{19}$$

Then by lemma 13 from Section 5.2.4 we obtain

$$\left(\frac{a_\alpha^{+1}}{h_\alpha^2}\mathring{y}^2,1\right)_{\omega_\alpha} \le b_\alpha(x_\beta)(a_\alpha\mathring{y}_{\bar{x}_\alpha}^2,1)_{\omega_\alpha^+}, \quad \alpha = 1,2.$$

We multiply this inequality by $\theta_\alpha(x_\beta)$ and take its inner product over ω_β. Then by (18) we will have

$$\left(\frac{\theta_\alpha a_\alpha^{+1}}{h_\alpha^2}\mathring{y}^2,1\right) \le (b_\alpha a_\alpha\theta_\alpha\mathring{y}_{\bar{x}_\alpha}^2,1)_\alpha, \quad \alpha = 1,2. \tag{20}$$

Suppose that $c_\alpha(x_\beta) = \max_{x_\alpha \in \omega_\alpha} w^\alpha(x)$, $\alpha = 1, 2$, $x_\beta \in \omega_\beta$, where $w^\alpha(x)$ is, for fixed x_β, the solution of the following three-point boundary-value problem:

$$\begin{gathered}\left(a_\alpha w^\alpha_{\bar{x}_\alpha}\right)_{x_\alpha} = -\frac{|a_{\alpha x_\alpha}|}{2h_\alpha}, \quad h_\alpha \le x_\alpha \le l_\alpha - h_\alpha, \\ w^\alpha(x) = 0, \quad x_\alpha = 0, l_\alpha, \quad x_\beta \in \omega_\beta.\end{gathered} \tag{21}$$

Analogously to the way in which (20) was obtained, by (1) we will have that:

$$\left(\frac{\kappa_\alpha \theta_\alpha |a_{\alpha x_\alpha}|}{2h_\alpha} \mathring{y}^2, 1\right) \le \left(\kappa_\alpha \theta_\alpha c_\alpha a_\alpha \mathring{y}^2_{\bar{x}_\alpha}, 1\right)_\alpha, \quad \alpha = 1, 2, \tag{22}$$

$$\left(\frac{|a_{\alpha x_\alpha}|}{2h_\alpha \theta_\alpha \kappa_\alpha} \mathring{y}^2, 1\right) \le \left(\frac{c_\alpha}{\kappa_\alpha \theta_\alpha} a_\alpha \mathring{y}^2_{\bar{x}_\alpha}, 1\right)_\alpha, \quad \alpha = 1, 2. \tag{23}$$

We now add (20) and (22) and sum then over α. Then by (16) we obtain

$$(d\mathring{y}^2, 1) = (\mathcal{D}y, y) \le \left((\kappa_\alpha c_\alpha + b_\alpha)\theta_\alpha a_\alpha \mathring{y}^2_{\bar{x}_\alpha}, 1\right)_\alpha.$$

Choosing θ_α from the formula

$$\theta_\alpha(x_\beta) = \frac{1}{b_\alpha(x_\beta) + c_\alpha(x_\beta)\kappa_\alpha(x_\beta)}, \quad \beta = 3 - \alpha, \ \alpha = 1, 2, \tag{24}$$

and taking into account (4), we then find that $(\mathcal{D}y, y) \le (Ay, y)$. Consequently, in (11) it is possible to set $\delta = 1$.

We now estimate Δ. To do this, we substitute (23) in (17) and take into account the choice of θ_α in (24). As a result we obtain the following estimate:

$$(R_1 \mathcal{D}^{-1} R_2 y, y) \le \sum_{\alpha=1}^{2} \left((1 + c_\alpha/\kappa_\alpha)(b_\alpha + c_\alpha \kappa_\alpha) a_\alpha \mathring{y}^2_{\bar{x}_\alpha}, 1\right)_\alpha.$$

We now choose the optimal κ_α in order to minimize the expression $(1 + c_\alpha/\kappa_\alpha)(b_\alpha + c_\alpha\kappa_\alpha)$ as a function of κ_α. We obtain $\kappa_\alpha(x_\beta) = \sqrt{b_\alpha(x_\beta)}$, $\beta = 3 - \alpha$, $\alpha = 1, 2$, and thus

$$(R_1 \mathcal{D}^{-1} R_2 y, y) \le \sum_{\alpha=1}^{2} \left((c_\alpha + \sqrt{b_\alpha})^2 a_\alpha \mathring{y}^2_{\bar{x}_\alpha}, 1\right)_\alpha.$$

Comparing this estimate with (4), we find that in (11) it is possible to set

$$\Delta = 4 \max_{\alpha=1,2} \left(\max_{x_\beta \in \omega_\beta} \left(c_\alpha(x_\beta) + \sqrt{b_\alpha(x_\beta)} \right)^2 \right), \quad \beta = 3 - \alpha. \tag{25}$$

Substituting in (16) the expressions for κ_α and θ_α, we obtain the following representation for the function $d(x)$

$$d(x) = \sum_{\alpha=1}^{2} \left(\frac{a_\alpha^{+1}}{h_\alpha^2 \sqrt{b_\alpha}} + \frac{|a_{\alpha x_\alpha}|}{2h_\alpha} \right) \frac{1}{c_\alpha + \sqrt{b_\alpha}}, \quad x \in \omega. \tag{26}$$

Thus, the function $d(x)$ and the constants δ and Δ have been found. Now all that remains is to apply theorem 1. Notice that, since $\delta = 1$, we have that $\omega_0 = 2/\sqrt{\Delta}$, and the iteration count can be estimated using

$$n \geq n_0(\epsilon), \quad n_0(\epsilon) = \frac{\sqrt[4]{\Delta} \ln(2/\epsilon)}{2\sqrt{2}}.$$

Further, using (2), from (19) and (21) we obtain that $b_\alpha = O(1/h_\alpha^2)$ and $c_\alpha = O(1/h_\alpha)$, if the number of points at which $a_{\alpha x_\alpha} = O(h_\alpha^{-1})$ is finite. From this it follows that $n_0(\epsilon) = O(\sqrt{N} \ln(2/\epsilon))$.

We look now at an implementation of this variant of the alternate-triangular method (9). First of all, for fixed x_β, $h_\beta \leq x_\beta \leq l_\beta - h_\beta$, the three-point boundary-value problems (19) and (21) are solved using the elimination method, and the values of $b_\alpha(x_\beta)$ and $c_\alpha(x_\beta)$, $\alpha = 1, 2$ are found. These four one-dimensional grid functions are stored and used in the iterative process to compute $d(x)$ using formula (26). The simplicity of the formula (26) allows us to avoid storing the two-dimensional grid function $d(x)$, and instead we compute it as needed.

Further, from formula (25) we find Δ and set $\delta = 1$. The values of the iterative parameters ω and τ_k for the scheme (9) are defined using theorem 1.

In order to find y_{k+1} given y_k, we use the first of the algorithms described for the alternate-triangular method in Section 10.1.1:

$$\begin{gathered} (\mathcal{D} + \omega_0 R_1) v = \varphi_k, \quad (\mathcal{D} + \omega_0 R_2) y_{k+1} = \mathcal{D} v, \\ \varphi_k = (\mathcal{D} + \omega_0 R_1) \mathcal{D}^{-1} (\mathcal{D} + \omega_0 R_2) y_k - \tau_{k+1}(A y_k - f). \end{gathered} \tag{27}$$

Ignoring the details, we give the difference form of algorithm (27):

$$\begin{gathered} v(i,j) = \alpha_1(i,j) v(i-1,j) + \beta_1(i,j) v(i,j-1) + \kappa(i,j) \varphi_k(i,j), \\ i = 1, 2, \ldots, N_1 - 1, \quad j = 1, 2, \ldots, N_2 - 1, \\ v(0,j) = 0, \quad 1 \leq j \leq N_2 - 1, \quad v(i,0) = 0, \quad 1 \leq i \leq N_1 - 1. \end{gathered} \tag{28}$$

$$\begin{gathered} \mathring{y}_{k+1}(i,j) = \alpha_2(i,j) \mathring{y}_{k+1}(i+1,j) + \beta_2(i,j) \mathring{y}_{k+1}(i,j+1) \\ + \kappa(i,j) d(i,j) v(i,j), \quad i = N_1 - 1, \ldots, 1, \quad j = N_2 - 1, \ldots, 1, \end{gathered} \tag{29}$$

where

$$\alpha_1 = \frac{\omega_0 a_1 \kappa}{h_1^2}, \quad \beta_1 = \frac{\omega_0 a_2 \kappa}{h_2^2}, \quad \alpha_2 = \frac{\omega_0 a_1^{+1} \kappa}{h_1^2}, \quad \beta_2 = \frac{\omega_0 a_2^{+1} \kappa}{h_2^2},$$

$$\frac{1}{\kappa} = d + \omega_0 \left[\frac{a_1^{+1} + a_1}{2h_1^2} + \frac{a_2^{+1} + a_2}{2h_2^2} \right].$$

The right-hand side $\varphi_k(i,j)$ is computed using the formulas

$$\begin{aligned}
\varphi_k(i,j) = {} & [P(i-1,j) + Q(i,j-1) + S(i,j)]\mathring{y}_k(i,j) \\
& + R_1(i,j)\mathring{y}_k(i+1,j) + R_1(i-1,j)\mathring{y}_k(i-1,j) \\
& + R_2(i,j)\mathring{y}_k(i,j+1) + R_2(i,j-1)\mathring{y}_k(i,j-1) \\
& + G(i-1,j)\mathring{y}_k(i-1,j+1) + G(i,j-1)\mathring{y}(i+1,j-1) + \tau_{k+1} f(i,j), \\
& 1 \le i \le N_1 - 1, \quad 1 \le j \le N_2 - 1,
\end{aligned} \tag{30}$$

where

$$G = \frac{\omega_0 a_1^{+1} a_2^{+1}}{h_1^2 h_2^2 d}, \quad R_\alpha = \left(\tau_{k+1} - \frac{\omega_0}{d\kappa} \right) \frac{a_\alpha^{+1}}{h_\alpha^2}, \quad \alpha = 1, 2,$$

$$S = \frac{1}{\omega_0 \kappa} \left[\frac{\omega_0}{d\kappa} - 2\tau_{k+1}/1 - \kappa d) \right], \quad P = \frac{\omega_0^2 (a_1^{+1})^2}{h_1^4 d}, \quad Q = \frac{\omega_0^2 (a_2^{+1})^2}{h_2^4 d},$$

and $P(0,j) = 0$, $1 \le j \le N_2 - 1$, $Q(i,0) = 0$, $1 \le i \le N_1 - 1$. Notice that, by (25) and (26), we have

$$c_\alpha + \sqrt{b_\alpha} \le \frac{\sqrt{\Delta}}{2} = \frac{1}{\omega_0}, \quad d \ge \omega_0 \sum_{\alpha=1}^{2} \frac{|a_{\alpha x_\alpha}|}{2h_\alpha}.$$

From this we obtain

$$\begin{aligned}
\frac{1}{\kappa} &= d + \omega_0 \sum_{\alpha=1}^{2} \frac{a_\alpha^{+1} + a_\alpha}{2h_\alpha^2} \ge \omega_0 \sum_{\alpha=1}^{2} \left(\frac{|a_{\alpha x_\alpha}|}{2h_\alpha} + \frac{a_\alpha^{+1} + a_\alpha}{2h_\alpha^2} \right) \\
&= \omega_0 \left(\max\left(\frac{a_1^{+1}}{h_1^2}, \frac{a_1}{h_1^2} \right) + \max\left(\frac{a_2^{+1}}{h_2^2}, \frac{a_2}{h_2^2} \right) \right)
\end{aligned}$$

or

$$\max(\alpha_1, \alpha_2) + \max(\beta_1, \beta_2) \le 1.$$

From this it follows that $\alpha_1 + \beta_1 \le 1$ and $\alpha_2 + \beta_2 \le 1$. Therefore the computations in (28), (30) are stable.

10.2.3 A comparison of the variants of the method. Above we constructed two variants of the alternate-triangular method for solving the difference problem (1). The variant (5), (7) was constructed on the basis of the regularizer R, and the variant (9), (10) uses the operator $\mathcal{D}$, which is chosen in a special way. These variants are characterized by one and the same asymptotic dependence of the iteration count on the number of nodes of the grid. However, the estimate for the iteration count in the first variant depends on the extrema of the coefficients $a_\alpha(x)$, $\alpha = 1, 2$ in the difference equation (1), whereas in the second variant it is determined from their integral characteristics.

We now compare these variants of the method on the following traditional model problem. Suppose that, on a square grid with $N_1 = N_2 = N$ introduced into the unit square ($l_1 = l_2 = l$), we are given the difference equation (1) in which

$$a_1(x) = 1 + c\left[(x_1 - 0.5)^2 + (x_2 - 0.5)^2\right],$$
$$a_2(x) = 1 + c\left[0.5 - (x_1 - 0.5)^2 - (x_2 - 0.5)^2\right], \quad x \in \bar{\omega}.$$

Then in (2) we have $c_1 = 1$, $c_2 = 1 + 0.5c$. By varying the parameter c, we will obtain coefficients $a_\alpha(x)$ with different extrema.

In table 10 for $\epsilon = 10^{-4}$, we give the iteration counts for the two variants as a function of the number of nodes N in one direction and of the ratio c_2/c_1. It is clear that for large values of c_2/c_1 the modified alternate-triangular method requires fewer iterations, and the iteration count only depends weakly on this ratio.

Table 10

	$N = 32$		$N = 64$		$N = 128$	
c_2/c_1	(5), (7)	(9), (10)	(5), (7)	(9), (10)	(5), (7)	(9), (10)
2	23	18	32	26	45	36
8	46	21	64	30	90	43
32	92	23	128	34	180	49
128	184	24	256	36	360	53
512	367	24	512	36	720	54

10.2.4 A boundary-value problem of the third kind. We now use the alternate-triangular method to solve a boundary-value problem of the third kind for an elliptic equation in the rectangle $\bar{G} = \{0 \le x_\alpha \le l_\alpha,\ \alpha = 1,2\}$:

$$\begin{aligned}
&\sum_{\alpha=1}^{2} \frac{\partial}{\partial x_\alpha}\left(k_\alpha(x)\frac{\partial u}{\partial x_\alpha}\right) = -\varphi(x), && x \in G,\\
&k_\alpha \frac{\partial u}{\partial x_\alpha} = \kappa_{-\alpha}(x)u - g_{-\alpha}(x), && x_\alpha = 0,\\
&-k_\alpha \frac{\partial u}{\partial x_\alpha} = \kappa_{+\alpha}(x)u - g_{+\alpha}(x), && x_\alpha = l_\alpha,
\end{aligned} \tag{31}$$

On the rectangular grid $\bar{\omega} = \{x_{ij} = (ih_1, jh_2) \in \bar{G},\ 0 \le i \le N_1,\ 0 \le j \le N_2,\ h_\alpha N_\alpha = l_\alpha,\ \alpha = 1,2\}$, the problem (31) corresponds to the difference problem

$$\begin{aligned}
&\Lambda y = -f(x), \quad x \in \bar{\omega},\\
&\Lambda = \Lambda_1 + \Lambda_2, \quad f(x) = \varphi(x) + \frac{2}{h_1}\varphi_1(x) + \frac{2}{h_2}\varphi_2(x),
\end{aligned} \tag{32}$$

where

$$\Lambda_\alpha y = \begin{cases} \dfrac{2}{h_\alpha}\left(a_\alpha^{+1} y_{x_\alpha} - \kappa_{-\alpha} y\right), & x_\alpha = 0,\\ (a_\alpha y_{\bar{x}_\alpha})_{x_\alpha}, & h_\alpha \le x_\alpha \le l_\alpha - h_\alpha,\\ \dfrac{2}{h_\alpha}\left(-a_\alpha y_{\bar{x}_\alpha} - \kappa_{+\alpha} y\right), & x_\alpha = l_\alpha,\end{cases}$$

$$\varphi_\alpha(x) = \begin{cases} g_{-\alpha}(x_\alpha), & x_\alpha = 0,\\ 0, & h_\alpha \le x_\alpha \le l_\alpha - h_\alpha,\\ g_{+\alpha}(x_\alpha), & x_\alpha = l_\alpha.\end{cases}$$

We will assume that the coefficients $a_\alpha(x)$ satisfy the conditions (2) and that there are a finite number of points at which $a_{\alpha x_\alpha} = O(h_\alpha^{-1})$. We will also assume that $\kappa_{-\alpha}(x_\beta)$ and $\kappa_{+\alpha}(x_\beta)$ do not simultaneously vanish for any x_β $(\kappa_{-\alpha} \ge 0,\ \kappa_{+\alpha} \ge 0,\ \kappa_{-\alpha} + \kappa_{+\alpha} > 0)$.

It is convenient to first reduce the difference problem (32) to a Dirichlet problem in the expanded region $\bar{\omega}^* = \{x_{ij} = (ih_1, jh_2),\ -1 \le i \le N_1 + 1,\ -1 \le j \le N_2 + 1\}$ containing the grid $\bar{\omega}$. We will denote by γ^* the boundary of the grid $\bar{\omega}^*$ and extend the function $y(x)$ by setting it equal to zero on γ^*.

If we denote

$$\bar{a}_\alpha(x) = \begin{cases} \rho(x_\beta)h_\alpha\kappa_{-\alpha}(x_\beta), & x_\alpha = 0, \\ \rho(x_\beta)a_\alpha(x), & h_\alpha \le x_\alpha \le l_\alpha, \\ \rho(x_\beta)h_\alpha\kappa_{+\alpha}(x_\beta), & x_\alpha = l_\alpha + h_\alpha, \quad 0 \le x_\beta \le l_\beta, \end{cases}$$

$$\bar{f}(x) = \rho(x_1)\rho(x_2)f(x), \quad x \in \bar{\omega},$$

$$\rho(x_\beta) = \begin{cases} 0.5, & x_\beta = 0, l_\beta, \\ 1, & h_\beta \le x_\beta \le l_\beta - h_\beta, \end{cases}$$

$$\beta = 3 - \alpha, \quad \alpha = 1, 2,$$

then the problem (32) can be written in the form

$$\begin{aligned} \bar{\Lambda} y = \sum_{\alpha=1}^{2} (\bar{a}_\alpha y_{\bar{x}_\alpha})_{x_\alpha} &= -\bar{f}(x), \quad x \in \bar{\omega}, \\ y(x) = 0, \quad x \in \gamma^*. \end{aligned} \tag{33}$$

Recall that a modified alternate-triangular method was constructed in Section 10.2.2 for a difference problem of the form (33). Consequently, in the formulas of Section 10.2.2, it is only necessary to change $a_\alpha(x)$ to $\bar{a}_\alpha(x)$, and we obtain a method for solving a boundary-value problem of the third kind for an elliptic equation in a rectangle.

For the case considered here, the three-point boundary-value problem (19) is written in the form

$$\begin{aligned} \left(\bar{a}_\alpha v^\alpha_{\bar{x}_\alpha}\right)_{x_\alpha} &= -\frac{\bar{a}_\alpha^{+1}}{h_\alpha^2}, \quad 0 \le x_\alpha \le l_\alpha, \\ v^\alpha(x) = 0, &\quad x_\alpha = -h_\alpha, l_\alpha + h_\alpha. \end{aligned} \tag{34}$$

Using the notation introduced above for $\bar{a}_\alpha$, we obtain that (34) can posed in another way

$$\begin{aligned} \left(a_\alpha v^\alpha_{\bar{x}_\alpha}\right)_{x_\alpha} &= -\frac{a_\alpha^{+1}}{h_\alpha^2}, \quad h_\alpha \le x_\alpha \le l_\alpha - h_\alpha, \\ a_\alpha^{+1} x^\alpha_{x_\alpha} - \kappa_{-\alpha} v^\alpha &= -\frac{a_\alpha^{+1}}{h_\alpha}, \quad x_\alpha = 0, \\ -a_\alpha v^\alpha_{\bar{x}_\alpha} - \kappa_{+\alpha} v^\alpha &= -\kappa_{+\alpha}, \quad x_\alpha = l_\alpha. \end{aligned} \tag{35}$$

By the assumptions made above concerning $a_\alpha(x)$, $\kappa_{-\alpha}$, and $\kappa_{+\alpha}$, the difference problem is soluble. Then

$$b_\alpha(x_\beta) = \max_{0\le x_\alpha\le l_\alpha} v^\alpha(x) = O\left(\frac{1}{h_\alpha^2}\right), \quad 0\le x_\beta \le l_\beta.$$

Analogously, problem (21), which in this case has the form

$$\begin{gathered} \left(\bar{a}_\alpha w^\alpha_{\bar{x}_\alpha}\right)_{x_\alpha} = -\frac{|\bar{a}_{\alpha x_\alpha}|}{2h_\alpha}, \quad 0\le x_\alpha\le l_\alpha, \\ w^\alpha(x) = 0, \quad x_\alpha = -h_\alpha, l_\alpha + h_\alpha, \end{gathered}$$

in this notation reduces to a boundary-value problem of the third kind

$$\begin{aligned} \left(a_\alpha w^\alpha_{\bar{x}_\alpha}\right)_{x_\alpha} &= -\frac{|a_{\alpha x_\alpha}|}{2h_\alpha}, & h_\alpha \le x_\alpha \le l_\alpha - h_\alpha, \\ a_\alpha^{+1} w^\alpha_{x_\alpha} - \kappa_{-\alpha} w^\alpha &= -\frac{|a_\alpha^{+1} - h_\alpha\kappa_{-\alpha}|}{2h_\alpha}, & x_\alpha = 0, \\ -a_\alpha w^\alpha_{\bar{x}_\alpha} - \kappa_{+\alpha} w^\alpha &= -\frac{|a_\alpha - h_\alpha\kappa_{+\alpha}|}{2h_\alpha}, & x_\alpha = l_\alpha. \end{aligned} \tag{36}$$

From this we obtain that

$$c_\alpha(x_\beta) = \max_{0\le x_\alpha\le l_\alpha} w^\alpha(x) = O\left(\frac{1}{h_\alpha}\right),$$

and, consequently,

$$\Delta = 4\max_{\alpha=1,2}\left(\max_{0\le x_\beta\le l_\beta}\left(c_\alpha(x_\beta) + \sqrt{b_\alpha(x_\beta)}\right)^2\right) = O\left(\frac{1}{|h|^2}\right).$$

Therefore, for the modified alternate-triangular method applied to the solution of the boundary-value problem of the third kind (32), the iteration count depends on the number of nodes in the same way as for a boundary-value problem of the first kind.

Obviously, the above transformation to a Dirichlet problem can also be carried out in the case when first-, second-, or third-kind boundary conditions are given on the sides of the rectangle.

10.2.5 A Dirichlet difference problem for an equation with mixed derivatives. Suppose that, in the rectangle $\bar{G} = \{0 \le x_\alpha \le l_\alpha,\ \alpha = 1,2\}$ with boundary Γ, it is necessary to find the solution to a Dirichlet problem for an equation of elliptic type with mixed derivatives

$$
\begin{aligned}
&Lu = \sum_{\alpha,\beta=1}^{2} \frac{\partial}{\partial x_\alpha}\left(k_{\alpha\beta}(x)\frac{\partial u}{\partial x_\beta}\right) = -\varphi(x), \quad x \in G,\\
&u(x) = g(x), \quad x \in \Gamma, \quad k_{\alpha\beta}(x) = k_{\beta\alpha}(x).
\end{aligned}
\tag{37}
$$

We will require that the following conditions are satisfied

$$
k_{\alpha\alpha}(x) \ge c > 0, \quad k_{12}^2(x) \le \rho^2 k_{11}(x)k_{22}(x), \quad x \in \bar{G}, \quad 0 \le \rho < 1. \tag{38}
$$

Notice that the conditions (38) guarantee the uniform ellipticity of equation (37). In fact, for fixed $x \in G$, we examine the eigenvalue problem for the matrix pencil

$$
\left\| \begin{matrix} k_{11} & k_{12} \\ k_{12} & k_{22} \end{matrix} \right\| - \lambda \left\| \begin{matrix} k_{11} & 0 \\ 0 & k_{22} \end{matrix} \right\| = 0.
$$

We obtain the quadratic equation $(1-\lambda)^2 k_{11}k_{22} - k_{12}^2 = 0$ for λ. From this we find

$$
|1-\lambda| = \frac{|k_{12}|}{\sqrt{k_{11}k_{12}}} \le \rho, \quad 1-\rho \le \lambda \le 1+\rho.
$$

Consequently, we obtain the inequality

$$
c_1 \sum_{\alpha=1}^{2} k_{\alpha\alpha}(x)\xi_\alpha^2 \le \sum_{\alpha,\beta=1}^{2} k_{\alpha\beta}(x)\xi_\alpha\xi_\beta \le c_2 \sum_{\alpha=1}^{2} k_{\alpha\alpha}(x)\xi_\alpha^2, \tag{39}
$$

where $\xi = (\xi_1, \xi_2)$ is an arbitrary vector. Then, the uniform ellipticity of the operator L follows from the condition $k_{\alpha\alpha}(x) \ge c > 0$.

On the rectangular grid $\bar{\omega} = \{x_{ij} = (ih_1, jh_2) \in \bar{G},\ 0 \le i \le N_1,\ 0 \le j \le N_2,\ h_\alpha N_\alpha = l_\alpha,\ \alpha = 1,2\}$, the problem (37), (38) can be placed in correspondence with a Dirichlet difference problem

$$
\begin{aligned}
&\Lambda y = \frac{1}{2}\sum_{\alpha,\beta=1}^{2} \left[(k_{\alpha\beta}y_{\bar{x}_\beta})_{x_\alpha} + (k_{\alpha\beta}y_{x_\beta})_{\bar{x}_\alpha}\right] = -\varphi(x),\ x \in \omega,\\
&y(x) = g(x), \quad x \in \gamma.
\end{aligned}
\tag{40}
$$

In the space H of grid functions defined on ω with inner product

$$(u,v) = \sum_{x\in\omega} u(x)v(x)h_1h_2$$

we define the operator A as follows: $Ay = -\Lambda\mathring{y}$, $y \in H$ and $y(x) = \mathring{y}(x)$ for $x \in \omega$, $\mathring{y}(x) = 0$ for $x \in \gamma$; we also define the operator R: $Ry = -\mathcal{R}\mathring{y}$, where

$$\mathcal{R}y = \sum_{\alpha=1}^{2}(a_\alpha y_{\bar{x}_\alpha})_{x_\alpha}, \quad x \in \omega, \quad a_\alpha(x) = \frac{k_{\alpha\alpha} + k_{\alpha\alpha}^{-1}}{2}.$$

Then problem (40) can be written in the form of equation (3), where $f(x)$ only differs from $\varphi(x)$ at the near-boundary nodes. Since $k_{\alpha\beta}(x) = k_{\beta\alpha}(x)$, the operators A and R are self-adjoint. We shall show that the following inequalities are satisfied

$$c_1R \le A \le c_2R, \quad c_1 = 1-\rho, \quad c_2 = 1+\rho, \tag{41}$$

where ρ is given in (38). From Green's difference formulas we obtain

$$(Ay,y) = -(\Lambda\mathring{y},\mathring{y}) = \sum_{\alpha,\beta=1}^{2} \frac{1}{2}\left[(k_{\alpha\beta}\mathring{y}_{\bar{x}_\beta},\mathring{y}_{\bar{x}_\alpha})_\alpha + {}_\alpha(k_{\alpha\beta}\mathring{y}_{x_\beta},\mathring{y}_{x_\alpha})\right],$$

where the inner product (u,v) is defined in Section 10.2.1, and

$${}_\alpha(u,v) = \sum_{x_\alpha=0}^{l_\alpha-h_\alpha}\sum_{x_\beta=h_\beta}^{l_\beta-h_\beta} u(x)v(x)h_1h_2, \quad \beta = 3-\alpha, \quad \alpha = 1,2.$$

Notice that, if one of the function $u(x)$ or $v(x)$ vanishes for $x_\beta = 0$ (or for $x_\beta = l_\beta$), then

$${}_\alpha(u,v) = [u,v) = \sum_{x_1=0}^{l_1-h_1}\sum_{x_2=0}^{l_2-h_2} u(x)v(x)h_1h_2,$$

$$(u,v)_\alpha = (u,v] = \sum_{x_1=h_1}^{l_1}\sum_{x_2=h_2}^{l_2} u(x)v(x)h_1h_2, \quad \alpha = 1,2.$$

This at once gives

$$(Ay,y) = \sum_{\alpha,\beta=1}^{2}\frac{1}{2}\left\{(k_{\alpha\beta}\mathring{y}_{\bar{x}_\beta},\mathring{y}_{\bar{x}_\alpha}] + [k_{\alpha\beta}\mathring{y}_{x_\beta},\mathring{y}_{x_\alpha})\right\}. \tag{42}$$

Further, since $\mathcal{R}y$ can be written in the form

$$\mathcal{R}y = \frac{1}{2}\sum_{\alpha=1}^{2}\left\{(k_{\alpha\alpha}y_{\bar{x}_\alpha})_{x_\alpha} + (k_{\alpha\alpha}y_{x_\alpha})_{\bar{x}_\alpha}\right\},$$

then

$$\begin{aligned}(Ry, y) = -(\mathcal{R}\mathring{y}, \mathring{y}) &= \sum_{\alpha=1}^{2}\frac{1}{2}\{(k_{\alpha\alpha}\mathring{y}_{\bar{x}_\alpha}, \mathring{y}_{\bar{x}_\alpha})_\alpha + {}_\alpha(k_{\alpha\alpha}\mathring{y}_{x_\alpha}, \mathring{y}_{x_\alpha})\} \\ &= \sum_{\alpha=1}^{2}\frac{1}{2}\left\{\left(k_{\alpha\alpha}\mathring{y}^2_{\bar{x}_\alpha}, 1\right] + \left[k_{\alpha\alpha}\mathring{y}^2_{x_\alpha}, 1\right)\right\}.\end{aligned} \tag{43}$$

From (42), (43) and the inequalities (39) we obtain

$$c_1\left(\sum_{\alpha=1}^{2}k_{\alpha\alpha}\mathring{y}^2_{\bar{x}_\alpha}, 1\right] \le \left(\sum_{\alpha,\beta=1}^{2}k_{\alpha\beta}\mathring{y}_{\bar{x}_\beta}\mathring{y}_{\bar{x}_\alpha}, 1\right] \le c_2\left(\sum_{\alpha=1}^{2}k_{\alpha\alpha}\mathring{y}^2_{\bar{x}_\alpha}, 1\right]$$

and analogously

$$c_1\left[\sum_{\alpha=1}^{2}k_{\alpha\alpha}\mathring{y}^2_{x_\alpha}, 1\right) \le \left[\sum_{\alpha,\beta=1}^{2}k_{\alpha\beta}\mathring{y}_{x_\beta}\mathring{y}_{x_\alpha}, 1\right) \le c_2\left[\sum_{\alpha=1}^{2}k_{\alpha\alpha}\mathring{y}^2_{x_\alpha}, 1\right),$$

and consequently, the estimates (41) have been established.

Thus the operator R defined above can be used as a regularizer in the alternate-triangular method

$$B\frac{y_{k+1} - y_k}{\tau_{k+1}} + Ay_k = f, \quad k = 0, 1, \ldots,$$

$$B = (\mathcal{D} + \omega R_1)\mathcal{D}^{-1}(\mathcal{D} + \omega R_2), \quad R_1 = R_1^*, \quad R_1 + R_2 = R,$$

where the operators R_1, R_2, and $\mathcal{D}$ were defined in Section 10.2.2. There we also found the constants δ and Δ for the inequalities $\delta\mathcal{D} \le R$, $R_1\mathcal{D}^{-1}R_2 \le \frac{\Delta}{4}R$, $\delta > 0$. The application of theorem 2 completes the construction of the alternate-triangular method for the difference problem (40).

10.3 The alternate-triangular method for elliptic equations in arbitrary regions

10.3.1 The statement of the difference problem. We now construct a modified alternate-triangular method for the solution of a Dirichlet difference problem in an arbitrary bounded region $\bar{G}$ with boundary Γ in the case of an elliptic equation with variable coefficients

$$\sum_{\alpha=1}^{2} \frac{\partial}{\partial x_\alpha}\left(k_\alpha(x)\frac{\partial u}{\partial x_\alpha}\right) = -\varphi(x), \quad x \in G,$$
$$u(x) = g(x), \quad x \in \Gamma, \quad k_\alpha(x) \geq c_1 > 0, \quad \alpha = 1, 2. \tag{1}$$

We will assume that the boundary Γ is sufficiently smooth. In addition, for simplicity we will suppose that the intersection of the region with a straight line passing through any point $x \in G$ parallel to a coordinate axis Ox_α, $\alpha = 1, 2$, consists of a single interval.

In the region $\bar{G}$ we construct a non-uniform grid $\bar{\omega}$ in the following way. We introduce the family of straight lines $x_\alpha = x_\alpha(i_\alpha)$, $i_\alpha = 0, \pm 1, \pm 2, \ldots$, $\alpha = 1, 2$. Then the points $x_i(x_1(i_1), x_2(i_2))$, $i = (i_1, i_2)$ form the basic grid on the surface. A lattice point x_i belonging to G will be called an interior node of the grid $\bar{\omega}$. The set of all interior nodes will, as usual, be denoted by ω.

The intersection with the region G of any straight line passing through a point $x_i \in \omega$ parallel to a coordinate axis Ox_α is an interval $\Delta_\alpha(x_i)$. The ends of this interval will be called the boundary nodes in the direction x_α. The set of all boundary nodes corresponding to x_α will be denoted by y_α. The boundary of the grid $\bar{\omega}$ is $\gamma = \gamma_1 \cup \gamma_2$, so that $\bar{\omega} = \omega \cup \gamma$. The grid $\bar{\omega}$ has been constructed.

We now introduce some notation. We denote by $\omega_\alpha(x_\beta)$, $\beta = 3 - \alpha$, $\alpha = 1, 2$ the set of nodes of the grid ω lying on the interval Δ_α; $\omega_\alpha^+(x_\beta)$ is the set consisting of $\omega_\alpha(x_\beta)$ and the right endpoint of the interval Δ_α; $\bar{\omega}_\alpha(x_\beta)$ consists of $\omega_\alpha(x_\beta)$ and the endpoints of the interval Δ_α.

We denote by $x^{(+1_\alpha)}$ and $x^{(-1_\alpha)}$ the nodes adjacent to $x \in \omega_\alpha(x_\beta)$ to the right and to the left belonging to $\bar{\omega}_\alpha(x_\beta)$. Notice that if, for example, $x^{(\pm 1_\alpha)} \in \gamma_\alpha$ then this node cannot be a node of the basic grid.

We define $h_\alpha^+(x) = x^{(+1_\alpha)} - x$, $h_\alpha^-(x) = x - x^{(-1_\alpha)}$, $x \in \omega_\alpha$, $x^{(\pm 1_\alpha)} \in \bar{\omega}_\alpha$. At all the interior nodes of the grid ω we also define the average step $\hbar_\alpha(x_\alpha) = 0.5(x_\alpha(i_\alpha+1) - x_\alpha(i_\alpha-1))$ as the distance between the corresponding straight lines of the basic grid.

We place the problem (1) on the grid $\bar{\omega}$ in correspondence with the difference problem

$$\Lambda y = \sum_{\alpha=1}^{2} (a_\alpha y_{\bar{x}_\alpha})_{\hat{x}_\alpha} = -\varphi(x), \quad x \in \omega,$$
$$y(x) = g(x), \quad x \in \gamma. \tag{2}$$

Here we have used the following notation:

$$(a_\alpha y_{\bar{x}_\alpha})_{\hat{x}_\alpha} = \frac{1}{\hbar_\alpha}\left(a_\alpha^{+1} y_{x_\alpha} - a_\alpha y_{\bar{x}_\alpha}\right), \quad a_\alpha^{+1} = a_\alpha\left(x^{(+1_\alpha)}\right),$$
$$y_{x_\alpha} = \frac{1}{h_\alpha^+}\left(y(x^{(+1_\alpha)}) - y(x)\right), \quad y_{\bar{x}_\alpha} = \frac{1}{h_\alpha^-}\left(y(x) - y(x^{(-1_\alpha)})\right).$$

The coefficients $a_\alpha(x)$ and $\varphi(x)$ are chosen so that the scheme (2) on a uniform grid has local second-order approximation.

We now introduce H — the set of grid functions defined on ω with inner product

$$(u, v) = \sum_{x \in \omega} u(x) v(x) \hbar_1(x_1) \hbar_2(x_2).$$

The operator A is defined in the usual way: $Ay = -\Lambda \mathring{y}$, $y \in H$ and $y(x) = \mathring{y}(x)$ for $x \in \omega$, $\mathring{y}(x) = 0$ for $x \in \gamma$. Then the difference problem (2) can be written in the form of the equation

$$Au = f, \tag{3}$$

where $f(x)$ only differs from $\varphi(x)$ at the near-boundary nodes.

10.3.2 The construction of an alternate-triangular method. For equation (3) we look at a modified alternate-triangular method

$$B\frac{y_{k+1} - y_k}{\tau_{k+1}} + Ay_k = f, \quad k = 0, 1, \ldots,$$
$$B = (\mathcal{D} + \omega R_1)\mathcal{D}^{-1}(\mathcal{D} + \omega R_2), \quad R_1 = R_1^*, \quad R_1 + R_2 = A. \tag{4}$$

We now define the operators R_1, R_2 and $\mathcal{D}$. As in the case of a rectangle, we choose a very simple operator $\mathcal{D}$:

$$\mathcal{D}y = d(x)y, \quad d(x) > 0 \quad \text{for} \quad x \in \omega, \tag{5}$$

and set $R_\alpha y = -\mathcal{R}_\alpha y$, $y \in H$ and $y(x) = \mathring{y}(x)$, $x \in \omega$, where

$$\mathcal{R}_1 y = -\sum_{\alpha=1}^{2}\left[\frac{a_\alpha}{\hbar_\alpha} y_{\bar{x}_\alpha} + \frac{1}{2\hbar_\alpha}\left(\frac{a_\alpha^{+1}}{h_\alpha^+} - \frac{a_\alpha}{h_\alpha^-}\right) y\right],$$
$$\mathcal{R}_2 y = \sum_{\alpha=1}^{2}\left[\frac{a_\alpha^{+1}}{\hbar_\alpha} y_{x_\alpha} + \frac{1}{2\hbar_\alpha}\left(\frac{a_\alpha^{+1}}{h_\alpha^+} - \frac{a_\alpha}{h_\alpha^-}\right) y\right]. \tag{6}$$

Since $\mathcal{R}_1 + \mathcal{R}_2 = \Lambda$, we obtain that $R_1 + R_2 = A$.

We will show that the operators R_1 and R_2 are conjugate. First we introduce some notation that will be used later. We define the following summations:

$$(u,v)_{\omega_\alpha(x_\beta)} = \sum_{x_\alpha \in \omega_\alpha(x_\beta)} u(x)v(x)\hbar_\alpha(x_\alpha),$$

$$(u,v)_{\omega_\alpha^+(x_\beta)} = \sum_{x_\alpha \in \omega_\alpha^+(x_\beta)} u(x)v(x)h_\alpha^-(x),$$

$$(u,v)_\alpha = \left((u,v)_{\omega_\alpha^+(x_\beta)}, 1\right)_{\omega_\beta(x_\alpha)}$$
$$= \sum_{x_\beta \in \omega_\beta} \sum_{x_\alpha \in \omega_\alpha^+} u(x)v(x)h_\alpha^-(x)\hbar_\beta(x_\beta),$$
$$\beta = 3-\alpha, \quad \alpha = 1,2.$$

Using this notation, the inner product in H can be written in the following form:

$$(u,v) = \left((u,v)_{\omega_1(x_2)}, 1\right)_{\omega_2(x_1)} = \left((u,v)_{\omega_2(x_1)}, 1\right)_{\omega_1(x_2)}. \tag{7}$$

First we prove one auxiliary result. Suppose that y_i and v_i are grid functions, defined for $0 \le i \le N$, where $y_0 = y_N = 0$ and $v_0 = v_N = 0$. Also assume that u_i is a grid function defined for $1 \le i \le N$. Then we have that

$$\sum_{i=1}^{N-1}(u_{i+1} - u_i)y_i v_i = -\sum_{i=1}^{N-1}(v_{i+1} - v_i)u_{i+1}y_i - \sum_{i=1}^{N-1}(y_i - y_{i-1})u_i v_i. \tag{8}$$

In fact, we have

$$\sum_{i=1}^{N-1}\left[(u_{i+1} - u_i)y_i v_i + (v_{i+1} - v_i)u_{i+1}y_i + (y_i - y_{i-1})u_i v_i\right]$$
$$= \sum_{i=1}^{N-1}(u_{i+1}v_{i+1}y_i - u_i v_i y_{i-1}) = u_N v_N y_{N-1} - u_1 v_1 y_0 = 0.$$

The assertion (8) has been proved. Using (8) it is easy to show that, for functions $\mathring{y}(x)$ and $\mathring{v}(x)$ defined on $\bar{\omega}$ and vanishing on γ, the following equation is valid

$$(u_{\hat{x}_\alpha}\mathring{y},\mathring{v})_{\omega_\alpha} = -\left(\mathring{y},\frac{h_\alpha^+}{\hbar_\alpha}u^{+1_\alpha}\mathring{v}_{x_\alpha}\right)_{\omega_\alpha} - \left(\frac{h_\alpha^-}{\hbar_\alpha}u\mathring{y}_{\bar{x}_\alpha},\mathring{v}\right)_{\omega_\alpha}. \tag{9}$$

Here the following notation has been used

$$u^{+1_\alpha} = u\left(x^{(+1_\alpha)}\right), \quad u_{\hat{x}_\alpha} = \frac{u^{+1_\alpha}-u}{\hbar_\alpha(x_\alpha)}.$$

Substituting the expression $u(x) = a_\alpha(x)/h_\alpha^-(x)$ in (9), and taking into account the equation $h_\alpha^-(x^{(+1_\alpha)}) = h_\alpha^+(x)$, we obtain

$$\left(\frac{1}{\hbar_\alpha}\left(\frac{a_\alpha^{+1}}{h_\alpha^+}-\frac{a_\alpha}{h_\alpha^-}\right)\mathring{y},\mathring{v}\right)_{\omega_\alpha} = -\left(\mathring{y},\frac{a_\alpha^{+1}}{\hbar_\alpha}\mathring{v}_{x_\alpha}\right)_{\omega_\alpha} - \left(\frac{a_\alpha}{\hbar_\alpha}\mathring{y}_{\bar{x}_\alpha},\mathring{v}\right)_{\omega_\alpha}.$$

Multiplying this relation by $\hbar(x_\beta)$, summing over ω_β, and using (7), we find

$$-\left(\frac{a_\alpha}{\hbar_\alpha}\mathring{y}_{\bar{x}_\alpha},\mathring{v}\right) = \left(\frac{a_\alpha^{+1}}{\hbar_\alpha}\mathring{v}_{x_\alpha},\mathring{y}\right) + \left(\frac{1}{\hbar_\alpha}\left(\frac{a_\alpha^{+1}}{h_\alpha^+}-\frac{a_\alpha}{h_\alpha^-}\right)\mathring{y},\mathring{v}\right). \tag{10}$$

We now show that the operators R_1 and R_2 are conjugate. From (6) and (10) we obtain

$$\begin{aligned}(R_1y,v) &= -(\mathcal{R}_1\mathring{y},\mathring{v}) \\ &= \sum_{\alpha=1}^{2}\left[\left(\frac{a_\alpha}{\hbar_\alpha}\mathring{y}_{\bar{x}_\alpha},\mathring{v}\right) + \left(\frac{1}{2\hbar_\alpha}\left(\frac{a_\alpha^{+1}}{h_\alpha^+}-\frac{a_\alpha}{h_\alpha^-}\right)\mathring{y},\mathring{v}\right)\right] \\ &= -\sum_{\alpha=1}^{2}\left[\left(\mathring{y},\frac{a_\alpha^{+1}}{\hbar_\alpha}\mathring{v}_{x_\alpha}\right) + \left(\frac{1}{2\hbar_\alpha}\left(\frac{a_\alpha^{+1}}{h_\alpha^+}-\frac{a_\alpha}{h_\alpha^-}\right)\mathring{y},\mathring{v}\right)\right] \\ &= -(\mathring{y},\mathcal{R}_2\mathring{v}) = (y,R_2v).\end{aligned}$$

The result has been proved. Incidentally, the self-adjointness of the operator A also follows from this result.

In order to construct the function $d(x)$ defining the operator $\mathcal{D}$, all that remains for us to do is to find the constants δ and Δ in the inequalities

$$\delta\mathcal{D} \le A, \quad R_1\mathcal{D}^{-1}R_2 \le \frac{\Delta}{4}A, \quad \delta > 0 \tag{11}$$

and to use the theorem 1. We do all of this in the same way as in Section 10.2.2, where we considered a Dirichlet problem for an elliptic equation in a rectangle on a uniform grid.

First of all we remark that, by Green's difference formulas, we have that

$$(Ay,y) = \sum_{\alpha=1}^{2} \left(a_\alpha \mathring{y}_{\bar{x}_\alpha}^2, 1\right)_\alpha, \quad \mathring{y}(x) = 0, \quad x \in \gamma.$$

Further, from (5) and (6) we find

$$(R_1 \mathcal{D}^{-1} R_2 y, y) = (\mathcal{D}^{-1} \mathcal{R}_2 \mathring{y}, \mathcal{R}_2 \mathring{y})$$
$$= \left(\frac{1}{d} \sum_{\alpha=1}^{2} \left[\frac{a_\alpha^{+1}}{\hbar_\alpha} \mathring{y}_{x_\alpha} + \frac{1}{2\hbar_\alpha}\left(\frac{a_\alpha^{+1}}{h_\alpha^+} - \frac{a_\alpha}{h_\alpha^-}\right) y\right]^2, 1\right).$$

We now use lemma 2, setting

$$p_\alpha = \frac{a_\alpha^{+1}}{\hbar_\alpha}, \quad q_\alpha = \frac{h_\alpha^{+1}}{2\hbar_\alpha}\left(\frac{a_\alpha^{+1}}{h_\alpha^+} - \frac{a_\alpha}{h_\alpha^-}\right),$$
$$u_\alpha = \mathring{y}_{x_\alpha}, \quad v_\alpha = \frac{1}{h_\alpha^+}\mathring{y}, \quad \alpha = 1, 2.$$

As a result, we obtain the inequality

$$(R_1 \mathcal{D}^{-1} R_2 y, y) \le \left(\frac{(1+\epsilon)}{d\hbar_1^2}\left[a_1^{+1} + \frac{\kappa_1 h_1^+}{2}\left|\frac{a_1^{+1}}{h_1^+} - \frac{a_1}{h_1^-}\right|\right]\left[a_1^{+1} \mathring{y}_{x_1}^2\right.\right.$$
$$\left.\left. + \frac{1}{2\kappa_1 h_1^+}\left|\frac{a_1^{+1}}{h_1^+} - \frac{a_1}{h_1^-}\right| \mathring{y}^2\right], 1\right) + \left(\frac{(1+\epsilon)}{d\epsilon\hbar_2^2}\left[a_2^{+1} + \frac{\kappa_2 h_2^+}{2}\left|\frac{a_2^{+1}}{h_2^+} - \frac{a_2}{h_2^-}\right|\right]\right.$$
$$\left. \times \left[a_2^{+1} \mathring{y}_{x_2}^2 + \frac{1}{2\kappa_2 h_2^+}\left|\frac{a_2^{+1}}{h_2^+} - \frac{a_2}{h_2^-}\right| \mathring{y}^2\right], 1\right).$$

We set here

$$\epsilon = \epsilon(x) = \frac{a_2^{+1} + 0.5\kappa_2 h_2^+ \left|\frac{a_2^{+1}}{h_2^+} - \frac{a_2}{h_2^-}\right|}{a_1^{+1} + 0.5\kappa_1 h_1^+ \left|\frac{a_1^{+1}}{h_1^+} - \frac{a_1}{h_1^-}\right|} \; \frac{\hbar_1 h_1^+}{\hbar_2 h_2^+} \; \frac{\theta_2}{\theta_1}$$

and define $d(x)$ as follows:

$$d(x) = \sum_{\alpha=1}^{2} \left(a_\alpha^{+1} + \frac{\kappa_\alpha h_\alpha^+}{2}\left|\frac{a_\alpha^{+1}}{h_\alpha^+} - \frac{a_\alpha}{h_\alpha^+}\right|\right) \frac{\theta_\alpha}{\hbar_\alpha h_\alpha^+}, \quad x \in \omega.$$

Here it is assumed that $\kappa_\alpha = \kappa_\alpha(x_\beta) > 0$, $\theta_\alpha = \theta_\alpha(x_\beta) > 0$, $\beta = 3 - \alpha$, $\alpha = 1, 2$. As a result we obtain that

$$(R_1 \mathcal{D}^{-1} R_2 y, y)$$
$$\leq \sum_{\alpha=1}^{2} \left(\frac{a_\alpha^{+1} h_\alpha^+}{\theta_\alpha \hbar_\alpha} \mathring{y}_{x_\alpha}^2, 1 \right) + \sum_{\alpha=1}^{2} \left(\frac{1}{2\hbar_\alpha \theta_\alpha \kappa_\alpha} \left| \frac{a_\alpha^{+1}}{h_\alpha^+} - \frac{a_\alpha}{h_\alpha^-} \right| \mathring{y}^2, 1 \right).$$

Since θ_α does not depend on x_α, then

$$\left(\frac{a_\alpha^{+1}}{\theta_\alpha} \frac{h_\alpha^+}{\hbar_\alpha} \mathring{y}_{x_\alpha}^2, 1 \right)_{\omega_\alpha} = \frac{1}{\theta_\alpha} \sum_{x_\alpha \in \omega_\alpha} a_\alpha^{+1} h_\alpha^+ \mathring{y}_{x_\alpha}^2 \leq \frac{1}{\theta_\alpha} \sum_{x_\alpha \in \omega_\alpha^+} a_\alpha h_\alpha^- \mathring{y}_{\bar{x}_\alpha}^2.$$

Consequently,

$$\left(\frac{a_\alpha^{+1}}{\theta_\alpha} \frac{h_\alpha^+}{\hbar_\alpha} \mathring{y}_{x_\alpha}^2, 1 \right) \leq \left(\frac{a_\alpha}{\theta_\alpha} \mathring{y}_{\bar{x}_\alpha}^2, 1 \right)_\alpha,$$

and therefore we finally have

$$(R_1 \mathcal{D}^{-1} R_2 y, y) \leq \sum_{\alpha=1}^{2} \left(\frac{a_\alpha}{\theta_\alpha} \mathring{y}_{\bar{x}_\alpha}^2, 1 \right)_\alpha + \sum_{\alpha=1}^{2} \left(\frac{1}{2\hbar_\alpha \theta_\alpha \kappa_\alpha} \left| \frac{a_\alpha^{+1}}{h_\alpha^+} - \frac{a_\alpha}{h_\alpha^-} \right| \mathring{y}^2, 1 \right).$$

The rest of the derivation is completely analogous to the transformations and estimates obtained in Section 10.2.2. We give here the results: in the inequalities (11)

$$\delta = 1, \quad \Delta = 4 \max_{\alpha=1,2} \left(\max_{x_\beta \in \omega_\beta} \left(c_\alpha(x_\beta) + \sqrt{b_\alpha(x_\beta)} \right)^2 \right), \quad \beta = 3 - \alpha,$$

where

$$b_\alpha(x_\beta) = \max_{x_\alpha \in \omega_\alpha} v^\alpha(x), \quad c_\alpha(x_\beta) = \max_{x_\alpha \in \omega_\alpha} w^\alpha(x), \quad x_\beta \in \omega_\beta.$$

The function $v^\alpha(x)$ is the solution of the three-point boundary-value problem

$$\left(a_\alpha v_{\bar{x}_\alpha}^\alpha \right)_{\hat{x}_\alpha} = -\frac{a_\alpha^{+1}}{\hbar_\alpha h_\alpha^+}, \quad x_\alpha \in \omega_\alpha(x_\beta), \qquad v^\alpha(x) = 0, \quad x_\alpha \in \gamma_\alpha, \tag{12}$$

and the function $w^\alpha(x)$ is the solution of the problem

$$\left(a_\alpha w_{\bar{x}_\alpha}^\alpha \right)_{\hat{x}_\alpha} = -\frac{1}{2\hbar_\alpha} \left| \frac{a_\alpha^{+1}}{h_\alpha^+} - \frac{a_\alpha}{h_\alpha^-} \right|, \quad x_\alpha \in \omega_\alpha(x_\beta), \qquad w^\alpha(x) = 0, \quad x_\alpha \in \gamma_\alpha. \tag{13}$$

The function $d(x)$ is then computed using the formula

$$d(x) = \sum_{\alpha=1}^{2} \left(\frac{a_\alpha^{+1}}{\hbar_\alpha h_\alpha^+ \sqrt{b_\alpha}} + \frac{1}{2\hbar_\alpha} \left| \frac{a_\alpha^{+1}}{h_\alpha^+} - \frac{a_\alpha}{h_\alpha^-} \right| \right) \frac{1}{c_\alpha + \sqrt{b_\alpha}}, \qquad x \in \omega.$$

The iterative parameters ω and $\{\tau_k\}$ are computed using the formulas in theorem 1. To find y_{k+1}, it is possible to use the algorithm

$$v(x) = \alpha_1(x) v^{(-1_1)} + \beta_1(x) v^{(-1_2)} + \kappa(x)\varphi_k(x), \quad x \in \omega,$$
$$v(x) = 0, \quad x \in \gamma,$$
$$\overset{\circ}{y}_{k+1}(x) = \alpha_2(x)\overset{\circ}{y}{}_{k+1}^{(+1_1)} + \beta_2(x)\overset{\circ}{y}{}_{k+1}^{(+1_2)} + \kappa(x) d(x) v(x), \quad x \in \omega,$$
$$\overset{\circ}{y}_{k+1}(x) = 0, \quad x \in \gamma,$$

where

$$\alpha_1 = \frac{\omega_0 a_1 \kappa}{\hbar_1 h_1^-}, \quad \beta_1 = \frac{\omega_0 a_2 \kappa}{\hbar_2 h_2^-}, \quad \alpha_2 = \frac{\omega_0 a_1^{+1} \kappa}{\hbar_1 h_1^+}, \quad \beta_2 = \frac{\omega_0 a_2^{+1} \kappa}{\hbar_2 h_2^+},$$
$$\frac{1}{\kappa} = d + \frac{1}{2} \sum_{\alpha=1}^{2} \frac{\omega_0}{\hbar_\alpha} \left(\frac{a_\alpha^{+1}}{h_\alpha^+} + \frac{a_\alpha}{h_\alpha^-} \right), \quad \varphi_k(x) = By_k - \tau_{k+1}(Ay_k - f).$$

It should be noted that in cases where the computation of By_k requires a great deal of computation, and where there is no restriction on the volume of intermediate information that can be stored, the second algorithm described in Section 10.1.1 should be used.

10.3.3 A Dirichlet problem for Poisson's equation in an arbitrary region. We now look at an application of this method to the solution of a Dirichlet problem for Poisson's equation

$$\frac{\partial^2 u}{\partial x_1^2} + \frac{\partial^2 u}{\partial x_2^2} = -\varphi(x), \quad x \in G, \quad u(x) = g(x), \quad x \in \Gamma.$$

We assume that the grid is square, i.e. $x_\alpha = x_\alpha(i_\alpha)$, $x_\alpha(i_\alpha + 1) = x_\alpha(i_\alpha) + h$, $i_\alpha = 0, \pm1, \pm2, \ldots$, and

$$\omega = \{x_i = (i_1 h, i_2 h) \in G, \quad i_\alpha = 0, \pm1, \pm2, \ldots\}.$$

Then $\hbar_\alpha \equiv h$, and the steps $h_\alpha^\pm(x)$ are only different from h at the near-boundary nodes of the grid $\bar{\omega}$.

We use the difference scheme (2), in which we set $a_\alpha(x) \equiv 1$ and $\hbar_\alpha \equiv h$. In order to apply the alternate-triangular method constructed in Section 10.3.2, it is necessary to find the solution to the homogeneous three-point boundary-value problems (12) and (13), which in this case have the form

$$\Lambda_\alpha v^\alpha = v^\alpha_{\bar{x}_\alpha \hat{x}_\alpha} = -\frac{1}{hh_\alpha^+}, \quad x_\alpha \in \omega_\alpha(x_\beta), \quad v^\alpha(x) = 0, \quad x_\alpha \in \gamma_\alpha, \quad (14)$$

$$\Lambda_\alpha w^\alpha = w^\alpha_{\bar{x}_\alpha \hat{x}_\alpha} = -\frac{1}{2h}\left|\frac{1}{h_\alpha^+} - \frac{1}{h_\alpha^-}\right|, \quad (15)$$

$$x_\alpha \in \omega_\alpha(x_\beta), \quad w^\alpha(x) = 0, \quad x_\alpha \in \gamma_\alpha.$$

We consider the interval Δ_α containing $\omega_\alpha(x_\beta)$, and denote by $l_\alpha(x_\beta)$ and $L_\alpha(x_\beta)$ its left and right endpoints. Then $h_\alpha^-(x) \le h$ if x is the closest node to l_α in the grid ω_α, and $h_\alpha^+(x) \le h$ if x is the closest node to L_α in the grid ω_α. All the remaining steps $h_\alpha^\pm$ are equal to the basic step h of the grid.

Here it is convenient to write the operator Λ_α on the grid ω_α in the following form:

$$\Lambda_\alpha y = \begin{cases} \dfrac{1}{h}\left(\dfrac{y^{+1_\alpha} - y}{h} - \dfrac{y - y^{-1_\alpha}}{h_\alpha^-}\right), & x_\alpha = l_\alpha + h_\alpha^-, \\ \dfrac{1}{h^2}\left(y^{+1_\alpha} - 2y + y^{-1_\alpha}\right), & l_\alpha + h_\alpha^- + h \le x_\alpha \le L_\alpha - h_\alpha^+ - h, \\ \dfrac{1}{h}\left(\dfrac{y^{+1_\alpha} - y}{h_\alpha^+} - \dfrac{y - y^{-1_\alpha}}{h}\right), & x_\alpha = L_\alpha - h_\alpha^+. \end{cases}$$

The solution of equations (14), (15) can be found explicitly. To do this, we substitute the boundary conditions into the equations at the points $x_\alpha = l_\alpha + h_\alpha^-$ and $x_\alpha = L_\alpha - h_\alpha^+$. These equations are then transformed into two-point equations, and they can be considered as new boundary conditions for the three-point equations with constant coefficients written at the points $l_\alpha + h_\alpha^- + h \le x_\alpha \le L_\alpha - h_\alpha^+ - h$. Consequently, we will have the following problem for $v(x)$ and $w(x)$ (we temporarily drop the superscript $^\alpha$ on v and w):

$$\begin{aligned} & v^{+1_\alpha} - 2v + v^{-1_\alpha} = -1, \qquad l_\alpha + h_\alpha^- + h \le x_\alpha \le L_\alpha - h_\alpha^+ - h, \\ & \left(1 + \frac{h}{h_\alpha^-}\right) v = v^{+1_\alpha} + 1, \quad x_\alpha = l_\alpha + h_\alpha^-, \\ & \left(1 + \frac{h_\alpha^+}{h}\right) v = \frac{h_\alpha^+}{h} v^{-1_\alpha} + 1, \quad x_\alpha = L_\alpha - h_\alpha^+, \end{aligned} \quad (16)$$

$$w^{+1_\alpha} - 2w + w^{-1_\alpha} = 0, \qquad l_\alpha + h_\alpha^- + h \le x_\alpha \le L_\alpha - h_\alpha^+ - h,$$

$$\left(1 + \frac{h}{h_\alpha^-}\right) w = w^{+1_\alpha} - \frac{1}{2}\left(1 - \frac{h}{h_\alpha^-}\right), \quad x_\alpha = l_\alpha + h_\alpha^-, \tag{17}$$

$$\left(1 + \frac{h_\alpha^+}{h}\right) w = \frac{h_\alpha^+}{h} w^{-1_\alpha} + \frac{1}{2}\left(1 - \frac{h_\alpha^+}{h}\right), \quad x_\alpha = L_\alpha - h_\alpha^+.$$

Using the methods for solving difference equations with constant coefficients described in Section 1.4, we find an explicit solution to the boundary-value problems (16) and (17):

$$v_\alpha(x)$$
$$= \frac{1}{2h^2}\left[(x_\alpha - l_\alpha)\left(L_\alpha - x_\alpha + \frac{2h^2 - (h_\alpha^+ + h_\alpha^-)(h_\alpha^+ + h - h_\alpha^-)}{L_\alpha - l_\alpha}\right) + h_\alpha^-(h - h_\alpha^-)\right]$$

$$w^\alpha(x) = \frac{1}{2} - \frac{h_\alpha^-(L_\alpha - x_\alpha) + h_\alpha^+(x_\alpha - l_\alpha)}{2h(L_\alpha - l_\alpha)}$$

for $l_\alpha + h_\alpha^- \le x_\alpha \le L_\alpha - h_\alpha^+$. Since $h_\alpha^\pm \le h$, then

$$(h_\alpha^+ + h_\alpha^-)\,(h_\alpha^+ + h - h_\alpha^-) \ge h_\alpha^-\,(h - h_\alpha^-),$$

therefore

$$v^\alpha(x) \le \frac{1}{2h^2}(x_\alpha - l_\alpha)(L_\alpha - x_\alpha) + 1 \le \frac{1}{2h^2}\left(\frac{L_\alpha - l_\alpha}{2}\right)^2 + 1,$$

$$w^\alpha(x) \le \frac{1}{2}, \quad \alpha = 1, 2.$$

Consequently,

$$b_\alpha(x_\beta) = \max_{x_\alpha \in \omega_\alpha} v^\alpha(x) \le \frac{1}{2h^2}\left(\frac{L_\alpha - l_\alpha}{2}\right)^2 + 1,$$

$$c_\alpha(x_\beta) = \max_{x_\alpha \in \omega_\alpha} w^\alpha(x) \le \frac{1}{2}.$$

Consequently, $\Delta = O(l_0^2/h^2)$ where l_0 is the diameter of the region G. Therefore by theorem 1, the iteration count can be bounded using

$$n \ge n_0(\epsilon) = \frac{\ln(2/\epsilon)}{2\sqrt{2\sqrt[4]{\eta}}} = \frac{\ln(2/\epsilon)}{2\sqrt{2\sqrt[4]{2}\sqrt{h/l_0}}} \approx 0.298\sqrt{N}\ln\frac{2}{\epsilon}, \tag{18}$$

where N is the maximal number of nodes in the direction x_1 or x_2. Thus, the iteration count for this model problem depends only on the basic step h of the grid and not the steps at the near-boundary nodes of the grid $\bar{\omega}$.

We now compare the estimate (18) with the estimate for the iteration count in the case of a Dirichlet problem in a square with side l_0 and with N nodes in each direction on the square grid $\bar{\omega}$. The corresponding estimate for the iteration count was obtained in Section 10.1.4, and it has the form

$$n \geq n_0(\epsilon) = 0.28\sqrt{N}\ln(2/\epsilon).$$

From this it follows that, for an arbitrary region G, the iteration count for the modified alternate-triangular method is practically the same as for the same Dirichlet problem for Poisson's equation in a square whose side is equal to the diameter of the region G.

Remark 1. The procedure laid out here for constructing an alternate-triangular method can obviously also be used in the case where it is necessary to solve an elliptic equation in a rectangle but on a non-uniform grid.

Remark 2. The construction of the method for the case of an equation with mixed derivatives can be performed with the aid of a regularizer R following the procedure outlined in Section 10.2.5.

Chapter 11

The Alternating-Directions Method

In this chapter we consider special iterative methods for solving grid elliptic equations of the form $Au = f$ where the operator A possesses a definite structure. In Section 11.1 the alternating-directions method is studied in the commutative case; an optimal set of parameters is constructed. In Section 11.2 the method is illustrated with examples involving the solution of boundary-value problems for elliptic equations with separable variables. In Section 11.3 the alternating-directions method is examined in the non-commutative case.

11.1 The alternating-directions method in the commutative case

11.1.1 The iterative scheme for the method. In Chapter 10 we studied the general alternate-triangular iterative method, where the operator B was chosen using the decomposition of the operator A into the sum of two mutually-conjugate operators. Most often we use a decomposition of the operator A into the sum of two triangular operators, and then B is the product of triangular operators that depend on an auxiliary iterative parameter. Taking into account the structure of the operator B allows us to optimally choose the iterative parameters and to construct a method which converges significantly faster than an explicit method. When applied to the solution of grid elliptic equations, this method is also economical since, to perform one iteration, the number of arithmetic operations required is proportional to the number of unknowns in the problem.

As we know, the operator A corresponding to the grid elliptic equations has a specific structure. Therefore, when choosing the operators B for an implicit iterative scheme, it is natural to try to use this peculiarity of the operator A. Obviously, such iterative methods will not be universal, however the restriction to a class of problems where the operator A has a definite structure allows us to construct rapidly converging iterative methods oriented to the solution of these particular grid equations.

In this chapter we will study a special method — the *alternating-directions iterative method.* First, we will give a description of this method in operator form, and then we will demonstrate this method on examples involving the approximate solution of various grid elliptic equations.

Our description of the method begins with the iterative scheme. Suppose that it is necessary to find the solution to a linear operator equation

$$Au = f \tag{1}$$

with a non-singular operator A defined in a Hilbert space H. Assume that the operator A is represented in the form of a sum of two operators A_1 and A_2, i.e. $A = A_1 + A_2$. To approximately solve equation (1), we consider an implicit two-level iterative scheme

$$B_{k+1}\frac{y_{k+1}-y_k}{\tau_{k+1}} + Ay_k = f, \quad k = 0,1,\ldots, \qquad y_0 \in H, \tag{2}$$

$$B_k = \left(\omega_k^{(1)}E + A_1\right)\left(\omega_k^{(2)}E + A_2\right), \quad \tau_k = \omega_k^{(1)} + \omega_k^{(2)}, \tag{3}$$

in which the operator B_{k+1} depends on the iteration number k. Here $\omega_k^{(1)}$ and $\omega_k^{(2)}$ are iterative parameters which also depend on the iteration number k and which will be defined later.

We examine first the method of finding y_{k+1} given y_k. One possible algorithm for realizing the scheme (2) is as follows:

$$\begin{aligned} \left(\omega_{k+1}^{(1)}E + A_1\right) y_{k+1/2} &= \left(\omega_{k+1}^{(1)}E - A_2\right) y_k + f, \\ \left(\omega_{k+1}^{(2)}E + A_2\right) y_{k+1} &= \left(\omega_{k+1}^{(2)}E - A_1\right) y_{k+1/2} + f, \quad k = 0,1,\ldots, \end{aligned} \tag{4}$$

where $y_{k+1/2}$ is an intermediate iterative approximation.

We will show that the system (4) is algebraically equivalent to the scheme (2). To do this, we eliminate $y_{k+1/2}$ from (4). Taking into account that $A = A_1 + A_2$, we rewrite (4) in the following form:

$$\left(\omega_{k+1}^{(1)}E + A_1\right)(y_{k+1/2} - y_k) + Ay_k = f, \tag{5}$$

$$\left(\omega_{k+1}^{(2)}E + A_2\right)(y_{k+1} - y_k) - \left(\omega_{k+1}^{(2)}E - A_1\right)(y_{k+1/2} - y_k) + Ay_k = f$$

and we subtract the second equation from the first. We obtain

$$y_{k+1/2} - y_k = \left(\omega_{k+1}^{(2)} E + A_2\right) \frac{y_{k+1} - y_k}{\omega_{k+1}^{(1)} + \omega_{k+1}^{(2)}} = \left(\omega_{k+1}^{(2)} E + A_2\right) \frac{y_{k+1} - y_k}{\tau_{k+1}}.$$

Substituting this expression in (5), we obtain the scheme (2). The reverse path is obvious.

To find y_{k+1} it is possible to use another algorithm that treats (2) as a scheme with a correction w_k,

$$\left(\omega_{k+1}^{(1)} E + A_1\right) v = r_k, \qquad r_k = Ay_k - f,$$

$$\left(\omega_{k+1}^{(2)} E + A_2\right) w_k = v,$$

$$y_{k+1} = y_k - \tau_{k+1} w_k, \quad k = 0, 1, \ldots.$$

A comparison with (4) shows that this algorithm is more economical, however it requires that more intermediate information be stored, i.e. it requires more computer memory, which is not always convenient.

Notice that, both when constructing the iterative scheme (3) and when constructing the algorithms, no conditions were placed on the operators A_1 and A_2, except for the natural assumption that the operators $\omega_k^{(\alpha)} E + A_\alpha$, $\alpha = 1, 2$ were non-singular. All the auxiliary requirements on the operators A_1 and A_2 are connected with the problem of optimally choosing the parameters $\omega_k^{(1)}$ and $\omega_k^{(2)}$.

11.1.2 The choice of the parameters. In the alternating-directions method, we have to deal with two sequences of parameters $\{\omega_k^{(1)}\}$ and $\{\omega_k^{(2)}\}$, which we will choose from the condition that the norm of the solution operator be a minimum in the original space H.

In order to solve this problem of choosing the iterative parameters, it is necessary to make definite assumptions of a functional character concerning the operators A_1 and A_2, and also to give some *a priori* information. We now formulate these assumptions.

We will assume that the operator A can be represented in the form of a sum of two self-adjoint and commutative operators A_1 and A_2:

$$A = A_1 + A_2, \quad A_1 = A_1^*, \quad A_2 = A_2^*, \qquad A_1 A_2 = A_2 A_1. \tag{6}$$

Suppose that the *a priori* information is given on the form of the bounds δ_α and Δ_α for the operator A_α, $\alpha = 1, 2$, i.e.

$$\delta_1 E \leq A_1 \leq \Delta_1 E, \quad \delta_2 E \leq A_2 \leq \Delta_2 E, \tag{7}$$

where the following condition is satisfied

$$\delta_1 + \delta_2 > 0. \tag{8}$$

Notice that it follows from (6)–(8) that the operator A is self-adjoint and positive-definite.

If the assumption about the commutativity of the operators A_1 and A_2 is satisfied, then we will say that we are considering the *commutative case*, and otherwise the *general case*. The conditions (6) guarantee the self-adjointness of the operators B_k for any k. In fact, from (6), the operators $\omega_k^{(1)} E + A_1$ and $A_2 + \omega_k^{(2)} E$ are self-adjoint and commutative, and the product of self-adjoint and commutative operators is a self-adjoint operator.

We move on now to the study of the convergence of the iterative scheme (2). Setting $y_k = z_k + u$ in (2), where z_k is the error and u is the solution of equation (1), we obtain the following homogeneous equation for z_k

$$z_{k+1} = S_{k+1} z_k, \quad k = 0, 1, \ldots, \quad z_0 = y_0 - u, \tag{9}$$

where

$$\begin{gathered} S_k = E - \tau_k B_k^{-1} A \\ \left(\omega_k^{(2)} E + A_2\right)^{-1} \left(\omega_k^{(1)} E + A_1\right)^{-1} \left(\omega_k^{(2)} E - A_1\right) \left(\omega_k^{(1)} E - A_2\right). \end{gathered} \tag{10}$$

Using (9), we express z_n in terms of z_0. We obtain

$$z_n = T_{n,0} z_0, \quad T_{n,0} = \prod_{j=1}^{n} S_j = S_n S_{n-1} \cdots S_1, \tag{11}$$

where $T_{n,0}$ is the resolving operator. Since the operators A_1 and A_2 are commutative, the ordering of the factors in (10) is unimportant; all the operators S_k are self-adjoint and pairwise commutative, and consequently, the operator $T_{n,0}$ is self-adjoint in H: $T_{n,0} = R_n(A_1, A_2)$, where $R_n(x, y)$ is the product of fractionally-rational functions of x and y:

$$R_n(x, y) = \prod_{j=1}^{n} \frac{\omega_j^{(2)} - x}{\omega_j^{(1)} + x} \frac{\omega_j^{(1)} - y}{\omega_j^{(2)} + y}. \tag{12}$$

From (11) we obtain

$$\| z_n \| \leq \| T_{n,0} \| \, \| z_0 \| . \tag{13}$$

By the self-adjointness of the operator $T_{n,0}$ we have $\| T_{n,0} \| = \max_k |\lambda_k(T_{n,0})|$, where $\lambda_k(T_{n,0})$ are the eigenvalues of the operator $T_{n,0}$. Further, by the conditions (6) (cf. Section 5.1.5), the operators A_1, A_2, and $T_{n,0}$ have a common system of eigenfunctions. Therefore

$$\lambda_k(T_{n,0}) = R_n\left(\lambda_{k_1}^{(1)}, \lambda_{k_1}^{(2)}\right),$$

where $\lambda_k^{(1)}$ and $\lambda_k^{(2)}$ are the eigenvalues of the operators A_1 and A_2 respectively, and where from (7) we have $\delta_1 \leq \lambda_k^{(1)} \leq \Delta_1$, $\delta_2 \leq \lambda_k^{(2)} \leq \Delta_2$. Consequently,

$$\| T_{n,0} \| = \max_{k_1,k_2} \left| R_n\left(\lambda_{k_1}^{(1)}, \lambda_{k_2}^{(2)}\right)\right| \leq \max_{\substack{\delta_1 \leq x \leq \Delta_1 \\ \delta_2 \leq y \leq \Delta_2}} |R_n(x,y)|.$$

Substituting this estimate in (13), we obtain

$$\| z_n \|_D \leq \max_{\substack{\delta_1 \leq x \leq \Delta_1 \\ \delta_2 \leq y \leq \Delta_2}} |R_n(x,y)| \, \| z_0 \|_D, \tag{14}$$

where $R_n(x,y)$ is defined in (12), and $D = E$. Notice that, by the commutativity of the operators A_1 and A_2, the operator $T_{n,0}$ will also be self-adjoint in the energy space H_D for $D = A, A^2$. Therefore by lemma 5 of Section 5.1, we will have $\| T_{n,0} \| = \| T_{n,0} \|_A = \| T_{n,0} \|_{A^2}$, and consequently the estimate (14) is valid for $D = A$, $D = A^2$.

Thus, the problem of estimating the error for the iterative scheme (2) leads to the problem of estimating the maximum modulus of a function of two variables $R_n(x,y)$ in the rectangle $G = \{\delta_1 \leq x \leq \Delta_1, \, \delta_2 \leq y \leq \Delta_2\}$, and to the choice of the iterative parameters from the condition that the maximum modulus of this function be a minimum. This problem is sufficiently complex that in Section 11.1.3 it will be reduced to the simpler problem of finding the fractionally-rational function of one variable that deviates least from zero on an interval.

11.1.3 A fractionally-linear transformation. We now study the function $R_n(x,y)$. With the aid of a fractionally-linear transformation of the unknowns, we map the rectangle G onto the square $\{\eta \leq u \leq 1, \, \eta \leq v \leq 1, \, \eta > 0\}$, where the transformation is chosen so that it does not change the form of the function $R_n(x,y)$. The desired transformation has the form

$$x = \frac{ru - s}{1 - tu}, \quad y = \frac{rv + s}{1 + tv}, \quad \eta \leq u, \, v \leq 1, \tag{15}$$

where the constants r, s, t and η have yet to be defined.

Substituting (15) in (12) and introducing the new parameters $\kappa_j^{(1)}$ and $\kappa_j^{(2)}$,

$$\kappa_j^{(1)} = \frac{\omega_j^{(1)} - s}{r - t\omega_j^{(1)}}, \quad \kappa_j^{(2)} = \frac{\omega_j^{(2)} + s}{r + t\omega_j^{(2)}}, \quad j = 1, 2, \ldots, n, \tag{16}$$

we obtain

$$R_n(x, y) = P_n(u, v) = \prod_{j=1}^{n} \frac{\kappa_j^{(2)} - u}{\kappa_j^{(1)} + u} \; \frac{\kappa_j^{(1)} - v}{\kappa_j^{(2)} + v}.$$

From (16) we find the correspondences which allow us to express the parameters $\omega_j^{(1)}$ and $\omega_j^{(2)}$ in terms of the new parameters $\kappa_j^{(1)}$ and $\kappa_j^{(2)}$:

$$\omega_j^{(1)} = \frac{r\kappa_j^{(1)} + s}{1 + t\kappa_j^{(1)}}, \quad \omega_j^{(2)} = \frac{r\kappa_j^{(2)} - s}{1 - t\kappa_j^{(2)}}, \quad j = 1, 2, \ldots, n. \tag{17}$$

Thus, if we can find the parameters $\kappa_j^{(1)}$ and $\kappa_j^{(2)}$, the formulas (17) define the parameters $\omega_j^{(1)}$ and $\omega_j^{(2)}$.

From the change of variables (15) we transform to the problem of finding those values of the parameters $\kappa_j^{(1)}$ and $\kappa_j^{(2)}$ for which

$$\min_{\kappa^{(1)}, \kappa^{(2)}} \max_{\eta \le u, v \le 1} |P_n(u, v)|.$$

Notice that, if certain restrictions are placed on the choice of the parameters $\kappa_j^{(1)}$ and $\kappa_j^{(2)}$, for example $\kappa_j^{(1)} \equiv \kappa_j^{(2)} \equiv \kappa_j$, then it is clear that the minimum can only be increased. Therefore

$$\min_{\kappa^{(1)}, \kappa^{(2)}} \max_{\eta \le u, v \le 1} |P_n(u, v)| \le \min_{\kappa} \max_{\eta \le u, v \le 1} \left| \prod_{j=1}^{n} \frac{\kappa_j - u}{\kappa_j + u} \; \frac{\kappa_j - v}{\kappa_j + v} \right|$$

$$= \min_{\kappa} \max_{\eta \le u \le 1} |r_n(u, \kappa)|^2, \quad r_n(u, \kappa) = \prod_{j=1}^{n} \frac{\kappa_j - u}{\kappa_j + u}.$$

Thus, the problem posed above of choosing the optimal iterative parameters $\omega_j^{(1)}$ and $\omega_j^{(2)}$ has been reduced to finding the fractionally-rational function $r_n(u, x)$ which deviates least from zero on the interval $[\eta, 1]$. In other words, it is necessary to find those κ_j^* for which

$$\max_{\eta \le u \le 1} |r_n(u, \kappa^*)| = \min_{\kappa} \max_{\eta \le u \le 1} |r_n(u, \kappa)| = \rho.$$

If these parameters have been found, then the error z_n in (14) can be estimated by $\| z_n \|_D \leq \rho^2 \| z_0 \|_D$, and accuracy of ϵ can be achieved if we set $\rho^2 = \epsilon$.

The desired set of iterative parameters will be given in Section 11.1.4, but here we will find the constants r, s, t and η in the transfomation (15).

If $r \neq ts$, then the transformation (15) is monotonic with respect to u and v, and consequently the inverse transformation $u = (x+s)/(r+tx)$, $v = (y-s)/(r-ty)$ will be monotonic with repsect to x and y. Therefore, to map the rectangle $\{\delta_1 \leq x \leq \Delta_1,\ \delta_2 \leq y \leq \Delta_2\}$ onto the square $\{\eta \leq u,\ v \leq 1\}$ it is sufficient that the endpoints of the intervals $[\delta_\alpha, \Delta_\alpha]$ map to the endpoints of the interval $[\eta, 1]$. This gives four relations for determining the constants in the transformation (15):

$$\delta_1 = \frac{r\eta - s}{1 - t\eta}, \quad \delta_2 = \frac{r\eta + s}{1 + t\eta}, \quad \Delta_1 = \frac{r-s}{1-t}, \quad \Delta_2 = \frac{r+s}{1+t}. \tag{18}$$

We now find the solution of the non-linear system (18). We first notice that, by the assumption (8), the following inequalities are valid

$$\Delta_2 + \delta_1 \geq \delta_1 + \delta_2 > 0, \quad \Delta_1 + \delta_2 \geq \delta_1 + \delta_2 > 0. \tag{19}$$

Further, from (18) we obtain

$$\begin{aligned} \Delta_1 - \delta_1 &= \frac{(1-\eta)(r-st)}{(1-t)(1-t\eta)}, \quad \Delta_2 - \delta_2 = \frac{(1-\eta)(r-st)}{(1+t)(1+t\eta)}, \\ \Delta_2 + \delta_1 &= \frac{(1+\eta)(r-st)}{(1+t)(1-t\eta)}, \quad \Delta_1 + \delta_2 = \frac{(1+\eta)(r-st)}{(1-t)(1+t\eta)}. \end{aligned} \tag{20}$$

From this we find

$$\left(\frac{1-\eta}{1+\eta}\right)^2 = \frac{(\Delta_1 - \delta_1)(\Delta_2 - \delta_2)}{(\Delta_1 + \delta_2)(\Delta_2 + \delta_1)} < 1,$$

and since by (19) the denominator does not vanish, then

$$\eta = \frac{1-a}{1+a}, \quad a = \sqrt{\frac{(\Delta_1 - \delta_1)(\Delta_2 - \delta_2)}{(\Delta_1 + \delta_2)(\Delta_2 + \delta_1)}}, \quad \eta \in [0,1]. \tag{21}$$

We now find t. From (20) we obtain

$$\frac{\Delta_2 + \delta_1}{\Delta_1 - \delta_1} = \frac{1+\eta}{1-\eta}\,\frac{1-t}{1+t} = \frac{1}{a}\,\frac{1-t}{1+t}.$$

From this we have

$$t = \frac{1-b}{1+b}, \quad b = \frac{\Delta_2 + \delta_1}{\Delta_1 - \delta_1} a. \tag{22}$$

From the last two equations of the system (18) we find

$$r = \frac{1}{2}[\Delta_1(1-t) + \Delta_2(1+t)] = \frac{1+t}{2}[\Delta_2 + \Delta_1 b] = \frac{\Delta_2 + \Delta_1 b}{1+b}, \tag{23}$$

$$s = \frac{1}{2}[\Delta_2(1+t) - \Delta_1(1-t)] = \frac{1+t}{2}[\Delta_2 - \Delta_1 b] = \frac{\Delta_2 - \Delta_1 b}{1+b}. \tag{24}$$

Since

$$r - st = \frac{2b(\Delta_1 + \Delta_2)}{(1+b)^2} > 0, \quad |t| < 1,$$

the transformation (15) is in fact monotonic. In Section 11.1.4 it will be shown that $\eta < \kappa_j \equiv \kappa_j^{(1)} \equiv \kappa_j^{(2)} < 1$. Therefore in (17) the denominator does not vanish.

We look now at several examples. Suppose $\delta_1 = \delta_2 = \delta$ and $\Delta_1 = \Delta_2 = \Delta$, i.e. the bounds for the operators A_1 and A_2 are identical. Then $\eta = \delta/\Delta$, $t = s = 0$, $r = \Delta$, $\omega_j^{(1)} \equiv \omega_j^{(2)} = \Delta\kappa_j$. Assume now that $\delta_1 = 0$, $\delta_2 = \delta$, $\Delta_1 = \Delta_2 = \Delta$, i.e. the operator A_1 is singular. Then

$$\eta = \delta \Big/ \left(\Delta + \sqrt{\Delta^2 - \delta^2}\right), \quad t = \eta, \quad s = \Delta\eta, \quad r = \Delta,$$

$$\omega_j^{(1)} = \frac{\Delta\kappa_j + \Delta\eta}{1 + \eta\kappa_j}, \quad \omega_j^{(2)} = \frac{\Delta\kappa_j - \Delta\eta}{1 - \eta\kappa_j}, \quad j = 1, 2, \ldots, n.$$

11.1.4 The optimal set of parameters. We give now the solution to the problem of optimally choosing the iterative parameters. Unlike the case of finding the polynomial differing least from zero, which was considered in Section 6.2, here the iterative parameters κ_j are expressed, not in terms of trigonometric functions, but using the Jacobi elliptic functions.

We recall certain definitions. The integral

$$K(k) = \int_0^{\pi/2} \frac{d\varphi}{\sqrt{1 - k^2 \sin^2 \varphi}}$$

is called a *complete elliptic integral of the first kind*, the number k is the *modulus* of this integral, and the number $k' = \sqrt{1-k^2}$ is the *auxiliary modulus*. It is customary to denote $K(k') = K'(k)$.

If we denote by $u(z,k)$ the function

$$u(z,k) = \int_z^1 \frac{dy}{\sqrt{(1-y^2)(y^2-k^2)}},$$

then the function $z = \mathrm{dn}(u, k')$, the inverse to $u(z,k)$ is called the *Jacobi elliptic function* with argument u and modulus k'.

Using this notation, the exact solution to the problem of optimally choosing the iterative integral parameters κ_j can be written in the following form:

$$\kappa_j \in \mathcal{M}_n = \left\{\mu_i = \mathrm{dn}\left(\frac{2i-1}{2n}K'(\eta), \eta'\right), \quad i = 1,2,\ldots,n\right\}, \tag{25}$$
$$j = 1,2,\ldots,n,$$

where the iteration count n, sufficient to achieve an accuracy ϵ, is estimated using the formula

$$n \geq n_0(\epsilon) = \frac{1}{4}\frac{K'(\eta)}{K(\eta)} \quad \frac{K'(\epsilon)}{K(\epsilon)}. \tag{26}$$

Here, as in the Chebyshev method, the κ_j are sequentially chosen as the elements of the set $\mathcal{M}_n$. We formulate these results for the alternating-directions method in the commutative case in the form of a theorem.

Theorem 1. *Suppose that the conditions* (6)–(8) *are satisfied, and the parameters* $\omega_j^{(1)}$ *and* $\omega_j^{(2)}$ *are chosen using the formulas*

$$\omega_j^{(1)} = \frac{r\kappa_j + s}{1 + t\kappa_j}, \quad \omega_j^{(2)} = \frac{r\kappa_j - s}{1 - t\kappa_j}, \quad j = 1,2,\ldots,n, \tag{27}$$

where κ_j *and* n *are defined in* (25), (26), *and* r, s, t, *and* η *are defined in* (21)–(24). *The alternating-directions method* (2), (3) *converges in* H_D, *and after* n *iterations, the error* $z_n = y_n - u$ *can be estimated by* $\| z_n \|_D \leq \epsilon \| z_0 \|_D$, *where* $D = E$, A *or* A^2, *and* n *is determined using* (26). □

We turn now to the compuational side of realizing the alternating-directions method with the optimal set of parameters. We find approximate formulas for computing κ_j and n and indicate the order in which the parameters κ_j should be chosen from the set $\mathcal{M}_n$.

Using the asymptotic representation for the complete elliptic integrals for small values of k:

$$\frac{1}{K(k)} = \frac{2}{\pi} + O(k^2), \quad K'(k) = \ln\frac{4}{k} + O\left(k^2 \ln\frac{1}{k}\right),$$

from (26) we obtain the following approximate formula for the iteration count n:

$$n \geq n_0(\epsilon) = \frac{1}{\pi^2} \ln\frac{4}{\eta} \ln\frac{4}{\epsilon}. \tag{28}$$

We look now at the question of computing μ_i. The function $\mathrm{dn}(u, k')$ is monotonically increasing in u, and takes on the following values: $\mathrm{dn}(0, k') = 1$, $\mathrm{dn}(K'(k), k') = k$. Therefore $\eta < \mu_n < \mu_{n-1} < \ldots < \mu_1 < 1$. Further, from the property of the elliptic function $\mathrm{dn}(u, k')$:

$$\mathrm{dn}(u, k') = \frac{k}{\mathrm{dn}(K'(k) - u, k')}$$

it follows that we have

$$\mu_i = \eta/\mu_{n+1-i}, \quad i = 1, 2 \ldots. \tag{29}$$

Therefore it is sufficient to find half of the values μ_i, and the remaining values are determined from the relation (29).

We obtain an approximate formula for μ_i using the expansion of the function $\mathrm{dn}(u, k')$ in powers of k. To do this we express the function $\mathrm{dn}(\sigma K'(\eta), \eta')$ in terms of the Jacobi θ-function and then express these functions as series. We obtain

$$\mathrm{dn}(\sigma K'(\eta), \eta') = \frac{\sqrt{\eta}\theta_3\left(\frac{i\sigma\pi K'}{K}, \bar{q}\right)}{\theta_2\left(\frac{i\sigma\pi K'}{K'}, \bar{q}\right)} = \sqrt{\eta}\bar{q}^{-\frac{2\sigma-1}{4}} \frac{\sum\limits_{m=-\infty}^{\infty} \bar{q}^{m(m+\sigma)}}{\sum\limits_{m=-\infty}^{\infty} \bar{q}^{m(m-1+\sigma)}},$$

where

$$\bar{q} = \exp\left(-\frac{\pi K'(\eta)}{K(\eta)}\right) = \frac{\eta^2}{16}\left(1 + \frac{\eta^2}{2}\right) + O(\eta^6).$$

From this we find

$$\mathrm{dn}(\sigma K'(\eta),\eta') = \sqrt{\eta} q^{\frac{2\sigma-1}{4}} \frac{1+q^{1-\sigma}+q^{1+\sigma}}{1+q^{\sigma}+q^{2-\sigma}} + O(\eta'^{\nu}), \tag{30}$$

where

$$q = \eta^2(1+\eta^2/2)/16, \quad \nu = \begin{cases} 4+5\sigma, & 0 < \sigma < 1/2, \\ 8-3\sigma, & 1/2 \le \sigma < 1. \end{cases}$$

For $\sigma \ge 1/2$, the order of the remainder term in (30) is equal to 5 and is uniform in σ, and for $\sigma < 1/2$, the order is equal to 4. Therefore the approximate formula for $\mathrm{dn}(\sigma K'(\eta),\eta)$ will be more accurate for $\sigma \ge 1/2$ than for $\sigma < 1/2$.

From (25), (29), and (30) we obtain the following formulas for computing μ_i:

$$\mu_i = \sqrt{\eta} q^{\frac{2\sigma_i-1}{4}} \frac{1+q^{1-\sigma_i}+q^{1+\sigma_i}}{1+q^{\sigma_i}+q^{2-\sigma_i}}, \quad [n/2]+1 \le i \le n,$$

$$\mu_i = \eta/\mu_{n+1-i}, \quad 1 \le i \le [n/2], \quad \sigma_i = (2i-1)/(2n),$$

$$q = \eta^2(1+\eta^2/2)/16,$$

where $[a]$ is the integer part of a.

We look now at the question of the order in which the κ_j are chosen from the set $\mathcal{M}_n$. From the definition of the transformation operator S_j in the scheme (2), and from the properties (6), (7) we obtain

$$\| S_j \| = \max_k |\lambda_k(S_j)| \le \max_{\substack{\delta_1 \le x \le \Delta_1 \\ \delta_2 \le y \le \Delta_2}} \left| \frac{\omega_j^{(2)} - x}{\omega_j^{(1)} + x} \; \frac{\omega_j^{(1)} - y}{\omega_j^{(2)} + y} \right|$$

or, using the change of variables (15),

$$\| S_j \| \le \max_{\eta \le u \le 1} \left| \frac{\kappa_j - u}{\kappa_j + u} \right|^2 .$$

Since all the κ_j belong to the interval $(\eta, 1)$, then from this it follows that $\| S_j \| < 1$ for any j. Therefore the iterative method (2), (3) will be stable with respect to rounding errors for any ordering of the κ_j from the set $\mathcal{M}_n$, for example for $\kappa_j = \mu_j$, $j = 1, 2, \ldots, n$.

To conclude this section we show that, for this set of parameters $\omega_j^{(1)}$ and $\omega_j^{(2)}$, the operators $\omega_j^{(\alpha)} E + A_\alpha$, $\alpha = 1, 2$ are positive-definite in H for any j. In fact, from (27) we obtain

$$\frac{\partial \omega_j^{(1)}}{\partial \kappa_j} = \frac{r - st}{(1+t\kappa_j)^2} > 0, \quad \frac{\partial \omega_j^{(2)}}{\partial \kappa_j} = \frac{(r - st)}{(1-t\kappa_j)^2} > 0.$$

Since the denominators in (27) do not vanish, and since $\eta < \kappa_j < 1$, from this and from (18) we find

$$\delta_2 = \frac{r\eta + s}{1 + t\eta} \leq \omega_j^{(1)} \leq \frac{r + s}{1 + t} = \Delta_2, \quad \delta_1 = \frac{r\eta - s}{1 + t\eta} \leq \omega_j^{(2)} \leq \frac{r - s}{1 - t} = \Delta_1. \tag{31}$$

Consequently, using the assumption (7), from (31) we obtain

$$\omega_j^{(1)} E + A_1 \geq (\delta_1 + \delta_2) E, \quad \omega_j^{(2)} E + A_2 \geq (\delta_1 + \delta_2) E,$$

and since $\delta_1 + \delta_2 > 0$, the assertion follows from assumption (8).

11.2 Sample applications of the method

11.2.1 A Dirichlet difference problem for Poisson's equation in a rectangle. We begin our look at sample applications of the alternating-directions method with the solution of a Dirichlet difference problem for Poisson's equations in a rectangle.

Suppose that, on the rectangular grid $\bar{\omega} = \{x_{ij} = (ih_1, jh_2) \in \bar{G}, 0 \leq i \leq N_1, 0 \leq j \leq N_2, h_\alpha = l_\alpha / N_\alpha, \alpha = 1, 2\}$, in the rectangle $\bar{G} = \{0 \leq x_\alpha \leq l_\alpha, \alpha = 1, 2\}$, it is necessary to find the solution to the problem

$$\begin{aligned} &\Lambda y = (\Lambda_1 + \Lambda_2) y = -\varphi(x), \quad x \in \omega, \quad y(x) = g(x), \quad x \in \gamma, \\ &\Lambda_\alpha y = y_{\bar{x}_\alpha x_\alpha}, \quad \alpha = 1, 2. \end{aligned} \tag{1}$$

We denote by H the set of grid functions defined on ω with inner product

$$(u, v) = \sum_{x \in \omega} u(x) v(x) h_1 h_2.$$

The operators A, A_1 and A_2 are defined on H as follows:

$$Ay = -\Lambda \mathring{y}, \quad A_\alpha y = -\Lambda_\alpha \mathring{y}, \ \alpha = 1, 2, \quad \text{where} \quad y \in h, \ \mathring{y} \in \mathring{H}, \quad y(x) = \mathring{y}(x)$$

for $x \in \omega$, and $\mathring{H}$ is the set of grid functions defined on $\bar{\omega}$ and vanishing on γ.

The difference problem (1) can then be written in the form of the operator equation $Au = f$, where $A = A_1 + A_2$.

As we know (cf. Section 5.1.5), the operators A_α are self-adjoint in H and have the bounds δ_α and Δ_α:

$$\delta_\alpha = \frac{4}{h_\alpha^2}\sin^2\frac{\pi h_\alpha}{2l_\alpha}, \quad \Delta_\alpha = \frac{4}{h_\alpha^2}\cos^2\frac{\pi h_\alpha}{2l_\alpha}, \quad \alpha = 1,2,$$

which are the same as the smallest and largest eigenvalues of the difference operators Λ_α. It remains to verify that the commutativity conditions on the operators A_1 and A_2 are satisfied. Using the definition of the operators A_α and the difference operators Λ_α, we obtain

$$A_1 A_2 y = \mathring{y}_{\bar{x}_1 x_1 \bar{x}_2 x_2} = \mathring{y}_{\bar{x}_2 x_2 \bar{x}_1 x_1} = A_2 A_1 y,$$

which is what we were required to prove.

Thus the conditions for the alternating-directions method in the commutative case are satisfied for this example.

Using the definition of the operators A_1 and A_2, the algorithm for the alternating-directions method for this example can be written in the following form:

$$\omega_{k+1}^{(1)} y_{k+1/2} - \Lambda_1 y_{k+1/2} = \omega_{k+1}^{(1)} y_k + \Lambda_2 y_k + \varphi, \; h_1 \le x_1 \le l_1 - h_1, \tag{2}$$

$$y_{k+1/2}(x) = g(x), \quad x_1 = 0, l_1, \quad h_2 \le x_2 \le l_2 - h_2,$$

$$\omega_{k+1}^{(2)} y_{k+1} - \Lambda_2 y_{k+1} = \omega_{k+1}^{(2)} y_{k+1/2} + \Lambda_1 y_{k+1/2} + \varphi,$$

$$h_2 \le x_2 \le l_2 - h_2, \tag{3}$$

$$y_{k+1}(x) = g(x), \quad x_2 = 0, l_2, \quad h_1 \le x_1 \le l_1 - h_1,$$

where $y_k(x) = g(x)$ for $x \in \gamma$ for any $k \ge 0$. Thus, the algorithm for the method consists of the sequential solution for fixed x_2 of the three-point boundary-value problems (2) in the direction x_1 in order to determine $y_{k+1/2}$ on ω, and the solution for fixed x_1 of the boundary-value problems (3) in the direction x_2 in order to determine the new iterative approximation y_{k+1} on ω. The alternation of the directions along which the boundary-value problems (2) and (3) are solved gives the name to the method — the alternating-directions method.

To find the solution of the problems (2), (3) it is possible to use the elimination method. We now write out equations (2), (3) in the form of a three-point system and verify the sufficient conditions for the stability of the elimination method. The equations will have the form

$$\begin{aligned} -y_{k+1/2}(i+1,j) + (2+h_1^2\omega_{k+1}^{(1)})y_{k+1/2}(i,j) - y_{k+1/2}(i-1,j) \\ = \varphi_1(i,j), \quad 1 \le i \le N_1 - 1, \\ y_{k+1/2}(0,j) = g(0,j), \quad y_{k+1/2}(N_1,j) = g(N_1,j), \\ 1 \le j \le N_2 - 1, \end{aligned} \tag{4}$$

where

$$\begin{aligned} \varphi_1(i,j) = \frac{h_1^2}{h_2^2}\Big[y_k(i,j+1) - (2-h_2^2\omega_{k+1}^{(1)})y_k(i,j) \\ + y_k(i,j-1) + h_2^2\varphi(i,j)\Big]; \end{aligned}$$

$$\begin{aligned} -y_{k+1}(i,j+1) + (2+h_2^2\omega_{k+1}^{(2)})y_{k+1}(i,j) - y_{k+1}(i,j-1) = \varphi_2(i,j), \\ 1 \le j \le N_2 - 1, \\ y_{k+1}(i,0) = g(i,0), \quad y_{k+1}(i,N_2) = g(i,N_2), \quad 1 \le i \le N_1 - 1, \end{aligned} \tag{5}$$

where

$$\begin{aligned} \varphi(i,j) = \frac{h_2^2}{h_1^2}\Big[y_{k+1/2}(i+1,j) \\ - (2-h_1^2\omega_{k+1}^{(2)})y_{k+1/2}(i,j) + y_{k+1/2}(i-1,j) + h_1^2\varphi(i,j)\Big]. \end{aligned}$$

Since for this example $\delta_1 > 0$ and $\delta_2 > 0$, then by the inequalities (31) in Section 11.1.4, the parameters $\omega_k^{(1)}$ and $\omega_k^{(2)}$ are positive. Therefore in the three-point equations (4) and (5), the coefficients of $y_{k+1/2}(i,j)$ and $y_{k+1}(i,j)$ dominate the remaining coefficients. Consequently, the elimination method applied to the problems (4), (5) will be stable with respect to rounding errors.

We now count the number of arithmetic operations which must be expended to perform one iteration in the method (2), (3) for this example. It is sufficient to count the number of operations for problem (4); the count for problem (5) is carried out analogously.

The formulas of the elimination method for problem (4) have the form (j fixed):

$$y_{k+1/2}(i,j) = \alpha_i y_{k+1/2}(i+1,j) + \beta_i, \quad 1 \le i \le N_1 - 1,$$

$$y_{k+1/2}(N_1,j) = g(N_1,j),$$

$$\alpha_{i+1} = 1/(C-\alpha_i),\ i = 1,2,\ldots,N_1-1,\ \alpha_1 = 0,\ C = 2 + h_1^2\omega_{k+1}^{(1)},$$

$$\beta_{i+1} = \alpha_{i+1}(\varphi_1(i,j) + \beta_i), \qquad i = 1,2,\ldots,N_1-1, \quad \beta_1 = g(0,j).$$

Notice that the elimination coefficients α_i do not depend on j and therefore can be computed once at a cost of $2(N_1 - 1)$ arithmetic operations. Further, the computation of $\varphi_1(i,j)$ on the grid ω requires $6(N_1-1)(N_2-1)$ arithmetic operations. The elimination coefficients β_i and the solution $y_{k+1/2}$ must be computed anew for each j. To do this requires $4(N_1 - 1)(N_2 - 1)$ operations. In all, finding $y_{k+1/2}$ on the grid ω given y_k requires $Q_1 = 10(N_1-1)(N_2-1)+2(N_1-1)$ arithmetic operations. To find y_{k+1} from (15) using the computed $y_{k+1/2}$ requires $Q_2 = 10(N_1-1)(N_2-1)+2(N_2-1)$ operations. Thus, for this example, performing one iteration of the alternating-directions method costs

$$Q = 20(N_1 - 1)(N_2 - 1) + 2(N_1 - 1) + 2(N_2 - 1) \tag{6}$$

arithmetic operations.

We estimate now the iteration count n which is sufficient to obtain the solution to an accuracy ϵ. In the special case where the region $\bar{G}$ is a square with side l ($l_1 = l_2 = l$) and the grid $\bar{\omega}$ is square with $N_1 = N_2 = N$ ($h_1 = h_2 = l/N$), we will have

$$\delta_1 = \delta_2 = \delta = \frac{4}{h^2}\sin^2\frac{\pi h}{2l}, \quad \Delta_1 = \Delta_2 = \Delta = \frac{4}{h^2}\cos^2\frac{\pi h}{2l}.$$

From (21) and (28) of Section 11.1 we obtain the following estimate for the iteration count:

$$n \geq n_0(\epsilon) = 0.1\ln\frac{4}{\eta}\ln\frac{4}{\epsilon}, \quad \eta = \delta/\Delta = \tan^2\frac{\pi h}{2l}$$

or for small h

$$n_0(\epsilon) = 0.2\ln(4N/\pi)\ln(4/\epsilon), \tag{7}$$

i.e. the iteration count is proportional to the logarithm of the number of unknowns in one direction.

From (6), (7) we obtain the following estimate for the number of arithmetic operations $Q(\epsilon)$ required to find the solution to the difference problem (1) using the alternating-directions method to an accuracy ϵ:

$$Q(\epsilon) = nQ = 4N^2\ln(4N/\pi)\ln(4/\epsilon). \tag{8}$$

In order to compare this method with the direct cycilc-reduction method (cf. Section 3.3), we convert the natural logarithm to the logarithm base 2.

We obtain

$$Q(\epsilon) \approx 2.12N^2\log_2(4N/\pi)\log_2(4/\epsilon).$$

Since the approximation error of the difference scheme (1) is $O(h^2)$, ϵ should correspondingly be chosen to be $O(h^2)$.

If we take $\epsilon = 4/N^2$, then we obtain

$$Q(\epsilon) = 4.24N^2 \log_2 N \log_2(4N/\pi).$$

For $N = 64$ we obtain $\epsilon \approx 10^{-3}$ and

$$Q(\epsilon) \approx 27.6N^2 \log_2 N.$$

A comparison with the estimate of the operation count for the cyclic-reduction method shows that for this grid the alternating-directions method requires approximately 5.5 times as many arithmetic operations as the cyclic-reduction method. As N increases and ϵ decreases, this difference grows.

For this particular case, we give the iteration count n as a function of the number of nodes N in one direction for $\epsilon = 10^{-4}$.

For comparison, we also give the iteration counts for other methods considered earlier.

Table 11

N	Simple iteration method	Explicit Chebyshev method	Successive over-relaxation method	Alternate-triangular method	Alternating directions method
32	1909	101	65	16	8
64	7642	202	128	23	10
128	30577	404	257	32	11

From the table it follows that the smallest iteration count is obtained for the alternating-directions method. Based on the iteration count, it is better than the alternate-triangular method with the Chebyshev parameters which was considered in Chapter 10.

Remark. If, for the problem (1), we look at the alternating-directions method with constant parameters, i.e. $\omega_j^{(1)} \equiv \omega^{(1)}$, $\omega_j^{(2)} \equiv \omega^{(2)}$, $\tau_j = \omega^{(1)} + \omega^{(2)}$, then from the formulas (25) of Section 11.1 we obtain, using the equation $\mathrm{dn}(1/2K'(k), k') = \sqrt{k}$, that the parameter $\kappa_j \equiv \sqrt{\eta}$. In particular when $\delta_1 = \delta_2 = \delta$, $\Delta_1 = \Delta_2 = \Delta$, earlier in Section 11.1.3 we obtained the following relation among the parameters $\omega_j^{(1)}$, $\omega_j^{(2)}$, and κ_j: $\omega_j^{(1)} = \omega_j^{(2)} = \Delta\kappa_j$. Since here $\eta = \delta/\Delta$, then from this we find $\omega^{(1)} \equiv \omega^{(2)} = \sqrt{\delta\Delta}$.

11.2.2 A boundary-value problem of the third kind for an elliptic equation with separable variables. Suppose that, in the rectangle $\bar{G} = \{0 \le x_\alpha \le l_\alpha, \alpha = 1, 2\}$, it is necessary to find the solution to the following boundary-value problem:

$$
\begin{aligned}
Lu &= \sum_{\alpha=1}^{2} \frac{\partial}{\partial x_\alpha}\left(k_\alpha(x_\alpha)\frac{\partial u}{\partial x_\alpha}\right) - qu = -f(x), && x \in G, \\
k_\alpha(x_\alpha)\frac{\partial u}{\partial x_\alpha} &= \kappa_{-\alpha} u - g_{-\alpha}(x), && x_\alpha = 0, \\
-k_\alpha(x_\alpha)\frac{\partial u}{\partial x_\alpha} &= \kappa_{+\alpha} u - g_{+\alpha}(x), \quad x_\alpha = l_\alpha, && \alpha = 1, 2.
\end{aligned}
\tag{9}
$$

We will assume that the following conditions are satisfied:

$$
\begin{gathered}
0 < c_{1,\alpha} \le k_\alpha(x_\alpha) \le c_{2,\alpha}, \quad \kappa_{\pm\alpha} = \text{constant} \ge 0, \ \alpha = 1, 2, \\
q = \text{constant} \ge 0, \qquad \sum_{\alpha=1}^{2} \kappa_{\pm\alpha}^2 + q^2 \ne 0.
\end{gathered}
\tag{10}
$$

A Neumann boundary-value problem ($\kappa_{\pm\alpha} = 0$) for the $q = 0$ will be considered separately in Chapter 12. The conditions (10) guarantee the existence and uniqueness of the solution to problem (9).

On the rectangular grid $\bar{\omega} = \{x_{ij} = (ih_1, jh_2) \in \bar{G}, 0 \le i \le N_1, 0 \le j \le N_2, h_\alpha = l_\alpha/N_\alpha, \alpha = 1, 2\}$ the problem (9) corresponds to the boundary-value difference problem

$$
\Lambda y = (\Lambda_1 + \Lambda_2) y = -\varphi(x), \quad x \in \bar{\omega},
\tag{11}
$$

where the difference operators Λ_1 and Λ_2 and the right-hand side φ are defined in the following way:

$$
\Lambda_\alpha y = \begin{cases}
\dfrac{2}{h_\alpha} a_\alpha(h_\alpha) y_{x_\alpha} - \left(0.5q + \dfrac{2}{h_\alpha}\kappa_{-\alpha}\right) y, & x_\alpha = 0, \\
(a_\alpha(x_\alpha) y_{\bar{x}_\alpha})_{x_\alpha} - 0.5qy, & h_\alpha \le x_\alpha \le l_\alpha - h_\alpha, \\
-\dfrac{2}{h_\alpha} a_\alpha(l_\alpha) y_{\bar{x}_\alpha} - \left(0.5q + \dfrac{2}{h_\alpha}\kappa_{+\alpha}\right) y, & x_\alpha = l_\alpha
\end{cases}
$$

for $0 \le x_\beta \le l_\beta$, $\beta = 3 - \alpha$, $\alpha = 1, 2$ and $\varphi = f + \varphi_1 + \varphi_2$,

$$
\varphi_\alpha(x) = \begin{cases}
\dfrac{2}{h_\alpha} g_{-\alpha}(x), & x_\alpha = 0, \\
0, & h_\alpha \le x_\alpha \le l_\alpha - h_\alpha, \\
\dfrac{2}{h_\alpha} g_{+\alpha}(x), & x_\alpha = l_\alpha.
\end{cases}
$$

We denote by H the space of grid functions defined on $\bar{\omega}$ with the inner product defined by the formula

$$(u,v)=\sum_{x\in\bar{\omega}}u(x)v(x)\hbar_1(x_1)\hbar_2(x_2),$$

$$\hbar_\alpha(x_\alpha)=\begin{cases}0.5h_\alpha, & x_\alpha=0,\\ h_\alpha, & h_\alpha\le x_\alpha\le l_\alpha-l_\alpha.\end{cases}$$

The operators A, A_1, and A_2 are defined in H by the relations $Ay=-\Lambda y$, $A_\alpha y=-\Lambda_\alpha y$, $\alpha=1,2$. In Section 5.2 it was shown that the operators A_1 and A_2 defined in this way are self-adjoint and commutative. In addition, by the conditions (10) the operator A is positive-definite in H (i.e. $\delta_1+\delta_2>0$). It remains to find the bounds for the operators A_1 and A_2, i.e. the constants δ_α and Δ_α in the inequalities $\delta_\alpha E\le A_\alpha\le\Delta_\alpha E$, $\alpha=1,2$.

We begin with δ_α. From the definition of the operators A_α and the Green difference formulas we obtain

$$\begin{aligned}(A_\alpha y,y)=&-\sum_{x_\beta=0}^{l_\beta}\sum_{x_\alpha=h_\alpha}^{l_\alpha-h_\alpha}\left[(a_\alpha y_{\bar{x}_\alpha})_{x_\alpha}-0.5qy\right]yh_1\hbar_2\\
&-\sum_{x_\beta=0}^{l_\beta}\left[a_\alpha(h_\alpha)y_{x_\alpha}-\left(\kappa_{-\alpha}+\frac{h_1}{4}q\right)y\right]y\Big|_{x_\alpha=0}\hbar_2\\
&+\sum_{x_\beta=0}^{l_\beta}\left[a_\alpha(l_\alpha)y_{\bar{x}_\alpha}+\left(\kappa_{+\alpha}+\frac{h_1}{4}q\right)y\right]y\Big|_{x_\alpha=l_\alpha}\hbar_2\\
=&\sum_{x_\beta}^{l_\beta}\sum_{x_\alpha=h_\alpha}^{l_\alpha}a_\alpha y_{\bar{x}_\alpha}^2h_1\hbar_2\\
&+\sum_{x_\beta=0}^{l_\beta}\left(\kappa_{-\alpha}y^2\,|_{x_\alpha}+\kappa_{+\alpha}y^2\big|_{x_\alpha=l_\alpha}\right)\hbar_2+0.5q(y^2,1).\end{aligned}$$

From this we find that, if $q=\kappa_{-\alpha}=\kappa_{+\alpha}=0$, then $\delta_\alpha=0$. If even one of the quantities q, $\kappa_{-\alpha}$ or $\kappa_{+\alpha}$ is different from zero, then δ_α can be found as follows. By lemma 16 of Section 5.2, we will have

$$(y^2,1)_{\bar{\omega}_\alpha}\le\max_{x_\alpha\in\bar{\omega}_\alpha}v^\alpha(x_\alpha)(A_\alpha y,y)_{\bar{\omega}_\alpha},\tag{12}$$

where $v^\alpha(x_\alpha)$ is the solution of the three-point boundary-value problem

$$\begin{aligned}
&\left(a_\alpha(x_\alpha)v^\alpha_{\bar{x}_\alpha}\right)_{x_\alpha} - 0.5qv = -1, \quad h_\alpha \le x_\alpha \le l_\alpha - h_\alpha,\\
&\frac{2}{h_\alpha}a_\alpha(h_\alpha)v^\alpha_{x_\alpha} - \left(0.5q + \frac{2}{h_\alpha}\kappa_{-\alpha}\right)v^\alpha = -1, \quad x_\alpha = 0,\\
&-\frac{2}{h_\alpha}a_\alpha(l_\alpha)v^\alpha_{\bar{x}_\alpha} - \left(0.5q + \frac{2}{h_\alpha}\kappa_{+\alpha}\right)v^\alpha = -1, \quad x_\alpha = l_\alpha,
\end{aligned} \tag{13}$$

and the inner product is defined as follows:

$$(u,v)_{\bar{\omega}_\alpha} = \sum_{x_\alpha=0}^{l_\alpha} u(x)v(x)\hbar_\alpha(x_\alpha).$$

Multiplying (12) by $\hbar_\beta(x_\beta)$ and summing over x_β from 0 to l_β, we obtain

$$(y^2,1) \le \max_{x_\alpha \in \bar{\omega}_\alpha} v^\alpha(x_\alpha)(A_\alpha y, y)$$

and, consequently,

$$\delta_\alpha = \frac{1}{\max\limits_{x_\alpha \in \bar{\omega}_\alpha} v^\alpha(x_\alpha)}, \quad \alpha = 1,2.$$

We now find Δ_α. The operator A_α corresponds to a tridiagonal matrix A_α. We denote by $\mathcal{D}$ the diagonal part of the matrix A_α, i.e. $\mathcal{D}y = d_\alpha(x_\alpha)y$,

$$d_\alpha(x_\alpha) = \begin{cases} 0.5q + \dfrac{2}{h_\alpha}\kappa_{-\alpha} + \dfrac{2}{h_\alpha^2}a_\alpha(h_\alpha), & x_\alpha = 0,\\[2mm] 0.5q + \dfrac{1}{h_\alpha^2}\left(a_\alpha(x_\alpha) + a_\alpha(x_\alpha + h_\alpha)\right), & h_\alpha \le x_\alpha \le l_\alpha - h_\alpha,\\[2mm] 0.5 + \dfrac{2}{h_\alpha}\kappa_{+\alpha} + \dfrac{2}{h_\alpha^2}a_\alpha(l_\alpha), & x_\alpha = l_\alpha. \end{cases}$$

We look at the eigenvalue problem

$$A_\alpha y - \lambda\mathcal{D}y = 0, \quad x \in \bar{\omega}. \tag{14}$$

It is easy to show that, if λ is an eigenvalue for the problem (14), then $2-\lambda$ is also an eigenvalue. Consequently,

$$\lambda_{\min}\mathcal{D} \le A_\alpha \le (2 - \lambda_{\min})\mathcal{D}$$

or

$$(A_\alpha y, y) \le (2-\lambda_{\min})(\mathcal{D}y, y) \le (2-\lambda_{\min}) \max_{x_\alpha\in\bar\omega_\alpha} d_\alpha(x_\alpha)(y,y).$$

Therefore it is possible to take Δ_α to be

$$\Delta_\alpha = (2-\lambda_{\min}) \max_{x_\alpha\in\bar\omega_\alpha} d_\alpha(x_\alpha).$$

It remains to find $\lambda_{\min}$. If $q=\kappa_{-\alpha}=\kappa_{+\alpha}=0$, then the operator A_α is singular and $\lambda_{\min}=0$. Otherwise, by remark 2 to lemma 14 of Section 5.2, we will have

$$(d_\alpha y, y)_{\bar\omega_\alpha} \le \max_{x_\alpha\in\bar\omega_\alpha} w^\alpha(x_\alpha)(A_\alpha y, y)_{\bar\omega_\alpha}, \tag{15}$$

where $w^\alpha(x_\alpha)$ is the solution of the following boundary-value problem:

$$\begin{aligned} &\left(a_\alpha w^\alpha_{\bar x_\alpha}\right)_{x_\alpha} - 0.5qw^\alpha = -d_\alpha(x_\alpha), \quad h_\alpha \le x_\alpha \le l_\alpha - h_\alpha,\\ &\frac{2}{h_\alpha}a_\alpha(h_\alpha)w^\alpha_{x_\alpha} - \left(0.5q + \frac{2}{h_\alpha}\kappa_{-\alpha}\right)w^\alpha = -d_\alpha(0), \quad x_\alpha = 0,\\ &-\frac{2}{h_\alpha}a_\alpha(l_\alpha)w^\alpha_{\bar x_\alpha} - \left(0.5q + \frac{2}{h_\alpha}\kappa_{+\alpha}\right)w^\alpha = -d_\alpha(l_\alpha), \quad x_\alpha = l_\alpha. \end{aligned} \tag{16}$$

Multiplying (15) by $\hbar_\beta(x_\beta)$ and summing over x_β from 0 to l_β, we obtain

$$(\mathcal{D}y, y) \le \max_{x_\alpha\in\bar\omega_\alpha} w^\alpha(x_\alpha)(A_\alpha y, y)$$

and, consequently,

$$\lambda_{\min} \ge \frac{1}{\max\limits_{x_\alpha\in\bar\omega_\alpha} w^\alpha(x_\alpha)}.$$

Thus, if $q=\kappa_{-\alpha}=\kappa_{+\alpha}=0$, then

$$\delta_\alpha = 0, \quad \Delta_\alpha = 2\max_{x_\alpha\in\bar\omega_\alpha} d_\alpha(x_\alpha),$$

otherwise

$$\delta_\alpha = \frac{1}{\max\limits_{x_\alpha\in\bar\omega_\alpha} v^\alpha(x_\alpha)},$$

$$\Delta_\alpha = \left(2 - \frac{1}{\max\limits_{x_\alpha\in\bar\omega_\alpha} w^\alpha(x_\alpha)}\right)\max_{x_\alpha\in\bar\omega_\alpha} d_\alpha(x_\alpha),$$

where $v^\alpha(x_\alpha)$ and $w^\alpha(x_\alpha)$ are the solutions to the problems (13) and (16). All the necessary *a priori* information for applying the alternating-directions method has been found. Using the formulas of theorem 1, we find the iterative parameters for the method and estimate the required iteration count.

We give now the algorithmic formulas for the alternating-directions method for this example. Taking into account the definition of the operators A_1, A_2 and the right-hand side φ, we obtain

$$\omega_{k+1}^{(1)} y_{k+1/2} - \Lambda_1 y_{k+1/2} = \omega_{k+1}^{(2)} y_k + \Lambda_2 y_k + \varphi,$$
$$0 \le x_1 \le l_1, \quad 0 \le x_2 \le l_2,$$
$$\omega_{k+1}^{(2)} y_{k+1} - \Lambda_2 y_{k+1} = \omega_{k+1}^{(2)} y_{k+1/2} + \Lambda_1 y_{k+1/2} + \varphi,$$
$$0 \le x_2 \le l_2, \quad 0 \le x_1 \le l_1.$$

Here, unlike in a Dirichlet problem, the three-point boundary-value problems must also be solved on the boundary of the grid $\bar{\omega}$, and the initial approximation y_0 is an arbitrary grid function given on the whole grid $\bar{\omega}$.

Using the conditions (10), it is possible to show that, for this example as for the case of a Dirichlet problem, the iteration count n can be bounded by the following asymptotic-in-h estimate:

$$n \ge n_0(\epsilon) = O(\ln|h| \ln \epsilon), \quad |h|^2 = h_1^2 + h_2^2.$$

We remark that all the observations given here are also valid in the case where $\bar{\omega}$ is an arbitrary non-uniform rectangular grid in the region $\bar{G}$. It is only necessary to change the operators Λ_α introduced here to the operators on the non-uniform grid.

We emphasize that the assumptions about the constants q, $\kappa_{\pm\alpha}$ and the dependence of the coefficients a_α only on x_α are essential. If even one of these assumptions is not satisfied, the commutativity conditions for the operators A_1 and A_2 will not be satisfied.

In conclusion we remark that the alternating-directions method can be applied to the solution of various analogs of equation (9) and also to other boundary conditions. In particular, on each side of the rectangle $\bar{G}$ it is possible to give boundary conditions of first, second, or third kind with constants $\kappa_{\pm\alpha}$.

11.2.3 A high-accuracy Dirichlet difference problem. We will look at one more sample application of the alternating-directions method. Suppose that, on the rectangular grid $\bar{\omega} = \{x_{ij} = (ih_1, jh_2) \in \bar{G}, 0 \le i \le N_1, 0 \le j \le N_2, h_\alpha = l_\alpha/N_\alpha, \alpha = 1,2\}$ in the rectangle $\bar{G} = \{0 \le x_\alpha \le l_\alpha, \alpha = 1,2\}$, it is necessary to find the solution to a high-accuracy Dirichlet difference problem for Poisson's equation

$$\Lambda y = (\Lambda_1 + \Lambda_2 + (\kappa_1 + \kappa_2)\Lambda_1\Lambda_2)y = -\varphi(x), \quad x \in \omega, \qquad (17)$$
$$y(x) = g(x), \quad x \in \gamma$$

where $\Lambda_\alpha y = y_{\bar{x}_\alpha x_\alpha}$, $\kappa_\alpha = h_\alpha^2/12$, $\alpha = 1,2$.

Here

$$\varphi = \tilde{f} + \kappa_1\Lambda_1\tilde{f} + \kappa_2\Lambda_2\tilde{f},$$

where $\tilde{f}(x)$ is the right-hand side of the original differential equation

$$Lu = \frac{\partial^2 u}{\partial x_1^2} + \frac{\partial^2 u}{\partial x_2^2} = -\tilde{f}(x), \quad x \in G, \quad u(x) = g(x), \quad x \in \Gamma.$$

The difference scheme (17) for this choice of $\varphi(x)$ has accuracy $O(|h|^4)$, $|h|^2 = h_1^2 + h_2^2$, and on a square grid ($h_1 = h_2 = h$) for the corresponding choice of $\varphi(x)$

$$\varphi = \tilde{f} + \frac{h^2}{12}(\Lambda_1 + \Lambda_2)\tilde{f} + \frac{h^4}{360}\left(\Lambda_1^2 + 4\Lambda_1\Lambda_2 + \Lambda_2^2\right)\tilde{f}$$

it has accuracy $O(h^6)$.

Introducing the operators $A_\alpha y = -\Lambda_\alpha \mathring{y}$, where $y \in H$, $\mathring{y} \in \mathring{H}$, H is the space of grid functions defined on ω with inner product

$$(u, v) = \sum_{x\in\omega} u(x)v(x)h_1h_2,$$

and $\mathring{H}$ is the set of grid functions vanishing on γ, we write (17) in the operator form

$$Au = f, \qquad (18)$$

where $A = A_1 + A_2 - (\kappa_1 + \kappa_2)A_1A_2$.

As was already shown, the operators A_1 and A_2 possess the following properties: A_1 and A_2 are self-adjoint in H and commute

$$A_\alpha = A_\alpha^*, \quad \alpha = 1,2, \quad A_1A_2 = A_2A_1, \qquad (19)$$

the operator A_α has bounds δ_α and Δ_α, where

$$\delta_\alpha = \frac{4}{h_\alpha^2}\sin^2\frac{\pi h_\alpha}{2l_\alpha}, \quad \Delta_\alpha = \frac{4}{h_\alpha^2}\cos^2\frac{\pi h_\alpha}{2l_\alpha}, \qquad (20)$$
$$\delta_\alpha E \le A_\alpha \le \Delta_\alpha E, \quad \delta_\alpha > 0, \quad \alpha = 1,2.$$

We have

Lemma 1. *If the conditions* (19), (20) *are satisfied and* $\kappa_\alpha \Delta_\alpha < 1$, *then the operators*

$$\bar{A}_\alpha = (E - \kappa_\alpha A_\alpha)^{-1} A_\alpha, \quad \alpha = 1, 2, \tag{21}$$

are self-adjoint in H, *commute, and have the bounds* $\bar{\delta}_\alpha$ *and* $\bar{\Delta}_\alpha$, *i.e.*

$$\bar{\delta}_\alpha E \le \bar{A}_\alpha \le \bar{\Delta}_\alpha E, \quad \bar{\delta}_\alpha > 0, \quad \alpha = 1, 2,$$

where $\bar{\delta}_\alpha$ *and* $\bar{\Delta}_\alpha$ *are defined by the formulas*

$$\bar{\delta}_\alpha = \frac{\delta_\alpha}{1 - \kappa_\alpha \delta_\alpha}, \quad \bar{\Delta}_\alpha = \frac{\Delta_\alpha}{1 - \kappa_\alpha \Delta_\alpha}. \tag{22}$$

Proof. The existence of the operator $\bar{A}_\alpha$ follows from the positive-definiteness of the operator $E - \kappa_\alpha A_\alpha$, if the condition $\kappa_\alpha \Delta_\alpha < 1$ is satisfied. Further, representing $\bar{A}_\alpha$ in the form $\bar{A}_\alpha = (A_\alpha^{-1} - \kappa_\alpha E)^{-1}$ and taking into account the self-adjointness of the operators A_α, A_α^{-1}, and $A_\alpha^{-1} - \kappa_\alpha E$, we obtain

$$\left(\frac{1}{\Delta_\alpha} - \kappa_\alpha\right) E \le (A_\alpha^{-1} - \kappa_\alpha E) \le \left(\frac{1}{\delta_\alpha} - \kappa_\alpha\right) E.$$

From this follows the assertion of the lemma. The commutativity of the operators $\bar{A}_1$ and $\bar{A}_2$ follows from the commutativity of A_1 and A_2. The lemma has been proved. □

For this example, the conditions of lemma 1 are satisfied, since $\kappa_\alpha \Delta_\alpha < 1/3$.

We now transform equation (18). To do this, we write (18) in the form

$$A_1(E - \kappa_2 A_2)u + A_2(E - \kappa_1 A_1)u = f. \tag{23}$$

Applying the operator $(E - \kappa_1 A_1)^{-1}(E - \kappa_2 A_2)^{-1}$ to (23) and taking into account the commutativity of all the operators, we obtain from (23) an equation which is equivalent to (18)

$$\bar{A}u = (\bar{A}_1 + \bar{A}_2)u = \bar{f} \tag{24}$$

where $\bar{A}_1$ and $\bar{A}_2$ are defined in (21), and $\bar{f} = (E - \kappa_1 A_1)^{-1}(E - \kappa_2 A_2)^{-1} f$. Thus, the solution of equation (18) leads to the solution of equation (24) with self-adjoint commuting operators $\bar{A}_1$ and $\bar{A}_2$, whose bounds are given in (22).

To find an approximation solution to equation (24), we use the alternating-directions method

$$\bar{B}_{k+1}\frac{y_{k+1}-y_k}{\tau_{k+1}}+\bar{A}y_k=\bar{f},\quad k=0,1,\ldots,\quad y_0\in H, \tag{25}$$

where

$$\bar{B}_k=\left(\omega_k^{(1)}E+\bar{A}_1\right)\left(\omega_k^{(2)}E+\bar{A}_2\right),\quad \tau_k=\omega_k^{(1)}+\omega_k^{(2)}.$$

The iterative parameters $\omega_k^{(1)}$ and $\omega_k^{(2)}$ are found from the formulas in theorem 1, in which δ_α and Δ_α are changed to $\bar{\delta}_\alpha$ and $\bar{\Delta}_\alpha$. All the necessary conditions for applying the alternating-directions method are satisfied here.

It remains to examine the algorithm used to perform the iterative method (25). We rewrite (25) in the form

$$\begin{aligned}\left(\omega_{k+1}^{(1)}E+\bar{A}_1\right)\left(\omega_{k+1}^{(2)}E+\bar{A}_2\right)y_{k+1}&\\ =\left(\omega_{k+1}^{(2)}E-\bar{A}_1\right)\left(\omega_{k+1}^{(1)}E-\bar{A}_2\right)y_k+\tau_{k+1}\bar{f}.&\end{aligned} \tag{26}$$

In Section 11.1.4 it was shown that the iterative parameters $\omega_k^{(1)}$ and $\omega_k^{(2)}$ satisfy the inequalities $\bar{\delta}_2\le\omega_k^{(1)}\le\bar{\Delta}_2$, $\bar{\delta}_1\le\omega_k^{(2)}\le\bar{\Delta}_1$ for any k.

Since $\bar{\delta}_\alpha>0$ for this example, dividing the left- and right-hand sides of (26) by $\omega_{k+1}^{(1)}\omega_{k+1}^{(2)}$ and denoting $\tau_{k+1}^{(1)}=1/\omega_{k+1}^{(1)}$, $\tau_{k+1}^{(2)}=1/\omega_{k+1}^{(2)}$, we obtain

$$\begin{aligned}\left(E+\tau_{k+1}^{(1)}\bar{A}_1\right)\left(E+\tau_{k+1}^{(2)}\bar{A}_2\right)y_{k+1}&\\ =\left(E-\tau_{k+1}^{(2)}\bar{A}_1\right)\left(E-\tau_{k+1}^{(1)}\bar{A}_2\right)+\left(\tau_{k+1}^{(1)}+\tau_{k+1}^{(2)}\right)\bar{f}.&\end{aligned}$$

We now apply to both sides of this equation the operator

$$(E-\kappa_1A_1)(E-\kappa_2A_2),$$

take into account that all the operators commute, and that

$$\begin{gathered}(E-\kappa_\alpha A_\alpha)\left(E+\tau_{k+1}^{(\alpha)}\bar{A}_\alpha\right)=E+\left(\tau_{k+1}^{(\alpha)}-\kappa_\alpha\right)A_\alpha,\\ (E-\kappa_\alpha A_\alpha)\left(E-\tau_{k+1}^{(\beta)}\bar{A}_\alpha\right)=E-\left(\tau_{k+1}^{(\beta)}+\kappa_\alpha\right)A_\alpha,\\ \beta=3-\alpha,\quad \alpha=1,2.\end{gathered}$$

As a result we obtain

$$\left(E+\left(\tau_{k+1}^{(1)}-\kappa_1\right)A_1\right)\left(E+\left(\tau_{k+1}^{(2)}-\kappa_2\right)A_2\right)y_{k+1} \tag{27}$$
$$=\left(E-\left(\tau_{k+1}^{(2)}+\kappa_1\right)A_1\right)\left(E-\left(\tau_{k+1}^{(1)}+\kappa_2\right)A_2\right)y_k+\left(\tau_{k+1}^{(1)}+\tau_{k+1}^{(2)}\right)f.$$

The iterative scheme (27) is equivalent to the following scheme:

$$\left(E+\left(\tau_{k+1}^{(1)}-\kappa_1\right)A_1\right)y_{k+1/2}$$
$$=\left(E-\left(\tau_{k+1}^{(1)}+\kappa_2\right)A_2\right)y_k+\left(\tau_{k+1}^{(1)}-\kappa_1\right)f, \tag{28}$$
$$\left(E+\left(\tau_{k+1}^{(2)}-\kappa_2\right)A_2\right)y_{k+1}$$
$$=\left(E-\left(\tau_{k+1}^{(2)}+\kappa_1\right)A_1\right)y_{k+1/2}+\left(\tau_{k+1}^{(2)}+\kappa_1\right)f. \tag{29}$$

The equivalence of (27) and (28), (29) is shown as follows. Multiplying (28) by $\tau_{k+1}^{(2)}+\kappa_1$, (29) by $-(\tau_{k+1}^{(1)}-\kappa_1)$ and adding the results, we obtain

$$\left(\tau_{k+1}^{(1)}+\tau_{k+1}^{(2)}\right)y_{k+1/2}=\left(\tau_{k+1}^{(1)}-\kappa_1\right)\left(E+\left(\tau_{k+1}^{(2)}-\kappa_2\right)A_2\right)y_{k+1}$$
$$+\left(\tau_{k+1}^{(2)}+\kappa_1\right)\left(E-\left(\tau_{k+1}^{(1)}+\kappa_2\right)A_2\right)y_k. \tag{30}$$

Substituting (30) in (28), after a simple transformation we obtain (27). The reverse argument is obvious.

Using the definition of the operators A_1 and A_2, the scheme (28) and (29) can be written in the form of the usual difference scheme

$$\left(E-\left(\tau_{k+1}^{(1)}-\kappa_1\right)\Lambda_1\right)y_{k+1/2}$$
$$=\left(E+\left(\tau_{k+1}^{(1)}+\kappa_2\right)\Lambda_2\right)y_k+\left(\tau_{k+1}^{(1)}-\kappa_1\right)\varphi \tag{31}$$

for $h_1 \le x_1 \le l_1-h_1$,

$$y_{k+1/2}=g(x)+(\kappa_1+\kappa_2)\Lambda_2 g(x), \quad x_1=0,l_1.$$

The boundary-value problem (31) must be sequentially solved for $h_2 \le x_2 \le l_2-h_2$. We thus find $y_{k+1/2}$ for $0\le x_1\le l_1$, $h_2\le x_2\le l_2-h_2$. Further

$$\left(E-\left(\tau_{k+1}^{(2)}-\kappa_2\right)\Lambda_2\right)y_{k+1}$$
$$=\left(E+\left(\tau_{k+1}^{(2)}+\kappa_1\right)\Lambda_1\right)y_{k+1/2}+\left(\tau_{k+1}^{(2)}+\kappa_1\right)\varphi \tag{32}$$

for $h_2 \le x_2 \le l_2 - h_2$,

$$y_{k+1} = g(x), \quad x_2 = 0, l_2.$$

The boundary-value problem (32) must be sequentially solved for $h_1 \le x_1 \le l_1 - h_1$. As a result, we obtain y_{k+1}.

If we compare the iteration counts of the alternating-directions method for the second-order scheme considered in Section 11.2.1 and for the high-order scheme, then in the latter case the iteration count will be somewhat larger. In the special case $l_1 = l_2 = l$, $N_1 = N_1 = N$ for $N = 10$, the increase is about 1%, and for $N = 100$ about 4%. The volume of computation at each iteration for both schemes is practically identical, and the difference in iteration counts is insignificant. Since the high-order scheme allows us to use a coarser grid to achieve a given accuracy in the solution of the differential problem, its use is particularly profitable in those cases where the solution of the differential problem possesses sufficient smoothness.

Recall that in Section 3.3 we considered a direct method for solving problem (17) — the reduction method. Both for the second-order scheme and for the scheme examined here, the direct method will require fewer arithmetic operations than the alternating-directions method with the optimal parameters.

11.3 The alternating-directions method in the general case

11.3.1 The case of non-commuting operators. Suppose that it is necessary to find the solution to the linear operator equation

$$Au = f \tag{1}$$

with a non-singular operator A which is representable in the form of the sum of two self-adjoint non-commuting operators A_1 and A_2 with bounds δ_1, Δ_1 and δ_2, Δ_2:

$$\begin{aligned} &A_\alpha = A_\alpha^*, \quad \delta_\alpha E \le A_\alpha \le \Delta_\alpha E, \quad \alpha = 1, 2, \quad \delta_1 + \delta_2 > 0, \\ &A = A_1 + A_2. \end{aligned} \tag{2}$$

To approximately solve equation (1), we look at a two-level scheme for the alternating-directions method with two iterative parameters $\omega^{(1)}$ and $\omega^{(2)}$:

$$\begin{aligned} &B\frac{y_{k+1} - y_k}{\tau} + Ay_k = f, \quad k = 0, 1, \ldots, \quad y_0 \in H, \\ &B = \left(\omega^{(1)}E + A_1\right)\left(\omega^{(2)}E + A_2\right), \quad \tau = \omega^{(1)} + \omega^{(2)}. \end{aligned} \tag{3}$$

Here the iterative parameters and the operator B do not depend on the iteration number k.

As in the case of commuting operators A_1 and A_2, the iterative approximation y_{k+1} for the scheme (3) can be found using the following algorithm:

$$\left(\omega^{(1)}E + A_1\right) y_{k+1/2} = \left(\omega^{(1)}E - A_2\right) y_k + f,$$
$$\left(\omega^{(2)}E + A_2\right) y_{k+1} = \left(\omega^{(2)}E - A_1\right) y_{k+1/2} + f, \quad k = 0, 1, \ldots .$$

We now investigate the convergence of the iterative scheme (3) and find the optimal values of the parameters $\omega^{(1)}$ and $\omega^{(2)}$. Assuming that the operator $\omega^{(2)}E + A_2$ is not singular, we study the convergence of (3) in the energy space H_D, where $D = (\omega^{(2)}E + A_2)^2$. By (2), the operator D is self-adjoint in H, and it follows from the assumption above that D is positive-definite in H.

For the error $z_k = y_k - u$ we obtain from (3) the homogeneous equation

$$\begin{gathered} z_{k+1} = S z_k, \quad k = 0, 1, \ldots, \quad z_0 = y_0 - u, \\ S = \left(\omega^{(2)}E + A_2\right)^{-1} \left(\omega^{(1)}E + A_1\right)^{-1} \left(\omega^{(2)}E - A_1\right) \left(\omega^{(1)}E - A_2\right). \end{gathered} \tag{4}$$

In (4), we transform to the problem for the equivalent error

$$x_k = \left(\omega^{(2)}E + A_2\right) z_k. \tag{5}$$

We obtain

$$\begin{gathered} x_{k+1} = \bar{S} x_k, \quad k = 0, 1, \ldots, \quad \bar{S} = \bar{S}_1 \bar{S}_2, \\ \bar{S}_1 = \left(\omega^{(1)}E + A_1\right)^{-1} \left(\omega^{(2)}E - A_1\right), \\ \bar{S}_2 = \left(\omega^{(2)}E + A_2\right)^{-1} \left(\omega^{(1)}E - A_2\right). \end{gathered} \tag{6}$$

Since by the change of variables (5) we have $\| x_k \| = \| z_k \|_D$, it is sufficient to investigate the behavior of the norm of the equivalent error x_k in the space H: From (6) we find

$$\| x_{k+1} \| \le \| \bar{S} \| \, \| x_k \| \le \| \bar{S}_1 \| \, \| \bar{S}_2 \| \, \| x_k \|, \quad k = 0, 1, \ldots$$

and, consequently,

$$\| z_n \|_D = \| x_n \| \le \| \bar{S} \|^n \| x_0 \| \le \left(\| \bar{S}_1 \| \, \| \bar{S}_2 \| \right)^n \| z_0 \|_D . \tag{7}$$

We now estimate the norms of the operators $\bar{S}_1$ and $\bar{S}_2$. We assume that the operators $\omega^{(\alpha)}E + A_\alpha$, $\alpha = 1,2$ are non-negative. Then by (2), from Section 5.1.4 we obtain

$$\| \bar{S}_1 \| \leq \max_{\delta_1 \leq x \leq \Delta_1} \left| \frac{\omega^{(2)} - x}{\omega^{(1)} + x} \right|, \quad \| \bar{S}_2 \| \leq \max_{\delta_2 \leq y \leq \Delta_2} \left| \frac{\omega^{(1)} - y}{\omega^{(2)} + y} \right|$$

and, consequently,

$$\| \bar{S}_1 \| \| \bar{S}_2 \| \leq \max_{\substack{\delta_1 \leq x \leq \Delta_1 \\ \delta_2 \leq y \leq \Delta_2}} |R_1(x,y)|, \quad R_1(x,y) = \frac{\omega^{(2)} - x}{\omega^{(1)} + x} \frac{\omega^{(1)} - y}{\omega^{(2)} + y}.$$

Taking into account the estimate (7), we choose the parameters $\omega^{(1)}$ and $\omega^{(2)}$ from the condition

$$\min_{\omega^{(1)},\omega^{(2)}} \max_{\substack{\delta_1 \leq x \leq \Delta_1 \\ \delta_2 \leq y \leq \Delta_2}} |R_1(x,y)|.$$

This problem is a special case of the problem solved in Section 11.1. With the aid of the fractionally-linear representation of the variables x and y (cf. (15), (21)–(24) in Section 11.1), the statement of the problem leads to the problem of finding the parameter κ^* from the condition

$$\max_{\eta \leq u \leq 1} \left| \frac{\kappa^* - u}{\kappa^* + u} \right| = \min_{\kappa} \max_{\eta \leq u \leq 1} \left| \frac{\kappa - u}{\kappa + u} \right| = \rho. \tag{8}$$

Here the parameters $\omega^{(1)}$ and $\omega^{(2)}$ are expressed in terms of κ^* using the formulas

$$\omega^{(1)} = \frac{r\kappa^* + s}{1 + t\kappa^*}, \quad \omega^{(2)} = \frac{r\kappa^* - s}{1 - t\kappa^*},$$

and for the error z_n we have the estimate

$$\| z_n \|_D \leq \rho^{2n} \| z_0 \|_D .$$

In addition, in Section 11.1.4 it was shown that, for the optimal choice of κ^*, the operators $\omega^{(\alpha)}E + A_\alpha$ are positive-definite if $\delta_1 + \delta_2 > 0$. Consequently, by (2) our assumptions about the non-negativity of the operators $\omega^{(\alpha)}E + A_\alpha$, $\alpha = 1,2$ will be automatically satisfied.

We now derive the solution of problem (8) independently of the results in Section 11.1.4. We consider the function $\varphi(u) = (\kappa - u)/(\kappa + u)$ on the interval $0 < \eta \leq u \leq 1$ for $\kappa > 0$. This function increases monotonically in u, and consequently

$$\max_{\eta \leq u \leq 1} |\varphi(u)| = \max\left(\left|\frac{\eta - \kappa}{\eta + \kappa}\right|, \left|\frac{1-\kappa}{1+\kappa}\right|\right).$$

From ths it is easy to obtain that the minimum over κ of the expression is achieved for κ^* defined by the equation

$$\frac{\kappa^* - \eta}{\kappa^* + \eta} = \frac{1 - \kappa^*}{1 + \kappa^*}.$$

From this we find

$$\kappa^* = \sqrt{\eta}, \quad \min_{\kappa} \max_{\eta \leq u \leq 1} \left|\frac{\kappa - u}{\kappa + u}\right| = \rho = \frac{1 - \sqrt{\eta}}{1 + \sqrt{\eta}}.$$

Thus, we have proved

Theorem 2. *Suppose the conditions* (2) *are satisfied, and the parameters* $\omega^{(1)}$ *and* $\omega^{(2)}$ *are chosen using the formulas*

$$\omega^{(1)} = \frac{r\sqrt{\eta} + s}{1 + t\sqrt{\eta}}, \quad \omega^{(2)} = \frac{r\sqrt{\eta} - s}{1 - t\sqrt{\eta}},$$

where r, s, t, *and* η *are defined in* (21)–(24) *of Section* 11.1. *The alternating-directions method* (3) *converges in* H_D, *and the error* z_n *can be estimated using*

$$\| z_n \|_D \leq \rho^{2n} \| z_0 \|_D, \quad \rho = \frac{1 - \sqrt{\eta}}{1 + \sqrt{\eta}}$$

where $D = (\omega^{(2)} E + A_2)^2$. *For the iteration count* n *we have the estimate*

$$n = n_0(\epsilon) = \ln \epsilon / (2 \ln \rho) \approx \ln \frac{1}{\epsilon} \Big/ (4\sqrt{\eta}). \square$$

11.3.2 A Dirichlet difference problem for an elliptic equation with variable coefficients. We look now at a sample application of the alternating-directions method in the non-commutative case. Suppose that, on the rectangular grid $\bar{\omega} = \{x_{ij} = (ih_1, jh_2) \in \bar{G}, 0 \leq i \leq N_1, 0 \leq j \leq N_2, h_\alpha = l_\alpha / N_\alpha, \alpha = 1, 2\}$ in the rectangle $\bar{G} = \{0 \le x_\alpha \le l_\alpha,\ \alpha = 1, 2\}$, it is necessary to find the solution to the following difference problem:

$$\Lambda y = \sum_{\alpha=1}^{2} (a_\alpha(x) y_{\bar{x}_\alpha})_{x_\alpha} - q(x) y = -\varphi(x), \quad x \in \omega,$$
$$y(x) = g(x), \quad x \in \gamma, \tag{9}$$

where the coefficients of the difference scheme satisfy the conditions

$$0 < c_1 \le a_\alpha(x) \le c_2, \quad 0 \le d_1 \le q(x) \le d_2. \tag{10}$$

In the space H of grid functions defined on ω with the inner product

$$(u, v) = \sum_{x \in \omega} u(x) v(x) h_1 h_2,$$

the operators A_1 and A_2 are defined as follows: $A_\alpha y = -\Lambda_\alpha \mathring{y} = -(a_\alpha \mathring{y}_{\bar{x}_\alpha})_{x_\alpha} + 0.5 q \mathring{y}$, $\alpha = 1, 2$, $y \in H$, $\mathring{y} \in \mathring{H}$, where, as usual, $\mathring{y}(x) = 0$ on γ.

The operators A_α are self-adjoint in H, and if $a_\alpha(x)$ depends only on the variable x_α and $q(x)$ is constant, then the operators A_1 and A_2 will commute. In the general case, they do not commute, and it is possible to apply the alternating-directions method examined in Section 11.3.1 to solve equation (1) corresponding to the difference problem (9).

The algorithm for the method has the simple form

$$\omega^{(1)} y_{k+1/2} - \Lambda_1 y_{k+1/2} = \omega^{(1)} y_k + \Lambda_2 y_k + \varphi, \qquad h_1 \le x_1 \le l_1 - h_1,$$
$$y_{k+1/2}(x) = g(x), \quad x_1 = 0, l_1, \qquad h_2 \le x_2 \le l_2 - h_2,$$
$$\omega^{(2)} y_{k+1} - \Lambda_2 y_{k+1} = \omega^{(2)} y_{k+1/2} + \Lambda_1 y_{k+1/2} + \varphi, \quad h_2 \le x_2 \le l_2 - h_2,$$
$$y_{k+1}(x) = g(x), \quad x_2 = 0, l_2, \qquad h_1 \le x_1 \le l_1 - h_1.$$

It remains to find the bounds δ_α and Δ_α for the operators A_α, $\alpha = 1, 2$. Since the conditions (10) are satisfied, from lemma 14 of Section 5.2 we obtain

$$(y^2, 1)_{\omega_\alpha} \le \kappa_\alpha(x_\beta)(A_\alpha y, y)_{\omega_\alpha}, \quad \beta = 3 - \alpha,\ \alpha = 1, 2, \tag{11}$$

where $\kappa_\alpha(x_\beta) = \max\limits_{x_\alpha \in \omega_\alpha} v^\alpha(x)$, and $v^\alpha(x)$ is the solution of the following three-point boundary-value problem:

$$\left(a_\alpha v^\alpha_{\bar{x}_\alpha}\right)_{x_\alpha} - 0.5 q v^\alpha = -1, \quad h_\alpha \le x_\alpha \le l_\alpha - h_\alpha,$$
$$v^\alpha(x) = 0, \quad x_\alpha = 0, l_\alpha, \quad h_\beta \le x_\beta \le l_\beta - h_\beta.$$

The inner product over ω_α is defined in the following way:

$$(u,v)_{\omega_\alpha} = \sum_{x_\alpha=h_\alpha}^{l_\alpha-h_\alpha} u(x)v(x)h_\alpha = \sum_{x_\alpha\in\omega_\alpha} u(x)v(x)h_\alpha.$$

Multiplying (11) by h_β and summing over ω_β, we obtain

$$\left(\frac{1}{\kappa_\alpha}y^2,1\right) \le (A_\alpha y,y), \quad \alpha=1,2.$$

Consequently, it is possible to take δ_α to be

$$\delta_\alpha = \min_{h_\beta\le x_\beta\le l_\beta-h_\beta} \frac{1}{\kappa_\alpha(x_\beta)}, \quad \beta=3-\alpha,\ \alpha=1,2.$$

We now find Δ_α. We will proceed by analogy with Section 11.2.2. We denote by $\mathcal{D}$ the diagonal part of the matrix $\mathcal{A}_\alpha$, corresponding to the operator A_α:

$$\mathcal{D}y = d_\alpha(x)y,$$

$$d_\alpha(x) = \begin{cases} 0.5q(x) + \dfrac{1}{h_\alpha^2}(a_\alpha(x_\alpha,x_\beta)+a_\alpha(x_\alpha+h_\alpha,x_\beta)), \\ h_\alpha \le x_\alpha \le l_\alpha - h_\alpha, \\ h_\beta \le x_\beta \le l_\beta - h_\beta. \end{cases}$$

Then we have

$$(A_\alpha y,y) \le (2-\lambda_{\min})(\mathcal{D}y,y) \le (2-\lambda_{\min})\max_{x\in\omega} d_\alpha(x)(y,y),$$

where $\lambda_{\min}$ is the constant from the operator inequality $\lambda_{\min}\mathcal{D} \le A_\alpha$.

We now find $\lambda_{\min}$. From lemma 14 of Section 5.2 we obtain

$$(d_\alpha y^2,1)_{\omega_\alpha} \le \rho_\alpha(x_\beta)(A_\alpha y,y)_{\omega_\alpha},$$

where $\rho_\alpha(x_\beta) = \max_{x_\alpha\in\omega_\alpha} w^\alpha(x)$, and $w^\alpha(x)$ is the solution of the following three-point boundary-value problem:

$$\left(a_\alpha w^\alpha_{\bar{x}_\alpha}\right)_{x_\alpha} - 0.5qw^\alpha = -d_\alpha(x), \quad h_\alpha \le x_\alpha \le l_\alpha - h_\alpha,$$

$$w^\alpha(x)=0, \quad x_\alpha = 0, l_\alpha, \quad h_\beta \le x_\beta \le l_\beta - h_\beta.$$

Multiplying (12) by h_β and summing over ω_β, we obtain

$$\left(\frac{d_\alpha}{\rho_\alpha}y^2,1\right) \le (A_\alpha y,y), \quad \alpha=1,2.$$

Table 12

c_2/c_1	$N=32$	$N=64$	$N=128$
2	65	132	264
8	90	187	380
32	110	233	482
128	122	264	556
512	128	282	603

Consequently, it is possible to take $\lambda_{\min}$ to be

$$\lambda_{\min} = \min_{h_\beta \le x_\beta \le l_\beta - h_\beta} \frac{1}{\rho_\alpha(x_\beta)},$$

therefore Δ_α is

$$\Delta_\alpha = \left(2 - \frac{1}{\max\limits_{x_\beta} \rho_\alpha(x_\beta)}\right) \max_{x\in\omega} d_\alpha(x), \qquad \alpha = 1,2.$$

Thus, the *a priori* information needed to apply the alternating-directions method has been found. Using the conditions (10), it can be shown that the quantity η, determining the convergence rate of the method, for this example is $O(|h|^2)$, where $|h|^2 = h_1^2 + h_2^2$. Therefore by theorem 2, we have the following estimate for the iteration count

$$n = O\left(\frac{1}{|h|} \ln \frac{1}{\epsilon}\right).$$

We look now at a model problem. Suppose that the difference scheme (9) is given on a square grid in the unit square ($N_1 = N_2 = N$, $l_1 = l_2 = l$). The coefficients $a_1(x)$, $a_2(x)$ and $q(x)$ are chosen as follows:

$$a_1(x) = 1 + c\left[(x_1 - 0.5)^2 + (x_2 - 0.5)^2\right],$$

$$a_2(x) = 1 + c\left[0.5 - (x_1 - 0.5)^2 - (x_2 - 0.5)^2\right],$$

$$q(x) \equiv 0, \quad c > 0.$$

In this case, in the inequalities (10) $c_1 = 1$, $c_2 = 1 + 0.5c$, $d_1 = d_2 = 0$; changing the parameter c, we obtain coefficients for a difference scheme (9) with different extrema.

We give now the iteration count for this alternating-directions method as a function of the ratio c_2/c_1 and of the number of nodes N in one direction for $\epsilon = 10^{-4}$.

We compare this method with the successive over-relaxation method (cf. Section 9.2), the alternate-triangular method (cf. Section 10.2) and the implicit Chebyshev method (cf. Section 6.2.3). Based on the iteration count, this alternating-directions method is worse than the successive over-relaxation method and the alternate-triangular method, but it is better than the implicit Chebyshev method by a factor of 1.5–2. However, based on the volume of computation, the alternating-directions method is also worse than the implicit Chebyshev method.

Chapter 12

Methods for Solving Equations with Indefinite and Singular Operators

In this chapter we study direct and iterative methods for solving equations with indefinite non-singular operators, with complex operators, and also with singular operators. In Section 12.1, we consider a method with Chebyshev parameters and also a method of variational type for the solution of an equation with an indefinite operator. In Section 12.2 complex versions of the simple iteration method and the alternating-directions method are constructed for an equation with a complex operator of special form. In Section 12.3 we study general iterative methods for solving equations with singular operators, but where the preconditioning operator is non-singular. Section 12.4 is concerned with the construction of special direct and iterative methods for equations with singular operators.

12.1 Equations with real indefinite operators

12.1.1 The iterative scheme. The choice of the iterative parameters. Suppose that, in a Hilbert space H, we are given the equation

$$Au = f \tag{1}$$

with a linear non-singular operator A. To solve equation (1), we look at an implicit two-level iterative scheme

$$B\frac{y_{k+1} - y_k}{\tau_{k+1}} + Ay_k = f, \quad k = 0, 1, \ldots, \tag{2}$$

with a non-singular operator B and an arbitrary $y_0 \in H$.

Using on the properties of the operators A, B, and D, iterative schemes of the form (2) were studied in Chapters 6 and 8, where several choices of the parameters τ_k were examined. Recall that D is the self-adjoint positive-definite operator which generates the energy space H_D. It was proved that, for the convergence of these iterative methods in H_D, it was necessary that the operator

$$C = D^{-1/2}(DB^{-1}A)D^{-1/2}. \tag{3}$$

be positive-definite. For specific choices of the operator D, this requirement led to the following conditions on the operators A and B:

1) the operator A must be positive-definite in H, if $D = A$, B, or $A^*B^{-1}A$;

2) the operator B^*A must be positive-definite in H if $D = A^*A$ or B^*B.

There exist problems for which these requirements are not satisfied, i.e. either the operator A is not definite, or it is difficult to find an operator B so that B^*A will be positive-definite. As an example of such a problem, we could consider a Dirichlet problem for the Helmholtz equation in a rectangle

$$y_{\bar{x}_1 x_1} + y_{\bar{x}_2 x_2} + m^2 y = 0, \quad x \in \omega,$$
$$y(x) = g(x), \quad x \in \gamma,$$

where $m^2 > 0$.

This sub-section is concerned with the construction of implicit two-level iterative methods for the case where the operator C is a non-singular indefinite operator in H. Here we will only look at real operators C; the complex case will be studied in Section 12.2.

We move on now to the construction of the iterative methods. In the equation

$$z_{k+1} = (E - \tau_{k+1}B^{-1}A)z_k, \quad k = 0, 1, \ldots, 2n-1,$$

for the error $z_k = y_k - u$ in the iterative scheme (2), we make the change of variable $z_k = D^{-1/2}x_k$ and transform to an equation for the equivalent error x_k:

$$x_{k+1} = (E - \tau_{k+1}C)x_k, \quad k = 0, 1, \ldots, 2n-1, \tag{4}$$

where the operator C is defined in (3). Since the operator C is indefinite, it is obvious that the norm of the operator $E - \tau_{k+1}C$ will be greater than or equal to one for any τ_{k+1}.

We now consider the equation relating the errors at even iterations. From (4) we obtain

$$x_{2k+2} = (E - \tau_{2k+2}C)(E - \tau_{2k+1}C)x_{2k}, \quad k = 0, 1, \ldots, n-1. \tag{5}$$

If we denote

$$\omega_{k+1} = -\tau_{2k+2}\tau_{2k+1}, \quad k = 0, 1, \dots, n-1, \tag{6}$$

and require that, for any k, the iterative parameters τ_{2k+2} and τ_{2k+1} satisfy the condition

$$1/\tau_{2k+2} + 1/\tau_{2k+1} = 2\alpha, \quad k = 0, 1, \dots, n-1, \tag{7}$$

where α is a constant yet to be determined, then (5) can be written in the form

$$x_{2k+2} = (E - \omega_{k+1}\bar{C})x_{2k}, \quad k = 0, 1, \dots, \quad \bar{C} = C^2 - 2\alpha C. \tag{8}$$

If ω_{k+1} and α are known, then the parameters τ_{2k+2} and τ_{2k+1} are defined using (6) and (7) by the formulas

$$\begin{aligned} \tau_{2k+1} &= -\alpha\omega_{k+1} - \sqrt{\alpha^2\omega_{k+1}^2 + \omega_{k+1}}, \\ \tau_{2k+2} &= -\alpha\omega_{k+1} + \sqrt{\alpha^2\omega_{k+1}^2 + \omega_{k+1}}, \\ k &= 0, 1, \dots, n-1. \end{aligned} \tag{9}$$

From (8) we obtain

$$x_{2n} = \prod_{j=1}^{n}(E - \omega_j\bar{C})x_0,$$

$$\| x_{2n} \| \le \left\| \prod_{j=1}^{n}(E - \omega_j\bar{C}) \right\| \| x_0 \| . \tag{10}$$

Since the operator $\bar{C}$ depends on α, the requirement that $\bar{C}$ be positive-definite will be one of the conditions underlying the choice of the parameter α. In addition, from (10) it follows that the parameters ω_j, $1 \le j \le n$, and the parameter α must be chosen from the condition that the norm of the resolving operator $\prod_{j=1}^{n}(E - \omega_j\bar{C})$ be a minimum.

This problem of the optimal choice for the iterative parameters ω_j and α, and hence also for the parameters τ_k in the scheme (2), will be solved below. First we show the connection between the above derivation and the derivation based on a Gauss transformation for the case of a self-adjoint operator C.

Notice that the change of variable $u = D^{-1/2}x$, $f = BD^{-1/2}\varphi$ allows us to write the original equation (1) in the following form:

$$Cx = \varphi, \tag{11}$$

where the operator C is defined in (3). Using (11), we obtain

$$\bar{C}x = C^2x - 2\alpha Cx = (C - 2\alpha E)\varphi = \bar{\varphi}. \tag{12}$$

Further, if we denote $v_k = D^{1/2}y_k$, where y_k is the iterative approximation in the scheme (2), then it is easy to find

$$x_k = D^{1/2}z_k = D^{1/2}y_k - D^{1/2}u = v_k - x.$$

Substituting x_k in (8) and taking into account (12), we obtain the iterative scheme

$$\frac{v_{2k+2} - v_{2k}}{\omega_{k+1}} + \bar{C}v_{2k} = \bar{\varphi}, \quad k = 0, 1, \ldots. \tag{13}$$

Thus, the scheme (13) is an explicit two-level scheme for the transformed equation (12).

Assume that $C = C^*$. Recall that in this case the first Gauss transformation involves converting from equation (11) to the equation $\bar{\bar{C}}x = C^2x = C\varphi = \bar{\bar{\varphi}}$. Since C is a non-singular operator, the operator C^2 will be positive-definite in H. Therefore this transformation leads us to the case of a definite operator. To solve the equation with this operator, it is possible to use a two-level scheme of the form (13), after having changed $\bar{C}$ to $\bar{\bar{C}}$ and $\bar{\varphi}$ to $\bar{\bar{\varphi}}$. Obviously, such a method is a special case (for $\alpha = 0$) of the method examined here.

12.1.2 Transforming the operator in the self-adjoint case. We will assume that the operator C is self-adjoint in H. Then the operator $\bar{C} = C^2 - \alpha C$ is also self-adjoint in H. Our first goal is to choose the parameter α so that the operator $\bar{C}$ will be positive-definite, and to find the bounds $\gamma_1 = \gamma_1(\alpha)$ and $\gamma_2 = \gamma_2(\alpha)$ of this operator, i.e. the values from the inequalities

$$\gamma_1 E \le \bar{C} \le \gamma_2 E, \quad \gamma_1 > 0. \tag{14}$$

If such a value of α exists, then using the estimate

$$\left\| \prod_{j=1}^{n} (E - \omega_j \bar{C}) \right\| \le \max_{\gamma_1 \le t \le \gamma_2} \left| \prod_{j=1}^{n} (1 - \omega_j t) \right|$$

the problem of finding the parameters ω_j, $j = 1,2,\ldots,n$, reduces to the construction of the polynomial $P_n(t)$ of degree n, normalized by the condition $P_n(0) = 1$, that deviates least from zero on the interval $[\gamma_1, \gamma_2]$ in the positive semi-axis. This problem was already studied in Chapter 6 when constructing the Chebyshev method. The solution has the form

$$\omega_k = \frac{\omega_0}{1+\rho_0\mu_k}, \quad \mu_k \in \mathcal{M}_n^* = \left\{-\cos\frac{2i-1}{2n}\pi,\ i = 1,2,\ldots,n\right\},$$

where $k = 1,2,\ldots,n$,

$$\omega_0 = \frac{2}{\gamma_1+\gamma_2}, \quad \rho_0 = \frac{1-\xi}{1+\xi}, \quad \rho_1 = \frac{1-\sqrt{\xi}}{1+\sqrt{\xi}}, \quad \xi = \frac{\gamma_1}{\gamma_2}.$$

By (10), the error x_{2n} can be estimated using

$$\| x_{2n} \| \leq q_n \| x_0 \|, \quad q_n = 2\rho_1^n/(1+\rho_1^{2n}) \ .$$

From this it follows that the choice of the parameter α should be made subject to the condition that the ratio γ_1/γ_2 be maximized.

We now find the optimal value for the parameter α. Suppose that the eigenvalues μ of the operator C lie in the intervals $[\mathring{\gamma}_1, \mathring{\gamma}_2]$ and $[\mathring{\gamma}_3, \mathring{\gamma}_4]$. Since the operator C is indefinite and non-singular, then

$$\mathring{\gamma}_1 \leq \mathring{\gamma}_2 < 0 < \mathring{\gamma}_3 \leq \mathring{\gamma}_4. \tag{15}$$

We now find the eigenvalues λ of the operator $\bar{C} = C^2 - 2\alpha C$. It is easy to see that the eigenvalues of the operators $\bar{C}$ and C are related by the formula

$$\lambda = \mu^2 - 2\alpha\mu, \quad \mu \in \Omega, \tag{16}$$

where Ω consists of the two intervals $[\mathring{\gamma}_1, \mathring{\gamma}_2]$ and $[\mathring{\gamma}_3, \mathring{\gamma}_4]$.

We first find a bound on α which guarantees the positivity of the eigenvalues λ, i.e. the positive-definiteness of the operator $\bar{C}$. An analysis of the inequality $\mu^2 - 2\alpha\mu > 0$ shows that it is valid for μ lying out of the interval $[0, 2\alpha]$. Therefore this inequality will be satisfied for $\mu \in \Omega$ if α satisfies the condition

$$\mathring{\gamma}_2 < 2\alpha < \mathring{\gamma}_3. \tag{17}$$

We will assume that (17) is satisfied. From (16) we obtain that the transformation $\lambda = \lambda(\mu) = \mu^2 - 2\alpha\mu$ maps the interval $[\mathring{\gamma}_1, \mathring{\gamma}_2]$ onto the interval $[\lambda_2, \lambda_1]$, and the interval $[\mathring{\gamma}_3, \mathring{\gamma}_4]$ onto the interval $[\lambda_3, \lambda_4]$, where $\lambda_i = \lambda_i(\mathring{\gamma}_i)$, $1 \le i \le 4$. Thus, all the eigenvalues of the operator $\bar{C}$ are positive and lie in the intervals $[\lambda_2, \lambda_1] \cup [\lambda_3, \lambda_4]$. Therefore, in the inequalities (14), it follows that

$$\gamma_1 = \min(\lambda_2, \lambda_3), \quad \gamma_2 = \max(\lambda_1, \lambda_4). \tag{18}$$

We now choose $2\alpha \in (\mathring{\gamma}_2, \mathring{\gamma}_3)$ from the condition that the ratio γ_1/γ_2 be maximal. From (18) we obtain

$$\gamma_1 = \begin{cases} \lambda_2 = \mathring{\gamma}_2(\mathring{\gamma}_2 - 2\alpha), & \mathring{\gamma}_2 < 2\alpha \le \mathring{\gamma}_2 + \mathring{\gamma}_3, \\ \lambda_3 = \mathring{\gamma}_3(\mathring{\gamma}_3 - 2\alpha), & \mathring{\gamma}_2 + \mathring{\gamma}_3 \le 2\alpha < \mathring{\gamma}_3, \end{cases}$$

$$\gamma_2 = \begin{cases} \lambda_4 = \mathring{\gamma}_4(\mathring{\gamma}_4 - 2\alpha), & 2\alpha \le \mathring{\gamma}_1 + \mathring{\gamma}_4, \\ \lambda_1 = \mathring{\gamma}_1(\mathring{\gamma}_1 - 2\alpha), & \mathring{\gamma}_1 + \mathring{\gamma}_4 \le 2\alpha. \end{cases}$$

We introduce the following notation: $\Delta_1 = \mathring{\gamma}_2 - \mathring{\gamma}_1$, $\Delta_2 = \mathring{\gamma}_4 - \mathring{\gamma}_3$, and we consider two cases.

1) Suppose first that $\Delta_1 \le \Delta_2$, i.e. $\mathring{\gamma}_2 + \mathring{\gamma}_3 \le \mathring{\gamma}_1 + \mathring{\gamma}_4$. In this case we obtain the following expression for $\xi = \gamma_1/\gamma_2$:

$$\xi = \xi(\alpha) = \begin{cases} \dfrac{\mathring{\gamma}_2(\mathring{\gamma}_2 - 2\alpha)}{\mathring{\gamma}_4(\mathring{\gamma}_4 - 2\alpha)}, & \mathring{\gamma}_2 < 2\alpha \le \mathring{\gamma}_2 + \mathring{\gamma}_3, \text{ increasing in } \alpha, \\ \dfrac{\mathring{\gamma}_3(\mathring{\gamma}_3 - 2\alpha)}{\mathring{\gamma}_4(\mathring{\gamma}_4 - 2\alpha)}, & \mathring{\gamma}_2 + \mathring{\gamma}_3 \le 2\alpha \le \mathring{\gamma}_1 + \mathring{\gamma}_4, \text{ decreasing in } \alpha, \\ \dfrac{\mathring{\gamma}_3(\mathring{\gamma}_3 - 2\alpha)}{\mathring{\gamma}_1(\mathring{\gamma}_1 - 2\alpha)}, & \mathring{\gamma}_1 + \mathring{\gamma}_4 \le 2\alpha, \text{ decreasing in } \alpha. \end{cases}$$

Consequently, in this case the optimal value of α is

$$\alpha = \alpha_0 = (\mathring{\gamma}_2 + \mathring{\gamma}_3)/\,2, \tag{19}$$

if the condition (17) is satisfied. For $\alpha = \alpha_0$ we have

$$\gamma_1 = \lambda_2 = \lambda_3 = -\mathring{\gamma}_2\mathring{\gamma}_3, \tag{20}$$

$$\gamma_2 = \lambda_4 = \mathring{\gamma}_4(\Delta_2 - \Delta_1) - \mathring{\gamma}_1\mathring{\gamma}_4 \ge \lambda_1. \tag{21}$$

2) Suppose now that $\Delta_1 \geq \Delta_2$, i.e. $\mathring{\gamma}_2 + \mathring{\gamma}_3 \geq \mathring{\gamma}_1 + \mathring{\gamma}_4$. In this case we will have

$$\xi = \xi(\alpha) = \begin{cases} \dfrac{\mathring{\gamma}_2(\mathring{\gamma}_2 - 2\alpha)}{\mathring{\gamma}_4(\mathring{\gamma}_4 - 2\alpha)}, & 2\alpha \leq \mathring{\gamma}_1 + \mathring{\gamma}_4, \text{ increasing in } \alpha, \\ \dfrac{\mathring{\gamma}_2(\mathring{\gamma}_2 - 2\alpha)}{\mathring{\gamma}_1(\mathring{\gamma}_1 - 2\alpha)}, & \mathring{\gamma}_1 + \mathring{\gamma}_4 \leq 2\alpha \leq \mathring{\gamma}_2 + \mathring{\gamma}_3, \text{ increasing in } \alpha, \\ \dfrac{\mathring{\gamma}_3(\mathring{\gamma}_3 - 2\alpha)}{\mathring{\gamma}_1(\mathring{\gamma}_1 - 2\alpha)}, & \mathring{\gamma}_2 + \mathring{\gamma}_3 \leq 2\alpha, < \mathring{\gamma}_3 \text{ decreasing in } \alpha. \end{cases}$$

Consequently, in this case also the optimal value of the parameter α is defined by formula (19), the value of γ_1 is given in (20), and

$$\gamma_2 = \lambda_1 = \mathring{\gamma}_1(\Delta_2 - \Delta_1) - \mathring{\gamma}_1\mathring{\gamma}_4 \geq \lambda_4. \tag{22}$$

Thus, we have proved

Lemma 1. *Suppose that the eigenvalues of the operator C lie in the intervals $[\mathring{\gamma}_1, \mathring{\gamma}_2]$ and $[\mathring{\gamma}_3, \mathring{\gamma}_4]$, $\mathring{\gamma}_2 < 0 < \mathring{\gamma}_3$. Then if $\alpha = \alpha_0 = (\mathring{\gamma}_2 + \mathring{\gamma}_3)/2$ and $\bar{C} = C^2 - \alpha C$, we have that*

$$\gamma_1 E \leq \bar{C} \leq \gamma_2 E, \quad \gamma_1 > 0,$$

where

$$\gamma_1 = -\mathring{\gamma}_2\mathring{\gamma}_3, \quad \gamma_2 = \max\,[\mathring{\gamma}_4(\Delta_2 - \Delta_1), \mathring{\gamma}_1(\Delta_2 - \Delta_1)] - \mathring{\gamma}_1\mathring{\gamma}_4.$$

For this value of α the ratio γ_1/γ_2 is maximal. □

The assertions of the lemma follows from (19)–(22). Notice that $\alpha_0 = 0$ only in the case where $\mathring{\gamma}_2 = -\mathring{\gamma}_3$.

12.1.3 The iterative method with the Chebyshev parameters. Above we considered the two-level iterative scheme (2), where the parameters τ_k, $k = 1, 2, \ldots, 2n$, were expressed in terms of the ω_k, $1 \leq k \leq n$, and α using the formulas (9). Here the parameters ω_k are the iterative parameters for the Chebyshev method and are determined by the corresponding formulas, where the necessary *a priori* information and the optimal value of the parameter α are given in lemma 1.

Notice that we assumed that the eigenvalues μ of the self-adjoint operator C belonged to the intervals $[\mathring{\gamma}_1, \mathring{\gamma}_2]$ and $[\mathring{\gamma}_3, \mathring{\gamma}_4]$. From the definition (3) of the operator C, it follows that the eigenvalues of the operator C are also eigenvalues for the following problem:

$$Au - \mu Bu = 0. \tag{23}$$

In order to verify this, it is sufficient to multiply this expression on the left by the operator $D^{1/2}B^{-1}$ and make a change of variable, setting $u = D^{-1/2}v$. Notice that the operator C will be self-adjoint in H if the operator $DB^{-1}A$ is self-adjoint.

We formulate these results in the form of a theorem.

Theorem 1. *Suppose that the operator $DB^{-1}A$ is self-adjoint in H and that the eigenvalues of the problem (23) belong to the intervals $[\mathring{\gamma}_1, \mathring{\gamma}_2]$ and $[\mathring{\gamma}_3, \mathring{\gamma}_4]$, $\mathring{\gamma}_1 \le \mathring{\gamma}_2 < 0 < \mathring{\gamma}_3 \le \mathring{\gamma}_4$. For the iterative process (2) with the parameters*

$$\tau_{2k-1} = -\alpha_0\omega_k - \sqrt{\alpha_0^2\omega_k^2 + \omega_k}, \quad \tau_{2k} = -\alpha_0\omega_k + \sqrt{\alpha_0^2\omega_k^2 + \omega_k},$$
$$k = 1, 2, \ldots, n,$$

we have the estimate

$$\| z_{2n} \|_D \le q_n \| z_0 \|_D,$$

where

$$\omega_k = \frac{\omega_0}{1+\rho_0\mu_k}, \ \mu_k \in \mathcal{M}_n^* = \left\{-\cos\frac{(2i-1)\pi}{2n}, \ 1 \le i \le n\right\}, \ 1 \le k \le n,$$

$$\omega_0 = \frac{2}{\gamma_1+\gamma_2}, \ \rho_0 = \frac{1-\xi}{1+\xi}, \ \rho_1 = \frac{1-\sqrt{\xi}}{1+\sqrt{\xi}}, \ q_n = \frac{2\rho_1^n}{1+\rho_1^{2n}}, \ \xi = \frac{\gamma_1}{\gamma_2},$$

$$\alpha_0 = 0.5\,(\mathring{\gamma}_2 + \mathring{\gamma}_3), \quad \gamma_1 = -\mathring{\gamma}_2\mathring{\gamma}_3,$$
$$\gamma_2 = \max\,[\mathring{\gamma}_4(\Delta_2 - \Delta_1), \ \mathring{\gamma}_1(\Delta_2 - \Delta_1)] - \mathring{\gamma}_1\mathring{\gamma}_4,$$
$$\Delta_1 = \mathring{\gamma}_2 - \mathring{\gamma}_1, \quad \Delta_2 = \mathring{\gamma}_4 - \mathring{\gamma}_3. \ \square$$

The iterative method (2) with these parameters τ_k will be called the *Chebyshev method.*

We now consider some special cases. Suppose $\Delta_1 = \Delta_2$, i.e. the lengths of the intervals $[\mathring{\gamma}_1, \mathring{\gamma}_2]$ and $[\mathring{\gamma}_3, \mathring{\gamma}_4]$ are the same. In this case we have

$$\gamma_1 = -\mathring{\gamma}_2\mathring{\gamma}_3, \quad \gamma_2 = -\mathring{\gamma}_1\mathring{\gamma}_4, \quad \xi = \frac{\mathring{\gamma}_2\mathring{\gamma}_3}{\mathring{\gamma}_1\mathring{\gamma}_4}.$$

We will show in this case the set of parameters τ_k is optimal. In order to construct the set of parameters τ_k for the scheme (2), we imposed n conditions (7); consequently, the choice of the parameters is fixed by the auxiliary constraints.

From (5) and (8) we find that

$$x_{2n} = Q_{2n}(C) \quad x_0 = P_n(\bar{C})x_0,$$

where

$$Q_{2n}(C) = \prod_{j=1}^{2n}(E - \tau_j C) = P_n(\bar{C}) = \prod_{j=1}^{n}(E - \omega_j \bar{C}). \tag{24}$$

We look at the corresponding algebraic polynomials $Q_{2n}(\mu)$ and $P_n(\lambda)$ ($\lambda = \mu^2 - 2\alpha\mu$). If the parameters ω_j are chosen in the manner indicated in theorem 1, then the polynomial $P_n(\lambda)$ can be expressed in terms of the Chebyshev polynomials of the first kind in the following way (cf. Section 6.2.1):

$$P_n(\lambda) = q_n T_n\left(\frac{1 - \omega_0\lambda}{\rho_0}\right), \quad P_n(0) = 1,$$

$$\max_{\gamma_1 \le \lambda \le \gamma_2} |P_n(\lambda)| = q_n.$$

Notice that at the points $\gamma_1 = \lambda_0 < \lambda_1 < \ldots < \lambda_n = \gamma_2$ where

$$\lambda_k = \frac{1 - \rho_0 \cos\frac{k\pi}{n}}{\omega_0}, \quad k = 0, 1, \ldots, n,$$

the polynomial $P_n(\lambda)$ takes on its extreme values on the intervals $[\gamma_1, \gamma_2]$:

$$P_n(\lambda_k) = (-1)^k q_n, \quad k = 0, 1, \ldots, n. \tag{25}$$

Since by (24) $Q_{2n}(\mu) = P_n(\lambda)$, where λ and μ are related by the formula $\lambda = \mu^2 - 2\alpha\mu$, then from (25) we find

$$Q_{2n}(\mu_k^-) = Q_{2n}(\mu_k^+) = (-1)^k q_n, \quad k = 0, 1, \ldots, n, \tag{26}$$

where μ_k^- and μ_k^+ are the roots of the quadratic equation

$$\mu_k^2 - 2\alpha\mu_k - \lambda_k = 0, \quad k = 0, 1, \ldots, n. \tag{27}$$

Further, for this case the transformation $\lambda = \lambda(\mu) = \mu^2 - 2\alpha\mu$ maps each of the intervals $[\mathring{\gamma}_1, \mathring{\gamma}_2]$ and $[\mathring{\gamma}_3, \mathring{\gamma}_4]$ onto the one interval $[\gamma_1, \gamma_2]$. Here the points $\mu = \mathring{\gamma}_2$, $\mu = \mathring{\gamma}_3$ correspond to $\lambda = \gamma_1$, and $\mu = \mathring{\gamma}_1$ and $\mu = \mathring{\gamma}_4$ correspond to $\lambda = \gamma_2$. Therefore the roots of equation (27) are distributed as follows:

$$\mathring{\gamma}_1 = \mu_n^- < \mu_{n-1}^- < \cdots < \mu_0^- = \mathring{\gamma}_2,\ \mathring{\gamma}_3 = \mu_0^+ < \mu_1^+ < \cdots < \mu_n^+ = \mathring{\gamma}_4.$$

We will now assume that the set of parameters constructed in theorem 1 is not optimal. This indicates that there exists another polynomial of degree not greater than $2n$ of the form

$$\bar{Q}_{2n}(\mu) = \prod_{j=1}^{2n} (1 - \bar{\tau}_j \mu),$$

for which

$$\max_{\mu \in \Omega} |\bar{Q}_{2n}(\mu)| < q_n, \quad \Omega = [\mathring{\gamma}_1, \mathring{\gamma}_2] \cup [\mathring{\gamma}_3, \mathring{\gamma}_4].$$

We consider the difference $R_{2n}(\mu) = Q_{2n}(\mu) - \bar{Q}_{2n}(\mu)$, which is a polynomial of degree not greater than $2n$. Since the polynomial $R_{2n}(\mu)$ must then have $2n+2$ roots, we can conclude that the assumption made above is invalid.

To prove this fact, we look at the values of $R_{2n}(\mu)$ at the points μ_k^-, $0 \le k \le n$. Since, by assumption, $-q_n < \bar{Q}_{2n}(\mu) < q_n$, $\mu \in \Omega$, then

$$R_{2n}(\mu_k^-) = Q_{2n}(\mu_k^-) - \bar{Q}_{2n}(\mu_k^-) = (-1)^k q_n - \bar{Q}_{2n}(\mu_k^-)$$

and $R_{2n}(\mu_k^-) < 0$ if k is odd, $R_{2n}(\mu_k^-) > 0$ if k is even. Consequently, the polynomial $R_{2n}(\mu)$ changes sign as we move from μ_k^- to μ_{k+1}^-, $0 \le k \le n-1$. Therefore, on the interval $[\mathring{\gamma}_1, \mathring{\gamma}_2]$, there exist n roots of this polynomial. Analogously, by considering the values $R_{2n}(\mu)$ at the points μ_k^+, $0 \le k \le n$, we prove the existence of n roots on the interval $[\mathring{\gamma}_3, \mathring{\gamma}_4]$. Further, since

$$R_{2n}(\mathring{\gamma}_2) = R_{2n}(\mu_0^-) > 0, \quad R_{2n}(\mathring{\gamma}_3) = R_{2n}(\mu_0^+) > 0,$$
$$R_{2n}(0) = 0,$$

in the interval $(\mathring{\gamma}_2, \mathring{\gamma}_3)$ there are another two different roots of the polynomial $R_{2n}(\mu)$ (one of which is zero), or zero is a double root. Consequently, on the interval $[\mathring{\gamma}_1, \mathring{\gamma}_4]$ the polynomial $R_{2n}(\mu)$ has $2n+2$ roots, which is impossible.

Thus, for the case $\Delta_1 = \Delta_2$, the set of parameters τ_k constructed in theorem 1 is optimal.

Suppose now that $\Delta_1 \le \Delta_2$. In this case we have $\gamma_1 = -\mathring{\gamma}_2\mathring{\gamma}_3$, $\gamma_2 = \mathring{\gamma}_4(\Delta_2 - \Delta_1) - \mathring{\gamma}_1\mathring{\gamma}_4$. Since $\gamma_2 = \mathring{\gamma}_4(\Delta_2 - \Delta_1) - \mathring{\gamma}_1\mathring{\gamma}_4 = \mathring{\gamma}_4(\mathring{\gamma}_4 - \mathring{\gamma}_3 - \mathring{\gamma}_2)$, then γ_1 and γ_2 do not depend on $\mathring{\gamma}_1$. Consequently, for any $\mathring{\gamma}_1$ in the interval $\mathring{\gamma}_2 + \mathring{\gamma}_3 - \mathring{\gamma}_4 \le \mathring{\gamma}_1 \le \mathring{\gamma}_2$ we have the same set of parameters τ_k, and the iterative method (2) converges at the same rate for any $\mathring{\gamma}_1$ from this interval.

In conclusion we note that the set of parameters τ_k constructed in theorem 1 will also be optimal in the case where $n = 1$, but Δ_1 and Δ_2 are not necessarly equal. This is the case of a cyclic simple iteration method, for which in the scheme (2) $\tau_{2k-1} \equiv \tau_1$, $\tau_{2k} \equiv \tau_2$, $k = 1, 2, \ldots$, and τ_1 and τ_2 are found from the formulas in theorem 1 for $n = 1$ $(\omega_1 = \omega_0)$

$$\tau_1 = -\alpha_0\omega_0 - \sqrt{\alpha_0^2\omega_0^2 + \omega_0}, \quad \tau_2 = -\alpha_0\omega_0 + \sqrt{\alpha_0^2\omega_0^2 + \omega_0},$$

where $\omega_0 = 2/(\gamma_1 + \gamma_2)$. Since in this case we have

$$x_{2n} = \prod_{j=1}^{n}(E - \omega_0\bar{C})x_0 = (E - \omega_0\bar{C})^n x_0,$$

$$\| E - \omega_0\bar{C} \| \le \rho_0, \quad \rho_0 = \frac{1-\xi}{1+\xi}, \quad \xi = \frac{\gamma_1}{\gamma_2},$$

then the error z_{2n} of the iterative scheme (2) will be bounded by

$$\| z_{2n} \|_D \le \rho_0^n \| z_0 \|_D .$$

Since by (6) and (7) the two parameters τ_1 and τ_2 are replaced by the parameters ω_1 and α, and the sequence is chosen optimally ($\omega_1 = \omega_0$, $\alpha = \alpha_0$), then we have that the parameters τ_1 and τ_2 for the simple iteration method are also chosen optimally.

12.1.4 Iterative methods of variational type. Above we looked at iterative methods for the case of a self-adjoint operator $DB^{-1}A$ when not all of the eigenvalues for the problem (2) were of one sign. There the convergence of the iterative method (2) was guaranteed by the construction of a special set of iterative parameters. We look now at iterative methods of the form (2) whose convergence with the usual set of iterative parameters is guaranteed by the structure of the operator B. We have looked at such methods for constructing iterative schemes when studying the symmetrization method (cf. Section 6.4.4) and the minimal error and conjugate-error methods in Chapter 8.

Suppose that the operator B has the form

$$B = (A^*)^{-1}\tilde{B}, \tag{28}$$

where $\tilde{B}$ is an arbitrary self-adjoint and positive-definite operator. We take D to be the operator $\tilde{B}$. Then $DB^{-1}A = A^*A$, $C = \tilde{B}^{-1/2}A^*A\tilde{B}^{-1/2}$. If the operator B is non-singular and indefinite, the operator C will be positive-definite. In addition, the operator C is self-adjoint in H. Therefore, if γ_1 and γ_2 are given in the inequalities $\gamma_1 E \le C \le \gamma_2 E$, $\gamma_1 > 0$, or in the equivalent inequalities

$$\gamma_1\tilde{B} \le A^*A \le \gamma_2\tilde{B}, \quad \gamma_1 > 0, \tag{29}$$

then the parameters τ_k in (2) can be chosen using the formulas for the Chebyshev two-level method (cf. Section 6.2.1)

$$\begin{gathered}\tau_k = \frac{\tau_0}{1+\rho_0\mu_k}, \quad \mu_k \in \mathcal{M}_n^* = \left\{-\cos\frac{(2i-1)\pi}{2n},\ 1 \le i \le n\right\}, \\ k = 1, 2, \ldots, n, \\ \tau_0 = \frac{2}{\gamma_1+\gamma_2}, \qquad \rho_0 = \frac{1-\xi}{1+\xi}, \quad \xi = \frac{\gamma_1}{\gamma_2}.\end{gathered} \tag{30}$$

Thus, we have

Theorem 2. *Suppose that A is a non-singular operator. For the iterative method* (2), (28) *with the parameters* (30), *where γ_1 and γ_2 are given in* (29), *we have the estimate*

$$\| z_n \|_{\tilde{B}} \le q_n \| z_0 \|_{\tilde{B}}, \quad q_n = \frac{2\rho_1^n}{1+\rho_1^{2n}}, \quad \rho_1 = \frac{1-\sqrt{\xi}}{1+\sqrt{\xi}}. \square$$

If the constants γ_1 and γ_2 in (29) are unknown, or if they can only be crudely estimated, it is possible to use the iterative methods of variational type examined in Chapter 8.

If the parameter τ_k in the scheme (2), (28) are chosen using the formulas

$$\tau_{k+1} = \frac{(r_k, r_k)}{(Aw_k, r_k)}, \quad k = 0, 1, \ldots,$$

where $r_k = Ay_k - f$ is the residual, and w_k is the correction defined by the equation $\tilde{B}w_k = A^*r_k$, then we obtain the minimal-error method (cf. Section 8.2.4). As is well-known, the error z_n for this method can be estimated using $\| z_n \|_{\tilde{B}} \le \rho_0^n \| z_0 \|_{\tilde{B}}$, where ρ_0 is defined in (30).

If we look at the three-level iterative scheme

$$By_{k+1} = \alpha_{k+1}(B - \tau_{k+1}A)y_k + (1 - \alpha_{k+1})By_{k-1} + \tau_{k+1}\alpha_{k+1}f, \quad k \ge 1,$$
$$By_1 = (B - \tau_1 A)y_0 + \tau_1 f, \quad y_0 \in H,$$

where the operator B is defined in (28), and the iterative parameters α_{k+1} and τ_{k+1} are chosen using the formulas

$$\tau_{k+1} = \frac{(r_k, r_k)}{(Aw_k, r_k)}, \qquad k = 0, 1, \ldots,$$
$$\alpha_{k+1} = \left(1 - \frac{\tau_{k+1}}{\tau_k}\frac{(r_k, r_k)}{(r_{k-1}, r_{k-1})}\frac{1}{\alpha_k}\right)^{-1}, \quad k = 1, 2, \ldots, \ \alpha_1 = 1,$$

then we obtain the conjugate-error method (cf. Section 8.4.1). For the error in this method we have the estimate

$$\| z_n \|_{\tilde{B}} \le q_n \| z_0 \|_{\tilde{B}} .$$

12.1.5 Examples. We look now at an application of the methods constructed above to finding the solution of a Dirichlet difference problem for the Helmholtz equation in a rectangle

$$\begin{aligned} &y_{\bar{x}_1 x_1} + y_{\bar{x}_2 x_2} + m^2 y = -f(x), \quad x \in \omega, \\ &y(x) = g(x), \quad x \in \gamma, \end{aligned} \tag{31}$$

where $\bar{\omega} = \{x_{ij} = (ih_1, jh_2),\ 0 \le i \le N_1,\ 0 \le j \le N_2,\ h_\alpha N_\alpha = l_\alpha,\ \alpha = 1, 2\}$, and γ is the boundary of the grid $\bar{\omega}$.

We shall reduce the problem (31) to the operator equation (1). In this case H is the space of grid functions defined on ω with the inner product

$$(u, v) = \sum_{x \in \omega} u(x)v(x)h_1h_2, \quad u \in H,\ v \in H.$$

We define the operator R as follows: $Ry = -\Lambda\mathring{y}$, $y \in H$, $\mathring{y} \in \mathring{H}$ and $y(x) = \mathring{y}(x)$, $x \in \omega$, where $\mathring{H}$ is the set of grid functions defined on $\bar{\omega}$ and vanishing on γ, and where Λ is the Laplace difference operator $\Lambda y = y_{\bar{x}_1 x_1} + y_{\bar{x}_2 x_2}$. Then the operator A is defined by the equality $A = R - m^2 E$. Since the operator R is self-adjoint in H and has the eigenvalues

$$\lambda_k = \lambda_{k_1}^{(1)} + \lambda_{k_2}^{(2)},\ \lambda_{k_\alpha}^{(\alpha)} = \frac{4}{h_\alpha^2}\sin^2\frac{k_\alpha \pi h_\alpha}{2l_\alpha},\ 1 \le k_\alpha \le N_\alpha - 1,$$

the operator A is also self-adjoint in H, and its eigenvalues μ_k can be expressed in terms of the λ_k using the formula

$$\mu_k = \lambda_k - m^2, \quad k = (k_1, k_2), \quad 1 \le k_\alpha \le N_\alpha - 1, \quad \alpha = 1, 2. \tag{32}$$

Assume that m^2 is not equal to any λ_k. We denote by λ_{m_1} and λ_{m_2} the closest eigenvalues to m^2 from below and above respectively, i.e.

$$\lambda_{m_1} < m^2 < \lambda_{m_2}.$$

In this case the operator A is non-singular and indefinite.

To solve equation (1) with this operator A, we look at the explicit iterative scheme (2) ($B = E$). If we set $D = E$, then the operator $DB^{-1}A$ is the same as A and is self-adjoint in H. The choice of the iterative parameters in this case can be obtained from theorem 1. From (23) we obtain that the necessary ***a priori*** information is given by the bounds of the intervals $[\mathring{\gamma}_1, \mathring{\gamma}_2]$, $[\mathring{\gamma}_3, \mathring{\gamma}_4]$ on the positive and negative semi-axes which contain the eigenvalues of the operator A.

From (32) and (33) we find $\mathring{\gamma}_1 = \delta - m^2$, $\mathring{\gamma}_2 = \lambda_{m_1} - m^2$, $\mathring{\gamma}_3 = \lambda_{m_2} - m^2$, $\mathring{\gamma}_4 = \Delta - m^2$, where

$$\delta = \min_k \lambda_k = \sum_{\alpha=1}^{2} \frac{4}{h_\alpha^2} \sin^2 \frac{\pi h_\alpha}{2l_\alpha}, \quad \Delta = \max_k \lambda_k = \sum_{\alpha=1}^{2} \frac{4}{h_\alpha^2} \cos^2 \frac{\pi h_\alpha}{2l_\alpha}.$$

We now find γ_1, γ_2, and the quantity $\sqrt{\xi}$ which determines the iteration count for this method, since $n \ge n_0(\epsilon) = \ln(2/\epsilon)/(2\sqrt{\xi})$. From the formulas of theorem 1 we find

$$\gamma_1 = (m^2 - \lambda_{m_1})(\lambda_{m_2} - m^2),$$

$$\gamma_2 = \begin{cases} (\Delta - m^2)(\Delta + m^2 - \lambda_{m_1} - \lambda_{m_2}), & \lambda_{m_1} + \lambda_{m_2} \le (\Delta + \delta), \\ (m^2 - \delta)(\lambda_{m_1} + \lambda_{m_2} - m^2 - \delta), & \lambda_{m_1} + \lambda_{m_2} \ge (\Delta + \delta), \end{cases}$$

The ratio $\xi = \gamma_1/\gamma_2$ depends on m^2. In order to determine the quality of this iterative method, we now find the value of m^2 on the interval $(\lambda_{m_1}, \lambda_{m_2})$ for which ξ is maximal. We obtain

$$m^2 = 0.5(\lambda_{m_1} + \lambda_{m_2}),$$

for which

$$\gamma_1 = \left(\frac{\lambda_{m_2} - \lambda_{m_1}}{2}\right)^2,$$

$$\gamma_2 = \begin{cases} (\Delta - m^2)^2, & 2m^2 \le \Delta + \delta, \\ (m^2 - \delta)^2, & 2m^2 \ge \Delta + \delta. \end{cases}$$

If m^2 is small, i.e. λ_{m_1} and λ_{m_2} are close to δ, then $\gamma_1 = O(1)$, and $\gamma_2 = (\Delta - m^2)^2 = O(1/|h|^4)$. In this case $\xi = O(|h|^4)$. If λ_{m_1} and λ_{m_2} are close to Δ, then again we obtain $\xi = O(|h|^4)$. Only in the case where λ_{m_1} and λ_{m_2} are close to $0.5(\Delta + \delta)$ do we obtain

$$\gamma_1 = O\left(\frac{1}{|h|^2}\right) \quad \text{and} \quad \gamma_2 = O\left(\frac{1}{|h|^4}\right),$$

so that

$$\xi = O(|h|^2).$$

Notice that the difference problem (31) can be solved by one of the direct methods which we examined in Chapters 3 and 4: either by the cyclic-reduction method, or by the method of separation of variables. Here the three-point boundary-value problems which arise should be solved, unlike in the case $m = 0$, by the non-monotone elimination method.

12.2 Equations with complex operators

12.2.1 The simple iteration method. Suppose that, in the complex Hilbert space H, we are given the equation

$$Au + qu = f, \tag{1}$$

where A is an Hermitian operator, and $q = q_1 + iq_2$ is a complex scalar. To approximately solve equation (1) we look at an explicit two-level scheme

$$\frac{y_{k+1} - y_k}{\tau} + (A + qE)y_k = f, \quad k = 0, 1, \ldots, \quad y_0 \in H, \tag{2}$$

where $\tau = \tau_1 + i\tau_2$ is a complex iterative parameter.

We will assume that $q_1 \neq 0$, and that γ_1 and γ_2 are the constants in the inequalities

$$\gamma_1 E \le A \le \gamma_2 E. \tag{3}$$

We will investigate the convergence of the iterative scheme (2) in the energy space H $(D = E)$ and we shall find the optimal value of the iterative parameter τ. Using (1) and (2), we write the equation for the error $x_k = y_k - u$ in the form:

$$x_{k+1} = Sx_k, \quad k = 0, 1, \ldots, \quad S = E - \tau C, \tag{4}$$

where

$$C = A + qE.$$

From (4) we find

$$x_n = S^n x_0, \quad \| x_n \| \le \| S^n \| \, \| x_0 \| . \tag{5}$$

We now study the transformation operator from iteration to iteration. Since the operator A is a Hermitian,

$$C^* = A + \bar{q}E, \quad C^*C = CC^*,$$

i.e. the operator C is a normal operator. Therefore the operator S is also normal. It is known (cf. Section 5.1.2) that for a normal operator S the following relation is valid:

$$\| S^n \| = \| S \|^n, \quad \| S \| = \sup_{x \neq 0} \frac{|(Sx, x)|}{(x, x)}.$$

Therefore from (5) it follows that the problem of choosing the iterative parameter τ reduces to finding it from the condition that the norm of the operator S be a minimum.

We now solve this problem. From (3) it follows that

$$z = \frac{(Cx, x)}{(x, x)} \in \Omega,$$

$$\Omega = \{z = z_1 + a(z_2 - z_1),\ 0 \le a \le 1,\ z_1 = \gamma_1 + q,\ z_2 = \gamma_2 + q\},$$

where Ω is an interval in the complex plane connecting the points z_1 and z_2. Therefore

$$\| S \| = \sup_{x \neq 0} \frac{|(Sx, x)|}{(x, x)} = \sup_{z \in \Omega} |1 - \tau z|$$

and the parameter τ can be found using the condition $\min\limits_{\tau} \max\limits_{z \in \Omega} |1 - \tau z|$.

We now investigate the function $\varphi(z) = |1 - \tau z|$. Since the level lines $|1 - \tau z| = \rho_0$ are concentric circles centered at the point $1/\tau$ and with radius $R = \rho_0/|\tau|$, then for the optimal value of the parameter $\tau = \tau_0$, the points z_1 and z_2 must lie on one level line. Consequently, it must satisfy the equations

$$|1 - \tau_0 z_1| = \rho_0, \quad |1 - \tau_0 z_2| = \rho_0,$$

where $|1 - \tau_0 z| \le \rho_0$ for $z \in \Omega$.

We write these equations in the equivalent form

$$\left|\frac{1-\tau_0 z_2}{1-\tau_0 z_1}\right| = 1, \quad \rho_0 = \frac{|z_2 - z_1|}{|z_1|\left|\dfrac{z_2}{z_1} - \dfrac{1-\tau_0 z_2}{1-\tau_0 z_1}\right|}.$$

If τ_0 changes, then by the first equation the complex number

$$z = \frac{1-\tau_0 z_2}{1-\tau_0 z_1}$$

traverses a unit cirle in the complex plane with center at the origin; thus ρ_0 will be minimal if it satisfies the equation

$$\frac{1-\tau_0 z_2}{1-\tau_0 z_1} = \frac{-z_2}{z_1}\frac{|z_1|}{|z_2|}.$$

This condition gives the following value for τ_0:

$$\tau_0 = \frac{|z_2|/z_2 + |z_1|/z_1}{|z_1| + |z_2|}. \tag{6}$$

For this value of $\tau = \tau_0$, the norm of the operator S can be estimated using

$$\| S \| = \rho_0 = \frac{|z_2 - z_1|}{|z_1| + |z_2|}, \tag{7}$$

from which we obtain the following estimate for the error x_n in the iterative scheme (2)

$$\| x_n \|_B \leq \rho_0^n \| x_0 \|_B . \tag{8}$$

We now find the conditions which guarantee that $\rho_0 < 1$. Since

$$|z_2 - z_1| = |z_1|\left|\frac{z_2}{|z_1|} - \frac{z_1}{|z_1|}\right| \leq |z_1|\left(1 + \frac{|z_2|}{|z_1|}\right) = |z_1| + |z_2|,$$

and equality is achieved only when

$$\frac{z_1}{|z_1|} = -\frac{z_2}{|z_1|}\frac{|z_1|}{|z_2|} = -\frac{z_2}{|z_2|}, \tag{9}$$

then $\rho_0 < 1$ if (9) does not hold.

In the present situation, $z_1 = \gamma_1 + q$ and $z_2 = \gamma_2 + q$. From (9) it is easy to see that there are two cases where $\rho_0 < 1$: either $q_2 \neq 0$ and γ_1 and γ_2 are arbitrary, or $q_2 = 0$ but γ_1 and γ_2 must satisfy the condition $(\gamma_1 + q_1)(\gamma_2 + q_1) > 0$. We will assume from now on that these conditions are satisfied. Then the iterative process (2) will converge.

Theorem 3. *Suppose that A is an Hermitian operator and that the inequalities* (3) *are satisfied. The iterative process* (2) *with the parameters*

$$\tau = \tau_0 = \frac{1}{|\gamma_1 + q| + |\gamma_2 + q|} \left(\frac{|\gamma_1 + q|}{\gamma_1 + q} + \frac{|\gamma_2 + q|}{\gamma_2 + q} \right)$$

converges in H, and the error can be estimated using (8), *where*

$$\rho_0 = \frac{|\gamma_2 - \gamma_1|}{|\gamma_1 + q| + |\gamma_2 + q|}. \ \square$$

Remark. Above we solved the problem of finding the optimal value of the parameter τ from the condition $\min\limits_{\tau} \max\limits_{z \in \Omega} |1 - \tau z|$, where Ω is an interval in the complex plane connecting the two points z_1 and z_2. It is also easy to find the solution to this problem in the case where Ω is a circle with center at the point z_0 of radius $r_0 < |z_0|$, i.e. a circle which does not include the origin. The solution of this problem has the form

$$\tau_0 = \frac{1}{z_0}, \quad \sup_{z \in \Omega} |1 - \tau z| = \rho_0 = \frac{r_0}{|z_0|} < 1.$$

We look now at the use of this method for finding the solution of the following difference problem:

$$\begin{gathered} \Lambda u - qu = -\varphi(x), \qquad x \in \omega, \\ u(x) = g(x), \quad x \in \gamma, \quad q = q_1 + iq_2, \\ \Lambda = \Lambda_1 + \Lambda_2, \quad \Lambda_\alpha u = (a_\alpha u_{\bar{x}_\alpha})_{x_\alpha}, \quad \alpha = 1, 2, \end{gathered} \tag{10}$$

where $\bar{\omega} = \{x_{ij} = (ih_1, jh_2) \in \bar{G},\ 0 \le i \le N_1,\ 0 \le j \le N_2,\ h_\alpha N_\alpha = l_\alpha,\ \alpha = 1, 2\}$ is a grid in the rectangle $\bar{G} = \{0 \le x_\alpha \le l_\alpha,\ \alpha = 1, 2\}$, and the coefficients $a_\alpha(x)$ are real and satisfy the conditions

$$0 < c_1 \le a_\alpha(x) \le c_2, \quad x \in \bar{\omega}. \tag{11}$$

In this case H is the space of complex grid functions defined on ω with inner product

$$(u,v)=\sum_{x\in\omega}u(x)\bar{v}(x)h_1h_2.$$

The problem (10) can be written in the form of equation (1), where the operator A is defined in the usual way: $Ay=-\Lambda\mathring{y}$, where $\mathring{y}\in\mathring{H}$, $y(x)=\mathring{y}(x)$ for $x\in\omega$, $\mathring{y}(x)=0$, $x\in\gamma$.

To solve equation (1), we look at the explicit iterative scheme (2).

Using the Green difference formulas for complex functions, and also the inequalities (11), we ascertain that the operator A is Hermitian in H, and in the inequalities (3)

$$\gamma_1=c_1\sum_{\alpha=1}^{2}\frac{4}{h_\alpha^2}\sin^2\frac{\pi h_\alpha}{2l_\alpha},$$

$$\gamma_2=c_2\sum_{\alpha=1}^{2}\frac{4}{h_\alpha^2}\cos^2\frac{\pi h_\alpha}{2l_\alpha}.$$

If we choose the iterative parameter τ according to theorem 3, then the error $x_n=y_n-u$ will be estimated by (8), where ρ_0 is defined in theorem 3.

In the special case where $l_1=l_2=l$, $N_1=N_2=N$ and $q_1=O(1)$, $q_2=O(1)$, we obtain $\rho_0=1-O(N^{-2})$. Consequently, to achieve an accuracy ϵ it is necessary to perform $n_0(\epsilon)=O(N^2\ln(1/\epsilon))$ iterations.

12.2.2 The alternating-directions method. We look again at equation (1) and assume that the operator A can be represented in the form of the sum of two Hermitian commutative operators A_1 and A_2:

$$A=A_1+A_2,\quad A_1A_2=A_2A_1,\quad A_\alpha=A_\alpha^*,\quad \alpha=1,2. \tag{12}$$

Suppose that δ and Δ are bounds for the operators A_1 and A_2, i.e.

$$\delta E\le A_\alpha\le\Delta E,\quad \alpha=1,2. \tag{13}$$

To solve equation (1) we look at an implicit two-level iterative scheme (2), in which the operator B is given in the following form:

$$B=(\omega E+A_1+q_0E)(\omega E+A_2+q_0E),\quad q_0=0.5q, \tag{14}$$

and the parameters τ and ω are related by the equation $\tau=2\omega$. We obtained an analogous iterative scheme in Chapter 11 when constructing the alternating-directions method. Notice that, to find y_{k+1} in the scheme (2), (14), it is possible to use the following algorithm:

$$(\omega E + C_1)y_{k+1/2} = (\omega E - C_2)y_k + f,$$
$$(\omega E + C_2)y_{k+1} = (\omega E - C_1)y_{k+1/2} + f, \quad k = 0, 1, \ldots,$$

where, to preserve notation, we have set $C_\alpha = A_\alpha + q_0 E$, $\alpha = 1, 2$.

We move on now to the investigation of the convergence of the scheme (2), (14) in the norm of H. Using the commutativity of the operators A_1 and A_2, we obtain an equation for the error z_k

$$z_{k+1} = S_1 S_2 z_k, \quad k = 0, 1, \ldots, \tag{15}$$
$$S_\alpha = (\omega E + C_\alpha)^{-1}(\omega E - C_\alpha), \quad \alpha = 1, 2, \tag{16}$$

where the operators S_1 and S_2 commute. From (15) we find

$$z_n = S_1^n S_2^n z_0, \quad \| z_n \| \leq \| S_1^n \| \| S_2^n \| \| z_0 \| . \tag{17}$$

We now estimate the norm of the operator S_α^n, $\alpha = 1, 2$. Since C_α is a normal operator ($C_\alpha^* C_\alpha = C_\alpha C_\alpha^*$, $\alpha = 1, 2$), the operator S_α is also normal. Therefore $\| S_\alpha^n \| = \| S_\alpha \|^n$ and it is sufficient to estimate the norm of the operator S_α.

Since the norm of a normal operator is equal to its spectral radius (cf. Section 5.1.2), from (16) we obtain

$$\| S_\alpha \| = \max_{\lambda_\alpha} \left| \frac{\omega - \lambda_\alpha}{\omega + \lambda_\alpha} \right|, \tag{18}$$

where λ_α are the eigenvalues of the operator C_α. By the assumptions (12), (13) made above concerning the operators A_α, we obtain that $\lambda_\alpha \in \Omega = \{z = z_1 + a(z_2 - z_1),\ 0 \leq a \leq 1,\ z_1 = \delta + q_0,\ z_2 = \Delta + q_0\}$ for $\alpha = 1, 2$. Consequently, from (18) we obtain that

$$\| S_\alpha \| \leq \max_{z \in \Omega} \left| \frac{\omega - z}{\omega + z} \right|, \quad \alpha = 1, 2. \tag{19}$$

We now pose the problem of choosing the parameter ω from the condition that the right-hand side in (19) be a minimum.

We look at the fractionally-linear transformation

$$w = (\omega - z)/(\omega + z), \quad \omega \neq 0, \tag{20}$$

which establishes the correspondence between points of the z-plane and points of the w-plane. From the properties of the transformation (20) it follows that the circle $|w| = \rho_0$ in the w-plane corresponds to a circle in the z-plane for $\rho \neq 1$, and the unit circle corresponds to a straight line in the z-plane which passes through the origin. The points of this straight line have an argument which differs from the argument of ω by $\pm\pi/2$.

We now find the center and radius of the circle in the z-plane which corresponds to the circle $|w| = \rho_0 \neq 1$ in the w-plane. To do this we use (20) to express z in terms of w:

$$z = \omega(1-w)/(1+w)$$

and, using this expression, we compute

$$\left|\frac{1+|w|^2}{1-|w|^2}\omega - z\right| = \left|\frac{1+|w|^2}{1-|w|^2}\omega - \frac{1-w}{1+w}\omega\right| = \frac{2|\omega|}{|1-|w|^2|}\cdot\frac{|w+|w|^2|}{|1+w|}.$$

Since

$$|w+|w|^2| = |w+w\bar{w}| = |w|\,|-1+\bar{w}| = |w|\,|1+w|,$$

then finally we obtain

$$\left|\frac{1+|w|^2}{1-|w|^2}\omega - z\right| = \frac{2|w|\,|\omega|}{|1-|w|^2|}.$$

From this it follows that the circle $|w| = \rho_0 < 1$ corresponds to a circle of radius R in the z-plane with center at the point z_0, where

$$z_0 = \frac{1+\rho_0^2}{1-\rho_0^2}\omega, \quad R = \frac{2\rho_0|\omega|}{1-\rho_0^2}. \tag{21}$$

Notice in addition that, because the mapping (20) is one-to-one, the equations

$$\left|\frac{\omega - z}{\omega + z}\right| = \rho_0 < 1, \quad |z_0 - z| = R \tag{22}$$

are equivalent.

We return now to the original problem. We consider the function

$$\varphi(z) = |w| = \left|\frac{\omega - z}{\omega + z}\right|.$$

From the above dsicussion it follows that the level lines $\varphi(z) = \rho_0$ for $\rho_0 < 1$ are circles of radius R with center at the point z_0, where z_0 and R are defined in (21). For different ρ_0 these circles do not intersect, and a circle corresponding to a smaller value of ρ_0 lies inside a circle corresponding to a larger value of ρ_0. From this we obtain that, for the optimal value $\omega = \omega_0$, the points z_1 and z_2 must lie on one level line:

$$\left|\frac{\omega_0 - z}{\omega_0 + z_1}\right| = \rho_0 < 1, \quad \left|\frac{\omega_0 - z_2}{\omega_0 + z_2}\right| = \rho_0 < 1, \tag{23}$$

where we will have that

$$\max_{z\in\Omega}\left|\frac{\omega_0 - z}{\omega_0 + z}\right| = \rho_0.$$

The parameter ω_0 should be chosen in order to minimize ρ_0.

We now find the optimal ω_0 and compute ρ_0. Using (22), from (23) we obtain

$$|z_0 - z_1| = R_0, \quad |z_0 - z_2| = R_0,$$
$$z_0 = \frac{1+\rho_0^2}{1-\rho_0^2}\omega_0, \quad R_0 = \frac{2\rho_0|\omega_0|}{1-\rho_0^2}$$

or

$$\left|\frac{z_0 - z_2}{z_0 - z_1}\right| = 1, \quad \frac{2\rho_0}{1+\rho_0^2} = \frac{R_0}{|z_0|} = \frac{|z_2 - z_1|}{|z_1|\left|\frac{z_2}{z_1} - \frac{z_0 - z_2}{z_0 - z_1}\right|}. \tag{24}$$

Notice that ρ_0 is minimal when $2\rho_0/(1+\rho_0^2)$ is minimal, and this will occur if we require that

$$\frac{z_0 - z_2}{z_0 - z_1} = -\frac{z_2}{z_1}\frac{|z_1|}{|z_2|}. \tag{25}$$

Substituting this expression in (24), we obtain

$$\frac{2\rho_0}{1+\rho_0^2} = \frac{|z_2 - z_1|}{|z_1| + |z_2|}.$$

From this we easily obtain

$$\gamma_1 = \frac{(1-\rho_0)^2}{1+\rho_0^2} = \frac{|z_1| + |z_2| - |z_2 - z_1|}{|z_1| + |z_2|},$$
$$\gamma_2 = \frac{(1+\rho_0)^2}{1+\rho_0^2} = \frac{|z_1| + |z_2| + |z_2 - z_1|}{|z_1| + |z_2|}, \quad \xi = \frac{\gamma_1}{\gamma_2} = \left(\frac{1-\rho_0}{1+\rho_0}\right)^2. \tag{26}$$

Consequently

$$\rho_0 = \frac{1-\sqrt{\xi}}{1+\sqrt{\xi}}, \quad \frac{1-\rho_0^2}{1+\rho_0^2} = \sqrt{\gamma_1\gamma_2}$$

and, in addition,

$$z_0 = \frac{1+\rho_0^2}{1-\rho_0^2}\omega_0 = \frac{\omega_0}{\sqrt{\gamma_1\gamma_2}}.$$

Substituting this expression in (25), we find the optimal value of the parameter ω_0:

$$\omega_0 = \frac{|z_1| + |z_2|}{|z_1|/z_1 + |z_2|/z_2}\sqrt{\gamma_1\gamma_2}. \qquad (27)$$

Thus, for the optimal $\omega = \omega_0$, an estimate for the norm of the operator S_α has been obtained: $\| S_\alpha \| \le \rho_0$, $\alpha = 1, 2$. Substituting it in (17), we find an estimate for the error z_n:

$$\| z_n \| \le \rho_0^{2n} \| z_0 \|, \quad \rho_0 = \frac{1-\sqrt{\xi}}{1+\sqrt{\xi}}, \quad \xi = \frac{\gamma_1}{\gamma_2}. \qquad (28)$$

Reasoning as in the simple iteration method, we find that the inequality $\gamma_1 > 0$, and together with it the inequality $\rho_0 < 1$, will be valid in two cases: either for $q_2 \ne 0$, or for $q_2 = 0$ but with $(\delta + 0.5q_1)(\Delta + 0.5q_1) > 0$.

Thus, we have proved

Theorem 4. *Suppose that the conditions* (12) *are satisfied, δ and Δ are given in the inequalities* (13) *and either $q_2 \ne 0$, or $q_2 = 0$ and $(\delta + 0.5q_1)(\Delta + 0.5q_1) > 0$. For the alternating-directions method* (2), (14), *in which the iterative parameter $\omega = \omega_0$ is chosen using the formula* (27), *and $\tau = 2\omega_0$, the estimate* (28) *is valid, where γ_1 and γ_2 are defined in* (26), *and $z_1 = \delta + 0.5q$ and $z_2 = \Delta + 0.5q$.* □

Remark 1. The solution of the problem

$$\min_{\omega} \max_{z\in\Omega} \left| \frac{\omega - z}{\omega + z} \right|,$$

where Ω is a circle with center at the point z_0 and radius $r_0 < |z_0|$ has the form

$$\omega_0 = z_0\sqrt{\gamma_1\gamma_2}, \quad \rho_0 = \max_{z\in\Omega} \left| \frac{\omega - z}{\omega + z} \right| = \frac{1-\sqrt{\xi}}{1+\sqrt{\xi}}, \quad \xi = \frac{\gamma_1}{\gamma_2},$$

where $\gamma_1 = 1 - r_0/|z_0|$, $\gamma_2 = 1 + r_0/|z_0|$.

Remark 2. If, in place of the inequalities (13), we are given the inequalities $\delta_\alpha E \le A_\alpha \le \Delta_\alpha E$, $\alpha = 1, 2$, then in theorem 4 we should set $\delta = \min(\delta_1, \delta_2)$, $\Delta = \max(\Delta_1, \Delta_2)$.

12.3 General iterative methods for equations with singular operators

12.3.1 Iterative schemes in the case of a non-singular operator B. Suppose that, in the finite-dimensional Hilbert space $H = H_N$, we are given the equation

$$Au = f \tag{1}$$

with a linear singular operator A. This indicates that the equation $Au = 0$ has a solution $u \neq 0$. We recall here (cf. Section 5.1.6) the derivation which related to the problem of solving equation (1).

Assume that ker A is the null-space of the operator A, i.e. the set of elements $u \in H$ for which $Au = 0$. We denote by im A — the image of the operator A — the set of elements of the form $y = Au$, where $u \in H$. It is known that the space H can be decomposed orthogonally into the direct sum of two subspaces:

$$H = \ker A \oplus \operatorname{im} A^*, \quad H = \ker A^* \oplus \operatorname{im} A. \tag{2}$$

This indicates that any element $u \in H$ can be represented in the form $u = \bar{u} + \tilde{u}$, where $\bar{u} \in \operatorname{im} A^*$, $\tilde{u} \in \ker A$, and $(\bar{u}, \tilde{u}) = 0$. Analogously, $u = \bar{u} + \tilde{u}$, where $\bar{u} \in \operatorname{im} A$ and $\tilde{u} \in \ker A^*$, $(\bar{u}, \tilde{u}) = 0$.

In equation (1) assume that $f = \bar{f} + \tilde{f}$, where $\bar{f} \in \operatorname{im} A$, $\tilde{f} \in \ker A^*$. An element $u \in H$ for which $Au = \bar{f}$ will be called a generalized solution of (1); it minimizes the functional $\| Au - f \|$. A generalized solution is not unique and is only determined up to an element of ker A. The normal solution is the generalized solution having minimal norm. The normal solution is unique and belongs to im A^*.

Our problem is to construct methods which allow us to approximately find the normal solution to equation (1). To do this, we will require that an approximate solution, like the exact normal solution, belong to the subspace im A^*.

To solve this problem we will use an implicit two-level scheme

$$B\frac{y_{k+1} - y_k}{\tau_{k+1}} + Ay_k = f, \quad k = 0, 1, \ldots, \quad y_0 \in H. \tag{3}$$

First we study the case where B is a non-singular operator in H. The general requirements on the iterative process are:

a) the iteration is carried out using the scheme (3), the approximate solution $y_n \in \operatorname{im} A^*$, but the intermediate approximations y_k can belong to H;

b) the specific structure of the subspaces ker A, ker A^*, im A and im A^* is not used in the iterative process.

We now find conditions on the operator B, the initial guess y_0, and the parameters τ_k, $k = 1, 2, \ldots, n$, which guarantee that the above requirements are satisfied.

Conditions 1. Assume that the operator B is such that

$$Bu \in \ker A^*, \quad \text{if} \quad u \in \ker A, \tag{4}$$

$$Bu \in \operatorname{im} A, \quad \text{if} \quad u \in \operatorname{im} A^*. \tag{5}$$

We then have

Lemma 2. *If the operators A and B satisfy the equalities*

$$A^*B = CA, \qquad BA^* = AD, \tag{6}$$

where C and D are some operators, then the conditions (4) *and* (5) *are satisfied.*

Proof. Suppose that the equalities (6) are satisfied. If $u \in \ker A$, then $Au = 0$ and, consequently, $A^*Bu = CAu = 0$. Therefore $Bu \in \ker A^*$, and (4) is satisfied. Suppose now that $u \in \operatorname{im} A^*$, i.e. $u = A^*v$, where $v \in H$. Then $Bu = BA^*v = ADv \in \operatorname{im} A$. Consequently, the condition (5) is satisfied. The lemma is proved. □

Corollary. *In the case $A = A^*$ the conditions of lemma* 2 *will be satisfied if the operators A and B commute: $AB = BA$.*

We derive one more assertion arising from (4) and (5).

Lemma 3. *Suppose that the conditions* (4) *and* (5) *are satisfied. Then*

$$B^{-1}u \in \ker A, \quad \text{if} \quad u \in \ker A^*, \tag{7}$$

$$B^{-1}u \in \operatorname{im} A^*, \quad \text{if} \quad u \in \operatorname{im} A, \tag{8}$$

and the operator AB^{-1} is non-singular on im A.

Proof. Assume that $u \in \ker A^*$ and $u \neq 0$. We denote $v = B^{-1}u$ and assume that $v \in \operatorname{im} A^*$. Then by (5) $u = Bv \in \operatorname{im} A$. But since $u \neq 0$, and the spaces im A and ker A^* are orthogonal, the assumptions made above are invalid. Consequently, $v = B^{-1}u \in \ker A$, and (7) is proved. The proof of (8) is analogous.

We now prove that AB^{-1} is non-singular on im A. In fact, suppose $u \in \operatorname{im} A$. Then by (8) $B^{-1}u \in \operatorname{im} A^*$ and, consequently, $B^{-1}u \perp \ker A$. From this we obtain that $AB^{-1}u \neq 0$, and therefore $(AB^{-1}u, AB^{-1}u) > 0$. The lemma is proved. □

We return now to the scheme (3) and examine the consequences of conditions 1. Corresponding to the decomposition of H in the form (2), we represent f and y_k for any k in the form

$$\begin{aligned} f &= \bar{f} + \tilde{f}, & \bar{f} &\in \operatorname{im} A, & \tilde{f} &\in \ker A^*, \\ y_k &= \bar{y}_k + \tilde{y}_k, & \bar{y}_k &\in \operatorname{im} A^* & \tilde{y}_k &\in \ker A. \end{aligned} \tag{9}$$

Using (9), we write the scheme (3) in the following way:

$$B\frac{\bar{y}_{k+1} - \bar{y}_k}{\tau_{k+1}} + B\frac{\tilde{y}_{k+1} - \tilde{y}}{\tau_{k+1}} + A\bar{y}_k = \bar{f} + \tilde{f}, \quad k = 0, 1, \dots. \tag{10}$$

From (4) and (5) we obtain that the first term on the left-hand side of (10) belongs to im A, and the second to ker A^*. Therefore from (10) we obtain the equation

$$B\frac{\bar{y}_{k+1} - \bar{y}_k}{\tau_{k+1}} + A\bar{y}_k = \bar{f}, \quad k = 0, 1, \dots, \ \bar{y}_0 \in \operatorname{im} A^* \tag{11}$$

for the component $\bar{y}_k \in \operatorname{im} A^*$ and the equation

$$B\frac{\tilde{y}_{k+1} - \tilde{y}_k}{\tau_{k+1}} = \tilde{f}, \quad k = 0, 1, \dots, \ \tilde{y}_0 \in \ker A \tag{12}$$

for the component $\tilde{y}_k \in \ker A$.

We now find conditions which guarantee that $y_n \in \operatorname{im} A^*$. From (9) it follows that if $\tilde{y}_n = 0$, then $y_n = \bar{y}_n \in \operatorname{im} A^*$. We find from (12) an explicit expression for $\tilde{y}_n$ and set it equal to zero. Then the condition a) formulated above will be satisfied.

From (12) we obtain

$$\tilde{y}_{k+1} = \tilde{y}_k + \tau_{k+1}B^{-1}\tilde{f} = \cdots = \tilde{y}_0 + \sum_{j=1}^{k+1} \tau_j B^{-1}\tilde{f}.$$

From this follow

Conditions 2. Suppose $y_0 = A^*\varphi$, where $\varphi \in H$, and the parameters τ_k, $k = 1, 2, \ldots, n$ satisfy the requirement

$$\sum_{j=1}^{n} \tau_j = 0, \tag{13}$$

if $f \in H$. If $f \perp \ker A^*$, then we do not impose a restriction on the parameters τ_k.

We now examine the choice of the initial guess y_0. Since for any $\varphi \in H$ we have that $y_0 = A^*\varphi \in \operatorname{im} A^*$, then in the decomposition (9) $\tilde{y}_0 = 0$ and $\bar{y}_0 = y_0$. In particular, choosing $\varphi = 0$, we obtain the initial guess $y_0 = 0$.

Thus, if the condition 2 is satisfied, then $y_n = \bar{y}_n$. Therefore the iterative process (3) will converge and give an approximate normal solution of equation (1) if the iterative process (11) converges, i.e. if the sequence $\bar{y}_k$ converges to the normal solution $\bar{u}$.

Remark 1. The conditions 2 allow us to separate from the iterative approximation y_n its projection on im A^*, i.e. to find y_n without using the subspaces ker A, im A, ker A^*, or im A^*. Further, if it is known that

$$\sum_{j=1}^{n} \tau_j \parallel B^{-1}\tilde{f} \parallel$$

is small, i.e. that $\parallel \tilde{y}_n \parallel$ is small, then using the equation $\parallel y_n - \bar{u} \parallel = \parallel \bar{y}_n - \bar{u} \parallel + \parallel \tilde{y}_n \parallel$ it is possible to take y_n to be the approximate solution and avoid the restriction (13). In this case $y_n \notin \operatorname{im} A^*$.

Remark 2. With the condition that all the elements of the subspace ker A^* are known, it is possible to restrict ourselves to the case $f \perp \ker A^*$, computing as necessary the projection of f onto ker A^*. If we look at the non-stationary iterative process

$$B_{k+1}\frac{y_{k+1} - y_k}{\tau_{k+1}} + Ay_k = f, \quad k = 0, 1, \ldots, \quad y_0 = A^*\varphi,$$

and require that condition 1 is satisfied, where B is changed to B_k, $k = 1, 2, \ldots$, then all the $y_k \in \operatorname{im} A^*$ and no additional restrictions on the τ_k are required.

12.3.2 The minimum-residual iterative method. We look now at the problem of choosing the iterative parameters τ_k for the scheme (3). We will assume that the operator B satisfies the conditions 1, and restrict our choice of the initial guess y_0 and the parameters τ_k using the conditions 2.

Above it was shown that the parameters τ_k should be chosen from the condition that the iterative process (11) converge to the normal solution $\bar{u}$ of equation (1).

We study now the iterative scheme (11). First we remark that the operator $D = A^*A$ is positive-definite on $\operatorname{im} A^*$. In fact, suppose $u \in \operatorname{im} A^*$ and $u \neq 0$. Since $u \perp \ker A$, then $Au \neq 0$ and, consequently, $(Du, u) = \| Au \|^2 > 0$. The operator D generates the energy space H_D consisting of the elements of $\operatorname{im} A^*$, where the inner product is defined in the usual way: $(u, v)_D = (Du, v)$, $u \in \operatorname{im} A^*$, $v \in \operatorname{im} A^*$.

We pose now the problem of choosing the parameters τ_{k+1} in the scheme (11) from the condition of minimizing $\| \bar{z}_{k+1} \|_D$, where $\bar{z}_{k+1}$ is the error: $\bar{z}_{k+1} = \bar{y}_{k+1} - \bar{u}$, $A\bar{u} = \bar{f}$ and $\bar{u}$ is the normal solution of equation (1).

For the error $\bar{z}_k \in \operatorname{im} A^*$, from (11) we obtain the following equation:

$$\bar{z}_{k+1} = (E - \tau_{k+1}B^{-1}A)\bar{z}_k. \tag{14}$$

From this we find

$$\| \bar{z}_{k+1} \|_D^2 = \| \bar{z}_k \|_D^2 - 2\tau_{k+1}(AB^{-1}A\bar{z}_k, A\bar{z}_k) + \tau_{k+1}^2 \| AB^{-1}A\bar{z}_k \|^2 .$$

Notice that by lemma 3 $\| AB^{-1}A\bar{z}_k \| > 0$, ($A\bar{z}_k \in \operatorname{im} A$). Therefore the minimum of $\| \bar{z}_{k+1} \|_D^2$ is achieved for

$$\tau_{k+1} = \frac{(AB^{-1}A\bar{z}_k, A\bar{z}_k)}{(AB^{-1}A\bar{z}_k, AB^{-1}A\bar{z}_k)} \tag{15}$$

and is equal to

$$\| \bar{z}_{k+1} \|_D^2 = \rho_{k+1}^2 \| \bar{z}_k \|_D^2, \quad \rho_{k+1}^2 = 1 - \frac{(AB^{-1}A\bar{z}_k, A\bar{z}_k)^2}{\| AB^{-1}A\bar{z}_k \|^2 \| A\bar{z}_k \|^2}. \tag{16}$$

The formula (15) is inconvenient for computations since it contains unknown quantities. We shall transform it. Using the decomposition (9) we obtain

$$A\bar{z}_k = A\bar{y}_k - \bar{f} = Ay_k - \bar{f} = r_k + \tilde{f}, \tag{17}$$

where $r_k = Ay_k - f$ is the residual. Since $\tilde{f} \in \ker A^*$, then by lemma 3 $B^{-1}\tilde{f} \in \ker A$ and, consequently, $AB^{-1}A\bar{z}_k = AB^{-1}r_k$. Substituting this expression, and also (17), in (15) and taking into account the equation $A^*\tilde{f} = 0$, we obtain

$$\tau_{k+1} = \frac{(AB^{-1}r_k, r_k)}{(AB^{-1}r_k, AB^{-1}r_k)} = \frac{(Aw_k, r_k)}{(Aw_k, Aw_k)}, \tag{18}$$

where the correction w_k is found from the equation $Bw_k = r_k$.

Notice that (18) is the same as the formula for the iterative parameter τ_{k+1} for the minimum-residual method examined in Chapter 8 for an equation with a non-singular operator A.

We now obtain an estimate for the convergence rate for this method. We multiply (14) on the left by A, compute the norm of the left- and right-hand sides and, taking into account that $\| A\bar{z}_k \| = \| \bar{z}_k \|_D$, we obtain the following estimate:

$$\| \bar{z}_{k+1} \|_D \leq \| E - \tau_{k+1}AB^{-1} \|_{\text{im}\, A} \| \bar{z}_k \|_D \tag{19}$$

for any τ_{k+1}. From (16) and (19) we obtain for any τ_{k+1}

$$\rho_{k+1} \leq \| E - \tau_{k+1}AB^{-1} \|_{\text{im}\, A} . \tag{20}$$

If we denote

$$\rho_0 = \min_{\tau} \| E - \tau AB^{-1} \|_{\text{im}\, A},$$

then an estimate for the error follows from (16) and (20)

$$\| \bar{z}_{k+1} \|_D \leq \rho_0 \| \bar{z}_k \|_D . \tag{21}$$

Here we use the notation $\| S \|_{\text{im}\, A}$ to denote the norm of the operator S in the subspace im A.

If $\rho_0 < 1$, then the iterative method (11), (18) will converge in H_D and from (21) we obtain that

$$\| \bar{z}_k \|_D \leq \rho_0^k \| \bar{z}_0 \|_D, \quad k = 0, 1, \ldots . \tag{22}$$

It only remains to restrict the choice of the parameters τ_k using the condition (13) if $\tilde{f} \neq 0$. To do this we proceed as follows. We perform $(n-1)$ iterations using the scheme (3), choosing $y_0 = A^*\varphi$ where $\varphi \in H$, using the formula (18) to compute the parameters τ_{k+1}, $k = 0, 1, \ldots, n-2$. We then perform one more iteration, choosing

$$\tau_n = -\sum_{j=1}^{n-1} \tau_j .$$

Then the condition (13) will be satisfied and, consequently, $y_n = \bar{y}_n$. We now estimate the norm of the error $z_n = y_n - \bar{u}$ in H_D. Since $y_n = \bar{y}_n$, then from (11) we obtain

$$y_n = \bar{y}_{n-1} - \tau_n B^{-1}(A\bar{y}_{n-1} - \bar{f}) = \bar{y}_{n-1} - \tau_n B^{-1} A\bar{z}_{n-1}.$$

From this

$$z_n = \bar{z}_{n-1} - \tau_n B^{-1} A\bar{z}_{n-1}$$

and after multiplying by A we will have

$$Az_n = (E - \tau_n AB^{-1})A\bar{z}_{n-1}.$$

Computing the norm, we obtain the estimate

$$\| z_n \|_D \leq \| E - \tau_n AB^{-1} \|_{\text{im } A} \| \bar{z}_{n-1} \|_D .$$

Substituting here (22) and taking into account that the choice of y_0 gives $\bar{y}_0 = y_0$, we find

$$\| y_n - \bar{u} \|_D \leq \| E - \tau_n AB^{-1} \|_{\text{im } A}\, \rho_0^{n-1} \| y_0 - \bar{u} \|_D . \tag{23}$$

We look now at some special cases.

1) Suppose $B = E$, and that the operator A is self-adjoint in H. Assume that γ_1 and γ_2 are the constants in the inequalities

$$\gamma_1(x, x) \leq (Ax, x) \leq \gamma_2(x, x), \quad \gamma_1 > 0, \quad Ax \neq 0. \tag{24}$$

In this case, the conditions 1 are satisfied.

We find ρ_0 and estimate the norm of the operator in (23). Since the operator A is self-adjoint in H, then, using (24), we obtain

$$\| E - \tau A \|_{\text{im } A} = \sup_{Au \neq 0} \left| 1 - \tau \frac{(Au, u)}{(u, u)} \right| \leq \max_{\gamma_1 \leq t \leq \gamma_2} |1 - \tau t|.$$

In Chapter 6 we encountered this minimax problem for the choice of τ when studying the simple iteration method. There it was found that

$$\min_{\tau} \max_{\gamma_1 \leq t \leq \gamma_2} |1 - \tau t| = \rho_0 = \frac{1 - \xi}{1 + \xi}, \quad \xi = \frac{\gamma_1}{\gamma_2}.$$

Thus ρ_0 has been found. Further, for $B = E$ the formula (15) for the parameter τ_{k+1} can be written in the form

$$\tau_{k+1} = \frac{(Ax, x)}{(Ax, Ax)}, \quad x = A\bar{z}_k \in \text{im } A.$$

Since $A = A^*$ and $\gamma_1 > 0$, then the inequalities (24) are equivalent to the following inequalities (cf. Section 5.1.3):

$$\gamma_1(Ax, x) \leq (Ax, Ax) \leq \gamma_2(Ax, x), \quad Ax \neq 0.$$

Therefore the parameters τ_k for $k \leq n-1$ satisfy the inequalities $1/\gamma_2 \leq \tau_k \leq 1/\gamma_1$. From this we obtain the estimate

$$0 < -\tau_n = \sum_{j=1}^{n-1} \tau_j \leq \frac{n-1}{\gamma_1}. \qquad (25)$$

We now estimate the norm of the operator in (23). Taking into account (24) and (25), we obtain

$$\begin{aligned} \| E - \tau_n A \|_{\mathrm{im}\, A} &\leq \max_{\gamma_1 \leq t \leq \gamma_2} |1 - \tau_n t| \\ &= 1 - \tau_n \gamma_2 \leq 1 + (n-1)\frac{\gamma_2}{\gamma_1} = 1 + (n-1)\frac{1+\rho_0}{1-\rho_0}. \end{aligned}$$

We substitute this estimate in (23) and find

$$\| y_n - \bar{u} \|_D \leq \rho_0^{n-1} \left[1 + (n-1)\frac{1+\rho_0}{1-\rho_0} \right] \| y_0 - \bar{u} \|_D. \qquad (26)$$

2) Suppose $B = B^*$, $A = A^*$ and $AB = BA$. Assume that γ_1 and γ_2 are the constants in the inequalities

$$\gamma_1(Bx, x) \leq (Ax, x) \leq \gamma_2(Bx, x), \quad \gamma_1 > 0, \ Ax \neq 0. \qquad (27)$$

In this case the conditions 1 are satisfied, the operator AB^{-1} is self-adjoint in H, and it can be shown that the estimate (26) is valid for the error in the method (3), (18).

3) Suppose that the operators B^*A and AB^* are self-adjoint in H, and that γ_1 and γ_2 are the constants in (27). In this case by lemma 2 the conditions 1 are satisfied. In addition, the operator AB^{-1} will be self-adjoint in H. It is possible to show that in this case also the estimate (26) holds.

12.3.3 A method with the Chebyshev parameters. We look now at the iterative methods (3) where the parameters τ_k are chosen using the *a priori* information about the operators A and B.

First we derive some auxiliary results that we will require later.

Lemma 4. *Suppose that the conditions*

$$A = A^* \geq 0, \quad B = B^* > 0, \quad AB = BA \tag{28}$$

are satisfied and that we are given the constants γ_1 *and* γ_2 *in the inequalities*

$$\gamma_1(Bx, x) \leq (Ax, x) \leq \gamma_2(Bx, x), \quad \gamma_1 > 0,\ Ax \neq 0. \tag{29}$$

We denote by D *one of the operators* A, B, *or* $AB^{-1}A$ *and define the operator* C *on the subspace* im A

$$C = D^{-1/2}(DB^{-1}A)D^{-1/2}.$$

The operator C *is self-adjoint in* im A *and satisfies the inequalities*

$$0 < \gamma_1(x, x) \leq (Cx, x) \leq \gamma_2(x, x), \quad x \in \operatorname{im} A. \tag{30}$$

Proof. From (28) and the corollary to lemma 2 it follows that the conditions 1 are satisfied. Further, the operator D is self-adjoint in H and is positive-definite on im A. As an example, we prove the positive-definiteness of the operator $D = AB^{-1}A$. Suppose $u \in \operatorname{im} A$ and $u \neq 0$. Since $(Du, u) = (B^{-1}Au, Au)$, and the operator B^{-1} is positive-definite by the boundedness and positive-definiteness of the operator B, then $(Du, u) \geq 0$, where equality is possible only if $Au = 0$. But this contradicts the above-made assumptions.

The operator D maps im A onto im A, therefore there exists $D^{-1/2}$ which also maps this subspace onto itself. Consequently, on im A it is possible to define the operator C indicated in the lemma. The transformation from (29) to (30) is proved in the same way as in Section 6.2.3. The lemma is proved. □

Lemma 5. *Suppose that the conditions*

$$B^*A = A^*B, \qquad AB^* = BA^* \tag{31}$$

are satisfied, and that we are given γ_1 *and* γ_2 *in* (29). *We denote* $C_1 = AB^{-1}$ *and* $C_2 = B^{-1}A$. *The operators* C_1 *and* C_2 *are self-adjoint in* H *and satisfy the inequalities*

$$\gamma_1(x, x) \leq (C_1x, x) \leq \gamma_2(x, x),\ \gamma_1 > 0, \quad x \in \operatorname{im} A, \tag{32}$$

$$\gamma_1(x, x) \leq (C_2x, x) \leq \gamma_2(x, x),\ \gamma_1 > 0, \quad x \in \operatorname{im} A^*, \tag{33}$$

Proof. The self-adjointness of the operators C_1 and C_2 follows immediately from (31). We shall prove (32) as an example. We look at the eigenvalue problem

$$AB^{-1}v - \lambda v = 0, \quad v \in H. \tag{34}$$

Since the operator AB^{-1} is self-adjoint in H, there exists an orthonormal system of eigenfunctions for the problem (34) $\{v_1, v_2, \dots, v_p, v_{p+1}, \dots, v_N\}$. Suppose that $v_1, \dots, v_p$ are functions corresponding to the eigenvalue $\lambda = 0$, and $v_{p+1}, \dots, v_N$ correspond to the non-zero λ. It is easy to see that $v_i \in \ker A^*$, $1 \le i \le p$, $v_i \in \operatorname{im} A$, $p+1 \le i \le N$, and by the decomposition of H into subspaces (2), the functions $v_{p+1}, \dots, v_N$ from a basis for im A. Then for $x \in \operatorname{im} A$ we have

$$x = \sum_{k=p+1}^{N} a_k v_k, \quad C_1 x = \sum_{k=p+1}^{N} \lambda_k a_k v_k,$$

and by the orthogonality of the eigenfunctions

$$(x, x) = \sum_{k=p+1}^{N} a_k^2, \quad (C_1 x, x) = \sum_{k=p+1}^{N} \lambda_k a_k^2.$$

From this we obtain the inequalities

$$\min_{p+1 \le k \le N} \lambda_k (x, x) \le (C_1 x, x) \le \max_{p+1 \le k \le N} \lambda_k (x, x).$$

It remains to find the smallest and largest eigenvalues corresponding to the eigenfunctions for the problem (34) belonging to im A. We write (34) in the form

$$Au_k - \lambda_k B u_k = 0, \quad p+1 \le k \le N, \tag{35}$$

where $u_k = B^{-1} v_k \in \operatorname{im} A^*$ and, consequently, $Au_k \ne 0$. Taking the inner product of (35) with u_k and using (29), we obtain that

$$\min_{p+1 \le k \le N} \lambda_k = \gamma_1, \quad \max_{p+1 \le k \le N} \lambda_k = \gamma_2.$$

The inequalities (32) have been proved. The validity of (33) is established analogously. The lemma is proved. □

We turn now to the problem of the choice of the iterative parameters for the scheme (3). Using the conditions 2 we write it in the following form:

$$B \frac{y_{k+1} - y_k}{\tau_{k+1}} + Ay_k = f, \quad y_0 \in A^* \varphi, \quad \sum_{j=1}^{n} \tau_j = 0. \tag{36}$$

If the conditions 1 are satisfied, then the parameters τ_k must be chosen from the condition that the scheme (11) converges with the restriction on the sum of the τ_j.

We look at equation (14) for the error in the scheme (11). If the conditions of lemma 4 are satisfied, then, setting $\bar{z}_k = D^{-1/2}x_k$ in (14), where D is one of the operators in lemma 4, we obtain the following equivalent error:

$$x_{k+1} = (E - \tau_{k+1}C)x_k, \quad k = 0, 1, \ldots, \quad x_k \in \operatorname{im} A. \tag{37}$$

The operator C is also defined in lemma 4.

If the conditions of lemma 5 are satisfied, then, denoting $B\bar{z}_k = x_k$ or $A\bar{z}_k = x_k$, we obtain the equation

$$x_{k+1} = (E - \tau_{k+1}C_1)x_k, \quad k = 0, 1, \ldots, \quad x_k \in \operatorname{im} A. \tag{38}$$

In this case $\| x_k \| = \| \bar{z}_k \|_D$, where $D = B^*B$ or A^*A. If we denote $\bar{z}_k = x_k$, then we obtain the equation

$$x_{k+1} = (E - \tau_{k+1}C_2)x_k, \quad k = 0, 1, \ldots, \quad x_k \in \operatorname{im} A^*, \tag{39}$$

and in this case $\| x_k \| = \| \bar{z}_k \|_D$, where $D = E$. The operators C_1 and C_2 are defined in lemma 5.

Thus, in all these cases we have obtained an equation of the form

$$x_{k+1} = (E - \tau_{k+1}C)x_k, \quad k = 0, 1, \ldots, \quad x_k \in H_1 \tag{40}$$

in the subspace H_1, where by lemmas 4 and 5 the operator C is self-adjoint in H_1, maps into H_1, and satisfies the inequalities

$$\gamma_1(x, x) \le (Cx, x) \le \gamma_2(x, x), \quad \gamma_1 > 0, \quad x \in H_1, \tag{41}$$

where γ_1 and γ_2 are taken from the inequalities (29).

From (40) we find

$$x_n = \prod_{j=1}^{n}(E - \tau_j C)x_0, \tag{42}$$

$$\| x_n \| \le \| P_n(C) \| \, \| x_0 \|, \quad P_n(C) = \prod_{j=1}^{n}(E - \tau_j C).$$

Taking into account the self-adjointness of C and the inequalities (41), we obtain

$$\| P_n(C) \| \le \max_{\gamma_1 \le t \le \gamma_2} |P_n(t)|.$$

It is easy to see that

$$\sum_{j=1}^{n} \tau_j = -P_n'(0),$$

therefore the polynomial $P_n(t)$ is normalized by the two conditions

$$P_n(0) = 1, \quad P_n'(0) = 0. \tag{43}$$

Consequently, we have transformed to the problem of constructing the polynomial of degree n that satisfies the conditions (43) and that deviates least from zero on the interval $0 < \gamma_1 \leq t \leq \gamma_2$. The construction of such a polynomial completely solves the problem of choosing the iterative parameters τ_k for the scheme (3).

The exact solution of this problem is unknown, and we now derive another solution to this problem. As in the minimum-residual method examined above, we choose the parameters $\tau_1, \tau_2, \ldots, \tau_{n-1}$ arbitrarily, and the condition

$$\sum_{j=1}^{n} \tau_j = 0$$

will be satisfied if we choose τ_n from the formula

$$\tau_n = -\sum_{j=1}^{n-1} \tau_j.$$

From (42) we obtain the following estimate

$$\| x_n \| \leq \| P_{n-1}(C) \| \, \| E - \tau_n C \| \, \| x_0 \|, \quad P_{n-1}(C) = \prod_{j=1}^{N-1} (E - \tau_j C). \tag{44}$$

We now choose the parameters $\tau_1, \tau_2, \ldots, \tau_{n-1}$ from the condition that the norm of the operator polynomial $P_{n-1}(C)$ be a minimum. Since no other conditions have been imposed on $P_{n-1}(C)$, the solution of this problem has the form (cf. Section 6.2):

$$\tau_k = \frac{\tau_0}{1 + \rho_0 \mu_k}, \quad \mu_k \in \mathcal{M}_{n-1}^* = \left\{ \cos \frac{(2i-1)\pi}{2(n-1)}, \ 1 \leq i \leq n-1 \right\}, \tag{45}$$

$k = 1, 2, \ldots, n-1$, where we have denoted

$$\tau_0 = \frac{2}{\gamma_1 + \gamma_2}, \quad \rho_0 = \frac{1-\xi}{1+\xi}, \quad \xi = \frac{\gamma_1}{\gamma_2}.$$

Then

$$P_{n-1}(t) = q_{n-1}T_{n-1}\left(\frac{1-\tau_0 t}{\rho_0}\right), \quad \| P_{n-1}(C) \| \leq q_{n-1}, \tag{46}$$

where $T_{n-1}(x)$ is the Chebyshev polynomial of the first kind of degree $n-1$,

$$q_k = 2\rho_1^k / (1+\rho_1^{2k}), \quad \rho_1 = \left(1-\sqrt{\xi}\right) / \left(1+\sqrt{\xi}\right).$$

It remains to find an explicit expression for τ_n. From (46) we find

$$\tau_n = -\sum_{j=1}^{n-1} \tau_j = P'_{n-1}(0) = -\frac{(n-1)\tau_0}{\rho_0} q_{n-1} U_{n-2}\left(\frac{1}{\rho_0}\right), \tag{47}$$

where $U_{n-2}(x)$ is the Chebyshev polynomial of the second kind of degree $n-2$. Here we have used the relation $T'_m(x) = mU_{m-1}(x)$. We compute $U_{n-2}(1/\rho_0)$. Since $\rho_0 < 1$, then from the explicit form for $U_{n-2}(x)$ (cf. Section 1.4.2):

$$U_{n-2}(x) = \frac{(x+\sqrt{x^2-1})^{n-1} - (x+\sqrt{x^2-1})^{-(n-1)}}{2\sqrt{x^2-1}}, \quad |x| \geq 1,$$

we obtain after a simple calculation

$$U_{n-2}\left(\frac{1}{\rho_0}\right) = \frac{1-\rho_1^{2(n-1)}}{2\rho_1^{n-1}} \frac{\rho_0}{\sqrt{1-\rho_0^2}}.$$

We substitute this expression in (47) and find

$$\tau_n = -\frac{(n-1)\tau_0}{\sqrt{1-\rho_0^2}} \frac{1-\rho_1^{2(n-1)}}{1+\rho_1^{2(n-1)}}. \tag{48}$$

Taking into account the self-adjointness of C and the inequalities (41), the formula (48) and the equation $\tau_0\gamma_2 = 1+\rho_0$, we obtain

$$\begin{aligned} \| E - \tau_n C \| &\leq \max_{\gamma_1 \leq t \leq \gamma_2} |1-\tau_n t| = 1 - \tau_n \gamma_2 \\ &= 1 + (n-1)\sqrt{\frac{1+\rho_0}{1-\rho_0}} \frac{1-\rho_1^{2(n-1)}}{1+\rho_1^{2(n-1)}} \leq 1 + (n-1)\sqrt{\frac{1+\rho_0}{1-\rho_0}}. \end{aligned} \tag{49}$$

Substituting (49) and (46) in (44), we obtain the following estimate for the norm of the equivalent error x_n:

$$\| x_n \| \leq \left(1 + (n-1)\sqrt{\frac{1+\rho_0}{1-\rho_0}}\right) q_{n-1} \| x_0 \|$$

under the condition that the parameters $\tau_1, \tau_2, \ldots, \tau_n$ are chosen using the formulas (45) and (48).

Theorem 5. *Suppose that the iterative parameters τ_k, $k = 1, \ldots, n$, for the scheme* (3) *are chosen using the formulas* (45) *and* (48) *and $y_0 = A^*\varphi$. Then we have the following estimate for the error*

$$\| y_n - \bar{u} \|_D \leq \left(1 + (n-1)\sqrt{\frac{1+\rho_0}{1-\rho_0}}\right) q_{n-1} \| y_0 - \bar{u} \|_D,$$

where $\bar{u}$ is the normal solution of equation (1) *and D is defined as follows: $D = B^*B$, A^*A, or E, if the conditions of lemma* 5 *are satisfied. The a priori information for this method with the Chebyshev parameters consists of the constants γ_1 and γ_2 from the inequalities* (29). □

12.4 Special methods

12.4.1 A Neumann difference problem for Poisson's equation in a rectangle. We will now illustrate the application of the iterative scheme with variable operators B_k for solving an equation with a singular operator A.

Suppose that, in the rectangle $\bar{G} = \{0 \leq x_\alpha \leq l_\alpha, \ \alpha = 1, 2\}$, it is necessary to find the solution to Poisson's equation

$$\frac{\partial^2 u}{\partial x_1^2} + \frac{\partial^2 u}{\partial x_2^2} = -\varphi(x), \quad x \in G, \tag{1}$$

satisfying the following boundary conditions:

$$\begin{aligned} \frac{\partial u}{\partial x_\alpha} &= -g_{-\alpha}(x_\beta), \quad x_\alpha = 0, \quad \beta = 3 - \alpha, \\ -\frac{\partial u}{\partial x_\alpha} &= -g_{+\alpha}(x_\beta), \quad x_\alpha = l_\alpha, \quad \alpha = 1, 2. \end{aligned} \tag{2}$$

On the rectangular grid $\bar{\omega} = \{x_{ij} = (ih_1, jh_2) \in \bar{G}, \ 0 \leq i \leq N_1, \ 0 \leq j \leq N_2, \ h_\alpha N_\alpha = l_\alpha, \ \alpha = 1, 2\}$ the problem (1), (2) corresponds to the following difference problem:

$$\begin{aligned} &\Lambda y = -f(x), \qquad x \in \bar{\omega}, \\ &\Lambda = \Lambda_1 + \Lambda_2, \quad f(x) = \varphi(x) + \frac{2}{h_1}\varphi_1(x) + \frac{2}{h_2}\varphi_2(x), \end{aligned} \tag{3}$$

where

$$\Lambda_\alpha y = \begin{cases} \dfrac{2}{h_\alpha} y_{x_\alpha}, & x_\alpha = 0, \\ y_{\bar{x}_\alpha x_\alpha}, & h_\alpha \le x_\alpha \le l_\alpha - h_\alpha, \\ -\dfrac{2}{h_\alpha} y_{\bar{x}_\alpha}, & x_\alpha = l_\alpha, \end{cases} \tag{4}$$

$$\varphi_\alpha(x) = \begin{cases} g_{-\alpha}(x_\beta), & x_\alpha = 0, \\ 0, & h_\alpha \le x_\alpha \le l_\alpha - h_\alpha, \\ g_{+\alpha}(x_\beta), & x_\alpha = l_\alpha. \end{cases}$$

The space H consists of the grid functions defined on the grid $\bar{\omega}$ with the inner product

$$(u, v) = \sum_{x \in \bar{\omega}} u(x) v(x) \hbar_1(x_1) \hbar_2(x_2),$$

where $\hbar_\alpha(x_\alpha)$ is the average step

$$\hbar_\alpha(x_\alpha) = \begin{cases} h_\alpha, & h_\alpha \le x_\alpha \le l_\alpha - h_\alpha, \\ 0.5h_\alpha, & x_\alpha = 0, l_\alpha, \quad \alpha = 1, 2. \end{cases}$$

The operator A is defined as the sum of the operators A_1 and A_2, where $A_\alpha = -\Lambda_\alpha$, $\alpha = 1, 2$. Then the problem (30) can be written in the form of the operator equation

$$Au = f \tag{5}$$

with this operator A.

We indicate the following properties of the operators A_1 and A_2. The operators A_1 and A_2 are self-adjoint in H and commute, i.e.

$$A_\alpha = A_\alpha^*, \quad \alpha = 1, 2, \quad A_1 A_2 = A_2 A_1.$$

These properties allow us to use the method of separation of variables to solve the eigenvalue problem for the operator A: $Au = \lambda u$. Working by analogy with the case of the Dirichlet problem which was considered in detail in Section 4.2.1, we obtain the solution to the problem in the form

$$\lambda_{k_1 k_2} = \lambda_{k_1}^{(1)} + \lambda_{k_2}^{(2)}, \quad \lambda_{k_\alpha}^{(\alpha)} = \frac{4}{h_\alpha^2} \sin^2 \frac{k_\alpha \pi}{2N_\alpha}, \quad k_\alpha = 0, 1, \ldots, N_\alpha,$$

$$\mu_{k_1 k_2}(i, j) = \mu_{k_1}^{(1)}(i) \mu_{k_2}^{(2)}(j), \quad 0 \le k_\alpha \le N_\alpha, \ \alpha = 1, 2,$$

$$\mu_{k_\alpha}^{(\alpha)}(m) = \begin{cases} \sqrt{\dfrac{2}{l_\alpha}} \cos \dfrac{k_\alpha \pi m}{N_\alpha}, & 1 \le k_\alpha \le N_\alpha - 1, \\ \sqrt{\dfrac{1}{l_\alpha}} \cos \dfrac{k_\alpha \pi m}{N_\alpha}, & k_\alpha = 0, N_\alpha, \ \alpha = 1, 2. \end{cases}$$

Here we have

$$A_\alpha \mu_{k_1 k_2} = \lambda_{k_\alpha}^{(\alpha)} \mu_{k_1 k_2}, \quad A\mu_{k_1 k_2} = \lambda_{k_1 k_2} \mu_{k_1 k_2}.$$

From this it follows that the operator A has a simple eigenvalue equal to zero which corresponds to the eigenfunction $\mu_{00}(i,j) \equiv 1/\sqrt{l_1 l_2}$. This function forms a basis for the subspace ker A. The functions $\mu_{k_1 k_2}(i,j)$ for $0 \le k_\alpha \le N_\alpha$ and $k_1 + k_2 \ne 0$ form a basis for the subspace im A.

To solve equation (5), we look at an iterative scheme for the alternating-directions method

$$\begin{gathered} B_{k+1} \frac{y_{k+1} - y_k}{\tau_{k+1}} + A y_k = f, \quad k = 0, 1, \ldots, \quad y_0 \in H, \\ B_k = \left(\omega_k^{(1)} E + A_1\right)\left(\omega_k^{(2)} E + A_2\right), \quad \tau_k = \omega_k^{(1)} + \omega_k^{(2)}. \end{gathered} \tag{6}$$

In order to avoid imposing auxiliary conditions on $\{\tau_k\}$ and $\{B_k\}$ connected with the separation of the component $\bar{y}_n \in \operatorname{im} A$, we require that the right-hand side f be orthogonal to ker A. If the given f does not satisfy this condition, then we change it to $f_1 = f - (f, \mu_{00})\mu_{00}$ in (6).

Notice that the operators B_k and A commute for any k. Therefore by the corollary to lemma 2, the conditions 1 will be satisfied (there we must change B to the operator B_k). In addition, by lemma 3, the operator B_k^{-1} maps im A onto im A.

We use the above facts to study the convergence of the scheme (6). Since $f \in \operatorname{im} A$, then, assuming that $y_k \in \operatorname{im} A$, we obtain from (6) that

$$y_{k+1} = y_k - \tau_{k+1} B_{k+1}^{-1}(A y_k - f) \in \operatorname{im} A.$$

Therefore, if we choose $y_0 = 0$, then $y_0 \in \operatorname{im} A$; consequently, for any $k \ge 0$ the iterative approximations $y_k \in \operatorname{im} A$. Thus, the scheme (6) can be considered only on the subspace im A.

We now investigate the convergence of the scheme (6) in im A using the norm of the space H_D, where D is taken to be one of the operators E, A, or A^2. Each of these operators will be positive-definite in im A. The means for studying the convergence of the scheme (6) is precisely the same as that used in Chapter 11 for constructing the alternating-directions method in the non-singular case. Therefore we limit ourselves only to the formulation of the problem of choosing the optimal set of parameters, omitting all the necessary computations.

The optimal parameters $\omega_j^{(1)}$ and $\omega_j^{(2)}$ for the scheme (6) must be chosen from the condition

$$\min_{\{\omega_j\}} \max_{x,y\in\Omega} \left| \prod_{j=1}^{n} \frac{\omega_j^{(2)} - x}{\omega_j^{(1)} + x} \frac{\omega_j^{(1)} - y}{\omega_j^{(2)} + y} \right| = \rho_n,$$

where $\Omega = \Omega_1 \cup \Omega_2 \cup \Omega_3$, $\Omega_1 = \{\lambda_1^{(1)} \le x \le \lambda_{N_1}^{(1)}, \lambda_1^{(2)} \le y \le \lambda_{N_2}^{(2)}\}$,

$$\Omega_2 = \left\{\lambda_1^{(1)} \le x \le \lambda_{N_2}^{(1)}, y = 0\right\} \text{ and } \Omega_3 = \left\{x = 0, \lambda_1^{(2)} \le y \le \lambda_{N_2}^{(2)}\right\}.$$

Here for the error $z_n = y_n - \bar{u}$, where $\bar{u}$ is the normal solution of equation (5), we have the estimate

$$\| z_n \|_D \le \rho_n \| z_0 \|_D .$$

Notice that for this example the condition that $\bar{u}$ is orthogonal to ker A can be written in the form $(u, 1) = 0$. Any other solution of equation (5) differs from the normal solution $\bar{u}$ by a function equal to a constant on the grid $\bar{\omega}$. Therefore one possible solution of problem (3) can be determined by fixing its value at one point of the grid $\bar{\omega}$.

The problem formulated above for the parameters differs from the problem we considered in Section 11.1, but it can be reduced to it by means of certain simplifications, at the cost of a decrease in the possible convergence rate of the iterative method. We denote

$$\delta = \min_\alpha \lambda_1^{(\alpha)}, \quad \Delta = \max_\alpha \lambda_{N_\alpha}^{(\alpha)}, \quad \eta = \frac{\delta}{\Delta},$$

$$\kappa_j = \frac{1}{\Delta}\omega_j^{(1)} = \frac{1}{\Delta}\omega_j^{(2)}, \quad j = 1, 2, \ldots, n.$$

Using this notation and the structure of the region Ω, the problem of choosing the parameters can be formulated as: choose κ_j, $1 \le j \le n$, from the condition

$$\min_{\{\kappa_j\}} \max_{\eta \le u \le 1} |r_n(u, \kappa)| = \bar{\rho}_n, \quad r_n(u, \kappa) = \prod_{j=1}^{n} \frac{\kappa_j - u}{\kappa_j + u}.$$

Here, obviously, $\bar{\rho}_n > \rho_n$.

This problem was also looked at in Section 11.1. Recall that there we obtained formulas for the κ_j and for the iteration count $n = n_0(\epsilon)$ which guaranteed that the inequality $\bar{\rho}_n^2 \leq \epsilon$ was satisfied. Since here we must insure that $\bar{\rho}_n \leq \epsilon$, then in the formulas for κ_j and $n_0(\epsilon)$ of Chapter 11 it is necessary to replace ϵ by ϵ^2. Then for the error in the method (6) we will have the estimate $\| z_n \|_D \leq \epsilon \| z_0 \|_D$. We give now the form of the estimate for the iteration count: $n \geq n_0(\epsilon)$,

$$n_0(\epsilon) = \frac{1}{\pi^2} \ln \frac{4}{\eta} \ln \frac{4}{\epsilon^2}.$$

For example, if $l_1 = l_2 = l$ and $h_1 = h_2 = h$, then

$$\delta = \frac{4}{h^2} \sin^2 \frac{\pi h}{2l}, \quad \Delta = \frac{4}{h^2}, \quad \eta = \sin^2 \frac{\pi h}{2}, \quad n_0(\epsilon) = O(\ln h \ln \epsilon).$$

Consequently, for the Neumann problem, the alternating-directions method requires practically twice as many iterations as for the case of a Dirichlet problem if the estimate for the iteration count is of the same order.

Notice that since the iterative parameters κ_j satisfy the estimate (cf. Section 11.1) $\eta < \kappa_j < 1$, then the parameters $\omega_j^{(1)}$ and $\omega_j^{(2)}$ belong to the interval (δ, Δ). Therefore the operators $\omega_k^{(\alpha)} E + A_\alpha$ are positive-definite in H, and they can be inverted using the usual three-point elimination algorithm.

12.4.2 A direct method for the Neumann problem. We look now at a direct method — a combination of the separation of variables method and the reduction method — for solving the difference problem (3). Recall that such a method was constructed in Section 4.3.2 for the following boundary-value problem: in the region G we are given equation (1), on the sides $x_2 = 0$ and $x_2 = l_2$ the boundary conditions (2) are given, but on the sides $x_1 = 0$ and $x_1 = l_1$, instead of the second-kind boundary conditions (2), we are given boundary conditions of third kind

$$\frac{\partial u}{\partial x_1} = \kappa_{-1} u - g_{-1}(x_1), \quad x_1 = 0,$$

$$-\frac{\partial u}{\partial x_2} = \kappa_{+1} u - g_{+1}(x_1), \quad x_1 = l_1,$$

where κ_{-1} and κ_{+1} are non-negative constants which are not both zero. The corresponding difference problem differs from the problem (3) only in the definition of the operator Λ_1. There we dealt with the operator Λ_1:

$$\Lambda_1 y = \begin{cases} \frac{2}{h_1}(y_{x_1} - \kappa_{-1} y), & x_1 = 0, \\ y_{\bar{x}_1 x_1}, & h_1 \le x_1 \le l_1 - h_1, \\ \frac{2}{h_1}(-y_{\bar{x}_1} - \kappa_{+1} y), & x_1 = l_1. \end{cases}$$

The requirement that κ_{-1} and κ_{+1} did not simultaneously vanish guaranteed the solubility of the difference problem and the uniqueness of the solution. In the algorithm for this method, this requirement was used only for the solution of the three-point boundary-value problems for the Fourier coefficients of the desired solution. Therefore to solve the problem (3), we can formally use the algorithm developed in Section 4.3.2, setting $\kappa_{-1} = \kappa_{+1} = 0$, and then separately consider the question of the solution of the three-point boundary-value problems which arise.

We return now to the problem (3). We will assume that $f \perp \ker A$, i.e. that $(f, 1) = 0$. Then the problem is soluble, the normal solution $\bar{u}$ is orthogonal to ker A, and one of the possible solutions can be determined by fixing its value at one node of the grid $\bar{\omega}$. In this algorithm, it is convenient to determine one of the possible solutions, not by fixing the value at a node, but by fixing the value of one of the Fourier coefficients. Suppose $y(i, j)$ is the solution of the problem (3). Then the normal solution $\bar{u}$ can be found using the formula

$$\bar{u} = y - (y, \mu_{00})\mu_{00}, \quad \mu_{00}(i, j) = 1 \Big/ \sqrt{l_1 l_2}. \tag{7}$$

We give now the algorithm for a direct method of solving the Neumann problem (3) for Poisson's equation in a rectangle.

1) For $0 \le i \le N_1$, compute the values of the function

$$\varphi(i,j) = \begin{cases} 2[f(i,0) + f(i,1)] - h_2^2 \Lambda_1 f(i,0), & j=0, \\ f(i,2j-1) + f(i,2j+1) + 2f(i,2j) - h_2^2 \Lambda_1 f(i,2j), & 1 \le j \le M_2 - 1, \\ 2[f(i,N_2) + f(i,N_2-1)] - h_2^2 \Lambda_1 f(i,N_2), & j = M_2, \end{cases}$$

$\kappa_{-1} = \kappa_{+1} = 0.$

2) Using the algorithm for the fast Fourier transform, compute the Fourier coefficients of the function $\varphi(i, j)$:

$$z_{k_2}(i) = \sum_{j=0}^{M_2} \rho_j \varphi(i, j) \cos \frac{k_2 \pi j}{M_2}, \quad 0 \le k_2 \le M_2, \; 0 \le i \le N_1.$$

3) Solve the three-point boundary-value problems

$$\begin{aligned} 4\sin^2\frac{k_2\pi}{2N_2}w_{k_2}(i) - h_2^2\Lambda_1 w_{k_2}(i) &= h_2^2 z_{k_2}(i), \quad 0 \le i \le N_1, \\ 4\cos^2\frac{k_2\pi}{2N_2}y_{k_2}(i) - h_2^2\Lambda_1 y_{k_2}(i) &= w_{k_2}(i), \quad 0 \le i \le N_1 \end{aligned} \tag{8}$$

for $0 \le k_2 \le M_2$, and as a result find the Fourier coefficients $y_{k_2}(i)$ of the function $y(i,j)$.

4) Use the algorithm for the fast Fourier transform to find the solution of the problem on the even rows of the grid $\bar{\omega}$

$$y(i,2j) = \sum_{k_2=0}^{M_2} \rho_{k_2} y_{k_2}(i) \cos\frac{k_2\pi j}{M_2},$$

$$0 \le j \le M_2, \quad 0 \le i \le N_1,$$

and solve the three-point boundary-value problems

$$\begin{aligned} &2y(i,2j-1) - h_2^2\Lambda_1 y(i,2j-1) \\ &\qquad = h_2^2 f(i,2j-1) + y(i,2j-2) + y(i,2j), \\ &0 \le i \le N_1, \quad 1 \le j \le M_2, \end{aligned}$$

to find the solution on the odd rows.

Here we have used the notation

$$M_2 = 0.5N_2, \ \rho_j = \begin{cases} 1, & 1 \le j \le M_2 - 1, \\ 0.5, & j = 0, M_2, \end{cases}$$

the operator Λ_1 is defined in (4), and we have assumed that N_2 is a power of 2. The operation count for this method will be equal to $O(N^2 \log_2 N)$ for $N_1 = N_2 = N$.

We separate one solution from the totality of solutions to problem (3) in the following way. Of all the three-point boundary-value problems which must be solved, only one problem (8) for $k_2 = 0$ has a non-unique solution. We choose here one of these solutions to guarantee the solution of the problem posed above. The difference problem (8) for $k_2 = 0$ has the form

$$\Lambda_1 w_0(i) = -z_0(i), \qquad 0 \le i \le N_1,$$

or

$$\begin{aligned} (w_0)_{\bar{x}_1 x_1} &= -z_0(i), & 1 \le i \le N_1 - 1, \\ \frac{2}{h_1}(w_0)_{x_1} &= -z_0(0), & i = 0, \\ -\frac{2}{h_1}(w_0)_{\bar{x}_1} &= -z_0(N_1), & i = N. \end{aligned} \tag{9}$$

It is not difficult to show, using the orthogonality of $f(i,j)$ to $\mu_{00}(i,j)$, that the grid function $z_0(i)$ is orthogonal to the function $\mu_0(i) = 1/\sqrt{l_1}$ in the sense of the inner product

$$(u,v)_1 = \sum_{x_1=0}^{l_1} u(x_1)v(x_1)\hbar_1(x_1).$$

And since $\mu_0(i)$ is a basis for the subspace $\ker \Lambda_1$, the problem (9) has a solution. We choose one of the solutions by fixing the value of $w_0(i)$ for some i, $0 \le i \le N_1$. We set, for example, $w_0(N_1) = 0$ and eliminate from (9) the boundary conditions for $i = N_1$. The difference problem obtained as a result of this change is easy to solve by the elimination method.

After one of the solutions $y(i,j)$ to the problem (3) has been found using the algorithm described above, the normal solution $\bar{u}$, if it is needed, can be determined using formula (7).

In conclusion we remark that an analogous procedure of separating out one of the possible solutions can also be used in the cyclic-reduction method when it is being used to solve a Neumann difference problem.

12.4.3 Iterative schemes with a singular operator B. The existence of direct methods for inverting the Laplace operator in a rectangle in the case of Neumann boundary conditions allows us to use such operators as the operator B in an implicit iterative scheme for solving singular equations. Since in this case the operator B is singular, it is necessary to again study the problem of choosing the iterative parameters.

We examine iterative methods for solving the equation (5) under the following assumptions: 1) the operator A is self-adjoint and singular; 2) the null space of the operator A is known, i.e. we are given a basis for $\ker A$; 3) the right-hand side f of equation (5) belongs to im A, i.e. $f = \bar{f} \in \operatorname{im} A$. It is easy to verify this condition, since a basis for ker A is known. Here the normal solution $\bar{u}$ of equation (5) is classical; it satisfied the relation

$$A\bar{u} = f. \tag{10}$$

Notice that by the self-adjointness of the operator A, we have the following orthogonal decomposition of the space H:

$$H = \ker A \oplus \operatorname{im} A. \tag{11}$$

To solve equation (5), we consider an implicit two-level scheme

$$B\frac{y_{k+1} - y_k}{\tau_{k+1}} + Ay_k = f, \quad k = 0, 1, \ldots, \quad y_0 \in H, \tag{12}$$

with a singular operator B. We are interested in using (12) to find an approximation to one of the solutions of equation (5).

We formulate now auxiliary assumptions concerning the operators A and B. Suppose that B is a self-adjoint operator in H, and that ker B = ker A. In additon, assume that for any $x \in \operatorname{im} A$

$$\gamma_1(Bx, x) \le (Ax, x) \le \gamma_2(Bx, x), \quad \gamma_1 > 0, \ Ax \ne 0, \tag{13}$$
$$(Bx, x) > 0.$$

Notice that, from the conditions $B = B^*$, $\ker B = \ker A$, and (11), it follows that im A and im B are the same.

We now study the scheme (12). Corresponding to (11), we write y_k in the form of a sum

$$y_k = \bar{y}_k + \tilde{y}_k, \quad \bar{y}_k \in \operatorname{im} A, \quad \tilde{y}_k \in \ker A.$$

From (12) we obtain the following equation for y_{k+1}:

$$By_{k+1} = \varphi_k, \tag{14}$$

where $\varphi_k = By_k - \tau_{k+1}(Ay_k - f)$.

Since $f \in \operatorname{im} A$ and $\operatorname{im} B = \operatorname{im} A$, then $\varphi_k \in \operatorname{im} A$ for any y_k. Consequently, $\varphi_k \perp \ker B$, and equation (14) has a totality of solutions in the usual sense; its solution $\bar{y}_{k+1}$ satisfies the equation

$$B\bar{y}_{k+1} = \varphi_k. \tag{15}$$

Notice that by the equation $B\tilde{y}_k = A\tilde{y}_k = 0$ we have

$$\varphi_k = B\bar{y}_k - \tau_{k+1}(A\bar{y}_k - f). \tag{16}$$

Therefore the component $\tilde{y}_k$ of the iterative approximation y_k, $\tilde{y} \in \ker A$, does not have any influence on $\bar{y}_{k+1}$. From this it follows that, in order to solve equation (14), it is sufficient to find some solution and then only at the end of the iterative process compute its projection y_n on im A, i.e. find $\bar{y}_n$.

We look now at the question of the choice of the iterative parameters τ_k. From the discussion above, it follows that we should choose them so that the sequence $\bar{y}_k$ converges to the normal solution $\bar{u}$ of equation (5). From (10), (15), and (16) we obtain the following problem for the error $\bar{z}_k = \bar{y}_k - u$:

$$B\bar{z}_{k+1} = (B - \tau_{k+1}A)\bar{z}_k, \quad k = 0, 1, \ldots, \tag{17}$$

where $\bar{z}_k \in \operatorname{im} A$ for any $k \geq 0$.

Since by (13) the operators A and B are positive-definite in the subspace im A, it is possible to investigate the convergence of the scheme (17) in the usual way using the norm of the energy space H_D, where $D = A$, B, or $AB^{-1}A$. Since in this case the operator $DB^{-1}A$ is self-adjoint, the parameters τ_k can be chosen using the formulas for the Chebyshev iterative method (cf. Section 6.2)

$$\tau_k = \frac{\tau_0}{1 + \rho_0 \mu_k}, \quad \mu_k \in \mathcal{M}_n^* = \left\{\cos\frac{(2i-1)\pi}{2n}, \ 1 \leq i \leq n\right\}, \ 1 \leq k \leq n,$$
$$\tau_0 = \frac{2}{\gamma_1 + \gamma_2}, \quad \rho_0 = \frac{1-\xi}{1+\xi}, \quad \rho_1 = \frac{1-\sqrt{\xi}}{1+\sqrt{\xi}}, \quad \xi = \frac{\gamma_1}{\gamma_2}, \tag{18}$$
$$n \geq n_0(\epsilon) = \ln(0.5\epsilon)/\ln \rho_1,$$

using γ_1 and γ_2 from the inequalities (13). Then for the error $\bar{z}_n$ after n iteration, we have the estimate

$$\| \bar{z}_n \|_D \leq \epsilon \| \bar{z}_0 \|_D .$$

The motivation for considering iterative methods with a singular operator B can be described as follows. If the operator B is such that finding the solution of equation (14) is sufficiently simple, then the original problem (5) can be approximately solved effectively by such a method, whenever the ratio ξ is not too small.

We give an example of one difference problem as an illustration of this method. Suppose that on the rectangular grid

$$\bar{\omega} = \{x_{ij} = (ih_1, jh_2) \in \bar{G}, \ 0 \leq i \leq N_1, \ 0 \leq j \leq N_2,$$
$$h_\alpha N_\alpha = l_\alpha, \ \alpha = 1, 2\},$$

in the rectangle $\bar{G}$, it is necessary to find the solution to a Neumann problem for an elliptic equation with variable coefficients

$$\begin{aligned} &\Lambda y = -f(x), \quad x \in \bar{\omega}, \\ &\Lambda = \Lambda_1 + \Lambda_2, \quad f(x) = \varphi(x) + \frac{2}{h_1}\varphi_1(x) + \frac{2}{h_2}\varphi_2(x), \end{aligned} \tag{19}$$

where

$$\Lambda_\alpha y = \begin{cases} \dfrac{2}{h_\alpha} a_\alpha^{+1} y_{x_\alpha}, & x_\alpha = 0, \\ (a_\alpha y_{\bar{x}_\alpha})_{x_\alpha}, & h_\alpha \le x_\alpha \le l_\alpha - h_\alpha, \\ -\dfrac{2}{h_\alpha} a_\alpha y_{\bar{x}_\alpha}, & x_\alpha = l_\alpha, \end{cases}$$

$$\varphi_\alpha(x_\alpha) = \begin{cases} g_{-\alpha}(x_\beta), & x_\alpha = 0, \\ 0, & h_\alpha \le x_\alpha \le l_\alpha - h_\alpha, \\ g_{+\alpha}(x_\beta), & x_\alpha = l_\alpha, \end{cases}$$

$a_1^{+1}(x) = a_1(x_1 + h_1, x_2)$, $a_2^{+1}(x) = a_2(x_1, x_2 + h_2)$.

It is assumed that the coefficients $a_1(x)$ and $a_2(x)$ satisfy the conditions

$$0 < c_1 \le a_\alpha(x) \le c_2, \quad \alpha = 1, 2, \quad x \in \bar{\omega}. \tag{20}$$

The scheme (19) is a difference analog of a Neumann problem for an elliptic equation

$$\begin{aligned} \frac{\partial}{\partial x_1}\left(k_1(x)\frac{\partial u}{\partial x_1}\right) + \frac{\partial}{\partial x_2}\left(k_2(x)\frac{\partial u}{\partial x_2}\right) &= -\varphi(x), \quad x \in G, \\ k_\alpha \frac{\partial u}{\partial x_\alpha} &= -g_{-\alpha}(x_\beta), \quad x_\alpha = 0, \quad \beta = 3 - \alpha, \\ -k_\alpha \frac{\partial u}{\partial x_\alpha} &= -g_{+\alpha}(x_\beta), \quad x_\alpha = l_\alpha, \quad \alpha = 1, 2. \end{aligned}$$

The space H was defined in Section 12.4.1. Introducing the operator $A = -\Lambda$, we write the difference problem (19) in the form of equation (5). It is easy to verify that $A = A^*$, and the Green difference formulas give

$$(Ay, y) = \sum_{\alpha=1}^{2} \left(a_\alpha y_{\bar{x}_\alpha}^2, 1\right)_\alpha, \tag{21}$$

where we have used the following notation

$$(u,v)_\alpha = \sum_{x_\beta=0}^{l_\beta} \sum_{x_\alpha=h_\alpha}^{l_\alpha} u(x)v(x)\hbar_\beta(x_\beta)h_\alpha, \ \beta = 3-\alpha, \ \alpha = 1,2.$$

It is not difficult to show that the operator A is singular and that for any coefficients $a_\alpha(x)$ satisfying (20), the kernel of the operator A consists of the grid functions which are constant on $\bar{\omega}$. Therefore, as a basis for ker A, it is possible to use the familiar function $\mu_{00}(i,j) = 1/\sqrt{l_1 l_2}$.

We now define the operator $B = -\mathring{\Lambda}$, where $\mathring{\Lambda} = \mathring{\Lambda}_1 + \mathring{\Lambda}_2$,

$$\mathring{\Lambda}_\alpha y = \begin{cases} \dfrac{2}{h_\alpha} y_{x_\alpha}, & x_\alpha = 0, \\ y_{\bar{x}_\alpha x_\alpha}, & h_\alpha \le x_\alpha \le l_\alpha - h_\alpha, \\ -\dfrac{2}{h_\alpha} y_{\bar{x}_\alpha}, & x_\alpha = l_\alpha, \quad \alpha = 1,2. \end{cases}$$

The operator B is self-adjoint in H, and in Section 12.4.1 it was remarked that the function $\mu_{00}(i,j)$ forms a basis for ker B. Consequently, ker A is known and ker A = ker B. If we take the projection of f on im A and use it if necessary to replace the right-hand side in the scheme (19), then all the requirements for the scheme (12) and for equation (5) will be satisfied.

To apply the iterative method (12), (18), all that remains is to indicate γ_1 and γ_2 in the inequalities (13). Since

$$(By,y) = \sum_{\alpha=1}^{2} \left(y_{\bar{x}_\alpha}^2, 1\right)_\alpha, \tag{22}$$

and the subspaces im A and im B are the same and consist of the grid functions which are not constant on the grid $\bar{\omega}$, then from (20)–(22) we obtain that $\gamma_1 = c_1$, $\gamma_2 = c_2$. The necessary ***a priori*** information has been found.

From the estimate (18) for the iteration count, it is clear that it does not depend on the number of unknowns in the problem, but that it is completely determined by the ratio c_1/c_2. Further, by the choice of the operator B, equation (14) for y_{k+1} is a Neumann difference problem for Poisson's equation in a rectangle. Its solution can be found by the direct method outlined in Section 12.4.2 at a cost of $O(N^2 \log_2 N)$ arithmetic operations. Then the total number of operations for this method, sufficient to obtain the solution of equation (19) to an accuracy ϵ, is equal to $Q(\epsilon) = O(N^2 \log_2 N |\ln \epsilon|)$.

Chapter 13

Iterative Methods for Solving Non-Linear Equations

In this chapter we study iterative methods for solving non-linear difference schemes. In Section 13.1 we outline the general theory of iterative methods for abstract non-linear operator equations in a Hilbert space; various assumptions concerning the operators are considered. In Section 13.2, we look at the application of the general theory to the solution of difference analogs of boundary-value problems for quasi-linear second-order elliptic equations.

13.1 Iterative methods. The general theory

13.1.1 The simple iteration method for equations with a monotone operator. In the preceding chapters we studied iterative methods for solving a linear first-kind operator equation

$$Au = f, \tag{1}$$

defined in a Hilbert space H. The majority of these methods were linear and converged at a geometric rate.

We move on now to the study of methods for solving equation (1) in the case where A is an arbitrary non-linear operator acting in H. This chapter is concerned with the construction of iterative methods for solving the non-linear equations (1). The construction of such methods is based as a rule on using an implicit iterative scheme with a linear operator B which is close in some sense to the non-linear operator A. Below, under various assumptions concerning the operators A, B, and D, we will prove general theorems about the convergence of an implicit two-level scheme

$$B\frac{y_{k+1} - y_k}{\tau} + Ay_k = f, \quad k = 0, 1, \ldots, \quad y_0 \in H \tag{2}$$

to the solution in H_D. The study of the iterative scheme (2) begins with the case of a monotone operator A. Recall that an operator A defined in a real Hilbert space is called *monotone* if

$$(Au - Av, u - v) \geq 0, \quad u, v \in H,$$

and *strongly monotone* if there exists a $\delta > 0$ such that for any $u, v \in H$

$$(Au - Av, u - v) \geq \delta \parallel u - v \parallel^2 . \tag{3}$$

We obtain from theorem 11 of Chapter 5 that the solution to equation (1) exists and is unique in the sphere $\parallel u \parallel \leq \frac{1}{\delta} \parallel A0 - f \parallel$ for a strongly-monotonic continuous operator in a finite-dimensional space H.

We will assume that B is a linear, bounded, and positive-definite operator in H, and that D is a self-adjoint positive-definite operator in H. Suppose in addition that we are given the constants γ_1 and γ_2 in the inequalities

$$(DB^{-1}(Au - Av), B^{-1}(Au - Av)) \leq \gamma_2(DB^{-1}(Au - Av), u - v), \tag{4}$$

$$(DB^{-1}(Au - Av), u - v) \geq \gamma_1(D(u - v), u - v), \tag{5}$$

where $\gamma_1 > 0$.

Lemma 1. *Assume that the conditions* (4), (5) *are satisfied. Then equation* (1) *will have a unique solution for any right-hand side.*

Proof. We write equation (1) in the equivalent form

$$u = Su, \tag{6}$$

where the non-linear operator S is defined as follows:

$$Su = u - \tau B^{-1}Au + \tau B^{-1}f, \quad \tau > 0.$$

For $\tau < 2/\gamma_2$, we will show that the operator S is uniformly contractive in H_D, i.e. for any $u, v \in H$ we have that

$$\parallel Su - Sv \parallel_D \leq \rho(\tau) \parallel u - v \parallel_D, \quad \rho(\tau) < 1, \tag{7}$$

where $\rho(\tau)$ does not depend on u and v. Then the assertion of the lemma will follow from theorem 8 of Chapter 5 concerning contractive mappings.

We have

$$\| Su - Sv \|_D^2 = (D(Su - Sv), (Su - Sv)) = \| u - v \|_D^2$$
$$- 2\tau(DB^{-1}(Au - Av), u - v) + \tau^2(DB^{-1}(Au - Av), B^{-1}(Au - Av)).$$

From (4), (5) we find that for $\tau < 2/\gamma_2$

$$\| Su - Sv \|_D^2 \le \| u - v \|_D^2 - \tau(2 - \tau\gamma_2)(DB^{-1}(Au - Av), u - v)$$
$$\le \rho^2(\tau) \| u - v \|_D^2,$$

where

$$\rho^2(\tau) = 1 - \tau(2 - \tau\gamma_2)\gamma_1. \qquad (8)$$

Since $\tau < 2/\gamma_2$, then $\rho(\tau) < 1$. The lemma is proved. □

We now investigate the convergence of the iterative scheme (2) under the assumption that the conditions (4), (5) are satisfied.

From (2) we find

$$y_{k+1} = y_k - \tau B^{-1}Ay_k + \tau B^{-1}f = Sy_k, \qquad (9)$$

where the non-linear operator S is defined above. Since the solution of equation (1) satisfies equation (6), from (6)–(9) we obtain

$$y_{k+1} - u = Sy_k - Su, \quad k = 0, 1, \ldots,$$
$$\| y_{k+1} - u \|_D^2 = \| Sy_k - Su \|_D^2 \le \rho^2(\tau) \| y_k - u \|_D^2,$$

where $\rho^2(\tau)$ is defined in (8). It is not difficult to show that the optimal value of the convergence rate is achieved when $\rho(\tau)$ is minimal, i.e. for $\tau = \tau_0 = 1/\gamma_2$. Then $\rho_0 = \rho(\tau_0) = \sqrt{1 - \xi}$, $\xi = \gamma_1/\gamma_2$. Thus we have proved

Theorem 1. *Suppose that the conditions* (4), (5) *are satisfied. The iterative method* (2) *with* $\tau = \tau_0 = 1/\gamma_2$ *converges in* H_D *and the error can be estimated using*

$$\| y_n - u \|_D \le \rho_0^n \| y_0 - u \|_D, \quad \rho = \sqrt{1 - \xi}, \quad \xi = \gamma_1/\gamma_2,$$

where u *is the solution of equation* (1). *For the iteration count we have the estimate*

$$n \ge n_0(\epsilon) = \ln \epsilon / \ln \rho_0. \ \square$$

Notice that, for a linear operator A, the conditions (4), (5) can be written in the form

$$(DB^{-1}Ay, B^{-1}Ay) \leq \gamma_2(DB^{-1}Ay, y), \quad (DB^{-1}Ay, y) \geq \gamma_1(Dy, y).$$

Consequently, in this case they are the same as the conditions imposed on the operators A, B, and D when the operator $DB^{-1}A$ is non-self-adjoint in H. Thus the method constructed here reduces to the first variant of the simple iteration method in the non-self-adjoint case (cf. Section 6.4.2).

Notice that, in place of (4), it is possible to require that the condition

$$\| B^{-1}(Au - Av) \|_D \leq \bar{\gamma}_2 \| u - v \|_D, \tag{10}$$

be satisfied; for $D = B = E$, this is a Lipschitz condition for the operator A. From (10) and (5) follows the inequality (4) with $\gamma_2 = \bar{\gamma}_2^2/\gamma_1$.

If the operator B is self-adjoint and positive-definite in H, then it is possible to take the operator D to be B. Then the conditions (4), (5) will have the form

$$(B^{-1}(Au - Av), Au - Av) \leq \gamma_2((Au - Av), u - v),$$
$$(Au - Av, u - v) \geq \gamma_1(B(u - v), u - v), \quad \gamma_1 > 0.$$

If B is non-self-adjoint and non-singular, then for $D = B^*B$ the conditions (4), (5) have the form

$$(Au - Av, Au - Av) \leq \gamma_2(Au - Av, B(u - v)),$$
$$(Au - Av, B(u - v)) \geq \gamma_1(B(u - v), B(u - v)), \quad \gamma_1 > 0.$$

For $D = B = E$ the condition (5) indicates that the operator A must be strongly-monotonic in H.

13.1.2 Iterative methods for the case of a differentiable operator. A better estimate for the convergence rate of the simple iteration method for equation (1) can be obtained if stronger assumptions are made concerning the operator A. Namely, we will assume that the operator A has a Gateaux derivative. Recall that a linear operator $A'(u)$ is called the *Gateaux derivative* of the operator A at the point $u \in H$ if for any $v \in H$ it satisfies

$$\lim_{t \to 0} \left\| \frac{A(u + tv) - A(u)}{t} - A'(u)v \right\| = 0.$$

If the operator A has a Gateaux derivative at each point of the space H, then the Lagrange inequality

$$\| Au - Av \| \leq \sup_{0 \leq t \leq 1} \| A'(u + t(u - v)) \| \, \| u - v \|, \quad u, v \in H,$$

is valid, and for any u, v, and $w \in H$ there exists a $t \in [0,1]$ such that

$$(Au - Av, w) = (A'(u + t(v-u))z, w), \quad z = u - v. \tag{11}$$

We return to the investigation of the convergence of the iterative scheme (2). We have

Theorem 2. *Assume that the operator A has a Gateaux derivative $A'(v)$ in the sphere $\Omega(r) = \{v : \| u - v \|_D \le r\}$, and that for any $v \in \Omega(r)$*

$$\begin{aligned}(DB^{-1}A'(v)y, B^{-1}A'(v)y) \le \gamma_2(DB^{-1}A'(v)y, y),\\ (DB^{-1}A'(v)y, y) \ge \gamma_1(Dy, y), \quad \gamma_1 > 0\end{aligned} \tag{12}$$

for any $y \in H$. The iterative method (2) with $\tau = 1/\gamma_2$ and $y_0 \in \Omega(r)$ converges in H_D, and the error can be estimated by

$$\| y_n - u \|_D \le \rho^n \| y_0 - u \|_D, \tag{13}$$

where u is the solution of equation (1), and $\rho = \sqrt{1-\xi}$, $\xi = \gamma_1/\gamma_2$. If the operator $DB^{-1}A'(v)$ is self-adjoint in H for $v \in \Omega(r)$ and the inequalities

$$\gamma_1(Dy, y) \le (DB^{-1}A'(v)y, y) \le \gamma_2(Dy, y), \quad \gamma_1 > 0 \tag{14}$$

are satisfied for any $v \in \Omega(r)$ and $y \in H$, then for $\tau = \tau_0 = 2/(\gamma_1 + \gamma_2)$ the estimate (13) is valid in the iterative process (2) with $\rho = \rho_0(1-\xi)/(1+\xi)$.

Proof. From the equation for the error

$$y_{k+1} - u = Sy_k - Su, \quad Sv = v - \tau B^{-1}Av + \tau B^{-1}f$$

and the Lagrange inequalities we obtain

$$\| y_{k+1} - u \|_D = \| Sy_k - Su \|_D \le \sup_{0 \le t \le 1} \| S'(v_k) \|_D \| y_k - u \|_D, \tag{15}$$

where $v_k = y_k + t(u - y_k) \in \Omega(r)$, if $y_k \in \Omega(r)$. Since $S'(v_k) = E - \tau B^{-1}A'(v_k)$, then the problem reduces to estimating the norm in H_D of the linear operator

$E - \tau B^{-1}A'(v_k)$. From the definition of the norm of an operator we have

$$\begin{aligned}\| S'(v_k) \|_D^2 &= \sup_{y\neq 0} \frac{(S'(v_k)y, S'(v_k)y)_D}{(y,y)_D} \\ &= \sup_{y\neq 0} \frac{(DS'(v_k)y, S'(v_k)y)}{(Dy,y)} \\ &= \sup_{z\neq 0} \frac{((E-\tau C(v_k))z, (E-\tau C(v_k))z)}{(z,z)} \\ &= \| E - \tau C(v_k) \|^2,\end{aligned}$$

where $C(v_k) = D^{-1/2}(DB^{-1}A'(v_k))D^{-1/2}$ and where we have made the change of variable $y = D^{-1/2}z$.

Substituting this relation in (15), we obtain

$$\| y_{k+1} - u \|_D \leq \sup_{0\leq t\leq 1} \| E - \tau C(v_k) \| \, \| y_k - u \|_D .$$

From (12) we find that for any $v_k \in \Omega(r)$ the operator $C(v_k)$ satisfies the inequalities

$$(C(v_k)y, C(v_k)y) \leq \gamma_2(C(v_k)y,y),$$
$$(C(v_k)y,y) \geq \gamma_1(y,y).$$

Recalll that the necessary estimate for the norm of the linear operator $E - \tau C(v_k)$ under these assumptions was obtained in Section 6.4.2. Namely, for $\tau = 1/\gamma_2$ we have $\| E - \tau C(v_k) \| \leq \rho$, where $\rho = \sqrt{1-\xi}$, $\xi = \gamma_1/\gamma_2$. The first assertion of the theorem has been proved. The proof of the second is analogous. In this case the operator $C(v_k)$ is self-adjoint in H, and the estimate for the norm of the operator $E - \tau C(v_k)$ was obtained earlier in Section 6.3.2. Theorem 2 has been proved. □

In Chapter 6, in addition to the estimate used here for the norm of the operator $E - \tau C(v_k)$ in the non-self-adjoint case, we obtained a second estimate under the assumption that we were given the three scalars $\bar\gamma_1$, $\bar\gamma_2$, and $\bar\gamma_3$ in the inequalities

$$\bar\gamma_1 E \leq C(v_k) \leq \bar\gamma_2 E, \quad \| C_1(v_k) \| \leq \bar\gamma_3, \quad \bar\gamma_1 > 0,$$

where $C_1 = 0.5(C - C^*)$ is the skew-symmetric part of the operator C. In this case for $\tau = \tau_0(1-\kappa\bar\rho)$ we have the estimate $\| E - \tau C(v_k) \| \leq \bar\rho$, where

$$\kappa = \frac{\bar\gamma_3}{\sqrt{\bar\gamma_1\bar\gamma_2 + \bar\gamma_3^2}}, \quad \tau_0 = \frac{2}{\bar\gamma_1 + \bar\gamma_2}, \quad \bar\rho = \frac{1-\bar\xi}{1+\bar\xi}, \quad \bar\xi = \frac{1-\kappa\bar\gamma_1}{1+\kappa\bar\gamma_2}. \tag{16}$$

Theorem 3. *Assume that the operator A has a Gateaux derivative in the sphere $\Omega(r)$, which for any $v \in \Omega(r)$ satisfies the inequalities*

$$\begin{aligned} &\bar{\gamma}_1(Dy,y) \le (DB^{-1}A(v)y,y) \le \bar{\gamma}_2(Dy,y), \quad \gamma_1 > 0, \\ &\| 0.5(DB^{-1}A'(v) - A'^*(v)(B^*)^{-1}D)y \|^2_{D^{-1}} \le \bar{\gamma}_3^2(Dy,y). \end{aligned} \tag{17}$$

Then for $\tau = \tau_0(1-\kappa\bar{\rho})$ and $y_0 \in \Omega(r)$ the iterative method (2) *converges in H_D, and the estimate* (13) *is valid for the error, where $\rho = \bar{\rho}$ is defined in* (16).

Proof. We will show now that, if the operator $A'(w)$ satisfies the conditions (17) for $w \in \Omega(r)$, then for any $u, v \in \Omega(r)$ the inequalities (4), (5) are valid with constants $\gamma_1 = \bar{\gamma}_1$, $\gamma_2 = (\bar{\gamma}_2 + \bar{\gamma}_3)^2/\bar{\gamma}_1$. Then from lemma 1 it will follow that equation (1) has a unique solution.

By (11) we have that for $u, v \in \Omega(r)$ and $t \in [0,1]$

$$(DB^{-1}Au - DB^{-1}Av, u-v) = (Ry,y), \quad R = DB^{-1}A'(w),$$

where $y = u - v$, $w = u + t(v-u) \in \Omega(r)$. From (17) we obtain $(Ry,y) \ge \bar{\gamma}_1(Dy,y)$, i.e. the inequality (5) is satisfied with $\gamma_1 = \bar{\gamma}_1$.

Further we have $(DB^{-1}Au - DB^{-1}Av, z) = (Ry,z)$. We represent the operator R in the form of the sum $R = R_0 + R_1$, where $R_0 = 0.5(R+R^*)$ is the symmetric and $R_1 = 0.5(R - R^*) = 0.5(DB^{-1}A'(w) - A'^*(w)(B^*)^{-1}D)$ is the skew-symmetric part of the operator R.
By the Cauchy-Schwartz-Bunyakovskij inequality and the conditions (17)

$$\begin{aligned} (R_1y,z) &= (D^{-1/2}R_1y, D^{1/2}z) \le (D^{-1}R_1y, R_1y)^{1/2}(Dz,z)^{1/2} \\ &= \| R_1y \|_{D^{-1}} (Dz,z)^{1/2} \le \bar{\gamma}_3(Dy,y)^{1/2}(Dz,z)^{1/2}. \end{aligned}$$

From the generalized Cauchy-Schwartz-Bunyakovskij inequality we find

$$\begin{aligned} (R_0y,z) &\le (R_0y,y)^{1/2}(R_0z,z)^{1/2} \\ &= (Ry,y)^{1/2}(Rz,z)^{1/2} \le \bar{\gamma}_2(Dy,y)^{1/2}(Dz,z)^{1/2}. \end{aligned}$$

Thus, we obtain the inequality

$$(Ry,z) \le (\bar{\gamma}_2 + \bar{\gamma}_3)(Dy,y)^{1/2}(Dz,z)^{1/2}.$$

Setting $z = B^{-1}(Au - Av)$ and using (5), we will have

$$(DB^{-1}(Au-Av), B^{-1}(Au-Av)) \le \frac{(\bar{\gamma}_2 + \bar{\gamma}_3)^2}{\bar{\gamma}_1}(DB^{-1}(Au-Av), u-v).$$

The assertion is proved. □

13.1.3 The Newton-Kantorovich method. In theorems 2 and 3 we assumed that the Gateaux derivative $A'(v)$ existed and satisfied the corresponding inequalities for $v \in \Omega(r) = \{v : \| u - v \|_D \leq r\}$, where u is the solution of equation (1).

From the proof of the theorem it follows that it is sufficient to require at each iteration $k = 0, 1, \ldots$ only that these inequalities be satisfied for $v \in \Omega(r_k)$, where $r_k = \| u - y_k \|_D$.

In this case γ_1 and γ_2 (and also $\bar{\gamma}_1$, $\bar{\gamma}_2$, and $\bar{\gamma}_3$) can depend on the iteration number k. If we choose the iterative parameter τ from the formulas of theorems 2 and 3, then we obtain the non-stationary iterative process (2) with $\tau = \tau_{k+1}$.

In addition, it is possible to consider the iterative process

$$B_{k+1} \frac{y_{k+1} - y_k}{\tau_{k+1}} + A y_k = f, \quad k = 0, 1, \ldots, \quad y_0 \in H, \tag{18}$$

in which the operator $B = B_{k+1}$ also depends on the iteration number. How should the operators B_k be chosen? If the operator A is linear, then $A'(v) = A$ for any $v \in H$. From theorems 2 and 3 it follows that for $B = A'(v) = A$ the convergence rate of the iterative method (2) will be maximal. Namely, for any initial approximation y_0, we obtain that $y_1 = u$.

We now choose the operator B_{k+1} for the case of a non-linear operator A in the following way: $B_{k+1} = A'(y_k)$. We obtain the iterative scheme

$$A'(y_k) \frac{y_{k+1} - y_k}{\tau_{k+1}} + A y_k = f, \quad k = 0, 1, \ldots, \quad y_0 \in H. \tag{19}$$

In correspondence with the remainder of the terminology, the iterative process (19) will be called non-linear. For $\tau_k \equiv 1$ it is referred to as the *Newton-Kantorovich method.* To estimate the convergence rate of the iterative process (19), it is possible to use theorems 2 and 3, in which B is changed to $A'(y_k)$. In particular, for the case $D = E$ for $\tau_{k+1} = 1/\gamma_2$ we have the estimate

$$\| y_{k+1} - u \| \leq \rho \| y_k - u \|, \quad \rho = \sqrt{1 - \gamma_1/\gamma_2} < 1, \tag{20}$$

where γ_1 and γ_2 are taken from the inequalities (12) in theorem 2

$$\| (A'(y_k))^{-1} A'(v) y \|^2 \leq \gamma_2((A'(y_k))^{-1} A'(v) y, y),$$
$$((A'(y_k))^{-1} A'(v) y, y) \geq \gamma_1(y, y), \quad \gamma_1 > 0$$

and $y \in H$, $v \in \Omega(r_k)$ and $r_k = \| u - y_k \|$. From (20) follows that $r_{k+1} = \| y_{k+1} - u \| \leq \rho r_k < r_k$ and, consequently, $r_k \to 0$ as $k \to \infty$. Therefore, if the derivative $A'(v)$ is continuous in a neighborhood of the solution as a function of v with values in the space of linear operators, then as $k \to \infty$ we have that $\gamma_1 \to 1$ and $\gamma_2 \to 1$. This leads to an acceleration in the convergence of the iterative method (19) as the iteration number k increases.

The discussion here shows that methods of the form (19) have, under certain auxiliary assumptions about the smoothness of the operator $A'(v)$, a convergence rate that is better than geometric.

We now consider the Newton-Kantorovich method (19) with $\tau \equiv 1$. We investigate the convergence of this method under the following assumptions: 1) the inequalities

$$\| A'(v) - A'(w) \| \leq \alpha \| v - w \|, \quad \alpha \geq 0, \tag{21}$$

$$\| A'(v)y \| \geq \frac{1}{\beta} \| y \|, \qquad y \in H, \quad \beta > 0 \tag{22}$$

are satisfied for $v, w \in \Omega(r)$; 2) the initial guess y_0 belongs to the sphere $\Omega(\bar{r})$, where $\bar{r} = \min(r, 1/(\alpha\beta))$.

Theorem 4. *If the assumptions* 1) *and* 2) *are satisfied, then for the error of the iterative method* (19) *with* $\tau_k \equiv 1$ *we have the estimate*

$$\| y_n - u \| \leq \frac{1}{\alpha\beta}(\alpha\beta \| y_0 - u \|)^{2^n}. \tag{23}$$

Proof. From (19) we obtain the following relation:

$$A'(y_k)(y_{k+1} - u) = A'(y_k)(y_k - u) - (Ay_k - Au) = Ty_k - Tu,$$
$$Tu = A'(y_k)u - Au,$$

where u is the solution of equation (1). From this, using the Lagrange inequalities for a non-linear operator T, we obtain

$$\| A'(y_k)(y_{k+1} - u) \| = \| Ty_k - Tu \| \leq \sup_{0 \leq t \leq 1} \| T'(v_k) \| \, \| y_k - u \|,$$

where $v_k = y_k + t(u - y_k)$. From the definition of the operator T we will have

$$T'(v_k) = A'(y_k) - A'(v_k).$$

We assume that $y_k \in \Omega(\bar{r})$. Since $\bar{r} \leq r$, then $y_k \in \Omega(r)$, and, consequently, $v_k \in \Omega(r)$. From the inequalities (21) we will have

$$\| T'(v_k) \| = \| A'(y_k) - A'(v_k) \| \leq \alpha \| y_k - v_k \| = \alpha t \| y_k - u \|,$$

$$\sup_{0 \leq t \leq 1} \| T'(v_k) \| \leq \alpha \| y_k - u \| .$$

Thus, we have found the estimate

$$\| A'(y_k)(y_{k+1} - u) \| \leq \alpha \| y_k - u \|^2 .$$

Using the inequality (22), from this we obtain

$$\| y_{k+1} - u \| \leq \alpha\beta \| y_k - u \|^2 . \tag{24}$$

Notice that, since $\| y_k - u \| \leq \bar{r}$ and $\alpha\beta\bar{r} \leq 1$, then

$$\| y_{k+1} - u \| \leq \alpha\beta\bar{r} \| y_k - u \| \leq \| y_k - u \| \leq \bar{r}.$$

Consequently, from the condition $y_k \in \Omega(\bar{r})$ it follows that $y_{k+1} \in \Omega(\bar{r})$. Since $y_0 \in \Omega(\bar{r})$, then by induction we find that $y_k \in \Omega(\bar{r})$ for any $k \geq 0$. Therefore the estimate (24) is valid for any $k \geq 0$.

We now solve inequality (24). We multiply it by $\alpha\beta$ and denote $q_k = \alpha\beta \| y_k - u \|$. For q_k we obtain the inequality $q_{k+1} \leq q_k^2$, $k = 0, 1, \ldots$. By induction it is easy to show that its solution has the form $q_n \leq q_0^{2^n}$, $n \geq 0$. Consequently, we have the estimate,

$$\alpha\beta \| y_n - u \| \leq (\alpha\beta \| y_0 - u \|)^{2^n} .$$

From this follows the assertion of the theorem. □

Remark 1. If the initial guess y_0 is chosen so that $\bar{r} \leq \rho/(\alpha\beta)$, $\rho < 1$, then from (23) we obtain the estimate

$$\| y_n - u \| \leq \rho^{2^n - 1} \| y_0 - u \|$$

and the estimate

$$n \geq n_0(\epsilon) = \log_2(\ln \epsilon / \ln \rho + 1)$$

for the iteration count.

Remark 2. If, instead of the condition (21), the inequality

$$\| A'(v) - A'(w) \| \leq \alpha \| v - w \|^p, \quad p \in (0, 1],$$

is satisfied, then for the error we have the estimate

$$\| y_n - u \| \leq \frac{1}{\sqrt[p]{\alpha\beta}} \left(\sqrt[p]{\alpha\beta} \| y_0 - u \| \right)^{(p+1)^n},$$

$$\left(\bar{r} = \min \left(r, \frac{1}{\sqrt[p]{\alpha\beta}} \right) \right).$$

In the proof of theorem 4, we obtained an estimate for the error (23). This estimate is useful from the point of view of its practical application, but it is also important for the theory of the method, since it demonstrates the convergence near to the solution u.

Theorem 4 enables us to find a region where a solution does not exist. In fact, the theorem asserts that y_k converged to u if $\| y_0 - u \| \leq \bar{r}$. Therefore, if the iteration does not converge, then there are no solutions to equation (1) in the sphere $\| y_0 - u \| \leq \bar{r}$ with center at the point y_0.

We remark that, if the operator A has a second Gateaux derivative in the sphere $\Omega(r)$, then in the inequality (21)

$$\alpha = \sup_{0 \leq t \leq 1} \| A''(v + t(w - v)) \| .$$

To realize the iterative scheme (19) for any k, it is necessary to solve the linear operator equation

$$A'(y_k)v = F(y_k), \tag{25}$$

where

$$F(y_k) = A'(y_k)y_k - \tau_{k+1}(Ay_k - f). \tag{26}$$

If v is the exact solution of equation (25), then in (19) $y_{k+1} = v$.

The operator $A'(y_k)$ must be computed at each iteration, and it can require a considerable computational expense. We consider an example. Suppose that the operator A corresponds to the system of non-linear equations

$$\varphi_i(u) = 0, \quad i = 1, 2, \ldots, m, \quad u = (u_1, u_2, \ldots, u_m).$$

The Gateaux derivative $A'(y)$ at the point $y = (y_1, y_2, \ldots, y_m)$ is a square matrix with elements $a_{ij}(y)$, where

$$a_{ij}(y) = \left. \frac{\partial \varphi_i(u)}{\partial u_j} \right|_{u=y}, \quad i, j = 1, 2, \ldots, m.$$

Consequently, at each iteration it is necessary to compute m^2 elements of the matrix $A'(y)$, where the number of unknowns in the problem is equal to m.

In order to avoid the computation of the derivative $A'(y_k)$ at each iteration, it is possible to use the following modification to the scheme (19):

$$A'(y_{km}) \frac{y_{km+i+1} - y_{km+i}}{\tau_{km+i+1}} + Ay_{km+i} = f,$$
$$i = 0, 1, \ldots, m-1, \quad k = 0, 1, \ldots .$$

Here the derivative A' is computed at the end of every m iterations and is used to find the intermediate approximations $y_{km+1}, y_{km+2}, \ldots, y_{(k+1)m}$. For $m = 1$ we obtain the iterative scheme (19).

13.1.4 Two-stage iterative methods. It is appropriate to use the iterative scheme (19) in the case where the operator $A'(y_k)$ is easy to invert. Then the exact solution v of equation (25) is taken as the new iterative approximation y_{k+1}, thus satisfying the scheme (19). We then obtain an iterative scheme whose operator B_{k+1} is given in explicit form: $B_{k+1} = A'(y_k)$.

If equation (25) is approximately solved, for example using an auxiliary (inner) iterative method and y_{k+1} is taken to be the m-th iterative approximation v_m, then y_{k+1} satisfies a general scheme(18) in which $B_{k+1} \neq A'(y_k)$. In this case, the explicit form of the operator B_{k+1} is not used; knowledge of its structure is needed only to investigate the convergence of the iterative scheme (18). The iterative method thus constructed is sometimes called a two-stage method, as implied by the special algorithm used to invert the operator B_{k+1}.

We shall describe in more detail a general scheme for constructing two-stage methods. Suppose that some implicit two-level iterative method

$$\bar{B}_{n+1}\frac{v_{n+1} - v_n}{\omega_{n+1}} + A'(y_k)v_n = F(y_k), \quad n = 0, 1, \ldots, m-1, \qquad (27)$$

is used to solve the linear equation (25), where $F(y_k)$ is defined in (26), $\{\omega_n\}$ is a set of iterative parameters, $\bar{B}_{n+1}$ are operators in H which can depend on y_k, and $v_0 = y_k$.

We now express v_m in terms of y_k. First we find an equation for the error $z_n = v_n - v$, where v is the solution of equation (25). From (25) and (27) we find

$$z_{n+1} = S_{n+1}z_n, \quad n = 0, 1, \ldots, \quad S_n = E - \omega_n \bar{B}_n^{-1} A'(y_k)$$

and, consequently,

$$\begin{aligned} z_m &= v_m - v = T_m z_0 = T_m(v_0 - v), \quad T_m = S_m S_{m-1} \cdots S_1, \\ v_m &= (E - T_m)v + T_m y_k. \end{aligned} \qquad (28)$$

From (25), (26) we obtain

$$v = [A'(y_k)]^{-1}F(y_k) = y_k - \tau_{k+1}[A'(y_k)]^{-1}(Ay_k - f).$$

Substituting this v in (28), we will have

$$y_{k+1} = v_m = y_k - \tau_{k+1}[A'(y_k)]^{-1}(Ay_k - f).$$

From this it follows that y_{k+1} satisfies the iterative scheme (18) if we denote

$$B_{k+1} = A'(y_k)(E - T_m)^{-1}. \tag{29}$$

Thus, performing one step of the two-stage method involves computing $F(y_k)$ using the formula (26) and completing m iterations of the scheme (27) with an initial guess $v_0 = y_k$. Then y_k is taken to be the resulting approximation v_m.

We look now at the iterative scheme (18), (29). To estimate the convergence rate it is possible to use theorems 2 and 3, in which B is changed to B_{k+1}, and τ to τ_{k+1}. A deficiency of such a choice of the parameter τ is that it is necessary to estimate γ_1, γ_2, and γ_3 quite accurately.

Notice that when constructing the two-stage method, it is possible to begin, not with equation (25), but with a "nearby" equation

$$Rv = F(y_k),$$

where the linear operator R is equivalent to the operator $A'(y_k)$ in some sense. In this case, in the iterative scheme (18) we have

$$B_{k+1} \equiv B = R(E - T_m)^{-1}.$$

We now investigate this case in more detail. Suppose that the following conditions are satisfied

$$R = R^* > 0, \quad T_m^* R = RT_m, \tag{30}$$

$$\| T_m \|_R \leq q < 1. \tag{31}$$

Lemma 2. *Assume that the conditions* (30), (31) *are satisfied. Then the operator* $B = R(E - T_m)^{-1}$ *is self-adjoint and positive-definite in* H, *and the following inequalities are valid*

$$(1 - q)B \leq R \leq (1 + q)B. \tag{32}$$

Proof. We examine the operator $B^{-1} = (E - T_m)R^{-1}$. From (30) we find $(E - T_m^*)R = R(E - T_m)$ or $R^{-1}(E - T_m^*) = (E - T_m)R^{-1}$. Consequently, the operator B^{-1} is self-adjoint in H.

Since by (30) the operator T_m is self-adjoint in H_R, then

$$\| T_m \|_R = \sup_{x\neq 0} \frac{|(T_n x, x)_R|}{(x,x)_R} = \sup_{x\neq 0} \frac{|(RT_m x, x)|}{(Rx, x)} \leq q < 1.$$

Consequently, for any $x \in H$ we have the inequality

$$|(RT_m x, x)| \leq q(Rx, x).$$

Setting here $x = R^{-1}y$, we obtain

$$|(T_m R^{-1} y, y)| \leq q(R^{-1}y, y),$$

therefore for $y \in H$ we find

$$(1-q)(R^{-1}y, y) \leq ((E - T_m)R^{-1}y, y) \leq (1+q)(R^{-1}y, y).$$

Thus we have obtained the estimate

$$(1-q)R^{-1} \leq B^{-1} \leq (1+q)R^{-1}. \qquad (33)$$

Since R^{-1} and B^{-1} are self-adjoint operators in H and $q < 1$, then from lemma 9 in Section 5.1 it follows that the inequalities (33) and (32) are equivalent. The lemma is proved. □

Lemma 3. *Suppose that the operator A has a Gateaux derivative $A'(v)$ in the sphere $\Omega(r)$, which for any $v \in \Omega(r)$ satisfies the inequalities*

$$c_1(Ry, y) \leq (A'(v)y, y) \leq c_2(Ry, y), \quad c_1 > 0, \qquad (34)$$

$$\| 0.5[A'(v) - (A'(v))^*]y \|_{R^{-1}}^2 \leq c_3^2(Ry, y), \quad c_3 \geq 0, \qquad (35)$$

and suppose that the conditions (30), (31) *are satisfied. Then the inequalities* (17) *of theorem* 3 *are valid, where*

$$\bar{\gamma}_1 = c_1(1-q), \quad \bar{\gamma}_2 = c_2(1+q), \quad \bar{\gamma}_3 = c_3(1+q)^2,$$
$$D = B = R(E - T_m)^{-1}.$$

Proof. By lemma 2, the operator D is self-adjoint and positive-definite in$H$. In addition, for $D = B$ the inequalities (17) have the form

$$\bar{\gamma}_1(By, y) \leq (A'(v)y, y) \leq \bar{\gamma}_2(By, y), \qquad (36)$$

$$\| 0.5[A'(v) - (A'(v))^*]y \|_{B^{-1}}^2 \leq \bar{\gamma}_3^2(By, y). \qquad (37)$$

The inequalities (36) with $\bar{\gamma}_1$ and $\bar{\gamma}_2$ as indicated in lemma 3 follow from (32) and (34), and (37) is a consequence of (32), (33) and (35) since

$$\| z \|^2_{B^{-1}} = (B^{-1}z, z) \le (1+q)(R^{-1}z, z) = (1+q) \| z \|^2_{R^{-1}},$$
$$(Rz, z) \le (1+q)(Bz, z). \square$$

Using lemma 3, it is possible to prove an analog of theorem 3 for the two-stage method.

Theorem 5. *Suppose that the conditions of lemma 3 are satisfied, and that the two-stage method is constructed on the basis of the equation $Rv = F(y_k)$ using the resolving operator T_m. If, in the iterative scheme (18) with $B_{k+1} \equiv B = R(E - T_m)^{-1}$ which describes this two-stage method, we choose $\tau_k \equiv \tau_0(1 - \kappa\bar{\rho})$ and $y_0 \in \Omega(r)$, then we have the following estimate for the error*

$$\| y_n - u \|_B \le \bar{\rho}^n \| y_0 - u \|_B,$$

where u is the solution of equation (1), *$\bar{\rho}$, κ and τ_0 are defined in* (16) *with $\bar{\gamma}_1$ and $\bar{\gamma}_2$ as indicated in lemma* 3.

13.1.5 Other iterative methods. In this subsection, we give a brief description of several iterative methods which are also used to solve equation (1) with a non-linear operator A.

Suppose that $\Phi(u)$ is a Gateaux-differentiable functional defined in H. An operator A acting in H is called *potential* if there exists a differentiable functional $\Phi(u)$ such that $Au = \operatorname{grad} \Phi(u)$ for all u. Here the gradient of the functional $\Phi(u)$ is defined by the equation

$$\left.\frac{d}{dt}\Phi(u + tv)\right|_{t=0} = (\operatorname{grad} \Phi(u), v).$$

An example of a potential operator is a bounded linear self-adjoint operator A acting in a Hilbert space H. It gives rise to the functional $\Phi(u) = 0.5(Au, u)$.

Assume that the operator A is continuously differentiable in H. The operator A is potential if and only if the Gateaux derivative $A'(v)$ is a self-adjoint operator in H.

If the operator A is potential, then the formula

$$\Phi(u) = \int_0^1 (A(u_0 + t(u - u_0)), u - u_0) dt,$$

where u_0 is an arbitrary but fixed element of H, gives a way of constructing the functional $\Phi(u)$ for the operator A.

If the operator A is generated by the gradient of a strictly convex functional, then the derivative $A'(v)$ is a positive-definite operator in H for any $v \in H$. In this case, it is possible to approximately solve equation (25) using iterative methods of variational type, for example in (27) the iterative parameters ω_{k+1} can be chosen using the formulas for the steepest-descent method, the minimal-residual method, and so forth.

As an example, we look at the two-stage method (18), (29) for which $\tau_{k+1} \equiv 1$, and in the scheme (27) $m = 1$ and $\bar{B}_1 = E$. Then $B_{k+1} = E/\omega_1$. If, in the auxiliary iterative process (27), the parameter ω_1 is chosen using the formulas for the minimal-residual method (or the minimal-correction method), then we obtain (cf. Sections 8.2.2, 8.2.3)

$$\omega_1 = \frac{(A'(y_k)r_k, r_k)}{\| A'(y_k)r_k \|^2}, \quad r_k = Ay_k - f. \tag{38}$$

In this case, the two-stage iterative method is described by the formula

$$\frac{y_{k+1} - y_k}{\omega_1} + Ay_k = f, \quad k = 0, 1, \ldots, \tag{39}$$

where ω_1 is defined in (38).

In the situation where the operator A is not potential, the parameter ω_1 can be chosen using the formulas for the minimal-error method, setting $\bar{B}_1 = [(A'(y_k))^*]^{-1}$ in (27) and

$$\omega_1 = \frac{(r_k, r_k)}{\| (A'(y_k))^* r_k \|^2}, \quad r_k = Ay_k - f. \tag{40}$$

In this case, the two-stage method has the form

$$\frac{y_{k+1} - y_k}{\omega_1} + (A'(y_k))^* Ay_k = (A'(y_k))^* f, \quad k = 0, 1, \ldots, \tag{41}$$

where ω_1 is defined in (40).

It is easy to see that in the method (38), (39) the parameter ω_1 is chosen in order to minimize $\| A'(y_k)(y_{k+1} - y_k) + Ay_k - f \|$, and in the method (40), (41) to minimize the norm $\| y_{k+1} - y_k + [A'(y_k)]^{-1}(Ay_k - f) \|$.

The problem of solving the equation $Au = f$ in the case of a potential operator can sometimes be changed to the problem of minimizing a functional generated by this operator. We remark that it is always possible to transform the problem of solving equation (1) to a minimization problem, even if the operator A is not potential.

In fact, suppose that $\Phi(u)$ is a functional defined in H and having a unique minimum at $u = 0$. As an example of such a functional it is possible to use $\Phi(u) = (Du, u)$, where D is a self-adjoint positive-definite operator in H. Further, for the given equation (1) we consider the functional

$$F(u) = \Phi(Au - f), \quad u \in H.$$

If equation (1) has the solution u, then obviously this gives the minimum of the functional $F(u)$.

We now describe a method for minimizing the function (a descent method). Suppose that equation (1) is generated by the gradient of strictly convex functional $\Phi(u)$. Assume that a minimizing sequence is constructed according to the iterative scheme (19), i.e. using the formula

$$y_{k+1} = y_k - \tau_{k+1}[A'(y_k)]^{-1}(Ay_k - f), \quad k = 0, 1, \ldots. \tag{42}$$

We denote

$$w_k = [A'(y_k)]^{-1} \operatorname{grad} \Phi(y_k), \tag{43}$$

where by our assumptions $\operatorname{grad} \Phi(y_k) = Ay_k - f$. We write (42) in the form

$$y_{k+1} = y_k - \tau_{k+1} w_k.$$

Notice that the operator $A'(y_k)$ is positive-definite and self-adjoint in H. Further, from the definition of the Gateaux derivative of a functional, we have

$$\lim_{\tau_{k+1} \to 0} \left[\frac{\Phi(y_k - \tau_{k+1} w_k) - \Phi(y_k)}{\tau_{k+1}} \right] + (\operatorname{grad} \Phi(y_k), w_k) = 0.$$

Since $A'(y_k) w_k = \operatorname{grad} \Phi(y_k)$, then

$$(\operatorname{grad} \Phi(y_k), w_k) = (A'(y_k) w_k, w_k) > 0.$$

Consequently, there exists a $\tau_{k+1} > 0$ such that $\Phi(y_{k+1})$ will be strictly less than $\Phi(y_k)$.

If the minimizing sequence $\{y_k\}$ is constructed using an explicit scheme (18) $(B_k \equiv E)$ i.e. according to the formulas

$$y_{k+1} = y_k - \tau_{k+1}(Ay_k - f),$$

then the move from y_k to y_{k+1} takes place along the direction of the gradient of the functional $\Phi(u)$ at the point y_k. Such methods are called *gradient-descent methods.* There exist several algorithms for choosing the iterative parameters τ_k, but we will not consider these questions here.

In conclusion we give a generalized explicit conjugate-gradient method which can be used to minimize the functional under the assumptions. The formulas for the Fletcher-Reeves algorithm have the form:

$$\begin{aligned} y_{k+1} &= y_k - a_{k+1} w_k, && k = 0,1,\ldots, \\ w_k &= \operatorname{grad}\Phi(y_k) + b_k w_{k-1}, && k = 1,2,\ldots, \\ w_0 &= \operatorname{grad}\Phi(y_0), \end{aligned}$$

where

$$b_k = \frac{\|\operatorname{grad}\Phi(y_k)\|^2}{\|\operatorname{grad}\Phi(y_{k-1})\|^2}, \qquad k = 1,2,\ldots,$$

and the parameter a_{k+1} is chosen in order to minimize $\Phi(y_k - a_{k+1}w_k)$. This one-dimensional minimization problem can be solved by a method from numerical analysis.

13.2 Methods for solving non-linear difference schemes

13.2.1 A difference scheme for a one-dimensional elliptic quasi-linear equation. The general theory of the iterative methods laid out in Section 13.1 will be applied to find an approximate solution of non-linear elliptic difference schemes. We begin with the simplest examples.

We look at a boundary-value problem of the third kind for a one-dimensional quasi-linear equation in divergent form

$$\begin{aligned} &Lu = \frac{d}{dx} k_1\left(x, u, \frac{du}{dx}\right) - k_0\left(x, u, \frac{du}{dx}\right) = -\varphi(x), \quad 0 < x < l, \\ &k_1\left(x, u, \frac{du}{dx}\right) = \kappa_0(u) - \mu_0, \quad x = 0, \\ &-k_1\left(x, u, \frac{du}{dx}\right) = \kappa_1(u) - \mu_1, \quad x = l. \end{aligned} \tag{1}$$

We will assume that the functions $k_1(x, p_0, p_1)$, $k_0(x, p_0, p_1)$, $\kappa_0(p_0)$ and $\kappa_1(p_0)$ are continuous in p_0 and p_1 and satisfy the strong ellipticity conditions

$$\sum_{\alpha=0}^{1} [k_\alpha(x, p_0, p_1) - k_\alpha(x, q_0, q_1)](p_\alpha - q_\alpha) \geq c_1 \sum_{\alpha=0}^{1} (p_\alpha - q_\alpha)^2, \tag{2}$$

$$[\kappa_\alpha(p_0) - \kappa_\alpha(q_0)](p_0 - q_0) \geq 0, \quad \alpha = 0,1, \tag{3}$$

where $c_1 > 0$ is a positive constant, $0 \leq x \leq 1$, $|p_0|$, $|q_0|$, $|p_1|$, $|q_1| < \infty$.

On the uniform grid $\bar{\omega} = \{x_i = ih,\, i = 0, 1, \ldots, N,\, hN = l\}$ the problem (1) can be put in correspondence with the difference scheme

$$\Lambda y_i = -f_i, \quad 0 \le i \le N, \tag{4}$$

where

$$f_i = \begin{cases} \varphi(0) + \dfrac{2}{h}\mu_0, & i = 0, \\ \varphi(x_i), & 1 \le i \le N-1, \\ \varphi(l) + \dfrac{2}{h}\mu_1, & i = N. \end{cases}$$

The difference operator Λ is defined by the formulas:

$$\begin{aligned} \Lambda y_i &= \frac{1}{2}\{[k_1(x, y, y_x)]_{\bar{x}} + [k_1(x, y, y_{\bar{x}})]_x \\ &\quad -k_0(x, y, y_x) - k_0(x, y, y_{\bar{x}})\}_i, \quad 1 \le i \le N-1, \\ \Lambda y_0 &= \frac{1}{h}[k_1(0, y_0, y_{x,0}) + k_1(h, y_1, y_{\bar{x},1})] \\ &\quad - k_0(0, y_0, y_{x,0}) - \frac{2}{h}\kappa_0(y_0), \quad i = 0, \\ \Lambda y_N &= -\frac{1}{h}[k_1(l-h, y_{N-1}, y_{x,N-1}) + k_1(l, y_N, y_{\bar{x},N})] \\ &\quad - k_0(l, y_N, y_{\bar{x},N}) - \frac{2}{h}\kappa_1(y_N), \quad i = N. \end{aligned}$$

If the non-linear operator A is defined in the space $H = H(\bar{\omega})$ by the relation $A = -\Lambda$, then the difference scheme (4) can be written in the form of the operator equation $Au = f$.

We investigate the properties of the non-linear operator A mapping from H into H. Recall that the inner product in $H(\bar{\omega})$ is defined by the formula

$$(u, v) = \sum_{i=1}^{N-1} u_i v_i h + 0.5h(u_0 v_0 + u_N v_N),$$

and we denote by $(u, v)_{\omega^+}$ and $(u, v)_{\omega^-}$ the sums

$$(u, v)_{\omega^+} = \sum_{i=1}^{N} u_i v_i h, \qquad (u, v)_{\omega^-} = \sum_{i=0}^{N-1} u_i v_i h,$$

so that

$$(u, v) = \frac{1}{2}\left[(u, v)_{\omega^+} + (u, v)_{\omega^-}\right].$$

We shall show that if the conditions (2), (3) are satisfied, then the operator A is strongly monotone in $H(\bar{\omega})$, i.e. the inequality

$$(Au - Av, u - v) \geq c_1 \parallel u - v \parallel^2, \quad c_1 > 0, \tag{5}$$

is satisfied, where c_1 is defined in (2).

We denote $p_0 = \bar{p}_0 = u_i$, $q_0 = \bar{q}_0 = v_i$, $p_1 = u_{x,i}$, $\bar{p}_1 = u_{\bar{x},i}$, $q_1 = v_{x,i}$, $\bar{q}_1 = v_{\bar{x},i}$. Using the definition of the operator A, the summation-by-parts formulas (cf. (7), (9) Section 5.2), and the conditions (2), (3), we obtain

$$\begin{aligned}
(Au-Av, u - v) &= (\Lambda v - \Lambda u, u - v) \\
&= \frac{1}{2}\sum_{i=1}^{N} h \left\{ \sum_{\alpha=0}^{1} [k_\alpha(x, \bar{p}_0, \bar{p}_1) - k_\alpha(x, \bar{q}_0, \bar{q}_1)](\bar{p}_\alpha - \bar{q}_\alpha) \right\}_i \\
&\quad + \frac{1}{2}\sum_{i=0}^{N-1} h \left\{ \sum_{\alpha=0}^{1} [k_\alpha(x, p_0, p_1) - k_\alpha(x, q_0, q_1)](p_\alpha - q_\alpha) \right\}_i \\
&\quad + [\kappa_1(\bar{p}_0) - \kappa_1(\bar{q}_0)](\bar{p}_0 - \bar{q}_0)\,|_{i=N} + [\kappa_0(p_0) - \kappa_0(q_0)](p_0 - q_0)|_{i=0} \\
&\geq \frac{c_1}{2} \sum_{i=1}^{N} h \sum_{\alpha=0}^{1} (\bar{p}_\alpha - \bar{q}_\alpha)_i^2 + \frac{c_1}{2} \sum_{i=0}^{N-1} h \sum_{\alpha=0}^{1} (p_\alpha - q_\alpha)_i^2 .
\end{aligned}$$

Taking into account the relation $u_{x,i} = u_{\bar{x},i+1}$, we write the resulting estimate in the form

$$\begin{aligned}
(Au - Av, u - v) &\geq \frac{c_1}{2}[(u - v, u - v)_{\omega^+} + (u - v, u - v)_{\omega^-} \\
&\quad + \sum_{i=1}^{N} h(u - v)_{\bar{x},i}^2 + \sum_{i=0}^{N-1} h(u - v)_{x,i}^2] \\
&= c_1 \left[\parallel u - v \parallel^2 + ((u - v)_{\bar{x}}^2, 1)_{\omega^+}\right] \geq c_1 \parallel u - v \parallel^2 .
\end{aligned}$$

From the remark 2 to lemma 12 of Chapter 5 it follows that this estimate is not optimal.

Thus, the strong monotonicity of the operator A has been established. By the continuity of the functions $k_\alpha(x, p_0, p_1)$ and $\kappa_\alpha(p_0)$, the operator A is continuous in H. Therefore from theorem 11 of Chapter 5 we obtain that the solution to equation $Au = f$ exists and is unique in the sphere $\parallel u \parallel \leq (1/c_1) \times \parallel A0 - f \parallel$, and, consequently, this is also true for the difference problem (4).

If $k_\alpha(x, p_0, p_1)$ and $\kappa_\alpha(p_0)$, $\alpha = 0, 1$ are continuously differentiable functions of their arguments, then in place of (2), (3) it is possible to use other sufficient conditions to guarantee the strong monotonicity of the operator A.

We will assume that the following conditions are satisfied

$$c_1 \sum_{\alpha=0}^{1} \xi_\alpha^2 \le \sum_{\alpha,\beta=0}^{1} a_{\alpha\beta}(x,p_0,p_1)\xi_\alpha\xi_\beta \le c_2 \sum_{\alpha=0}^{1} \xi_\alpha^2, \quad c_1 > 0, \tag{6}$$

$$0 \le \sigma_\alpha(p_0) \le c_3, \quad \alpha = 1,2, \tag{7}$$

where $\xi = (\xi_0, \xi_1)^T$ is an arbitrary vector and

$$a_{\alpha\beta}(x,p_0,p_1) = \frac{\partial k_\alpha(x,p_0,p_1)}{\partial p_\beta}, \quad \sigma_\alpha(p_0) = \frac{\partial \kappa_\alpha(p_0)}{\partial p_0}, \quad \alpha,\beta = 0,1.$$

We shall show that (2), (3) follow from the conditions (6), (7). In fact, we have that

$$\begin{aligned} &k_\alpha(x,p_0,p_1) - k_\alpha(x,q_0,q_1) \\ &\quad = \int_0^1 \frac{d}{dt} k_\alpha(x, tp_0 + (1-t)q_0, tp_1 + (1-t)q_1)dt \\ &\quad = (p_0 - q_0)\int_0^1 \frac{\partial k_\alpha(x,s_0,s_1)}{\partial s_0}dt + (p_1 - q_1)\int_0^1 \frac{\partial k_\alpha(x,s_0,s_1)}{\partial s_1}dt \\ &\quad = \sum_{\beta=0}^{1}(p_\beta - q_\beta)\int_0^1 a_{\alpha\beta}(x,s_0,s_1)dt, \quad \alpha = 0,1, \end{aligned}$$

where $s_0 = tp_0 + (1-t)q_0$, $s_1 = tp_1 + (1-t)q_1$. Multiplying this equation by $p_\alpha - q_\alpha$ and summing it for α between 0 and 1, we obtain using (6)

$$\begin{aligned} &\sum_{\alpha=0}^{1}[k_\alpha(x,p_0,p_1) - k_\alpha(x,q_0,q_1)](p_\alpha - q_\alpha) \\ &\qquad = \int_0^1 \sum_{\alpha,\beta=0}^{1} a_{\alpha\beta}(x,s_0,s_1)(p_\alpha - q_\alpha)(p_\beta - q_\beta)dt \\ &\qquad \ge c_1 \int_0^1 \sum_{\alpha=0}^{1}(p_\alpha - q_\alpha)^2 dt = c_1 \sum_{\alpha=0}^{1}(p_\alpha - q_\alpha)^2. \end{aligned}$$

Thus, the inequality (2) has been obtained. Analogously, from (7) we obtain inequality (3)

$$[\kappa_\alpha(p_0) - \kappa_\alpha(q_0)](p_0 - q_0) = \int_0^1 \frac{\partial \kappa_\alpha(s_0)}{\partial s_0} dt (p_0 - q_0)^2 \geq 0.$$

Thus, the conditions (6), (7) guarantee the existence and uniqueness of a solution to the difference problem (4).

We now find the Gateaux derivative of the operator A, assuming that the functions $k_\alpha(x, p_0, p_1)$ and $\kappa_\alpha(p_0)$, $\alpha = 0, 1$, have bounded derivatives of the necessary order in p_0 and p_1.

From the definition of the Gateaux derivative of a non-linear operator we will have

$$\begin{aligned} A'(u)y_i = &-\frac{1}{2}\{[a_{11}(x, u, u_x)y_x]_{\bar{x},i} + [a_{11}(x, u, u_{\bar{x}})y_{\bar{x}}]_{x,i} \\ &+ [a_{10}(x, u, u_x)y]_{\bar{x},i} + [a_{10}(x, u, u_{\bar{x}})y]_{\bar{x},i}\} \\ &+ \frac{1}{2}\{a_{01}(x, u, u_x)y_{x,i} + a_{01}(x, u, u_{\bar{x}})y_{\bar{x},i} \\ &+ [a_{00}(x, u, u_x) + a_{00}(x, u, u_{\bar{x}})]y_i\}, \quad 1 \leq i \leq N-1. \end{aligned}$$

For $i = 0$ we obtain

$$\begin{aligned} A'(u)y_0 = &-\frac{1}{h}[a_{11}(0, u_0, u_{x,0}) + a_{11}(h, u_1, u_{\bar{x},1}) \\ &- ha_{01}(0, u_0, u_{x,0}) + ha_{10}(h, u_1, u_{\bar{x},1})]y_{x,0} + \frac{2}{h}[\sigma_o(u_0) \\ &- \frac{1}{2}a_{10}(0, u_0, u_{x,0}) - \frac{1}{2}a_{10}(h, u_1, u_{\bar{x},1}) + \frac{h}{2}a_{00}(0, u_0, u_{x,0})]y_0, \end{aligned}$$

and for $i = N$ we will have

$$\begin{aligned} A'(u)y_N = &\frac{1}{h}[a_{11}(l-h, u_{N-1}, u_{x,N-1}) + a_{11}(l, u_N, u_{\bar{x},N}) \\ &+ ha_{01}(l, u_N, u_{\bar{x},N}) - ha_{10}(l-h, u_{N-1}, u_{x,N-1})]y_{\bar{x},N} + \frac{2}{h}[\sigma_1(u_N) \\ &+ \frac{1}{2}a_{10}(l, u_N, u_{\bar{x},N}) + \frac{1}{2}a_{10}(l-h, u_{N-1}, u_{x,N-1}) \\ &+ \frac{h}{2}a_{00}(l_1, u_N, u_{\bar{x},N})]y_N. \end{aligned}$$

Notice that, when computing $A'(u)y_0$ and $A'(u)y_N$, we used the equalities

$$y_1 = y_0 + hy_{x,0}, \quad y_{N-1} = y_N - hy_{\bar{x},N}. \tag{8}$$

We now investigate the properties of the Gateaux derivative $A'(u)$ of the operator A.

Lemma 4. *If the conditions*

$$\frac{\partial k_1(x,p_0,p_1)}{\partial p_0} = \frac{\partial k_0(x,p_0,p_1)}{\partial p_1}, \tag{9}$$

are satisfied, then $A'(u)$ is a self-adjoint operator in H. If the conditions (6), (7) *are satisfied, then it is positive-definite in H.*

Proof. Using the summation-by-parts formulas, and also the relations (8), we obtain

$$\begin{aligned}(A'(u)y,z) &= \frac{1}{2}\sum_{i=0}^{N-1} h[a_{11}(x,u,u_x)y_x z_x + a_{10}(x,u,u_x)yz_x \\ &\qquad + a_{01}(x,u,u_x)y_x z + a_{00}(x,u,u_x)yz]_i \\ &\quad + \frac{1}{2}\sum_{i=1}^{N} h[a_{11}(x,u,u_{\bar{x}})y_{\bar{x}} z_{\bar{x}} + a_{10}(x,u,u_{\bar{x}})yz_{\bar{x}} \\ &\quad + a_{01}(x,u,u_{\bar{x}})y_{\bar{x}} z + a_{00}(x,u,u_x)yz]_i + \sigma_0(u_0)y_0z_0 + \sigma_1(u_N)y_Nz_N.\end{aligned} \tag{10}$$

Equating this expression with the expression for $(y, A'(u)z)$, we obtain, using the condition $a_{10}(x,p_0,p_1) = a_{01}(x,p_0,p_1)$, which is just another form of (9), that the operator $A'(u)$ is self-adjoint in H for any $u \in H$.

Suppose now that the conditions (6), (7) are satisfied. Setting $z_i \equiv y_i$ in (10), we obtain

$$\begin{aligned}(A'(u)y,y) &\geq \frac{c_1}{2}\left[\sum_{i=0}^{N-1} h\left(y_i^2 + y_{x,i}^2\right) + \sum_{i=1}^{N} h\left(y_i^2 + y_{\bar{x},i}^2\right)\right] \\ &= c_1\left[(y,y) + \left(y_{\bar{x}}^2, 1\right)_{\omega^+}\right] \geq c_1(y,y),\end{aligned} \tag{11}$$

i.e. the operator $A'(u)$ is positive-definite in H. The lemma is proved. □

Notice that by theorem 2 of Chapter 5, the positive-definiteness of the Gateaux derivative of a continuous operator A implies that it is strongly monotone. Thus, if the conditions (6), (7) are satisfied, the operator A is strongly monotone.

Setting $z_i \equiv y_i$ in (10), we obtain from the conditions (6), (7) the upper bound

$$(A'(u)y,y) \le \frac{c_2}{2}\left[\sum_{i=0}^{N-1} h\left(y_i^2 + y_{x,i}^2\right) + \sum_{i=1}^{N} h\left(y_i^2 + y_{\bar{x},i}^2\right)\right]$$
$$+ c_3\left(y_0^2 + y_N^2\right) = c_2\left[(y,y) + \left(y_{\bar{x}}^2, 1\right)_{\omega+}\right] + c_3\left(y_0^2 + y_N^2\right).$$

From the inequalities (36) of lemma 15 Chapter 5 for $\epsilon = 1$ we find that

$$y_0^2 + y_N^2 \le c_4\left[(y,y) + \left(y_{\bar{x}}^2, 1\right)_{\omega+}\right], \quad c_4 = \frac{8 + l^2}{l\sqrt{16 + l^2}}. \tag{12}$$

Consequently, we have

$$(A'(u)y,y) \le \gamma_2\left[(y,y) + \left(y_{\bar{x}}^2, 1\right)_{\omega+}\right], \quad \gamma_2 = c_2 + c_3 c_4. \tag{13}$$

In the space $H = H(\bar{\omega})$ we define the linear operator R mapping H onto H by the formulas

$$Ry_i = \begin{cases} -\frac{2}{h} y_{x,0} + y_0, & i = 0, \\ -y_{\bar{x}x,i} + y_i, & 1 \le i \le N-1, \\ \frac{2}{h} y_{\bar{x},N} + y_N, & i = N. \end{cases}$$

From the first Green difference formula we find

$$(Ry,y) = (y,y) + \left(y_{\bar{x}}^2, 1\right)_{\omega+}. \tag{14}$$

Then from (11), (13), (14) it follows easily that, if the conditions (6), (7) are satisfied, the Gateaux derivative $A'(u)$ of the operator A can be bounded using

$$\gamma_1(Ry,y) \le (A'y,y) \le \gamma_2(Ry,y), \tag{15}$$

where $\gamma_1 = c_1 > 0$, $\gamma_2 = c_2 + c_3 c_4$, i.e. the operators R and A' are energy equivalent with constants that do not depend on the grid spacing h.

Recall that we obtained above that

$$(Au - Av, u - v) \ge c_1\left[\| u - v \|^2 + ((u-v)_{\bar{x}}^2, 1)_{\omega+}\right],$$

if the conditions (2), (3) are satisfied. From this and from (14) it follows that if the conditions (2), (3) are satisfied, then we have the estimate

$$(Au - Av, u - v) \ge \gamma_1(R(u-v), u - v), \quad \gamma_1 = c_1 > 0. \tag{16}$$

We will show now that, for any $u, v \in H$, the following estimate is valid

$$(R^{-1}(Au - Av), Au - Av) \le \gamma_2(Au - Av, u - v), \tag{17}$$

where $\gamma_2 = c_2(1 + c_4)$, if the conditions

$$\begin{aligned} &\sum_{\alpha=0}^{1}[k_\alpha(x, p_0, p_1) - k_\alpha(x, q_0, q_1)]^2 \\ &\quad \le c_2 \sum_{\alpha=0}^{1}[k_\alpha(x, p_0, p_1) - k_\alpha(x, q_0, q_1)](p_\alpha - q_\alpha), \\ &[\kappa_\alpha(p_0) - \kappa_\alpha(q_0)]^2 \le c_2[\kappa_\alpha(p_0) - \kappa_\alpha(q_0)](p_0 - q_0), \end{aligned} \tag{18}$$

are satisfied. In fact, to prove (17) it is sufficient to obtain the following estimate for any $u, v \in H$

$$(Au - Av, z)^2 \le \gamma_2(Au - Av, u - v)(Rz, z). \tag{19}$$

Then, setting here $z = R^{-1}(Au - Av)$, we will have (17).

We denote

$$\begin{aligned} &p_0 = \bar{p}_0 = u_i, \quad q_0 = \bar{q}_0 = v_i, \quad s_0 = \bar{s}_0 = z_i, \\ &p_1 = u_{x,i}, \quad \bar{p}_1 = u_{\bar{x},i}, \quad q_1 = v_{x,i}, \quad \bar{q}_1 = v_{\bar{x},i}, \\ &s_1 = z_{x,i}, \quad \bar{s}_1 = z_{\bar{x},i}. \end{aligned}$$

Using the definition of the operator A and the summation-by-parts formulas, we obtain

$$\begin{aligned} (Au - Av, z)^2 &= (\Lambda v - \Lambda u, z)^2 \\ &= \left\{ \frac{1}{2}\sum_{\alpha=0}^{1}([k_\alpha(x, p_0, p_1) - k_\alpha(x, q_0, q_1)], s_\alpha)_{\omega^-} \right. \\ &\quad + \frac{1}{2}\sum_{\alpha=0}^{1}([k_\alpha(x, \bar{p}_0, \bar{p}_1) - k_\alpha(x, \bar{q}_0, \bar{q}_1)], \bar{s}_\alpha)_{\omega^+} \\ &\quad \left. + [\kappa_1(\bar{p}_0) - \kappa_1(\bar{q}_0)]\, \bar{s}_0|_{i=N} + [\kappa_0(p_0) - \kappa_0(q_0)]\, s_0|_{i=0} \right\}^2. \end{aligned}$$

Using the Cauchy-Schwartz-Bunyakovskij inequality, we sequentially obtain

$$(Au - Av, z)^2 \le \left\{ \frac{1}{2} \sum_{\alpha=0}^{1} \left([k_\alpha(x, p_0, p_1) - k_\alpha(x, q_0, q_1)]^2, 1 \right)_{\omega^-}^{1/2} \left(s_\alpha^2, 1 \right)_{\omega^-}^{1/2} \right.$$

$$+ \frac{1}{2} \sum_{\alpha=0}^{1} \left([k_\alpha(x, \bar{p}_0, \bar{p}_1) - k_\alpha(x, \bar{q}_0, \bar{q}_1)]^2, 1 \right)_{\omega^+}^{1/2} \left(\bar{s}_\alpha^2, 1 \right)_{\omega^+}^{1/2}$$

$$\left. + [\kappa_1(\bar{p}_0) - \kappa_1(\bar{q}_0)] \, \bar{s}_0 |_{i=N} + [\kappa_0(p_0) - \kappa_0(q_0)] \, s_0 |_{i=0} \right\}^2$$

$$\le \left\{ \frac{1}{2} \sum_{\alpha=0}^{1} \left([k_\alpha(x, p_0, p_1) - k_\alpha(x, q_0, q_1)]^2, 1 \right)_{\omega^-} \right.$$

$$+ \frac{1}{2} \sum_{\alpha=0}^{1} \left([k_\alpha(x, \bar{p}_0, \bar{p}_1) - k_\alpha(x, \bar{q}_0, \bar{q}_1)]^2, 1 \right)_{\omega^+}$$

$$\left. + [\kappa_1(\bar{p}_0) - \kappa_1(\bar{q}_0)]_{i=N}^2 + [\kappa_0(p_0) - \kappa_0(q_0)]_{i=0}^2 \right\}$$

$$\times \left\{ \frac{1}{2} \sum_{\alpha=0}^{1} \left[\left(s_\alpha^2, 1 \right)_{\omega^-} + \left(\bar{s}_\alpha^2, 1 \right)_{\omega^+} \right] + \bar{s}_0^2 \big|_{i=N} + s_0^2 \big|_{i=0} \right\}.$$

Taking into account that

$$(Au - Av, u - v)$$

$$= \frac{1}{2} \sum_{\alpha=0}^{1} \left([k_\alpha(x, p_0, p_1) - k_\alpha(x, q_0, q_1)](p_\alpha - q_\alpha), 1 \right)_{\omega^-}$$

$$+ \frac{1}{2} \sum_{\alpha=0}^{1} \left([k_\alpha(x, \bar{p}_0, \bar{p}_1) - k_\alpha(x, \bar{q}_0, \bar{q}_1)(\bar{p}_\alpha - \bar{q}_\alpha), 1 \right)_{\omega^+}$$

$$+ [\kappa_1(\bar{p}_0) - \kappa_1(\bar{q}_0)] \, (\bar{p}_0 - \bar{q}_0)|_{i=N} + [\kappa_0(p_0) - \kappa_0(q_0)] \, (p_0 - q_0)|_{i=0} ,$$

and also from (12), (14) and the notation introduced above

$$\frac{1}{2} \sum_{\alpha=0}^{1} \left[\left(s_\alpha^2, 1 \right)_{\omega^-} + \left(\bar{s}_\alpha^2, 1 \right)_{\omega^+} \right] + \bar{s}_0^2 \big|_{i=N} + s_0^2 \big|_{i=0}$$

$$= \frac{1}{2} \left[(z^2, 1)_{\omega^+} + (z^2, 1)_{\omega^-} + \left(z_{\bar{x}}^2, 1 \right)_{\omega^+} + \left(z_x^2, 1 \right)_{\omega^-} \right] + z_N^2 + z_0^2$$

$$= (z^2, 1) + \left(z_{\bar{x}}^2, 1 \right)_{\omega^+} + z_N^2 + z_0^2 \le (1 + c_4)(Rz, z),$$

we obtain the estimate (19), if the conditions (18) are satisfied. The assertion is proved.

13.2.2 The simple iteration method. We look now at iterative methods for solving the non-linear difference scheme (4) that we constructed. We shall first show that the conditions (2), (3), and (18) are satisfied.

To solve equation (4), an implicit iteration method is used

$$B\frac{y_{k+1}-y_k}{\tau}+Ay_k=f, \quad k=0,1,\ldots, \quad y_0\in H, \tag{20}$$

where $A=-\Lambda$, $B=R$ and the operator R was defined above. From (20) it follows that finding y_{k+1} given y_k requires the solution of the linear equation

$$B_{y+1}=\varphi, \quad \varphi=By_k-\tau(Ay_k-f)$$

or in expanded form

$$\begin{aligned}
-y_{k+1}(i-1)+cy_{k+1}(i)-y_{k+1}(i+1)&=h^2\varphi(i), \quad 1\le i\le N-1,\\
cy_{k+1}(0)-2y_{k+1}(1)&=h^2\varphi(0), \quad i=0,\\
-2y_{k+1}(N-1)+cy_{k+1}(N)&=h^2\varphi(N), \quad i=N,
\end{aligned}$$

where $c=2+h^2$. Since $c>2$, the boundary-value difference problem can be solved using the monotone elimination method at a cost of $O(N)$ arithmetic operations.

It remains to indicate the value of the iterative parameter τ and to give an estimate for the number of iterations required. Since the conditions (2), (3), and (18) are satisfied, the estimates (16) and (17) are valid, and they can be written in the form

$$\begin{aligned}
(Au-Av,u-v)&\ge\gamma_1(B(u-v),u-v), \quad \gamma_1=c_1>0,\\
(B^{-1}(Au-Av),Au-Av)&\le\gamma_2(Au-Av,u-v), \quad \gamma_2=c_2(1+c_4),
\end{aligned} \tag{21}$$

where c_1 was given in (2), c_2 in (18), and c_4 in (12).

Since the operator B is self-adjoint and positive-definite, we investigate the convergence of the method (20) in the energy space H_D, where $D=B$. For this choice of the operator D, the inequalities (21) are the same as the inequalities (4), (5). Therefore it is possible to use theorem 1 to choose the iterative parameter τ. We obtain that, for $\tau=1/\gamma_2=1/(c_2(1+c_4))$, the iterative method (20) converges in H_D, and the error can be estimated using $\| y_n-u\|_B\le\rho^n\| y_0-u\|_B$, $\rho=\sqrt{1-\xi}$, $\xi=\gamma_1/\gamma_2$, for any initial guess y_0.

Thus, if the conditions (2), (3), (18) are satisfied, then the iterative simple iteration method (20) with the indicated value of the parameter τ

enables us to obtain the solution of the non-linear difference scheme (4) to an accuracy ϵ after $n \geq n_0(\epsilon)$ iterations, where

$$n_0(\epsilon) = \frac{\ln \epsilon}{\ln \rho} = \frac{2 \ln \epsilon}{\ln\left(1 - \dfrac{c_1}{c_2(1+c_4)}\right)}.$$

Since the constants c_1, c_2, and c_4 do not depend on the grid spacing h, the iteration count $n_0(\epsilon)$ only depends on ϵ and does not change when the grid is refined.

We look now at the iterative method (20) under the assumptions that (6) and (7) are satisfied for arbitrary $a_{\alpha\beta} = \partial k_\alpha / \partial p_\beta$ and $\sigma_\alpha = \partial \kappa_\alpha / \partial p_0$, and also assuming that the symmetry condition (9) is satisfied. Then the Gateaux derivative of the operator A will satisfy the inequalities (15), which, if we choose $B = R$, can be written in the form

$$\gamma_1(By, y) \leq (A'(v)y, y) \leq \gamma_2(By, y), \quad v, y \in H, \tag{22}$$

where $\gamma_1 = c_1$, $\gamma_2 = c_2 + c_3c_4$, c_1, c_2, and c_3 are defined in (6), (7), and c_4 in (12).

Suppose that $D = B$. Then the operator $DB^{-1}A'(v) = A'(v)$ will be self-adjoint in H by lemma 4, and, consequently, the conditions of theorem 2 will be satisfied, and the inequalities (22) will be the same as the inequalities (14). Therefore the parameter τ in (20) can be taken to be $\tau = \tau_0 = 2/(\gamma_1 + \gamma_2)$. Then the error $y_n - u$ and the iteration count can be estimated using

$$\| y_n - u \|_B \leq \rho_0^n \| y_0 - u \|_B, \quad \rho_0 = \frac{1-\xi}{1+\xi}, \quad \xi = \frac{\gamma_1}{\gamma_2} = \frac{c_1}{c_2 + c_3c_4},$$
$$n \geq n_0(\epsilon) = \ln \epsilon / \ln \rho_0.$$

Here, as for the preceding method, the iteration count does not depend on the grid spacing h. Using the first Green difference formula, we will have the following representation of the norm $\| z \|_B$ for this choice of the operator B:

$$\| z \|_B^2 = (z, z) + \left(z_{\bar{x}}^2, 1\right)_{\omega^+}.$$

We have examined methods for solving a non-linear difference scheme that approximates a quasi-linear one-dimensional equation on a uniform grid. It presents no difficulty to transfer these comments to the case of an arbitrary non-uniform grid, and also to difference schemes that approximate basic boundary-value problems for quasi-linear second-order elliptic equations in a rectangle.

13.2.3 Iterative methods for quasi-linear elliptic difference equations in a rectangle. In the rectangle $\bar{G} = \{0 \le x_\alpha \le l_\alpha,\ \alpha = 1,2\}$ with boundary Γ, it is necessary to find the solution of the equations

$$\sum_{\alpha=1}^{2} \frac{\partial}{\partial x_\alpha} k_\alpha \left(x, u, \frac{\partial u}{\partial x_1}, \frac{\partial u}{\partial x_2} \right) - k_0 \left(x, u, \frac{\partial u}{\partial x_1}, \frac{\partial u}{\partial x_2} \right) = -\varphi(x),\ x \in G, \quad (23)$$

satisfying the third-kind boundary conditions

$$\begin{aligned} k_\alpha \left(x, u, \frac{\partial u}{\partial x_1}, \frac{\partial u}{\partial x_2} \right) &= \kappa_{-\alpha}(x,u) - g_{-\alpha}(x), \quad x_\alpha = 0, \\ -k_\alpha \left(x, u, \frac{\partial u}{\partial x_1}, \frac{\partial u}{\partial x_2} \right) &= \kappa_{+\alpha}(x,u) - g_{+\alpha}(x), \quad x_\alpha = l_\alpha, \quad \alpha = 1,2. \end{aligned} \quad (24)$$

We will assume, as in the one-dimensional case, that the following conditions are satisfied. The functions $k_\alpha(x,p)$ and $\kappa_{\pm\alpha}(x,p_0)$ are continuous in $p = (p_0, p_1, p_2)$ and p_0, and, in addition,

$$\sum_{\alpha=0}^{2} [k_\alpha(x,p) - k_\alpha(x,q)](p_\alpha - q_\alpha) \ge c_1 \sum_{\alpha=0}^{2} (p_\alpha - q_\alpha)^2,$$

$$\sum_{\alpha=0}^{2} [k_\alpha(x,p) - k_\alpha(x,q)]^2 \le c_2 \sum_{\alpha=0}^{2} [k_\alpha(x,p) - k_\alpha(x,q)](p_\alpha - q_\alpha),$$

$$[\kappa_{\pm\alpha}(x,p_0) - \kappa_{\pm\alpha}(x,q_0)]^2 \le c_2 [\kappa_{\pm\alpha}(x,p_0) - \kappa_{\pm\alpha}(x,q_0)](p_0 - q_0), \quad \alpha = 1,2,$$

where $c_1 > 0$ and $c_2 > 0$, $x \in \bar{G}$ and $|p|$, $|q| < \infty$.

In the region $\bar{G}$, we introduce the rectangular grid

$$\bar{\omega} = \{x_{ij} = (ih_1, jh_2),\ 0 \le i \le N_1,\ 0 \le j \le N_2,\ h_\alpha N_\alpha = l_\alpha,\ \alpha = 1,2\}.$$

The simplest difference scheme corresponding to the problem (23), (24) has the form

$$\begin{aligned} \Lambda y &= -f, \qquad x \in \bar{\omega}, \\ \Lambda &= \Lambda_1 + \Lambda_2, \quad f = \varphi + 2\varphi_1/h_1 + 2\varphi_2/h_2, \end{aligned} \quad (25)$$

where

$$\varphi_\alpha(x) = \begin{cases} g_{-\alpha}(x), & x_\alpha = 0, \\ 0, & h_\alpha \le x_\alpha \le l_\alpha - h_\alpha, \\ g_{+\alpha}(x), & x_\alpha = l_\alpha, \quad 0 \le x_{3-\alpha} \le l_{3-\alpha}, \end{cases}$$

and the operators Λ_α, $\alpha = 1,2$ are defined by the formulas:

1) for $h_\beta \le x_\beta \le l_\beta - h_\beta$ we have

$$\begin{aligned}
\Lambda_\alpha y =& \frac{1}{2}\left\{[k_\alpha(x,y,y_{\bar{x}_1},y_{\bar{x}_2})]_{x_\alpha} + [k_\alpha(x,y,y_{x_1},y_{x_2})]_{\bar{x}_\alpha}\right\} \\
&- \frac{1}{4}[k_0(x,y,y_{\bar{x}_1},y_{\bar{x}_2}) + k_0(x,y,y_{x_1},y_{x_2})], \ h_\alpha \le x_\alpha \le l_\alpha - h_\alpha; \\
\Lambda_\alpha y =& \frac{1}{h_\alpha}\left[k_\alpha^{+1_\alpha}(x,y,y_{\bar{x}_1},y_{\bar{x}_2}) + k_\alpha(x,y,y_{x_1},y_{x_2})\right] \\
&- \frac{1}{2}k_0(x,y,y_{x_1},y_{x_2}) - \frac{2}{h_\alpha}\kappa_{-\alpha}(x,y), \ x_\alpha = 0; \\
\Lambda_\alpha y =& -\frac{1}{h_\alpha}\left[k_\alpha(x,y,y_{\bar{x}_1},y_{\bar{x}_2}) + k_\alpha^{-1_\alpha}(x,y,y_{x_1},y_{x_2})\right] \\
&- \frac{1}{2}k_0(x,y,y_{\bar{x}_1},y_{\bar{x}_2}) - \frac{2}{h_\alpha}\kappa_{+\alpha}(x,y), \ x_\alpha = l_\alpha;
\end{aligned}$$

2) for $x_\beta = 0$ we have

$$\Lambda_\alpha y = [k_\alpha(x,y,y_{x_1},y_{x_2})]_{\bar{x}_\alpha} - \frac{1}{2}k_0(x,y,y_{x_1},y_{x_2})$$

when $h_\alpha \le x_\alpha \le l_\alpha - h_\alpha$;

$$\Lambda_\alpha y = \frac{2}{h_\alpha}k_\alpha(x,y,y_{x_1},y_{x_2}) - k_0(x,y,y_{x_1},y_{x_2}) - \frac{2}{h_\alpha}\kappa_{-\alpha}(x,y)$$

when $x_\alpha = 0$;

$$\Lambda_\alpha y = -\frac{2}{h_\alpha}k_\alpha^{-1_\alpha}(x,y,y_{x_1},y_{x_2}) - \frac{2}{h_\alpha}\kappa_{+\alpha}(x,y), \quad x_\alpha = l_\alpha;$$

3) for $x_\beta = l_\beta$ we have

$$\Lambda_\alpha y = [k_\alpha(x,y,y_{\bar{x}_1},y_{\bar{x}_2})]_{x_\alpha} - \frac{1}{2}k_0(x,y,y_{\bar{x}_1},y_{\bar{x}_2})$$

when $h_\alpha \le x_\alpha \le l_\alpha - h_\alpha$;

$$\begin{aligned}
\Lambda_\alpha y &= \frac{2}{h_\alpha}k_\alpha^{+1_\alpha}(x,y,y_{\bar{x}_1},y_{\bar{x}_2}) - \frac{2}{h_\alpha}\kappa_{-\alpha}(x,y), \quad x_\alpha = 0; \\
\Lambda_\alpha y &= -\frac{2}{h_\alpha}k_\alpha(x,y,y_{\bar{x}_1},y_{\bar{x}_2}) - k_0(x,y,y_{\bar{x}_1},y_{\bar{x}_2}) - \frac{2}{h_\alpha}\kappa_{+\alpha}(x,y)
\end{aligned}$$

when $x_\alpha = l_\alpha$.

Here $\beta = 3 - \alpha$, $\alpha = 1, 2$ and we have used the notation

$$k_1^{+1_1}\,(x, y, y_{\bar{x}_1}, y_{\bar{x}_2})|_{x_{ij}}$$
$$= k_1\left(x_{i+1,j}, y(i+1,j), y_{\bar{x}_1}(i+1,j), y_{\bar{x}_2}(i+1,j)\right),$$

and also the analogous notation for $k_1^{-1_1}$ and $k_2^{\pm 1_2}$.

In the space H of grid functions defined on $\bar{\omega}$, we define the inner product

$$(u, v) = \sum_{i=0}^{N_1}\sum_{j=0}^{N_2} \hbar_1(i)\hbar_2(j)u(i,j)v(i,j),$$

$$\hbar_\alpha(k) = \begin{cases} h_\alpha, & 1 \le k \le N_\alpha - 1, \\ 0.5h_\alpha, & k = 0, N_\alpha \end{cases}$$

and the operators $A_\alpha = -\Lambda_\alpha$, $\alpha = 1, 2$, $A = A_1 + A_2$, $R = R_1 + R_2$, where

$$R_\alpha y = \begin{cases} -\dfrac{2}{h_\alpha} y_{x_\alpha} + \dfrac{1}{2} y, & x_\alpha = 0, \\[2ex] -y_{\bar{x}_\alpha x_\alpha} + \dfrac{1}{2} y, & h_\alpha \le x_\alpha \le l_\alpha - h_\alpha, \\[2ex] \dfrac{2}{h_\alpha} y_{\bar{x}_\alpha} + \dfrac{1}{2} y, & x_\alpha = l_\alpha, \ \alpha = 1, 2, \end{cases}$$

and $0 \le x_\beta \le l_\beta$. Then the difference scheme (25) can be written in the form of the operator equation

$$Au = f \tag{26}$$

with the non-linear operator A.

Using the assumptions made above concerning the coefficients $k_\alpha(x, p)$ and $\kappa_{\pm\alpha}(x, p_0)$, we obtain as in the one-dimensional case that the inequalities (16) and (17) are valid, where c_4 is the constant from the inequality

$$\sum_{j=0}^{N_2} \hbar_2(j)[y^2(0,j) + y^2(N_1,j)] + \sum_{i=0}^{N_1} \hbar_1(i)[y^2(i,0) + y^2(i,N_2)]$$
$$\le c_4 \left[\sum_{i=0}^{N_1}\sum_{j=0}^{N_2} y^2(i,j)\hbar_1(i)\hbar_2(j) + \sum_{j=0}^{N_2}\sum_{i=1}^{N_1} h_1\hbar_2(j) y_{\bar{x}_1}^2(i,j) \right. \tag{27}$$
$$\left. + \sum_{i=0}^{N_1}\sum_{j=1}^{N_2} \hbar_1(i) h_2 y_{\bar{x}_2}^2(i,j) \right].$$

We shall show that

$$c_4 = \sqrt{2}(16+l^2)/(l\sqrt{32+l^2}), \quad l = \min(l_1, l_2). \tag{28}$$

From the inequality (36) in lemma 15 of Chapter 5 for $\epsilon = \sqrt{2}$, we obtain

$$y^2(0,j) + y^2(N_1,j)$$
$$\le \frac{(16+l_1^2)\sqrt{2}}{l_1\sqrt{32+l_1^2}}\left[\sum_{i=1}^{N_1} h_y y_{\bar{x}_1}^2(i,j) + \frac{1}{2}\sum_{i=0}^{N_1} \hbar_1(i) y^2(i,j)\right].$$

Notice that, if here we change l_1 to l, then the inequality is only strengthened. We first multiply the left- and right-hand sides of this inequality by $\hbar_2(j)$ and sum for j between 0 and N_2. We will have

$$\sum_{j=0}^{N_2} \hbar_2(j)\left[y^2(0,j) + y^2(N_1,j)\right]$$
$$\le c_4\left[\sum_{j=0}^{N_2}\sum_{i=1}^{N_1} h_1\hbar_2(j) y_{\bar{x}_1}^2(i,j) + \frac{1}{2}\sum_{i=0}^{N_1}\sum_{j=0}^{N_2} y^2(i,j)\hbar_1(i)\hbar_2(j)\right], \tag{29}$$

where c_4 was defined in (28). Analogously we find

$$\sum_{i=0}^{N_1} \hbar_1(i)\left[y^2(i,0) + y^2(i,N_2)\right]$$
$$\le c_4\left[\sum_{i=0}^{N_1}\sum_{j=1}^{N_2} \hbar_1(i) h_2 y_{\bar{x}_2}^2(i,j) + \frac{1}{2}\sum_{i=0}^{N_1}\sum_{j=0}^{N_2} y^2(i,j)\hbar_1(i)\hbar_2(j)\right]. \tag{30}$$

Adding together (29) and (30), we obtain inequality (27).

To solve equation (26), it is possible to use an implicit simple iteration method (20) with $B = R$ and $\tau = 1/\gamma_2 = 1/(c_2(1+c_4))$. Then by theorem 1 the iterative method (20) will converge in H_B and the error can be estimated using

$$\| y_n - u \|_B \le \rho^n \| y_0 - u \|_B, \quad \rho = \sqrt{1-\xi}, \quad \xi = \gamma_1/\gamma_2 = c_1/(c_2(1+c_4)).$$

Consequently, the iteration count $n_0(\epsilon)$ that is required to achieve an accuracy ϵ does not depend on the number of nodes in the grid $\bar{\omega}$.

To find y_{k+1}, we have the problem

$$Ry_{k+1} = \varphi, \quad \varphi = Ry_k - \tau(Ay_k - f).$$

Since the operator R corresponds to a boundary-value problem of the second kind for a difference equation with constant coefficients, the problem can be solved by the direct methods described in Chapters 3 and 4 at a cost of $O(N^2 \log_2 N)$ arithmetic operations ($N_1 = N_2 = N = 2^n$). If the functions $k_\alpha(x,p)$ and $\kappa_{\pm\alpha}(x,p_0)$ are differentiable, then the operator A will have a Gateaux derivative that is a self-adjoint operator in H if the following conditions are satisfied

$$a_{\alpha\beta}(x,p) = a_{\beta\alpha}(x,p), \quad \alpha, \beta = 0,1,2, \tag{31}$$

where $a_{\alpha\beta}(x,p) = \partial k_\alpha(x,p)/\partial p_\beta$. It is possible to show that, if in addition to (31) the following conditions are satisfied

$$c_1 \sum_{\alpha=0}^{2} \xi_\alpha^2 \le \sum_{\alpha,\beta=0}^{2} a_{\alpha\beta}(x,p)\xi_\alpha\xi_\beta \le c_2 \sum_{\alpha=0}^{1} \xi_\alpha^2, \quad c_1 > 0,$$

$$0 \le \frac{\partial \kappa_{\pm\alpha}(x,p_0)}{\partial p_0} \le c_3, \quad \alpha = 1,2,$$

then the inequalities (15) are valid, where $\gamma_1 = c_1$, $\gamma_2 = c_2 + c_3 c_4$, and where c_4 was defined in (28). Then in the iterative method (20) with $B = R$, the parameter τ can be set equal to $\tau_0 = 2/(\gamma_1 + \gamma_2)$. By theorem 4, we have the following estimate for the error

$$\| y_n - u \|_B \le \rho_0^n \| y_0 - u \|_B, \quad \rho_0 = (1-\xi)/(1+\xi), \quad \xi = \gamma_1/\gamma_2.$$

Suppose now that we are required to find the solution to a boundary-value problem of the first kind in the rectangle $\bar{G}$

$$\sum_{\alpha=1}^{2} \frac{\partial}{\partial x_\alpha} k_\alpha \left(x, u, \frac{\partial u}{\partial x_1}, \frac{\partial u}{\partial x_2}\right) - k_0 \left(x, u, \frac{\partial u}{\partial x_1}, \frac{\partial u}{\partial x_2}\right) = -\varphi(x), \ x \in G, \tag{32}$$

$$u(x) = 0, \quad x \in \Gamma.$$

We will assume that the functions $k_\alpha(x,p)$ are continuous in $p = (p_0, p_1, p_2)$

and that the following conditions are satisfied

$$\sum_{\alpha=1}^{2}[k_\alpha(x,p)-k_\alpha(x,q)](p_\alpha-q_\alpha)\geq c_1\sum_{\alpha=1}^{2}(p_\alpha-q_\alpha)^2,\quad c_1>0,$$
$$[k_0(x,p)-k_0(x,q)](p_0-q_0)\geq 0,\tag{33}$$
$$\sum_{\alpha=0}^{2}[k_\alpha(x,p)-k_\alpha(x,q)]^2\leq c_2\sum_{\alpha=0}^{2}[k_\alpha(x,p)-k_\alpha(x,q)](p_\alpha-q_\alpha),$$

where $c_1>0$, $c_2>0$ for $x\in G$ and $|p|$, $|q|<\infty$.

The problem (32), on the rectangular uniform grid $\bar\omega=\omega\cup\gamma$ introduced earlier, will be put in correspondence with the difference scheme

$$\Lambda y=-f,\quad x\in\omega,\quad y(x)=0,\quad x\in\gamma,\tag{34}$$

where $f=\varphi$, and the difference operator Λ is defined as follows:

$$\Lambda y=\Lambda^- y=\frac{1}{2}\{[k_1(x,y,y_{\bar x_1},y_{\bar x_2})]_{x_1}+[k_1(x,y,y_{x_1},y_{x_2})]_{\bar x_1}$$
$$+[k_2(x,y,y_{\bar x_1},y_{\bar x_2})]_{x_2}+[k_2(x,y,y_{x_1},y_{x_2})]_{\bar x_2}$$
$$-k_0(x,y,y_{\bar x_1},y_{\bar x_2})-k_0(x,y,y_{x_1},y_{x_2})\}\,.$$

We give here two more possible approximations:

$$\Lambda y=\Lambda^+ y=\frac{1}{2}\{[k_1(x,y,y_{\bar x_1},y_{x_2})]_{x_1}+[k_1(x,y,y_{x_1},y_{\bar x_2})]_{\bar x_1}$$
$$+[k_2(x,y,y_{x_1},y_{\bar x_2})]_{x_2}+[k_2(x,y,y_{\bar x_1},y_{x_2})]_{\bar x_2}$$
$$-k_0(x,y,y_{\bar x_1},y_{x_2})-k_0(x,y,y_{x_1},y_{\bar x_2})\}$$

and $\Lambda=1/2(\Lambda^-+\Lambda^+)$.

For this example, H is the space of grid functions defined on ω with the inner product

$$(u,v)=\sum_{i=1}^{N_1-1}\sum_{j=1}^{N_2-1}h_1h_2u(i,j)v(i,j).$$

If we set $y|_\gamma=0$ in the equation for the scheme (34), then we obtain the difference scheme $\bar\Lambda y=-f$. Defining the operator $A=-\bar\Lambda$, we write the resulting scheme in the form of the operator equation (26) in the space H.

Using the conditions (33), we obtain for all three approximations that the corresponding operator A satisfies the inequalities (16), (17):

$$(Au - Av, u - v) \geq \gamma_1(R(u - v), u - v), \quad \gamma_1 = c_1 > 0,$$
$$(R^{-1}(Au - Av), Au - Av) \leq \gamma_2(Au - Av, u - v), \quad \gamma_2 = c_2(1 + c_4),$$

where

$$c_4 = \frac{1}{\delta}, \quad \delta = \frac{4}{h_1^2}\sin^2\frac{\pi h_1}{2l_1} + \frac{4}{h_2^2}\sin^2\frac{\pi h_2}{2l_2} \geq \frac{8}{l_1^2} + \frac{8}{l_2^2},$$

and the operator R corresponds to the Laplace difference operator $Ry = -\mathcal{R}\mathring{y}$, $y(x) = \mathring{y}(x)$ for $x \in \omega$, and $\mathring{y}(x) = 0$ for $x \in \gamma$, $\mathcal{R}u = u_{\bar{x}_1 x_1} + u_{\bar{x}_2 x_2}$.

To solve equation (26), we use the simple iteration method (20) with $B = R$ and $\tau = 1/\gamma_2$. By theorem 1, we will have the estimate

$$\| y_n - u \|_B \leq \rho^n \| y_0 - u \|_B, \quad \rho = \sqrt{1 - \xi}, \quad \xi = \gamma_1/\gamma_2.$$

As before, to solve the equation $Ry_{k+1} = Ry_k - \tau(Ay_k - f)$ it is possible to use the direct methods of cyclic reduction or separation of variables described in Chapters 3 and 4.

13.2.4 Iterative methods for weakly-nonlinear equations. In the rectangle $\bar{G} = \{0 \leq x_\alpha \leq l_\alpha, \ \alpha = 1, 2\}$ we consider a weakly-nonlinear second-order elliptic equation

$$Lu = \frac{\partial^2 u}{\partial x_1^2} + \frac{\partial^2 u}{\partial x_2^2} - k_0\left(x, u, \frac{\partial u}{\partial x_1}, \frac{\partial u}{\partial x_2}\right) = 0, \quad x \in G \tag{35}$$

with the first-order boundary conditions

$$u(x) = 0, \quad x \in \Gamma. \tag{36}$$

The *weak non-linearity* of equation (35) indicates that the functions $k_0(x, p_0, p_1, p_2)$ are defined for $x \in \bar{G}$ and $|p_0|$, $|p_1|$, $|p_2| < \infty$, that they are continuous in x for fixed p_0, p_1, p_2, and also that there exist derivatives of the functions $k_0(x, p_0, p_1, p_2)$ in p_0, p_1, and p_2, that satisfy the conditions

$$c_2 \geq \frac{\partial k_0}{\partial p_0} \geq 0, \quad \left|\frac{\partial k_0}{\partial p_\alpha}\right| \leq M, \quad \alpha = 1, 2. \tag{37}$$

On the rectangular uniform grid $\bar{\omega} = \omega \cup \gamma$ introduced earlier, the difference scheme corresponding to the problem (35), (36) has the form

$$\begin{aligned} &\Lambda y = 0, \quad x \in \omega, \quad y(x) = 0, \quad x \in \gamma, \\ &\Lambda y = \mathcal{R}y - \frac{1}{2}\left[k_0(x, y, y_{\bar{x}_1}, y_{\bar{x}_2}) + k_0(x, y, y_{x_1}, y_{x_2})\right], \end{aligned} \tag{38}$$

where $\mathcal{R}y = y_{\bar{x}_1 x_1} + y_{\bar{x}_2 x_2}$ is the Laplace difference operator.

We define now the difference operator $\Lambda'(v)$:

$$\begin{aligned} \Lambda'(v)y = \mathcal{R}y - \frac{1}{2}&\left[a_{01}(x, v, v_{\bar{x}_1}, v_{\bar{x}_2})y_{\bar{x}_1}\right. \\ &+ a_{01}(x, v, v_{x_1}, v_{x_2})y_{x_1} + a_{02}(x, v, v_{\bar{x}_1}, v_{\bar{x}_2})y_{\bar{x}_2} \\ &\left.+ a_{02}(x, v, v_{x_1}, v_{x_2})y_{x_2} + (a_{00}(x, v, v_{\bar{x}_1}, v_{\bar{x}_2}) + a_{00}(x, v, v_{x_1}, v_{x_2}))y\right], \end{aligned}$$

where

$$a_{0\alpha}(x, p_0, p_1, p_2) = \frac{\partial k_0(x, p_0, p_1, p_2)}{\partial p_\alpha}, \quad \alpha = 0, 1, 2.$$

In the space H of grid functions defined on ω, we define the operators:

$$Ay = -\Lambda \mathring{y}, \quad Ry = -\mathcal{R}\mathring{y}, \quad A'(v)y = -\Lambda'(\mathring{v})\mathring{y},$$

where

$$y(x) = \mathring{y}(x), \quad v(x) = \mathring{v}(x) \quad \text{for} \quad x \in \omega$$

and

$$\mathring{y}(x) = 0, \quad \mathring{v}(x) = 0, \quad \text{for} \quad x \in \gamma.$$

The operator $A'(v)$ is the Gateaux derivative of the operator A. Using this notation, we write the difference scheme in the form of the operator equation (26).

If $k_0(x, p_0, p_1, p_2)$ does not depend on p_1 and p_2, i.e.

$$k(x, p_0, p_1, p_2) = k_0(x, p_0),$$

then

$$a_{01}(x, p) = a_{02}(x, p) = 0.$$

In this case the operator $A'(v)$ is self-adjoint in H.

Using the lower bound for the difference operator $(-\mathcal{R})$

$$(-\mathcal{R}\mathring{y},\mathring{y}) = -(\mathring{y}_{\bar{x}_1x_1} + \mathring{y}_{\bar{x}_2x_2},\mathring{y}) \geq \delta(\mathring{y},\mathring{y}),$$

where

$$\delta = \frac{4}{h_1^2}\sin^2\frac{\pi h_1}{2l_1} + \frac{4}{h_2^2}\sin^2\frac{\pi h_2}{2l_2} \geq \frac{8}{l_1^2} + \frac{8}{l_2^2},$$

the condition (37) for $M = 0$, and the equation

$$-(\Lambda(\mathring{v})\mathring{y},\mathring{y}) = -(\mathcal{R}\mathring{y},\mathring{y}) + (a_{00}(x,\mathring{v})\mathring{y},\mathring{y}),$$

we obtain

$$\gamma_1(Ry,y) \leq (A'(v)y,y) \leq \gamma_2(Ry,y),$$

where

$$\gamma_1 = 1, \quad \gamma_2 = 1 + c_2/\delta.$$

Consequently, if for this "self-adjoint" case we use the iterative method (20) with $B = D = R$ and $\tau = \tau_0 = 2/(\gamma_1+\gamma_2)$, then by theorem 2 the error can be estimated using

$$\| y_n - u \|_B \leq \rho_0^n \| y_0 - u \|_B, \quad \rho_0 = (1-\xi)/(1+\xi), \quad \xi = \gamma_1/\gamma_2.$$

The operator R in the scheme (20) can be inverted using one of the direct methods.

Chapter 14

Example Solutions of Elliptic Grid Equations

In Section 14.1 we consider certain ways of constructing implicit iterative schemes, in particular ones based on the extraction of a regularizer. In Section 14.2 we construct direct and iterative methods for solving difference problems in the multi-dimensional case. We consider an iterative method for solving a boundary-value problem of the third kind for an elliptic equation with mixed derivatives in a rectangle. Methods for solving systems of elliptic equations are discussed in Section 14.3. Here we consider the application of the general theory to the solution of certain problems in elasticity theory. In Section 14.4 we illustrate methods for solving a Dirichlet difference problem for Poisson's equation in a region of complex form, based on the reduction of the given problem to the solution of a series of problems in a rectangle.

14.1 Methods for constructing implicit iterative schemes

14.1.1 The regularizer principle in the general theory of iterative methods. In Chapters 6–8, 12, and 13 we laid out the general theory of iterative methods used to solve the operator equation

$$Au = f. \tag{1}$$

In the general theory of iterative methods, we did not use the concrete structure of the operators of the iterative scheme — the theory uses minimal information of a general functional character concerning the operators. This enables us to indicate (for fixed operators in the scheme) general principles for constructing optimal iterative methods. For example, if the operators A

and B in the two-level iterative scheme

$$B\frac{y_{k+1}-y_k}{\tau_{k+1}}+Ay_k=f, \quad k=0,1,\ldots, \quad y_0\in H \tag{2}$$

satisfy the conditions

$$B=B^*>0, \quad A=A^*>0, \tag{3}$$

$$\gamma_1 B\le A\le \gamma_2 B, \quad \gamma_1>0, \tag{4}$$

then the set of Chebyshev iterative parameters τ_k:

$$\tau_k=\frac{\tau_0}{1+\rho_0\mu_k}, \quad \mu_k\in\mathcal{M}_n=\left\{-\cos\frac{(2i-1)\pi}{2n},\ 1\le i\le n\right\}, \quad 1\le k\le n,$$

where

$$\tau_0=\frac{2}{\gamma_1+\gamma_2}, \quad \rho_0=\frac{1-\xi}{1+\xi}, \quad \xi=\frac{\gamma_1}{\gamma_2}$$

is optimal.

How do such requirements guide us to a choice of the operator B? It was remarked in Section 5.3 that the choice of B should be made subject to two requirements: 1) guaranteeing the optimal convergence rate of the method; 2) economizing the cost of inverting this operator.

For the example given above, the first requirement is fulfilled if the energy of the operator B is close to the energy of the operator A, i.e. γ_1 and γ_2 are close in the inequalities (4). In order to satisfy the second requirement, it is necessary to choose easily invertible operators B from the class of operators that are close in energy to the operator A.

How do we construct easily invertible operators? It is obvious that, if $B^1, B^2, \ldots, B^p$ are easily invertible operators, then the operator $B = B^1B^2\ldots B^p$, their product, is also easily invertible.

Notice that, unlike the factors, the operator B can have a complex structure. For example, suppose that $B^\alpha = E+\omega R_\alpha$, $\alpha=1,2$, where R_α is the operator corresponding to the difference operator $(-\mathcal{R}_\alpha)$: $\mathcal{R}_\alpha y = y_{\bar{x}_\alpha x_\alpha}$, $\alpha=1,2$. The operator B^α corresponds to a three-point difference operator that can be inverted using the elimination method, whose operation count is proportional to the number of unknowns in the problem. The operator $B=B^1B^2$ has a nine-point stencil and it corresponds to the difference operator $\mathcal{B}$;

$$\mathcal{B}y = y-\omega\sum_{\alpha=1}^{2} y_{\bar{x}_\alpha x_\alpha}+\omega^2 y_{\bar{x}_1x_1\bar{x}_2x_2}.$$

The complex structure of the operator B allows us to increase the ratio $\xi = \gamma_1/\gamma_2$, and that leads to an increase in the convergence rate of the iterative method.

To construct the operator B, it is possible to begin with some operator $R = R^* > 0$ (a regularizer) that is energy equivalent to A and B:

$$c_1 R \le A \le c_2 R, \quad c_2 \ge c_1 > 0, \tag{5}$$

$$\mathring{\gamma}_1 B \le R \le \mathring{\gamma}_2 B, \quad \mathring{\gamma}_2 \ge \mathring{\gamma}_1 > 0. \tag{6}$$

Then the inequalities (4) are valid with constants $\gamma_1 = c_1\mathring{\gamma}_1$, $\gamma_2 = c_2\mathring{\gamma}_2$, where

$$\xi = \gamma_1/\gamma_2 = (c_1/c_2)\mathring{\xi}, \quad \mathring{\xi} = \mathring{\gamma}_1/\mathring{\gamma}_2.$$

What is the idea behind the introduction of the regularizer R? For grid elliptic boundary-value problems, the operator R is usually chosen so that the constants c_1 and c_2 in the inequalities (5) do not depend on the parameters of the grid (on the number of nodes of the grid). For example, if the operator A corresponds to a difference operator with variable coefficients

$$\Lambda y = (a_1 y_{\bar{x}_1})_{x_1} + (a_2 y_{\bar{x}_2})_{x_2}, \quad 0 < c_1 \le a_\alpha \le c_2,$$

defined on the uniform grid $\bar{\omega} = \{x_{ij} = (ih_1, jh_2),\ 0 \le i \le N_1,\ 0 \le j \le N_2,\ h_\alpha N_\alpha = l_\alpha,\ \alpha = 1,2\}$, in the rectangle $\bar{G} = \{0 \le x_\alpha \le l_\alpha,\ \alpha = 1,2\}$, so that $Ay = -\Lambda\mathring{y}$, where $y(x) = \mathring{y}(x)$ for $x \in \omega$ and $\mathring{y}(x) = 0$ for $x \in \gamma$, then R can be taken to be the operator corresponding the Laplace difference operator $\mathcal{R}y = (\mathcal{R}_1 + \mathcal{R}_2)y = y_{\bar{x}_1x_1} + y_{\bar{x}_2x_2}$, $Ry = -\mathcal{R}\mathring{y}$, and where the operators $\mathcal{R}_\alpha$ were defined above.

Using the Green difference formulas, it is easy to show (cf. Section 5.2.8) that the operators A and B are self-adjoint in H and satisfy the inequalities (5). Here H is the space of grid functions defined on ω, with inner product defined by the formula

$$(u,v) = \sum_{x\in\omega} u(x)v(x)h_1h_2.$$

Suppose now that the operator A corresponds to an elliptic difference operator containing mixed derivatives

$$\Lambda y = \sum_{\alpha,\beta=1}^{2} 0.5\left[(k_{\alpha\beta}y_{\bar{x}_\beta})_{x_\alpha} + (k_{\alpha\beta}y_{x_\beta})_{\bar{x}_\alpha}\right],$$

and that the strong ellipticity conditions are satisfied:

$$c_1 \sum_{\alpha=1}^{2} \xi_\alpha^2 \le \sum_{\alpha,\beta=1}^{2} k_{\alpha\beta}(x)\xi_\alpha\xi_\beta \le c_2 \sum_{\alpha=1}^{2} \xi_\alpha^2, \quad c_1 > 0.$$

We take as a regularizer the operator R defined above. In Section 5.2.8 it was proved that the inequalities (5) are satisfied for these operators A and R.

We give still another example. Suppose that the operator A corresponds to a Laplace difference operator of high order

$$\Lambda y = y_{\bar{x}_1 x_1} + y_{\bar{x}_2 x_2} + \frac{h_1^2 + h_2^2}{12} y_{\bar{x}_1 x_1 \bar{x}_2 x_2}.$$

We shall show that, if the operator R is chosen as the operator indicated above, then the inequalities (5) are valid with constants $c_1 = 2/3$, $c_2 = 1$.

Using the first Green difference formula and the equality $y_{\bar{x}_1 x_1 \bar{x}_2 x_2} = y_{\bar{x}_1 \bar{x}_2 x_1 x_2}$ (valid for grid functions defined on the rectangular grid $\bar{\omega}$), we obtain

$$\begin{aligned} -(\Lambda \mathring{y}, \mathring{y}) &= (\mathring{y}_{\bar{x}_1}^2, 1)_1 + (\mathring{y}_{\bar{x}_2}^2, 1)_2 - \frac{h_1^2 + h_2^2}{12} (\mathring{y}_{\bar{x}_1 \bar{x}_2}^2, 1)_{12}, \\ -(\mathcal{R}\mathring{y}, \mathring{y}) &= (\mathring{y}_{\bar{x}_1}^2, 1)_1 + (\mathring{y}_{\bar{x}_2}^2, 1)_2 . \end{aligned} \tag{7}$$

Here we have denoted

$$(u, v)_1 = \sum_{i=1}^{N_1} \sum_{j=1}^{N_2-1} u(i,j)v(i,j)h_1 h_2,$$

$$(u, v)_2 = \sum_{i=1}^{N_1-1} \sum_{j=1}^{N_2} u(i,j)v(i,j)h_1 h_2,$$

$$(u, v)_{12} = \sum_{i=1}^{N_1} \sum_{j=1}^{N_2} u(i,j)v(i,j)h_1 h_2.$$

From (7) follows the estimate $A \le R$, i.e. $c_2 = 1$ in (5). Further, taking into account that $\mathring{y}_{\bar{x}_2}(x) = 0$ for $x_1 = 0, l_1$ and $\mathring{y}_{\bar{x}_1} = 0$ for $x_2 = 0, l_2$, from lemma 12 of Chapter 5 we obtain the estimate

$$(\mathring{y}_{\bar{x}_1 \bar{x}_2}^2, 1)_{12} \le \frac{4}{h_2^2} (\mathring{y}_{\bar{x}_1}^2, 1)_1 \tag{8}$$

and in analogous manner

$$\left(\overset{\circ}{y}{}^2_{\bar{x}_1 x_2}, 1\right)_{12} = \left(\overset{\circ}{y}{}^2_{\bar{x}_2 \bar{x}_1}, 1\right)_{12} \leq \frac{4}{h_1^2} \left(\overset{\circ}{y}{}^2_{\bar{x}_2}, 1\right)_2 . \tag{9}$$

Multiplying (8) by $h_2^2/12$, (9) by $h_1^2/12$, and combining the resulting inequalities, we will have

$$\frac{h_1^2 + h_2^2}{12} \left(\overset{\circ}{y}{}^2_{\bar{x}_1 \bar{x}_2}, 1\right)_{12} \leq \frac{1}{3} \left[\left(\overset{\circ}{y}{}^2_{\bar{x}_1}, 1\right)_1 + \left(\overset{\circ}{y}{}^2_{\bar{x}_2}, 1\right)_2\right].$$

From this and from (7) follows the inequality $A \geq 2R/3$. The assertion is proved.

These examples show that, for various operators A, it is possible to choose the same regularizer R. Therefore the problem of constructing the operator B for an implicit iterative scheme is simplified. The operator B is constructed from the condition that it is near in energy to the regularizer R. The class of regularizers is significantly narrower than the class containing the operators A. If the operator B has already been chosen and, consequently, the constants $\overset{\circ}{\gamma}_1$ and $\overset{\circ}{\gamma}_2$ in the inequalities (6) have been found, then for each specific operator A we need only find the constants c_1 and c_2 in the inequalities (5).

The basic difficulty is using a regularizer is obtaining estimates for $\overset{\circ}{\gamma}_1$ and $\overset{\circ}{\gamma}_2$. Most often, the operator B has a factored form, where the factors depend on certain iterative parameters. This gives a family of operators B with a definite structure characterized by the indicated parameters. These parameters can be chosen from the condition that $\overset{\circ}{\xi}$ be a maximum. Certain examples involving factored operators will be considered in the next Section. Here we remark that the operator B can sometimes be chosen as the regularizer R ($\overset{\circ}{\gamma}_1 = \overset{\circ}{\gamma}_2 = 1$).

14.1.2 Iterative schemes with a factored operator. In Section 14.1.1 the principle of a regularizer was illustrated on an example with a self-adjoint operator A. In this case, the inequalities (4) are consequences of the inequalities (5) and (6).

In Chapter 6 it was shown that, if the operator A is not self-adjoint in H, and the energy space H_D is generated by a self-adjoint positive-definite operator D, where D is either B or $A^*B^{-1}A$, then the inequalities (4) can be changed to the inequalities

$$\gamma_1(Bx, x) \leq (Ax, x), \quad (B^{-1}Ax, Ax) \leq \gamma_2(Ax, x), \quad \gamma_1 > 0 \tag{10}$$

or the inequalities

$$\gamma_1 B \leq A \leq \gamma_2 B, \quad (B^{-1}A_1 x, A_1 x) \leq \gamma_3^2(Bx, x), \quad \gamma_1 > 0, \tag{11}$$

where $A_1 = 0.5(A - A^*)$ is the non-self-adjoint part of the operator A. Suppose that the operator $B = B^* > 0$ has been constructed starting from a regularizer R, and that the conditions (6) are satisfied. Then, if the operator R satisfies the conditions

$$c_1(Rx, x) \le (Ax, x), \quad (R^{-1}Ax, Ax) \le c_2(Ax, x), \quad c_1 > 0, \qquad (10')$$

then the inequalities (10) are valid with constants $\gamma_1 = c_1\mathring{\gamma}_1$, $\gamma_2 = c_2\mathring{\gamma}_2$.

From lemma 9 of Chapter 5 and the inequality (6) follow the inequalities $\mathring{\gamma}_1 R^{-1} \le B^{-1} \le \mathring{\gamma}_2 R^{-1}$. From this we obtain

$$(B^{-1}Ax, Ax) \le \mathring{\gamma}_2(R^{-1}Ax, Ax) \le c_2\mathring{\gamma}_2(Ax, x).$$

Analogously we prove that if the operator R satisfies the conditions

$$c_1 R \le A \le c_2 R, \quad (R^{-1}A_1x, A_1x) \le c_3^2(Rx, x), \quad c_1 > 0, \qquad (11')$$

then the inequalities (11) are valid with constants $\gamma_1 = c_1\mathring{\gamma}_1$, $\gamma_2 = c_2\mathring{\gamma}_2$, $\gamma_3 = c_3\mathring{\gamma}_3$.

Thus, in the case of a non-self-adjoint operator A as well, it is necessary to obtain estimates for $\mathring{\gamma}_1$ and $\mathring{\gamma}_2$ in the inequalities (6).

We look now that the inequalities (6) for self-adjoint operators R and B. We consider two cases:

1) The operator R is represented in the form of a sum $R = R_1 + R_2$ of mutually conjugate operators R_1 and R_2:

$$R_2 = R_1^*, \qquad (12)$$

so that $(R_1x, x) = (R_2x, x) = 0.5(Rx, x)$, $x \in H$, and the operator B has the form

$$B = (E + \omega R_1)(E + \omega R_2), \qquad (13)$$

where $\omega > 0$ is a parameter.

2) The operator R is represented in the form of a sum $R = R_1 + R_2 + \ldots + R_p$, $p \ge 2$, of self-adjoint pairwise-commuting operators R_α, $\alpha = 1, 2, \ldots, p$, so that

$$R_\alpha = R_\alpha^*, \quad R_\alpha R_\beta = R_\beta R_\alpha, \quad \alpha, \beta = 1, 2, \ldots, p, \qquad (14)$$

and the operator B is factored and has the form

$$B = \prod_{\alpha=1}^{p}(E + \omega R_\alpha), \qquad (15)$$

where $\omega > 0$ is a parameter.

In each of these cases the operator B is self-adjoint in H. We especially emphasize the universality of choosing the operator B in the form (13), where the operators R_1 and R_2 satisfy the condition (12).

Our problem consists of finding estimates for $\mathring{\gamma}_1$ and $\mathring{\gamma}_2$ in (6) and choosing the iterative parameter ω from the condition that the ratio $\mathring{\xi} = \mathring{\gamma}_1/\mathring{\gamma}_2$ be a maximum.

Each case will be examined separately. The first case was studied in detail in Chapter 10 in connection with the alternate-triangular method. Therefore we limit ourselves here to formulating the results.

Theorem 1. *Suppose that the conditions* (12) *are satisfied and that we are given the constants* δ *and* Δ *in the inequalities*

$$R \geq \delta E, \quad (R_2 x, R_2 x) \leq \frac{\Delta}{4}(Rx, x), \quad \delta > 0 \tag{16}$$

Then, for the optimal value of the parameter $\omega = \omega_0 = 2/\sqrt{\delta\Delta}$, *the operator* B *defined by equation* (13) *satisfies the inequalities* (6) *with constants*

$$\mathring{\gamma}_1 = \frac{\delta}{2(1+\sqrt{\eta})}, \quad \mathring{\gamma}_2 = \frac{\delta}{4\sqrt{\eta}}, \quad \eta = \frac{\delta}{\Delta}. \square$$

Notice that it is possible to consider a more general form for the operator B than (13), namely:

$$B = (\mathcal{D} + \omega R_1)\mathcal{D}^{-1}(\mathcal{D} + \omega R_2),$$

where $\mathcal{D} = \mathcal{D}^* > 0$. From lemma 1 of Chapter 10 it follows that theorem 1 remains valid, and it is only necessary to change (16) to the following inequalities

$$R \geq \delta\mathcal{D}, \quad (\mathcal{D}^{-1}R_2 x, R_2 x) \leq \frac{\Delta}{4}(Rx, x).$$

Here the operator $\mathcal{D}$ plays the role of an auxiliary iterative parameter.

The operator B is easy to invert, for example, in the case where the operator R_1 corresponds to a lower-triangular matrix, R_2 to an upper-triangular matrix, and $\mathcal{D}$ to a diagonal matrix. If the operator R corresponds to an elliptic difference operator, then these triangular matrices will only have a finite number of non-zero elements in each row which does not depend on the number of nodes in the grid. Therefore the cost of inverting each factor entering into the operator B is proportional to the number of unknowns in the problem.

We look now at the second case.

Theorem 2. *Suppose that the operator B has the form* (15), *the conditions* (14) *are satisfied, and that the bounds for the operator R_α are given:*

$$\delta_\alpha E \le R_\alpha \le \Delta_\alpha E, \quad \delta_\alpha > 0, \quad \alpha = 1, 2, \ldots, p.$$

Then for the optimal value of the parameter ω

$$\omega = \omega_0 = \frac{1}{\Delta} \frac{1 - \eta^{1/p}}{\eta^{1/p} - \eta}$$

the operator B satisfies the inequalities (6) *with the constants*

$$\mathring{\gamma}_1 = \frac{p\Delta}{(1 + \omega_0 \Delta)^p}, \quad \mathring{\gamma}_2 = \mathring{\gamma}_1 \frac{p - k(1 - \eta)}{p\eta^{k/p}}, \quad \eta = \frac{\delta}{\Delta},$$

where

$$\delta = \min_\alpha \delta_\alpha, \quad \Delta = \max_\alpha \Delta_\alpha, \quad k = \left[\frac{p}{1 - \eta} - \frac{\eta^{1/p}}{1 - \eta^{1/p}}\right],$$

and $[a]$ is the integer part of the number a. □

In view of the awkwardness of the proof, it will be omitted. We will only remark that, by the conditions (14), the operator B commutes with the operators R_α, $\alpha = 1, 2, \ldots, p$, and therefore

$$\mathring{\gamma}_1 = \min_{\delta \le x_\alpha \le \Delta} \frac{x_1 + x_2 + \cdots + x_p}{\prod_{\alpha=1}^p (1 + \omega x_\alpha)}, \quad \mathring{\gamma}_2 = \max_{\delta \le x_\alpha \le \Delta} \frac{x_1 + x_2 + \cdots + x_p}{\prod_{\alpha=1}^p (1 + \omega x_\alpha)}.$$

We indicate special cases of theorem 2. If $p = 2$, then

$$k = 1, \quad \omega_0 = \frac{1}{\sqrt{\delta\Delta}}, \quad \mathring{\gamma}_1 = \frac{2\delta}{(1 + \sqrt{\eta})^2}, \quad \mathring{\gamma}_2 = \frac{\delta}{\sqrt{\eta}} \frac{1 + \eta}{(1 + \sqrt{\eta})^2}.$$

If $p = 3$, then

$$k = 2, \ \omega_0 = \frac{1}{\sqrt[3]{\delta\Delta}(\sqrt[3]{\delta} + \sqrt[3]{\Delta})}, \quad \mathring{\gamma}_1 = 3\delta \left(\frac{1 - \eta^{2/3}}{1 - \eta}\right)^3,$$

$$\mathring{\gamma}_2 = \frac{\delta(1 + 2\eta)}{\eta^{2/3}} \left(\frac{1 - \eta^{2/3}}{1 - \eta}\right)^3.$$

For the case $p = 2$, it is possible to obtain better estimates for $\mathring{\gamma}_1$ and $\mathring{\gamma}_2$ by introducing into the operator

$$B = (E + \omega_1 R_1)(E + \omega_2 R_2) \tag{17}$$

two parameters ω_1 and ω_2 which take into account that the bounds of the operators R_1 and R_2 are different. We have

Theorem 3. *Suppose that the operator B has the form (17), the conditions*

$$R_\alpha = R_\alpha^*, \quad \alpha = 1,2, \quad R_1 R_2 = R_2 R_1,$$

are satisfied, and that the bounds for the operator R_1 and R_2 are given:

$$\delta_\alpha E \le R_\alpha \le \Delta_\alpha E, \quad \alpha = 1,2, \quad \delta_1 + \delta_2 > 0.$$

Then for the optimal values of the parameters ω_1 and ω_2

$$\omega_1 = \frac{1 + t\sqrt{\eta}}{r\sqrt{\eta} + s}, \quad \omega_2 = \frac{1 - t\sqrt{\eta}}{r\sqrt{\eta} - s}$$

the inequalities (6) are satisfied with constants

$$\mathring{\gamma}_1 = \frac{4\sqrt{\eta}}{(\omega_1 + \omega_2)(1 + \sqrt{\eta})^2}, \quad \mathring{\gamma}_2 = \frac{2(1+\eta)}{(\omega_1 + \omega_2)(1 + \sqrt{\eta})^2},$$

where

$$r = \frac{\Delta_2 + \Delta_1 b}{1 + b}, \quad s = \frac{\Delta_2 - \Delta_1 b}{1 + b}, \quad t = \frac{1 - b}{1 + b}, \quad \eta = \frac{1 - a}{1 + a},$$

$$a = \sqrt{\frac{(\Delta_1 - \delta_1)(\Delta_2 - \delta_2)}{(\Delta_1 + \delta_2)(\Delta_2 + \delta_1)}}, \quad b = \frac{\Delta_2 + \delta_1}{\Delta_1 - \delta_1} a.$$

Proof. For the proof of the theorem, we make a change of variable, setting

$$R_1 = (r\bar{R}_1 - sE)(E - t\bar{R}_1)^{-1}, \quad R_2 = (r\bar{R}_2 + sE)(E + t\bar{R}_2)^{-1}$$

with the indicated values for r, s, and t. It is possible to show that the operators

$$\bar{R}_1 = (R_1 + sE)(rE + tR_1)^{-1}, \quad \bar{R}_2 = (R_2 - sE)(rE - tR_2)^{-1}$$

satisfy the conditions $\bar{R}_\alpha = \bar{R}^*_\alpha$, $\alpha = 1, 2$, $\bar{R}_1\bar{R}_2 = \bar{R}_2\bar{R}_1$ and that they have the same bounds $\eta E \le \bar{R}_\alpha \le E$, $\eta > 0$, $\alpha = 1, 2$. Further, since the operators $E - t\bar{R}_1$ and $E + t\bar{R}_2$ are self-adjoint and positive-definite, then there exist commuting operators $(E - t\bar{R}_1)^{1/2}$ and $(E + t\bar{R}_2)^{1/2}$. We set

$$x = (E - t\bar{R}_1)^{1/2}(E - t\bar{R}_2)^{1/2} y.$$

We obtain

$$(Bx, x) = (1 - \omega_1 s)(1 + \omega_2 s)(\bar{B}y, y), \tag{18}$$

$$(Rx, x) = (r - st)(\bar{R}y, y), \tag{19}$$

where $\bar{B} = (E + \bar{\omega}\bar{R}_1)(E + \bar{\omega}\bar{R}_2)$, $\bar{R} = \bar{R}_1 + \bar{R}_2$,

$$\bar{\omega} = \frac{\omega_1 r - t}{1 - \omega_1 s} = \frac{\omega_2 r + t}{1 + \omega_2 s}. \tag{20}$$

From (20) we find

$$2\bar{\omega} = \frac{\omega_1 r - t}{1 - \omega_1 s} + \frac{\omega_2 r + t}{1 + \omega_2 s} = \frac{(r - st)(\omega_1 + \omega_2)}{(1 - \omega_1 s)(1 + \omega_2 s)}.$$

From this and from (18), (19) we obtain

$$\frac{(Rx, x)}{(Bx, x)} = \frac{2\bar{\omega}}{\omega_1 + \omega_2} \frac{(\bar{R}y, y)}{(\bar{B}y, y)}. \tag{21}$$

Using theorem 2, we obtain that for

$$\bar{\omega} = \omega_0 = 1/\sqrt{\eta} \tag{22}$$

we have the inequalities

$$\mathring{\gamma}_1(\bar{R}y, y) \le (\bar{B}y, y) \le \mathring{\gamma}_2(\bar{R}y, y), \tag{23}$$

where

$$\mathring{\gamma}_1 = \frac{2\eta}{(1 + \sqrt{\eta})^2}, \quad \mathring{\gamma}_2 = \frac{\sqrt{\eta}(1 + \eta)}{(1 + \sqrt{\eta})^2}.$$

Consequently, from (20) and (22) we obtain the optimal values for the parameters ω_1 and ω_2:

$$\omega_1 = \frac{1 + t\sqrt{\eta}}{r\sqrt{\eta} + s}, \quad \omega_2 = \frac{1 - t\sqrt{\eta}}{r\sqrt{\eta} - s},$$

and from (21) and (23) follow the inequalities (6) with constants $\mathring{\gamma}_1$ and $\mathring{\gamma}_2$ indicated in the formulation of theorem 3. Theorem 3 is proved. □

14.1.3 A method for implicitly inverting the operator B (a two-stage method). In Section 14.1.2 we studied a method for constructing implicit iterative schemes where the operator B is given constructively in the form of the product of easily invertible operators. We now look at still another method in which the iterative approximation y_{k+1} is found as the result of an auxiliary procedure that can be considered as the implicit inversion of some operator B.

Recall that the general idea of such a method was examined in Section 5.3.4. In Section 13.1.4 this method was used to construct an iterative method for solving an equation with a non-linear operator A. There we formulated conditions which enabled us to obtain estimates for $\mathring{\gamma}_1$ and $\mathring{\gamma}_2$ in the inequalities (6).

We give those results here. Suppose that the iterative approximation y_{k+1} is found using a scheme with a correction: $y_{k+1} = y_k - \tau_{k+1} w^p$, and the correction w^p is an approximate solution to the auxiliary equation

$$Rw = r_k, \quad r_k = Ay_k - f. \tag{24}$$

Here R is a regularizer satisfying the inequalities (5) (for the case of a self-adjoint operator A) or the inequalities (10′) or (11′) (for a non-self-adjoint A).

Suppose that equation (24) is solved using some two-level iterative scheme, so that the error $z^m = w^m - w$ satisfies the equation

$$z^{m+1} = S_{m+1} z^m, \quad m = 0, 1, \ldots, p-1, \quad z^0 = w^0 - w,$$

where S_{m+1} is the transformation operator from the m-th to the $(m+1)$-st iteration.

Choosing $w^0 = 0$, then from the equations

$$z^p = w^p - w = T_p(w^0 - w), \quad T_p = \prod_{m=1}^{p} S_m,$$

$$w = R^{-1} r_k,$$

we will have

$$w^p = B^{-1} r_k, \quad \text{where} \quad B = R(E - T_p)^{-1}.$$

Substituting the resulting expression for w^p in (23), we obtain an implicit iterative scheme (2) with the indicated operator B.

Theorem 4. *Suppose that the conditions*

$$R = R^* > 0, \quad T_p^* R = RT_p, \quad \| T_p \|_R \leq q < 1,$$

are satisfied. Then the operator $B = R(E - T_p)^{-1}$ *is self-adjoint and positive-definite in* H *and the inequalities* (6) *are valid with constants* $\mathring{\gamma}_1 = 1 - q$, $\mathring{\gamma}_2 = 1 + q$.

Proof. For the proof see lemma 2 of Chapter 13. □

Remark. If the operators R and T_p are self-adjoint, commute, and $\| T_p \| \leq q < 1$, then the assertions of theorem 4 are valid.

By the method described above, we constructed an implicit two-level iterative scheme. If we start from the formulas

$$\begin{aligned} y_{k+1} &= \alpha_{k+1} y_k + (1 - \alpha_{k+1}) y_{k-1} - \tau_{k+1} \alpha_{k+1} w_k^p, \quad k = 1, 2, \ldots, \\ y_1 &= y_0 - \tau_1 w_0^p, \end{aligned}$$

and we find the correction w_k^p for any $k = 0, 1, \ldots$ as the approximate solution to equation (24), then we obtain an implicit three-level iterative scheme

$$\begin{aligned} By_{k+1} &= \alpha_{k+1}(B - \tau_{k+1} A) y_k - (1 - \alpha_{k+1}) B y_{k-1} + \alpha_{k+1} \tau_{k+1} f, \\ By_1 &= (B - \tau_1 A) y_0 + \tau_1 f. \end{aligned} \tag{25}$$

In conclusion we remark that the iterative parameters τ_k for the scheme (2) and the parameters τ_k, α_k for the scheme (25) are chosen using the general theory of iterative methods. Here there arises the problem of choosing the optimal iteration count p for the auxiliary iterative process. We now clarify this issue. Suppose for simplicity that the auxiliary process is stationary ($S_m \equiv S$), the operators R and S are self-adjoint and commutative, and that the condition $\| S \| \leq \rho$ is satisfied. Then $q = \rho^p$, i.e.

$$p = \ln q / \ln \rho. \tag{26}$$

The operators A and B satisfy the inequalities (4) with constants

$$\gamma_1 = c_1(1 - q), \quad \gamma_2 = c_2(1 + q).$$

If the iterative parameters τ_k in the scheme (2) are chosen using the formulas for the Chebyshev method, then we have the following estimate for the iteration count

$$n \geq n_0(\epsilon), \quad n_0(\epsilon) = \ln(0.5\epsilon) / \ln \rho_1,$$

where

$$\rho_1 = \rho_1(q) = \frac{1-\sqrt{\xi}}{1+\sqrt{\xi}}, \quad \xi = \frac{\gamma_1}{\gamma_2} = \frac{c_1}{c_2}\frac{1-q}{1+q}.$$

Then the general iteration count $k = pn$ can be estimated using

$$k \geq k_0(\epsilon), \quad k_0(\epsilon) = \frac{\ln 0.5\epsilon}{\ln \rho}\, \frac{\ln q}{\ln \rho_1(q)}.$$

From this it follows that the quantity q that determines, via (26), the number of inner iterations should be chosen in order to minimize the function $\varphi(q) = \ln q / \ln \rho_1(q)$. This problem can be solved numerically.

14.2 Examples of solving elliptic boundary-value problems

14.2.1 Direct and iterative methods. We consider methods for solving multidimensional boundary-value difference problems for elliptic equations. We begin with a Dirichlet problem for Poisson's equation in an p-dimensional rectangular parallelipiped $\bar{G} = \{0 \leq x_\alpha \leq l_\alpha,\ \alpha = 1,2,\ldots,p,\ ,p \geq 3\}$

$$Lu = \sum_{\alpha=1}^{p} \frac{\partial^2 u}{\partial x_\alpha^2} = -f(x), \quad x = (x_1, x_2, \ldots, x_p) \in G, \quad u(x) = g(x),\, x \in \Gamma. \tag{1}$$

On a uniform rectangular grid $\bar{\omega} = \{x_i = (i_1 h_1, \ldots, i_p h_p) \in \bar{G},\ 0 \leq i_\alpha \leq N_\alpha,\ h_\alpha N_\alpha = l_\alpha,\ \alpha = 1,2,\ldots,p\}$ the problem (1) corresponds to a Dirichlet difference problem

$$\Lambda y = \sum_{\alpha=1}^{p} \Lambda_\alpha y = -\varphi(x),\ x \in \omega, \qquad y(x) = g(x),\ x \in \gamma, \tag{2}$$

where $\Lambda_\alpha y = y_{\bar{x}_\alpha x_\alpha}$, $1 \leq \alpha \leq p$. To solve (2) we use the method of separation of variables. As in the case $p = 2$, it is possible to show that the eigenvalue problem for the p-dimensional Laplace difference operator on the grid $\bar{\omega}$

$$\Lambda \mu + \lambda \mu = 0,\ x \in \omega,\ \mu(x) = 0,\ x \in \gamma,$$

has the following solution:

$$\mu_k(i_1, \ldots, i_p) = \prod_{\alpha=1}^{p} \mu_{k_\alpha}^{(\alpha)}(i_\alpha),\ \mu_{k_\alpha}^{(\alpha)}(i_\alpha) = \sqrt{\frac{2}{l_\alpha}} \sin \frac{k_\alpha \pi i_\alpha}{N_\alpha},$$

$$1 \leq k_\alpha \leq N_\alpha - 1,\ 0 \leq i_\alpha \leq N_\alpha,\ 1 \leq \alpha \leq p,$$

$$\lambda_k = \sum_{\alpha=1}^{p} \frac{4}{h_\alpha^2} \sin^2 \frac{k_\alpha \pi}{2N_\alpha},\ k = (k_1, \ldots, k_p).$$

The functions $\mu_k(i_1, \ldots, i_p)$ are orthonormal in the sense of the scalar product

$$(u, v) = \sum_{i_1=1}^{N_1-1} \cdots \sum_{i_p=1}^{N_p-1} u(i_1, \ldots, i_p) v(i_1, \ldots, i_p) \prod_{\alpha=1}^{p} h_\alpha.$$

Besides, any grid function $v(i_1, \ldots, i_p)$ that reduces to zero on γ can be represented in the form of a sum

$$v(i_1, \ldots, i_p) = \sum_{k_1=1}^{N_1-1} \cdots \sum_{k_p=1}^{N_p-1} c_{k_1 \cdots k_p} \mu_k(i_1, \ldots, i_p),$$

where the Fourier coefficients $C_{k_1 \cdots k_p} = C_k$ are defined by the formula

$$c_k = \sum_{i_1=1}^{N_1-1} \cdots \sum_{i_p=1}^{N_p-1} v(i_1, \ldots, i_p) \mu_k(i_1, \ldots, i_p) \prod_{\alpha=1}^{p} h_\alpha.$$

To apply the method of separation of variables it is necessary to reduce (2) to a problem with uniform boundary conditions. To do this it is sufficient to change the right hand side $\varphi(x)$ of (2) only at points near the boundary. We will assume that this has been done and that in (2) the function $g(x)$ is equal to zero.

The solution of the problem

$$\sum_{\alpha=1}^{p} y_{\bar{x}_\alpha x_\alpha} = -\varphi(x), \; x \in \omega, \; y(x) = 0, \; x \in \gamma, \tag{3}$$

will be found in the form of a sum

$$y(i_1, \ldots, i_p) = \sum_{k_1=1}^{N_1-1} \cdots \sum_{k_p=1}^{N_p-1} c_k \mu_k(i_1, \ldots, i_p). \tag{4}$$

Expanding the right-hand side $\varphi(x)$ in an analogous sum, we obtain from (3) that

$$c_k = \frac{\varphi_k}{\lambda_k}, \; k = (k_1, \ldots, k_p), \; 1 \le k_\alpha \le N_\alpha - 1, \tag{5}$$

where $\varphi_{k_1 \cdots k_p} = \varphi_k$ are the Fourier coefficients of the function $\varphi(x)$, which are computed from

$$\varphi_k = \sum_{i_1=1}^{N_1-1} \cdots \sum_{i_p=1}^{N_p-1} \varphi(i_1, \ldots, i_p) \mu_k(i_1, \ldots, i_p) \prod_{\alpha=1}^{p} h_\alpha. \tag{6}$$

Thus, initially the Fourier coefficients $\varphi_{k_1\cdots k_p}$ are computed for $1 \le k_\alpha \le N_\alpha - 1$, $1 \le \alpha \le p$; then $c_{k_1\cdots k_p}$ are found using (5); and finally, the desired solution $y(i_1,\dots,i_p)$ is computed for $1 \le i_\alpha \le N_\alpha - 1$, $1 \le \alpha \le p$, using (4).

Taking into account the structure of the functions $\mu_k(i_1,\dots,i_p)$, it is possible to offer the following algorithm:

1) compute the sums

$$\varphi_{k_1}(i_2,\dots,i_p) = \sum_{i_1=1}^{N_1-1} \varphi(i_1,\dots,i_p)\sin\frac{k_1\pi i_1}{N_1},\quad 1 \le k_1 \le N_1 - 1,$$

for $1 \le i_\alpha \le N_\alpha - 1$, $2 \le \alpha \le p$;

2) sequentially compute the sums for $m = 2, 3, \dots, p-1$

$$\varphi_{k_1\cdots k_m}(i_{m+1},\dots,i_p) = \sum_{i_m=1}^{N_m-1} \varphi_{k_1\cdots k_{m-1}}(i_m,\dots,i_p)\sin\frac{k_m\pi i_m}{N_m},$$
$$1 \le k_m \le N_m - 1,$$

for $1 \le i_\alpha \le N_\alpha - 1$, $m+1 \le \alpha \le p$, and $1 \le k_\beta \le N_\beta - 1$, $1 \le \beta \le m-1$;

3) for $1 \le k_\beta \le N_\beta - 1$, $1 \le \beta \le p-1$, compute the sums

$$\bar{c}_{k_1\cdots k_p} = \frac{1}{\lambda_k}\prod_{\alpha=1}^{p}\frac{2}{N_\alpha}\sum_{i_p=1}^{N_p-1}\varphi_{k_1\cdots k_{p-1}}(i_p)\sin\frac{k_p\pi i_p}{N_p}$$

for $1 \le k_p \le N_p - 1$. As a result the coefficients $c_{k_1 k_2\cdots k_p}$ will be found (up to a scale factor).

4) compute the sums

$$y_{k_2\cdots k_p}(i_1) = \sum_{k_1=1}^{N_1-1} \bar{c}_{k_1\cdots k_p}\sin\frac{k_1\pi i_1}{N_1},\quad 1 \le i_1 \le N_1 - 1,$$

for $1 \le k_\beta \le N_\beta - 1$, $2 \le \beta \le p$;

5) sequentially compute the sums for $m = 2, 3, \dots, p-1$

$$y_{k_{m+1}\cdots k_p}(i_1,\dots,i_m) = \sum_{k_m=1}^{N_m-1} y_{k_m\cdots k_p}(i_1,\dots,i_{m-1})\sin\frac{k_m\pi i_m}{N_m}$$

for $1 \le k_\beta \le N_\beta - 1$, $m+1 \le \beta \le p$, $1 \le i_\alpha \le N_\alpha - 1$, $1 \le \alpha \le m-1$, and $1 \le i_m \le N_m - 1$;

6) compute the sums

$$y(i_1,\ldots,i_p)=\sum_{k_p=1}^{N_p-1} y_{k_p}(i_1,\ldots,i_{p-1})\sin\frac{k_p\pi i_p}{N_p},\ 1\le i_p\le N_p-1,$$

for $1\le i_\alpha\le N_\alpha-1$, $1\le\alpha\le p-1$.

We now estimate the number of arithmetic operations required for the algorithm desribed above. If $N_\alpha=2^{n_\alpha}$, $n_\alpha\ge 1$, then these single sums can be computed using the fast Fourier transform, which for sums of the form

$$y_i=\sum_{k=1}^{N-1} a_k\sin\frac{k\pi i}{N},\ 1\le i\le N-1,\ N=2^n,$$

requires (cf. Section 4.1.2)

$$q=(2N-1)\log_2 N-3N+3$$

arithmetic operations. From this it follows that the total number of operations required to compute all the sums indicated in steps 1)–6) is estimated by

$$Q=2\sum_{m=1}^{p}\left(2\log_2 N_m-3-\frac{\log N_m-3}{N_m}\right)N_1N_2\cdots N_p$$
$$\le 4\prod_{\alpha=1}^{p}N_\alpha\log_2\left(\prod_{\alpha=1}^{p}N_\alpha\right).$$

If $N_1=N_2=\cdots N_p=N$, then $Q\approx 4pN^p\log_2 N$. In addition, it is necessary to include another N^p multiplications which are required in step 3).

Note that for $p=2$ the algorithm constructed here coincides with the algorithm described in Section 4.2.2.

Suppose now that we need to solve an elliptic equation with variable coefficients in the case of a Dirichlet problem

$$Lu=\sum_{\alpha=1}^{p}\frac{\partial}{\partial x_\alpha}\left(k_\alpha(x)\frac{\partial u}{\partial x_\alpha}\right)=-f(x),\ x\in G, \tag{7}$$
$$u(x)=g(x),\ x\in\Gamma.$$

On the uniform rectangular grid $\bar{\omega}$, introduced above, we will have the Dirichlet difference problem (2), where

$$\Lambda_\alpha y = (a_\alpha(x) y_{\bar{x}_\alpha})_{x_\alpha}, \; a_\alpha(x) = k_\alpha(x)|_{x_\alpha=(i_\alpha-0.5)h_\alpha}, \; 1 \le \alpha \le p.$$

We will use an iterative method to get an approximate solution of (2). As usual, we reduce (2) to the operator equation

$$Au = f \tag{8}$$

in the Hilbert space H of grid functions defined on ω, with the scalar product defined by

$$(u, v) = \sum_{x\in\omega} u(x)v(x) \prod_{\alpha=1}^{p} h_\alpha.$$

If we assume that the coefficients $k_\alpha(x)$ satisfy

$$0 < c_1 \le k_\alpha(x) \le c_2, \; x \in \bar{G}, \; 1 \le \alpha \le p, \tag{9}$$

then using Green's difference formula, it is possible to show that

$$c_1 R \le A \le c_2 R, \; c_1 > 0, \tag{10}$$

where the operator R corresponds to the p-dimensional Laplace difference operator $(-\mathcal{R})$:

$$\mathcal{R}y = \sum_{\alpha=1}^{p} y_{\bar{x}_\alpha x_\alpha}.$$

In addition, the operators A and R are self-adjoint in H. To solve (8) we use an implicit iterative scheme

$$B\frac{y_{k+1} - y_k}{\tau_{k+1}} + Ay_k = f, \; k = 0, 1, \ldots, \; y_0 \in H, \tag{11}$$

with the Chebyshev set of parameters $\{\tau_k\}$ and with operator $B = R$. With this choice of B the iterative method (11) converges in H_D where D is either A, or B, or $AB^{-1}A$, and with an upper bound $n \ge n_0(\epsilon)$ on the number of iterations, where

$$n_0(\epsilon) = \frac{\ln 0.5\epsilon}{\ln \rho_1}, \; \rho_1 = \frac{1 - \sqrt{\xi}}{1 + \sqrt{\xi}}, \; \xi = \frac{c_1}{c_2}.$$

To find y_{k+1} it is necessary to solve a Dirichlet problem for Poisson's equation. If this is done using the direct, separation-of-variables method described above, then the number of arithmetic operations required to solve (2) is estimated by

$$Q = 2pN^p \sqrt{\frac{c_2}{c_1}} \ln \frac{2}{\epsilon} \log_2 N$$

for the square grid ($N_1 = N_2 = \cdots = N_p = N$).

Analogously it follows that if it is necessary to solve a Dirichlet problem for an elliptic equation containing mixed derivatives

$$Lu = \sum_{\alpha,\beta=1}^{p} \frac{\partial}{\partial x_\alpha} \left(k_{\alpha\beta}(x) \frac{\partial u}{\partial x_\beta} \right) = -f(x),$$

whose coefficients satisfy a strong ellipticity condition

$$c_1 \sum_{\alpha=1}^{p} \xi_\alpha^2 \leq \sum_{\alpha,\beta=1}^{p} k_{\alpha\beta}(x)\xi_\alpha\xi_\beta \leq c_2 \sum_{\alpha=1}^{p} \xi_\alpha^2, \; c_1 > 0.$$

We consider now examples of elliptic boundary value problems, approximately solved using the iterative scheme (11) with factored operator B.

Suppose we must solve (7) on an arbitrary non-uniform grid $\hat{\bar{\omega}}$ introduced into the parallelipiped $\bar{G}$. The corresponding difference scheme (2) with operator

$$\Lambda_\alpha y = (a_\alpha(x) y_{\bar{x}_\alpha})_{\hat{x}_\alpha}, \; 1 \leq \alpha \leq p,$$

is written in the form of an operator equation (8), and the scalar product in H is given by

$$(u,v) = \sum_{i_1=1}^{N_1-1} \cdots \sum_{i_p=1}^{N_p-1} u(i_1,\ldots,i_p) v(i_1,\ldots,i_p) \hbar_1(i_1) \cdots \hbar_p(i_p).$$

We choose as a regularizer the operator $R = R_1 + R_2 + \cdots + R_p$, corresponding to the Laplace difference operator $(-\mathcal{R})$ on the nonuniform grid $\hat{\bar{\omega}}$

$$\mathcal{R} = \mathcal{R}_1 + \mathcal{R}_2 + \cdots + \mathcal{R}_p, \; \mathcal{R}_\alpha y = y_{\bar{x}_\alpha \hat{x}_\alpha}, \; 1 \leq \alpha \leq p.$$

Applying Green's difference formula, we convince ourselves that the operators R_α are self-adjoint in H. In addition they commute pairwise. From remark I to lemma 16 in Chapter 5 it follows that the operator inequality

$$R_\alpha \geq \delta_\alpha E,\ \delta_\alpha = 8/l_\alpha^2,\ 1 \leq \alpha \leq p$$

holds, giving a lower bound for the operator R_α. An estimate for an upper bound on R_α

$$R_\alpha \leq \Delta_\alpha E, \Delta_\alpha = \max_{1 \leq i_\alpha \leq N_\alpha - 1} \frac{4}{h_\alpha(i_\alpha)h_\alpha(i_\alpha + 1)},\ 1 \leq \alpha \leq p,$$

follows from lemma 17 of Chapter 5.

If (9) is satisfied then (10) is true. To solve (8) we use the iterative scheme (11), where B is the factored operator $B = \prod_{\alpha=1}^p (E + \omega R_\alpha)$.

From theorem 2 of Chapter 14 we obtain the optimal value for the iterative parameter ω, and also the constants γ_1° and γ_2° in the inequalities $\gamma_1^\circ B \leq R \leq \gamma_2^\circ B$.

From this and from (10) we find that $\gamma_1 = c_1\gamma_1^\circ$ and $\gamma_2 = c_2\gamma_2^\circ$ are the energy equivalence constants for the operators A and B. Using them it is possible to construct for the scheme (11) the Chebyshev set of iterative parameters $\{\tau_k\}$.

14.2.2 A high-accuracy Dirichlet difference problem in the multi-dimensional case. In this Section we consider a method for solving this problem for Poisson's equation in a p-dimensional rectangular parallelipiped. Recall that we constructed direct methods for $p = 2$ in Section 3.3.2 and in Section 4.3.3, and an alternating-direction iterative method in Section 11.2.3.

Suppose now that $p \geq 3$. In the p-dimensional parallelipiped $\bar{G} = \{0 \leq x_\alpha \leq l_\alpha,\ \alpha = 1, 2, \ldots, p\}$ on the grid $\bar{\omega} = \{x_i = (i_1, h_1, \ldots, i_p h_p) \in \bar{G},\ 0 \leq i_\alpha \leq N_\alpha,\ h_\alpha N_\alpha = l_\alpha,\ 1 \leq \alpha \leq p\}$ we pose, corresponding to (1), the following fourth-order accuracy difference scheme:

$$\Lambda y = \left(\sum_{\alpha=1}^p \Lambda_\alpha + \sum_{\alpha=1}^p \kappa_\alpha \sum_{\beta \neq \alpha}^{1/p} \Lambda_\alpha \Lambda_\beta \right) y = -\varphi(x),\ x \in \omega,\ y(x) = g(x), x \in \gamma, \tag{12}$$

where

$$\Lambda_\alpha y = y_{\bar{x}_\alpha x_\alpha},\ 1 \leq \alpha \leq p,\ \varphi = f + \sum_{\alpha=1}^p \kappa_\alpha \Lambda_\alpha f,\ \kappa_\alpha = \frac{h_\alpha^2}{12}.$$

We will solve (12) using separation of variables. As usual, we assume that $g(x) = 0$, otherwise we change the right-hand side $\varphi(x)$ of (12) at the near-boundary nodes and thereby take into account non-homogeneous boundary conditions.

Note that the functions $\mu_k(i_1, \ldots, i_p)$ indicated in Section 14.2.1 are eigenfunctions of the difference operator Λ defined in (12), and the numbers

$$\lambda_k = \sum_{\alpha=1}^{p} \lambda_{k_\alpha}^{(\alpha)} - \sum_{\alpha=1}^{p} \kappa_\alpha \sum_{\beta\neq\alpha}^{1/p} \lambda_{k_\alpha}^{(\alpha)} \lambda_{k_\beta}^{(\beta)}, \; k = (k_1, \ldots, k_p), \tag{13}$$

where

$$\lambda_{k_\alpha}^{(\alpha)} = \frac{4}{h_\alpha^2} \sin^2 \frac{k_\alpha \pi}{2N_\alpha}, \; 1 \leq k_\alpha \leq N_\alpha - 1, \; 1 \leq \alpha \leq p,$$

are the eigenvalues of this operator.

Expanding the solution $y(x)$ and the right-hand side $\varphi(x)$ as sums in $\mu_k(i_1, \ldots, i_p)$ (as is usually done in the method of separation of variables), we obtain

$$y(i_1, \ldots, i_p) = \sum_{k_1=1}^{N_1-1} \cdots \sum_{k_p=1}^{N_p-1} \frac{\varphi_{k_1\cdots k_p}}{\lambda_k} \mu_k(i_1, \ldots, i_p),$$

where $\varphi_{k_1\cdots k_p}$ are the Fourier coefficients of the function $\varphi(x)$ defined in (6).

An algorithm for computing similar sums was described in detail in Section 14.2.1. In comparison with the method for solving the second-order accuracy Dirichlet difference problem constructed in Section 14.2.1, the only difference is the value of the eigenvalues λ_k. Thus, these methods are practically equivalent in computational costs. At the same time the difference scheme (12) will guarantee a more accurate solution to the differential problem (1) than the scheme (2).

Note that because of the inequalities

$$\lambda_{k_\alpha}^{(\alpha)} < \frac{4}{h_\alpha^2}, \; 1 \leq k_\alpha \leq N_\alpha - 1, \; 1 \leq \alpha \leq p,$$

from (13) we obtain

$$\begin{aligned} \lambda_k &\geq \sum_{\alpha=1}^{p} \lambda_{k_\alpha}^{(\alpha)} - \tfrac{1}{3} \sum_{\alpha=1}^{p} \sum_{\beta\neq\alpha}^{1/p} \lambda_{k_\beta}^{(\beta)} \\ &= \sum_{\alpha=1}^{p} \lambda_{k_\alpha}^{(\alpha)} - \tfrac{1}{3} \sum_{\alpha=1}^{p} \left(\sum_{\beta=1}^{p} \lambda_{k_\beta}^{(\beta)} - \lambda_{k_\alpha}^{(\alpha)} \right) = \frac{4-p}{3} \sum_{\alpha=1}^{p} \lambda_{k_\alpha}^{(\alpha)}. \end{aligned}$$

For $p \geq 4$ the difference operator Λ defining the difference scheme (12) loses the property of sign-fixedness (ellipticity).

Without disturbing the order of the approximation, it is possible to choose another difference operator Λ' which preserves the property of ellipticity for any $p \geq 2$:

$$\Lambda' y = \sum_{\alpha=1}^{p} \prod_{\beta \neq \alpha}^{1/p} (E + \kappa_\beta \Lambda_\beta) \Lambda_\alpha y.$$

The difference operator Λ' has the same eigenfunctions $\mu_k(i_1, \ldots, i_p)$ as the operator Λ constructed above, and the numbers

$$\lambda'_k = \sum_{\alpha=1}^{p} \prod_{\beta \neq \alpha}^{1/p} (1 - \kappa_\beta \lambda_{k_\beta}^{(\beta)}) \lambda_{k_\alpha}^{(\alpha)}, \ k = (k_1, \ldots, k_p)$$

are its eigenvalues. Consequently, to solve the high-accuracy Dirichlet difference problem with the operator Λ', it is possible to use the algorithm constructed in Section 14.2.1, changing λ_k to λ'_k.

14.2.3 A boundary-value problem of the third kind for an equation with mixed derivatives in a rectangle. Suppose that, in the rectangle $\bar{G} = \{0 \leq x_\alpha \leq l_\alpha, \alpha = 1, 2\}$, it is necessary to solve

$$Lu = \sum_{\alpha,\beta=1}^{2} \frac{\partial}{\partial x_\alpha} \left(k_{\alpha\beta}(x) \frac{\partial u}{\partial x_\beta} \right) - q(x) u = -\bar{f}(x), \ x = (x_1, x_2) \in G, \quad (14)$$

satisfying on the boundary Γ the boundary conditions

$$\begin{gathered} k_{\alpha\alpha} \frac{\partial u}{\partial x_\alpha} + k_{\alpha\beta} \frac{\partial u}{\partial x_\beta} = \kappa_\alpha^- u - g_\alpha^-, \ x_\alpha = 0, \\ -k_{\alpha\alpha} \frac{\partial u}{\partial x_\alpha} - k_{\alpha\beta} \frac{\partial u}{\partial x_\beta} = \kappa_\alpha^+ u - g_\alpha^+, \ x_\alpha = l_\alpha, \\ \beta = 3 - \alpha, \ \alpha = 1, 2. \end{gathered} \quad (15)$$

We will assume that the strong ellipticity condition is satisfied

$$c_1 \sum_{\alpha=1}^{2} \xi_\alpha^2 \leq \sum_{\alpha,\beta=1}^{2} k_{\alpha\beta}(x) \xi_\alpha \xi_\beta \leq c_2 \sum_{\alpha=1}^{2} \xi_\alpha^2, \ c_1 > 0, \quad (16)$$

and also the conditions

$$\kappa_1^{\pm} = \kappa_1^{\pm}(x_2) \geq 0,\ 0 \leq x_2 \leq l_2,\ \kappa_2^{\pm} = \kappa_2^{\pm}(x_1) \geq 0,\ 0 \leq x_1 \leq l_1,\ q(x) \geq 0. \tag{17}$$

In the rectangle $\bar{G}$ we introduce the arbitrary non-uniform rectangular grid

$$\begin{aligned}\bar{\omega} = \{\, & x_i = (x_1(i_1), x_2(i_2)) \in \bar{G}, \\ & x_\alpha(i_\alpha) = x_\alpha(i_\alpha - 1) + h_\alpha(i_\alpha), \\ & 1 \leq i_\alpha \leq N_\alpha,\ x_\alpha(0) = 0,\ x_\alpha(N_\alpha) = l_\alpha,\ \alpha = 1, 2\},\end{aligned}$$

and define the average step $\hbar_\alpha(i_\alpha)$:

$$\hbar_\alpha(i_\alpha) = \begin{cases} 0.5h_\alpha(1), & i_\alpha = 0, \\ 0.5[h_\alpha(i_\alpha) + h_\alpha(i_\alpha + 1)], & 1 \leq i_\alpha \leq N_\alpha - 1, \\ 0.5h_\alpha(N_\alpha), & i_\alpha = N_\alpha. \end{cases}$$

We construct a difference scheme, approximating the problem (14), (15) on $\bar{\omega}$. We introduce the following notation for the flux

$$w_\alpha = k_{\alpha\alpha}\frac{\partial u}{\partial x_\alpha} + k_{\alpha\beta}\frac{\partial u}{\partial x_\beta},\ \beta = 3 - \alpha,\ \alpha = 1, 2,$$

and write the problem (14), (15) in the form

$$\begin{aligned} Lu &= \sum_{\alpha=1}^{2} \frac{\partial w_\alpha}{\partial x_\alpha} - qu = -\bar{f}(x),\ x \in G, \\ w_\alpha &= \kappa_\alpha^- u - g_\alpha^-,\ x_\alpha = 0, \\ -w_\alpha &= \kappa_\alpha^+ u - g_\alpha^+,\ x_\alpha = l_\alpha. \end{aligned} \tag{18}$$

On the grid $\bar{\omega}$ we define the following difference analog for the flux

$$\begin{aligned} w_\alpha^{(-)} &= k_{\alpha\alpha} y_{\bar{x}_\alpha} + k_{\alpha\beta} y_{\bar{x}_\beta},\ w_\alpha^{(+)} = k_{\alpha\alpha} y_{\bar{x}_\alpha} + k_{\alpha\beta} y_{x_\beta}, \\ z_\alpha^{(-)} &= k_{\alpha\alpha} y_{x_\alpha} + k_{\alpha\beta} y_{x_\beta},\ z_\alpha^{(+)} = k_{\alpha\alpha} y_{x_\alpha} + k_{\alpha\beta} y_{\bar{x}_\beta}. \end{aligned} \tag{19}$$

Using this notation, we pose for (5) the corresponding difference scheme

$$\Lambda y = (\Lambda_1 + \Lambda_2)y - dy = -\varphi(x),\ x \in \bar{\omega}, \tag{20}$$

where $\Lambda_\alpha = \Lambda_\alpha^-$, $\alpha = 1, 2$ is the difference operator defined by:

a) for $h_\beta(1) \le x_\beta \le l_\beta - h_\beta(N_\beta)$

$$\Lambda_\alpha^- y = \begin{cases} \frac{1}{2}[\frac{h_\beta}{\hbar_\beta}(w_\alpha^{(-)})_{\hat{x}_\alpha} + \frac{h_\beta^{+1}}{\hbar_\beta}(z_\alpha^{(-)})_{\hat{x}_\alpha}], & h_\alpha(1) \le x_\alpha \le l_\alpha - h_\alpha(N_\alpha), \\ \frac{1}{2\hbar_\alpha}[\frac{h_\beta}{\hbar_\beta}w_\alpha^{(-)+1} + \frac{h_\beta^{+1}}{\hbar_\beta}z_\alpha^{(-)}] - \frac{\kappa_\alpha^-}{\hbar_\alpha}y, & x_\alpha = 0, \\ -\frac{1}{2\hbar_\alpha}[\frac{h_\beta}{\hbar_\beta}w_\alpha^{(-)} + \frac{h_\beta^{+1}}{\hbar_\beta}z_\alpha^{(-)-1}] - \frac{\kappa_\alpha^+}{\hbar_\alpha}y, & x_\alpha = l_\alpha; \end{cases}$$

b) for $x_\beta = 0$

$$\Lambda_\alpha^- y = \begin{cases} (z_\alpha^{(-)})_{\hat{x}_\alpha}, & h_\alpha(1) \le x_\alpha \le l_\alpha - h_\alpha(N_\alpha), \\ \frac{1}{\hbar_\alpha}[z_\alpha^{(-)} - \kappa_\alpha^- y], & x_\alpha = 0, \\ -\frac{1}{\hbar_\alpha}[z_\alpha^{(-)-1} + \kappa_\alpha^+ y], & x_\alpha = l_\alpha; \end{cases}$$

c) for $x_\beta = l_\beta$

$$\Lambda_\alpha^- y = \begin{cases} (w_\alpha^{(-)})_{\hat{x}_\alpha}, & h_\alpha(1) \le x_\alpha \le l_\alpha - h_\alpha(N_\alpha), \\ \frac{1}{\hbar_\alpha}[w_\alpha^{(-)+1} - \kappa_\alpha^- y], & x_\alpha = 0, \\ -\frac{1}{\hbar_\alpha}[w_\alpha^{(-)} + \kappa_\alpha^+ y], & x_\alpha = l_\alpha. \end{cases}$$

Here we have used the notation

$$(w_\alpha^{(-)})_{\hat{x}_\alpha} = \frac{w_\alpha^{(-)+1} - w_\alpha^{(-)}}{\hbar_\alpha},$$

$$(z_\alpha^{(-)})_{\hat{x}_\alpha} = \frac{z_\alpha^{(-)} - z_\alpha^{(-)-1}}{\hbar_\alpha},$$

$$d(x) = q(x),$$

$$h_\beta^{+1} = h_\beta(i_\beta + 1),$$

$$w_\alpha^{(-)\pm 1} = w_\alpha^{(-)}(i_\alpha \pm 1, i_\beta),$$

$$z_\alpha^{(-)\pm 1} = z_\alpha^{(-)}(i_\alpha \pm 1, i_\beta).$$

The right-hand side $\varphi(x)$ of the difference scheme (20) is defined by

$$\varphi(x) = \bar{f}(x) + \frac{1}{\hbar_1}\varphi_1(x) + \frac{1}{\hbar_2}\varphi_2(x),$$

where

$$\varphi_\alpha(x) = \begin{cases} g_\alpha^-(x_\beta), & x_\alpha = 0, \\ 0, & h_\alpha(1) \le x_\alpha \le l_\alpha - h_\alpha(N_\alpha), \\ g_\alpha^+(x_\beta), & x_\alpha = l_\alpha,\ \beta = 3-\alpha,\ \alpha = 1,2, \end{cases}$$

for $0 \le x_\beta \le l_\beta$.

In place of the operator Λ_α in the difference scheme (20), it is also possible to use the operator Λ_α^+, which is defined by:

a) for $h_\beta(1) \le x_\beta \le l_\beta - h_\beta(N_\beta)$

$$\Lambda_\alpha^+ y = \begin{cases} \frac{1}{2}[\frac{h_\beta^{+1}}{\hbar_\beta}(w_\alpha^{(+)})_{\hat{x}_\alpha} + \frac{h_\beta}{\hbar_\beta}(z_\alpha^{(+)})_{\hat{x}_\alpha}], & h_\alpha(1) \le x_\alpha \le l_\alpha - h_\alpha(N_\alpha), \\ \frac{1}{2\hbar_\alpha}[\frac{h_\beta^{+1}}{\hbar_\beta} w_\alpha^{(+)+1} + \frac{h_\beta}{\hbar_\beta} z_\alpha^{(+)}] - \frac{\kappa_\alpha^-}{\hbar_\alpha} y, & x_\alpha = 0, \\ -\frac{1}{2\hbar_\alpha}[\frac{h_\beta^{+1}}{\hbar_\beta} w_\alpha^{(+)} + \frac{h_\beta}{\hbar_\beta} z_\alpha^{(+)-1}] - \frac{\kappa_\alpha^+}{\hbar_\alpha} y, & x_\alpha = l_\alpha; \end{cases}$$

b) for $x_\beta = 0$

$$\Lambda_\alpha^+ y = \begin{cases} (w_\alpha^{(+)})_{\hat{x}_\alpha}, & h_\alpha(1) \le x_\alpha \le l_\alpha - h_\alpha(N_\alpha), \\ \frac{1}{\hbar_\alpha}[w_\alpha^{(+)+1} - \kappa_\alpha^- y], & x_\alpha = 0, \\ -\frac{1}{\hbar_\alpha}[w_\alpha^{(+)} + \kappa_\alpha^+ y], & x_\alpha = l_\alpha; \end{cases}$$

c) for $x_\beta = l_\beta$

$$\Lambda_\alpha^+ y = \begin{cases} (z_\alpha^{(+)})_{\hat{x}_\alpha}, & h_\alpha(1) \le x_\alpha \le l_\alpha - h_\alpha(N_\alpha), \\ \frac{1}{\hbar_\alpha}[z_\alpha^{(+)} - \kappa_\alpha^- y], & x_\alpha = 0, \\ -\frac{1}{\hbar_\alpha}[z_\alpha^{(+)-1} + \kappa_\alpha^+ y], & x_\alpha = l_\alpha. \end{cases}$$

For the difference schemes considered above, the difference operator Λ has a stencil consisting of seven points for interior nodes of the grid $\bar{\omega}$. If the difference operator Λ_α is defined by $\Lambda_\alpha = \Lambda_\alpha^0 = 0.5(\Lambda_\alpha^- + \Lambda_\alpha^+)$, then the stencil of the operator Λ will consist of nine points. We give here the form of the operator Λ_α^0:

a) for $h_\beta(1) \le x_\beta \le l_\beta - h_\beta(N_\beta)$

$$\Lambda_\alpha^0 y =$$
$$\begin{cases} (a_{\alpha\alpha} y_{\bar{x}_\alpha})_{\hat{x}_\alpha} + (k_{\alpha\beta} y_{\mathring{x}_\beta})_{\mathring{x}_\alpha}, \\ \qquad h_\alpha(1) \le x_\alpha \le l_\alpha - h_\alpha(N_\alpha), \\ \dfrac{1}{\hbar_\alpha}(a_{\alpha\alpha}^{+1} y_{x_\alpha} + k_{\alpha\beta} y_{\mathring{x}_\beta} - \kappa_\alpha^- y) + (k_{\alpha\beta} y_{\mathring{x}_\beta})_{x_\alpha}, \\ \qquad x_\alpha = 0, \\ -\dfrac{1}{\hbar_\alpha}(a_{\alpha\alpha} y_{\bar{x}_\alpha} + k_{\alpha\beta} y_{\mathring{x}_\beta} + \kappa_\alpha^+ y) + (k_{\alpha\beta} y_{\mathring{x}_\beta})_{\bar{x}_\alpha}, \\ \qquad x_\alpha = l_\alpha; \end{cases}$$

b) for $x_\beta = 0$

$$\Lambda_\alpha^0 y =$$
$$\begin{cases} (a_{\alpha\alpha} y_{\bar{x}_\alpha})_{\hat{x}_\alpha} + (k_{\alpha\beta} y_{x_\beta})_{\mathring{x}_\alpha}, \\ \qquad h_\alpha(1) \le x_\alpha \le l_\alpha - h_\alpha(N_\alpha), \\ \dfrac{1}{\hbar_\alpha}(a_{\alpha\alpha}^{+1} y_{x_\alpha} + k_{\alpha\beta} y_{x_\beta} - \kappa_\alpha^- y) + (k_{\alpha\beta} y_{x_\beta})_{x_\alpha}, \\ \qquad x_\alpha = 0, \\ -\dfrac{1}{\hbar_\alpha}(a_{\alpha\alpha} y_{\bar{x}_\alpha} + k_{\alpha\beta} y_{x_\beta} + \kappa_\alpha^+ y) + (k_{\alpha\beta} y_{x_\beta})_{\bar{x}_\alpha}, \\ \qquad x_\alpha = l_\alpha; \end{cases}$$

c) for $x_\beta = l_\beta$

$$\Lambda_\alpha^0 y =$$
$$\begin{cases} (a_{\alpha\alpha} y_{\bar{x}_\alpha})_{\hat{x}_\alpha} + (k_{\alpha\beta} y_{\bar{x}_\beta})_{\mathring{x}_\alpha}, \\ \qquad h_\alpha(1) \le x_\alpha \le l_\alpha - h_\alpha(N_\alpha), \\ \dfrac{1}{\hbar_\alpha}(a_{\alpha\alpha}^{+1} y_{x_\alpha} + k_{\alpha\beta} y_{\bar{x}_\beta} - k_\beta^- y) + (k_{\alpha\beta} y_{\bar{x}_\beta})_{x_\alpha}, \\ \qquad x_\alpha = 0, \\ -\dfrac{1}{\hbar_\alpha}(a_{\alpha\alpha} y_{\bar{x}_\alpha} + k_{\alpha\beta} y_{\bar{x}_\beta} + \kappa_\alpha^+ y) + (k_{\alpha\beta} y_{\bar{x}_\beta})_{\bar{x}_\alpha}, \\ \qquad x_\alpha = l_\alpha; \end{cases}$$

where

$$a_{\alpha\alpha} = 0.5(k_{\alpha\alpha} + k_{\alpha\alpha}^{-1}),\ y_{\mathring{x}_\beta} = 0.5(y_{\hat{x}_\beta} + y_{\bar{x}_\beta}).$$

Remark. The flux $w_\alpha = k_{\alpha\alpha}\frac{\partial u}{\partial x_\alpha} + k_{\alpha\beta}\frac{\partial u}{\partial x_\beta}$ can be approximated in other ways, namely using one of the difference analogs:

$$w_\alpha^{(-)} = k_{\alpha\alpha}^{-0.5_\alpha-0.5_\beta} y_{\bar{x}_\alpha} + k_{\alpha\beta}^{-0.5_\alpha-0.5_\beta} y_{\bar{x}_\beta},$$

$$w_\alpha^{(+)} = k_{\alpha\alpha}^{-0.5_\alpha+0.5_\beta} y_{\bar{x}_\alpha} + k_{\alpha\beta}^{-0.5_\alpha+0.5_\beta} y_{x_\beta},$$

$$z_\alpha^{(-)} = k_{\alpha\alpha}^{+0.5_\alpha+0.5_\beta} y_{x_\alpha} + k_{\alpha\beta}^{+0.5_\alpha+0.5_\beta} y_{x_\beta},$$

$$z_\alpha^{(+)} = k_{\alpha\alpha}^{+0.5_\alpha-0.5_\beta} y_{x_\alpha} + k_{\alpha\beta}^{+0.5_\alpha-0.5_\beta} y_{\bar{x}_\beta},$$

where

$$k_{11}^{+0.5_1+0.5_2} = k_{11}(x_1(i_1) + 0.5h_1(i_1+1), x_2(i_2) + 0.5h_2(i_2+1)),$$

$$k_{11}^{-0.5_1-0.5_2} = k_{11}(x_1(i_1) - 0.5h_1(i_1), x_2(i_2) - 0.5h_2(i_2)),$$

and so forth. The definitions of the operators Λ_α^-, Λ_α^+, and Λ_α° are preserved.

We note that if (1) does not contain mixed derivatives ($k_{\alpha\beta}(x) \equiv 0$), then the difference schemes (20) constructed with the aid of the difference operators Λ_α^-, Λ_α^+, and Λ_α° will differ only in the definition of the coefficients of the difference equation.

14.2.4 Iterative methods for solving a difference problem. The boundary value difference problem (20) constructed above will be written in the form of an operator equation

$$Au = f \tag{21}$$

in the Hilbert space H of grid functions defined on $\bar{\omega}$, defining the operator A by: $Ay = -\Lambda y + dy$, $y \in H$, and letting $f = \varphi$.

We also define the operator R, acting in H as the sum of three operators, $R = R_1^- + R_2^- + aE$, where $a \geq 0$ is a constant, and R_1^- and R_2^- are given by:

$$R_\alpha^- y = \begin{cases} -\dfrac{1}{\hbar_\alpha} y_{x_\alpha}, & i_\alpha = 0, \\ -y_{\hat{x}_\alpha \bar{x}_\alpha}, & 1 \leq i_\alpha \leq N_\alpha - 1, \\ \dfrac{1}{\hbar_\alpha} y_{\bar{x}_\alpha}, & i_\alpha = N_\alpha, \end{cases} \qquad \alpha = 1, 2.$$

We will study the properties of the operators A and R in the space H, where the scalar product is defined in the usual way:

$$(u,v) = \sum_{i_1=0}^{N_1} \sum_{i_2=0}^{N_2} u(i_1,i_2)v(i_1,i_2)\hbar_1(i_1)\hbar_2(i_2), \quad u,v \in H.$$

Below we will use the following notation:

$$(u,v)_\gamma = \sum_{x\in\gamma} u(x)v(x)\tau(x),$$

where

$$\tau(x) = \tau(x_i) = \begin{cases} \hbar_\alpha(i_\alpha), & 1 \le i_\alpha \le N_\alpha - 1,\ i_\beta = 0, N_\beta,\ \beta \ne \alpha, \\ \hbar_\alpha(i_\alpha) + \hbar_\beta(i_\beta), & i_\alpha = 0, N_\alpha,\ i_\beta = 0, N_\beta, \end{cases}$$

$$_\alpha(u,v)_\beta = \sum_{i_\alpha=0}^{N_\alpha-1} \sum_{i_\beta=1}^{N_\beta} uvh_\alpha(i_\alpha+1)h_\beta(i_\beta),$$

$$(u,v)_{\alpha\beta} = \sum_{i_\alpha=1}^{N_\alpha} \sum_{i_\beta=1}^{N_\beta} uvh_\alpha(i_\alpha)h_\beta(i_\beta),$$

$$_{\alpha\beta}(u,v) = \sum_{i_\alpha=0}^{N_\alpha-1} \sum_{i_\beta=0}^{N_\beta-1} uvh_\alpha(i_\alpha+1)h_\beta(i_\beta+1),$$

$$(u,v)_\alpha = \sum_{i_\alpha=1}^{N_\alpha} \sum_{i_\beta=0}^{N_\beta} uvh_\alpha(i_\alpha)\hbar_\beta(i_\beta),$$

where $\beta = 3 - \alpha$, $\alpha = 1, 2$.

Using the definition of the difference operator Λ_α^-, the formulas (19), and Green's first difference formula, we obtain

$$\begin{aligned}(A_\alpha^- u,v) = &-\tfrac{1}{2}\sum_{i_\beta=1}^{N_\beta} h_\beta(i_\beta)\left[\sum_{i_\alpha=1}^{N_\alpha-1} \hbar_\alpha(i_\alpha)(w_\alpha^{(-)})_{\hat{x}_\alpha} v + w_\alpha^{(-)+1} v|_{i_\alpha=0} - w_\alpha^{(-)} v|_{i_\alpha=N_\alpha}\right] \\ &-\tfrac{1}{2}\sum_{i_\beta=0}^{N_\beta-1} h_\beta(i_\beta+1)\left[\sum_{i_\alpha=1}^{N_\alpha-1} \hbar_\alpha(i_\alpha)(z_\alpha^{(-)})_{\hat{x}_\alpha} v\right. \\ &\left. + z_\alpha^{(-)} v|_{i_\alpha=0} - z_\alpha^{(-)-1} v|_{i_\alpha=N_\alpha}\right] \\ &+\sum_{i_\beta=0}^{N_\beta} \hbar_\beta(i_\beta)\left[\kappa_\alpha^- uv|_{i_\alpha=0} + \kappa_\alpha^+ uv|_{i_\alpha=N_\alpha}\right]\end{aligned}$$

$$
\begin{aligned}
&=\tfrac{1}{2}\sum_{i_\beta=1}^{N_\beta}\sum_{i_\alpha=1}^{N_\alpha} h_\alpha(i_\alpha)h_\beta(i_\beta)w_\alpha^{(-)}v_{\bar{x}_\alpha}\\
&\quad+\tfrac{1}{2}\sum_{i_\beta=0}^{N_\beta-1}\sum_{i_\alpha=0}^{N_\alpha-1} h_\alpha(i_\alpha+1)h_\alpha(i_\beta+1)z_\alpha^{(-)}v_{x_\alpha}\\
&\quad+\sum_{i_\beta=0}^{N_\beta}\hbar_\beta(i_\beta)\left[\kappa_\alpha^- uv|_{i_\alpha=0}+\kappa_\alpha^+ uv|_{i_\alpha=N_\alpha}\right]\\
&=\tfrac{1}{2}(k_{\alpha\alpha}u_{\bar{x}_\alpha}+k_{\alpha\beta}u_{\bar{x}_\beta},v_{\bar{x}_\alpha})_{\alpha\beta}+\tfrac{1}{2}{}_{\alpha\beta}(k_{\alpha\alpha}u_{x_\alpha}+k_{\alpha\beta}u_{x_\beta},v_{x_\alpha})\\
&\quad+\sum_{i_\beta=0}^{N_\beta}\hbar_\beta(i_\beta)\left[\kappa_\alpha^- uv|_{i_\alpha=0}+\kappa_\alpha^+ uv|_{i_\alpha=N_\alpha}\right],\\
&\beta=3-\alpha,\alpha=1,2.
\end{aligned}
$$

From this we obtain for the operator A

$$
\begin{aligned}
(Au,v)&=\sum_{\alpha=1}^{2}(A_\alpha^- u,v)+(du,v)\\
&=0.5(k_{11}u_{\bar{x}_1}v_{\bar{x}_1}+k_{12}u_{\bar{x}_2}v_{\bar{x}_1}+k_{21}u_{\bar{x}_1}v_{\bar{x}_2}+k_{22}u_{\bar{x}_2}v_{\bar{x}_2},1)_{12}\\
&\quad+0.5_{12}(k_{11}u_{x_1}v_{x_1}+k_{12}u_{x_2}v_{x_1}+k_{21}u_{x_1}v_{x_2}+k_{22}u_{x_2}v_{x_2},1)\\
&\quad+(\kappa u,v)_\gamma+(du,v)\\
&=0.5\Big(\sum_{\alpha,\beta=1}^{2}k_{\alpha\beta}u_{\bar{x}_\beta}v_{\bar{x}_\alpha},1\Big)_{12}+0.5\Big(\sum_{\alpha,\beta=1}^{2}k_{\alpha\beta}u_{x_\beta}v_{x_\alpha},1\Big)\\
&\quad+(\kappa u,v)_\gamma+(du,v).
\end{aligned}
\tag{22}
$$

Here the grid function κ is defined on γ as follows:

$$
\kappa(x_i)=\begin{cases}
\kappa_\alpha^-, & i_\alpha=0,\ 1\le i_\beta\le N_\beta-1,\\
\kappa_\alpha^+, & i_\alpha=N_\alpha,\ 1\le i_\beta\le N_\beta-1,\\
\dfrac{\kappa_\alpha^\pm\hbar_\beta+k_\beta^\pm\hbar_\alpha}{\hbar_\beta+\hbar_\alpha}, & i_\alpha=0,N_\alpha \text{ and } i_\beta=0,N_\beta.
\end{cases}
$$

From (22) it follows that the operator A is self-adjoint in H, if

$$k_{12}=k_{21} \tag{23}$$

is satisfied. Setting $v = u$ in (22) and using (16), we have that

$$\begin{aligned}(Au, v) &\geq 0.5c_1 \sum_{\alpha=1}^{2}[(u_{\bar{x}_\alpha}^2, 1)_{12} + {}_{12}(u_{x_\alpha}^2, 1)] + (du^2, 1) + (\kappa u^2, 1)_\gamma \\ &= c_1 \sum_{\alpha=1}^{2}(u_{\bar{x}_\alpha}^2, 1)_\alpha + (du^2, 1) + (\kappa u^2, 1)_\gamma,\end{aligned} \tag{24}$$

and analogously

$$(Au, u) \leq c_2 \sum_{\alpha=1}^{2}(u_{\bar{x}_\alpha}^2, 1)_\alpha + (du^2, 1) + (\kappa u^2, 1)_\gamma. \tag{25}$$

If we use Λ_α^+ in place of Λ_α in (20), then for the corresponding operator A, instead of (22) we will have

$$\begin{aligned}(Au, v) = 0.5_2(&k_{11}u_{\bar{x}_1}v_{\bar{x}_1} + k_{12}u_{x_2}v_{\bar{x}_1} + k_{21}u_{\bar{x}_1}v_{x_2} + k_{22}u_{x_2}v_{x_2}, 1)_1 \\ &+ 0.5_1(k_{11}u_{x_1}v_{x_1} + k_{12}u_{\bar{x}_2}v_{x_1} + k_{21}u_{x_1}v_{\bar{x}_2} + k_{22}u_{\bar{x}_2}v_{\bar{x}_2}, 1)_2 \\ &+ (\kappa u, v)_\gamma + (du, v),\end{aligned}$$

from which the self-adjointness of A follows, if (23) is satisfied. The inequalities (23), (25) are valid. These assertions also remain true in the case when we use Λ_α° instead of Λ_α in (23).

The self-adjointness of the operator R in H, and the equality

$$(Ru, u) = \sum_{\alpha=1}^{2}(u_{\bar{x}_\alpha}^2, 1)_\alpha + a(u^2, 1), \tag{26}$$

for any $u \in H$ follow from the Green difference formulas.

Our problem is to find the constants m_1 and m_2 in

$$m_1(Ru, u) \leq (Au, u) \leq m_2(Ru, u), \tag{27}$$

i.e. the energy equivalence constants for the operators A and R. We solve this problem assuming that the conditions (17) are satisfied, and that the coefficients $q(x)$ and $\kappa_\alpha^\pm(x_\beta)$ are not identically equal to zero.

To find m_1 and m_2 we need two lemmas.*

---

* See lemmas 9 and 10 in §5, chapter V in the book by A.A. Samarskii and V.B. Andreev, *Difference methods for elliptic equations*, Science Press (Moscow, 1976).

Lemma 1. *Let ω^* be some set of nodes of the grid ω where*

$$\sum_{x\in\omega^*} \hbar_1(x_1)\hbar_2(x_2) = s > 0.$$

For any grid function $u(x)$ defined on $\bar{\omega}$, we have

$$\nu \sum_{\alpha=1}^{2} (u^2_{\bar{x}_\alpha}, 1)_\alpha + \mu(u^2, 1)_{\omega^*} \geq \mu M(u^2, 1), \tag{28}$$

where μ is an arbitrary constant, ν is a constant satisfying

$$\nu > \max[0, -0.5\mu(l_1^2 + l_2^2)],$$

and

$$M = \frac{2s\nu}{l_1 l_2[2\nu + \mu(l_1^2 + l_2^2)]} > 0.$$

Proof. We give here the basic steps in the proof of the lemma, referring the reader to the reference for details. Let $x = (x_1, x_2)$ and $y = (y_1, y_2)$ be two arbitrary nodes of the grid $\bar{\omega}$. For any grid functon u defined on $\bar{\omega}$, and for $\epsilon_1 > 0$ and $\epsilon_2 > 0$ we have

$$(1-\epsilon_1)u^2(x) \leq \left(\frac{1}{\epsilon_1} - 1\right) u^2(y) + (1+\epsilon_2)|x_1 - y_1| \sum_{x_1\in\omega_1^+} u^2_{\bar{x}_1}(x_1, x_2)h_1(x_1)$$
$$+ \left(1 + \frac{1}{\epsilon_2}\right) |x_2 - y_2| \sum_{x_2\in\omega_2^+} u^2_{\bar{x}_2}(y_1, x_2)h_2(x_2).$$

We multiply this inequality by $\hbar_1(x_1)\hbar_2(x_2)$, sum over $x \in \bar{\omega}$ and take into account that

$$\sum_{x_\alpha\in\bar{\omega}_\alpha} |x_\alpha - y_\alpha|\hbar_\alpha(x_\alpha) \leq 0.5l_\alpha^2, \quad \alpha = 1, 2.$$

As a result we obtain

$$(1-\epsilon_1)(u^2, 1) \leq l_1 l_2 \left(\frac{1}{\epsilon_1} - 1\right) u^2(y) + 0.5l_1^2(1+\epsilon_2)(u^2_{\bar{x}_1}, 1)_1$$
$$+ 0.5l_1 l_2 \left(1 + \frac{1}{\epsilon_2}\right) \sum_{x_2\in\omega_2^+} u^2_{\bar{x}_2}(y_1, x_2)h_2(x_2).$$

We multiply this inequality by $\hbar_1(y_1)\hbar_2(y_2)$ and sum over $y \in \omega^*$. Taking into account that $s \leq l_1 l_2$, we obtain

$$s(1-\epsilon_1)(u^2,1) \leq l_1 l_2 \left(\frac{1}{\epsilon_1} - 1\right)(u^2,1)_{\omega^*} + 0.5 l_1^3 l_2 (1+\epsilon_2)(u_{\bar{x}_1}^2, 1)_1 + 0.5 l_1 l_2^3 \left(1 + \frac{1}{\epsilon_2}\right)(u_{\bar{x}_2}^2, 1)_2. \tag{29}$$

Setting here

$$\epsilon_1 = \frac{2\nu}{2\nu + \mu(l_1^2 + l_2^2)}, \quad \epsilon_2 = \left(\frac{l_2}{l_1}\right)^2,$$

we obtain the estimate (28). Lemma 1 is proved. □

Lemma 2. *Let γ^* be some set of nodes of the boundary γ, where*

$$\sum_{x\in\gamma^*} \tau(x) \geq \sigma.$$

For any grid function $u(x)$ defined on $\bar{\omega}$, we have

$$\nu \sum_{\alpha=1}^{2} (u_{\bar{x}_\alpha}^2, 1)_\alpha + \mu(u^2,1)_{\gamma^*} \geq \mu M(u^2,1), \tag{30}$$

where μ is an arbitrary constant, ν is a constant satisfying $\nu > \max[0, -\mu(l_1^2 + l_2^2)(1/l_1 + 1/l_2)]$, and

$$M = \frac{\sigma\nu}{\nu l_1 l_2 + \mu(l_1^2 + l_2^2)(l_1 + l_2)} > 0.$$

Proof. In the proof of lemma 2, the basic role is played by an analog to (29)

$$\sigma(1-\epsilon_1)(u^2,1) \leq l_1 l_2 \left(\frac{1}{\epsilon_1} - 1\right)(u^2,1)_{\gamma^*} + l_1^2(l_1+l_2)(1+\epsilon_1)(u_{\bar{x}_1}^2,1)_1 + l_2^2(l_1+l_2)\left(1+\frac{1}{\epsilon_2}\right)(u_{\bar{x}_2}^2,1)_2.$$

Setting here

$$\epsilon_1 = \frac{\nu l_1 l_2}{\nu l_1 l_2 + \mu(l_1^2 + l_2^2)(l_1 + l_2)}, \quad \epsilon_2 = \left(\frac{l_2}{l_1}\right)^2,$$

we obtain the bound (30). □

We now use lemmas 1 and 2 for the bounds m_1 and m_2 in (27). We denote by ω^* the nodes of the grid $\bar{\omega}$ at which $d(x) \neq 0$, and by γ^* the nodes of the boundary γ at which $\kappa(x) \neq 0$; we set

$$d_* = \min_{x\in\omega^*} d(x), \quad d^* = \max_{x\in\omega^*} d(x),$$

$$\kappa_* = \min_{x\in\gamma^*} \kappa(x), \quad \kappa^* = \max_{x\in\gamma^*} \kappa(x).$$

Setting $\mu = d_*$, $\nu = \nu_1 > 0$ in (28), and $\mu = \kappa_*$, $\nu = \nu_2 > 0$ in (30), from (24) we obtain

$$\begin{aligned}(Au, u) &\geq c_1 \sum_{\alpha=1}^{2} (u_{\bar{x}_\alpha}^2, 1)_\alpha + (du^2, 1)_{\omega^*} + (\kappa u^2, 1)_{\gamma^*} \\ &\geq (c_1 - \nu_1 - \nu_2) \sum_{\alpha=1}^{2} (u_{\bar{x}_\alpha}^2, 1)_\alpha + \nu_1 \sum_{\alpha=1}^{2} (u_{\bar{x}_\alpha}^2, 1)_\alpha \\ &\quad + d_*(u^2, 1)_{\omega^*} + \nu_2 \sum_{\alpha=1}^{2} (u_{\bar{x}_\alpha}^2, 1)_\alpha + \kappa_*(u^2, 1)_{\gamma^*} \\ &\geq (c_1 - \nu_1 - \nu_2) \sum_{\alpha=1}^{2} (u_{\bar{x}_\alpha}^2, 1)_\alpha + M_1(u^2, 1),\end{aligned} \tag{31}$$

where

$$M_1 = \frac{2s\nu_1 d_*}{l_1 l_2[2\nu_1 + d_*(l_1^2 + l_2^2)]} + \frac{\sigma\nu_2\kappa_*}{\nu_2 l_1 l_2 + \kappa_*(l_1^2 + l_2^2)(l_1 + l_2)}.$$

Further, setting $\mu = -d^*$, $\nu = \nu_3 > 0.5d^*(l_1^2 + l_2^2)$ in (28), and $\mu = -\kappa^*$, $\nu = \nu_4 > \kappa^*(l_1^2 + l_2^2)(1/l_1 + 1/l_2)$ in (30), from (25) we obtain the bound

$$\begin{aligned}(Au, u) &\leq c_2 \sum_{\alpha=1}^{2} (u_{\bar{x}_\alpha}^2, 1)_\alpha + (du^2, 1)_{\omega^*} + (\kappa u^2, 1)_{\gamma^*} \\ &\leq (c_2 + \nu_3 + \nu_4) \sum_{\alpha=1}^{2} (u_{\bar{x}_\alpha}^2, 1)_\alpha - \left[\nu_3 \sum_{\alpha=1}^{2} (u_{\bar{x}_\alpha}^2, 1)_\alpha - d^*(u^2, 1)_{\omega^*}\right] \\ &\quad - \left[\nu_4 \sum_{\alpha=1}^{2} (u_{\bar{x}_\alpha}^2, 1)_\alpha - \kappa^*(u^2, 1)_{\gamma^*}\right] \\ &\leq (c_2 + \nu_3 + \nu_4) \sum_{\alpha=1}^{2} (u_{\bar{x}_\alpha}^2, 1)_\alpha + M_2(u^2, 1),\end{aligned} \tag{32}$$

where

$$M_2 = \frac{2s\nu_3 d^*}{l_1 l_2[2\nu_3 - d^*(l_1^2 + l_2^2)]} + \frac{\sigma\nu_4\kappa^*}{\nu_4 l_1 l_2 - \kappa^*(l_1^2 + l_2^2)(l_1 + l_2)}.$$

Let a be the constant in the definition of the operator R. We choose the parameters ν_2 and ν_4 so that

$$\frac{M_1}{c_1 - \nu_1 - \nu_2} = a, \quad \frac{M_2}{c_2 + \nu_3 + \nu_4} = a, \tag{33}$$

where M_1 and M_2 are defined above. From (33) we obtain that $\nu_2 = \nu_2(a, \nu_1)$ and $\nu_4 = \nu_4(a, \nu_3)$. We note that the region of admissible values of ν_1 and ν_3 is determined by the bounds

$$\nu_1 > 0,\ \nu_2 > 0,\ c_1 - \nu_1 - \nu_2 > 0,\ \nu_3 > 0.5d^*(l_1^2 + l_2^2),$$
$$\nu_4 > \kappa^*(l_1^2 + l_2^2)(1/l_1 + 1/l_2).$$

We choose the values of ν_1 and ν_3 in these regions so as to maximize $c_1 - \nu_1 - \nu_2(a, \nu_1)$ over ν_1, and minimize $c_2 + \nu_3 + \nu_4(a, \nu_3)$ over ν_3. We denote

$$\begin{aligned} m_1 &= m_1(a) = \max_{\nu_1}[c_1 - \nu_1 - \nu_2(a, \nu_1)], \\ m_2 &= m_2(a) = \min_{\nu_3}[c_2 + \nu_3 + \nu_4(a, \nu_3)]. \end{aligned} \tag{34}$$

From (31), (32) and (26) we obtain that (27) is valid for the values of m_1 and m_2 indicated in (34). It remains to choose the optimal value for the constant a so as to maximize the ratio $m_1(a)/m_2(a)$.

We note that m_1 and m_2 do not depend on the steps of the grid $\bar{\omega}$, if s and σ are bounded below by quantities not depending on the grid $\bar{\omega}$.

We use the chosen operator R as a regularizer to construct iterative schemes for solving (21). We limit ourselves to considering implicit two-level schemes

$$B\frac{y_{k+1} - y_k}{\tau_{k+1}} + Ay_k = f,\ k = 0, 1, \ldots,\ y_0 \in H, \tag{35}$$

with a self-adjoint operator B, and the Chebyshev set of parameters $\{\tau_k\}$. We recall that for the scheme (35) it is necessary to be given the constants γ_1 and γ_2 in

$$\gamma_1 B \le A \le \gamma_2 B,\ \gamma_1 > 0, \tag{36}$$

and a number of iterations n bounded by $n(\epsilon)$, where

$$n(\epsilon) = \frac{\ln 0.5\epsilon}{\ln \rho_1},\ \rho_1 = \frac{1 - \sqrt{\xi}}{1 + \sqrt{\xi}},\ \xi = \frac{\gamma_1}{\gamma_2}.$$

If the grid $\bar{\omega}$ is uniform in one direction, we set $B = R$. Then in (36) $\gamma_1 = m_1$ and $\gamma_2 = m_2$, and the equation $Ry_{k+1} = (R - \tau_{k+1}A)y_k + \tau_{k+1}f$ with known right-hand side can be solved by either the cyclic reduction method (cf. Chapter 3), or the separation of variables method (cf. Chapter 4) at a cost of $O(N^2 \log_2 N)$ arithmetic operations ($N_1 = N_2 = N = 2^n$). Since m_1 and m_2 do not depend on the number of nodes of the grid $\bar{\omega}$, then the total number of operations expended for the approximate solution of (21) with tolerance ϵ is estimated by $Q(\epsilon) = O(N^2 \log_2 N |\ln 0.5\epsilon|)$.

If the grid $\bar{\omega}$ is non-uniform in each direction, then we set $B = (E + \omega R_1)(E + \omega R_2)$, where $R_\alpha = R_\alpha^- + 0.5aE$, and R_α^- is defined above.

It is easy to see that the operators R_1 and R_2 are self-adjoint in H and commutative. Besides, in view of the non-negativity of the operator we have

$$R_\alpha \geq \delta_\alpha E, \ \alpha = 1, 2, \ \delta_1 = \delta_2 = 0.5a.$$

From lemma 17 of Chapter 5 we find the upper bounds

$$R_\alpha \leq \Delta_\alpha E,$$

$$\Delta_\alpha = 0.5a + \max\left[\frac{4}{h_\alpha(1)}, \frac{4}{h_\alpha(N_\alpha)}, \max_{1 \leq i_\alpha \leq N_\alpha - 1} \frac{4}{h_\alpha(i_\alpha)h_\alpha(i_\alpha + 1)}\right].$$

We are using theorem 3 of Chapter 14, and we obtain optimal values for the parameters ω_1, ω_2, and also the energy equivalence constants $\mathring{\gamma}_1$ and $\mathring{\gamma}_2$ for the operators B and R:

$$\mathring{\gamma}_1 B \leq R \leq \mathring{\gamma}_2 B, \ \mathring{\gamma}_1 > 0.$$

From this and (27) we find that in (36) $\gamma_1 = m_1 \mathring{\gamma}_1$ and $\gamma_2 = m_2 \mathring{\gamma}_2$.

It is possible to show for this method that $\gamma_1 = O(1)$ and $\gamma_2 = O(1/h)$, where h is the minimal step of the grid $\bar{\omega}$. Consequently, the number of iterations will be bounded by $O(h^{-\frac{1}{2}} |\ln 0.5\epsilon|)$. The number of arithmetic operations expended on one iteration is proportional to the number of unknowns in the problem, since the inversion of the operator $E + w_\alpha R_\alpha$, $\alpha = 1, 2$ can be done using the elimination method.

Thus, using the regularizer principle, we construct two iterative methods which can be used to find approximate solutions to the difference analog of a third-order boundary value problem for an equation with mixed derivatives in a rectangle.

14.3 Systems of elliptic equations

14.3.1 A Dirichlet problem for systems of elliptic equations in a p-dimensional parallelipiped. Let $u = (u^1(x), u^2(x), \dots, u^{m_0}(x))^T$ and $f = (f^1(x), f^2(x), \dots, f^{m_0}(x))^T$ be vectors of dimension m_0, $x = (x_1, x_2, \dots, x_p)$ be a point in p-dimensional space, and $k = (k_{\alpha\beta})$ be a block matrix of dimension $p \times p$ such that the block $k_{\alpha\beta} = (k_{\alpha\beta}^{sm}(x))$ is a matrix of size $m_0 \times m_0$:

$$k = \begin{Vmatrix} k_{11} & k_{12} & \dots & k_{1p} \\ k_{21} & k_{22} & \dots & k_{2p} \\ \dots & \dots & \dots & \dots \\ k_{p1} & k_{p2} & \dots & k_{pp} \end{Vmatrix}, \quad k_{\alpha\beta} = \begin{Vmatrix} k_{\alpha\beta}^{11} & k_{\alpha\beta}^{12} & \dots & k_{\alpha\beta}^{1m_0} \\ k_{\alpha\beta}^{21} & k_{\alpha\beta}^{22} & \dots & k_{\alpha\beta}^{2m_0} \\ \dots & \dots & \dots & \dots \\ k_{\alpha\beta}^{m_01} & k_{\alpha\beta}^{m_02} & \dots & k_{\alpha\beta}^{m_0m_0} \end{Vmatrix}.$$

In the p-dimensional parallelipiped $\bar{G} = \{0 \le x_\alpha \le l_\alpha,\ \alpha = 1, 2, \dots, p\}$ with boundary Γ, we consider a Dirichlet problem for a system of elliptic equations:

$$\begin{aligned} Lu &= \sum_{\alpha,\beta=1}^{p} \frac{\partial}{\partial x_\alpha}\left(k_{\alpha\beta}\frac{\partial u}{\partial x_\beta}\right) = -f(x), \quad x \in G, \\ u(x) &= g(x), \quad x \in \Gamma. \end{aligned} \tag{1}$$

If we transform from vector to scalar notation, then the problem (1) can be written in the form of the system

$$(Lu)^s = -f^s(x), \quad x \in G,$$
$$u^s(x) = g^s(x), \quad x \in \Gamma, \quad s = 1, 2, \dots, m_0,$$

where

$$(Lu)^s = \sum_{\alpha,\beta=1}^{p} \sum_{m=1}^{m_0} \frac{\partial}{\partial x_\alpha}\left(k_{\alpha\beta}^{sm}(x)\frac{\partial u^m}{\partial x_\beta}\right). \tag{2}$$

We will assume that the strong ellipticity condition is satisfied

$$c_1 \sum_{\alpha=1}^{p} |\xi_\alpha|^2 \le \sum_{\alpha,\beta=1}^{p} (k_{\alpha\beta}\xi_\alpha, \xi_\beta) \le c_2 \sum_{\alpha=1}^{p} |\xi_\alpha|^2, \tag{3}$$

where $c_1 > 0$, $c_2 > 0$ are constants not depending on x, $\xi_\alpha = (\xi_\alpha^1, \xi_\alpha^2, \dots, \xi_\alpha^{m_0})^T$, $\alpha = 1, 2, \dots, p$ are arbitrary vectors,

$$|\xi_\alpha|^2 = \sum_{s=1}^{m_0} (\xi_\alpha^s)^2, \quad (k_{\alpha\beta}\xi_\alpha, \xi_\beta) = \sum_{s,m=1}^{m_0} k_{\alpha\beta}^{sm}\xi_\alpha^s\xi_\beta^m.$$

Notice that the left-hand inequality in (3) demonstrates the positive-definiteness of the matrices k.

We shall construct a difference scheme approximating problem (1). To do this, we introduce into the region $\bar{G}$ a rectangular grid

$$\bar{\omega} = \{x_i = (i, h_1, \dots, i_p h_p) \in \bar{G}, \dots 0 \le i_\alpha \le N_\alpha, \\ h_\alpha N_\alpha = l_\alpha, \quad \alpha = 1, 2, \dots, p\}$$

with boundary γ, such that $\bar{\omega} = \omega \cup \gamma$. On the grid $\bar{\omega}$ we will examine vector grid functions whose components are grid functions of p discrete variables, for example $y = (y^1, y^2, \dots, y^{m_0})^T$ where $y^s = y^s(x_i) = y^s(i_1, i_2, \dots, i_p)$.

In vector notation, the Dirichlet difference problem for the system (1) on the grid $\bar{\omega}$ has the form

$$\Lambda^- y = \sum_{\alpha,\beta=1}^{p} 0.5 \left[(k_{\alpha\beta} y_{\bar{x}_\beta})_{x_\alpha} + (k_{\alpha\beta} y_{x_\beta})_{\bar{x}_\alpha} \right] = -\varphi(x), \quad x \in \omega,$$

$$y(x) = g(x), \quad x \in \gamma.$$

Transforming to scalar notation, we obtain the system

$$\begin{aligned} (\Lambda^- y)^s &= -\varphi^s(x), \quad x \in \omega, \\ y^s(x) &= g^s(x), \quad x \in \gamma, \quad s = 1, 2, \dots, m_0, \end{aligned} \tag{4}$$

where

$$(\Lambda^- y)^s = \sum_{\alpha,\beta=1}^{p} \sum_{m=1}^{m_0} 0.5 \left[\left(k_{\alpha\beta}^{sm} y_{\bar{x}_\beta}^m \right)_{x_\alpha} + \left(k_{\alpha\beta}^{sm} y_{x_\beta}^m \right)_{\bar{x}_\alpha} \right].$$

The operator Λ^-, as in the case of scalar elliptic equations, can be written in a different notation, namely:

$$\Lambda^- y = \sum_{\alpha=1}^{p} 0.5 \left[(k_{\alpha\alpha} y_{\bar{x}_\alpha})_{x_\alpha} + (k_{\alpha\alpha} y_{x_\alpha})_{\bar{x}_\alpha} \right] \\ + \sum_{\substack{\alpha,\beta=1 \\ \alpha \ne \beta}}^{p} 0.5 \left[\left(k_{\alpha\beta}^{sm} y_{\bar{x}_\beta}^m \right)_{x_\alpha} + \left(k_{\alpha\beta}^{sm} y_{x_\beta}^m \right)_{\bar{x}_\alpha} \right].$$

Notice that we can also approximate the differential operator L using the difference operators different from Λ^-, for example

$$\Lambda^+ y = \sum_{\alpha=1}^{p} 0.5\left[(k_{\alpha\alpha} y_{\bar{x}_\alpha})_{x_\alpha} + (k_{\alpha\alpha} y_{x_\alpha})_{\bar{x}_\alpha}\right] + \sum_{\substack{\alpha,\beta=1\\ \alpha\neq\beta}}^{p} 0.5\left[\left(k_{\alpha\beta} y_{x_\beta}\right)_{x_\alpha} + \left(k_{\alpha\beta} y_{\bar{x}_\beta}\right)_{\bar{x}_\alpha}\right].$$

or

$$\Lambda^0 y = 0.5(\Lambda^- + \Lambda^+) y = \sum_{\alpha=1}^{p} 0.5\left[(k_{\alpha\alpha} y_{\bar{x}_\alpha})_{x_\alpha} + (k_{\alpha\alpha} y_{x_\alpha})_{\bar{x}_\alpha}\right] + \sum_{\substack{\alpha,\beta=1\\ \alpha\neq\beta}}^{p} (k_{\alpha\beta} y_{\mathring{x}_\beta})_{\mathring{x}_\alpha}.$$

We introduce the space H of vector grid functions defined on ω, and define in it the inner product

$$(u,v) = \sum_{s=1}^{m_0} (u^s, v^s), \quad (u^s, v^s) = \sum_{x\in\omega} u^s(x) v^s(x) h_1 h_1 \dots h_p,$$
$$u = (u^1, u^2, \dots, u^{m_0})^T, \quad v = (v^1, v^2, \dots, v^{m_0})^T, \quad u, v \in H.$$

We define the Laplace difference operator:

$$\mathcal{R} y = \sum_{\alpha=1}^{p} y_{\bar{x}_\alpha x_\alpha}, \quad (\mathcal{R} y)^s = \sum_{\alpha=1}^{p} y^s_{\bar{x}_\alpha x_\alpha}.$$

In the space H, we define the operators A and R in the usual way:

$$Ay = -\Lambda^- \mathring{y}, \quad Ry = -\mathcal{R}\mathring{y}, \quad y \in H,$$

where $\mathring{y}(x) = y(x)$ for $x \in \omega$ and $\mathring{y}(x) = 0$ if $x \in \gamma$.

Using this notation and redefining the right-hand side φ of equation (4) at the near-boundary nodes in the usual way, we write the difference problem (4) in the form of an operator equation

$$Au = f, \tag{5}$$

defined in the Hilbert space H.

Using the Green difference formula for scalar grid functions, the conditions (3), and also assuming that the symmetry conditions are satisfied

$$k_{\alpha\beta}^{sm} = k_{\beta\alpha}^{ms}, \quad \alpha, \beta = 1, 2, \ldots, p, \quad s, m = 1, 2, \ldots, m_0, \tag{6}$$

we obtain that the operators R and A are self-adjoint in H and are energy equivalent with constants c_1 and c_2, i.e. we have the operator inequalities

$$c_1 R \leq A \leq c_2 R, \quad c_1 > 0. \tag{7}$$

To find an approximate solution to equation (5), we can use an implicit two-level iterative method with the Chebyshev parameters

$$B\frac{y_{k+1} - y_k}{\tau_{k+1}} + Ay_k = f, \quad k = 0, 1, \ldots, \quad y_0 \in H, \tag{8}$$

where

$$\tau_k = \frac{\tau_0}{1 + \rho_0 \mu_k}, \quad \mu_k \in \mathcal{M}_n = \left\{ -\cos\frac{(2i-1)\pi}{2n}, \ 1 \leq i \leq n \right\},$$

$$k = 1, 2, \ldots, n,$$

$$\tau_0 = \frac{2}{\gamma_1 + \gamma_2}, \quad \rho = \frac{1 - \xi}{1 + \xi}, \quad \rho_1 = \frac{1 - \sqrt{\xi}}{1 + \sqrt{\xi}}, \quad \xi = \frac{\gamma_1}{\gamma_2},$$

$$n \geq n_0(\epsilon) = \ln(0.5\epsilon)/\ln \rho_1,$$

and γ_1 and γ_2 are the constants of energy equivalence for the self-adjoint operators A and B:

$$\gamma_1 B \leq A \leq \gamma_2 B, \quad \gamma_1 > 0, \quad A = A^*, \quad B = B^*. \tag{9}$$

If the operator B is taken to be the above-defined operator R, then from (7) we obtain that $\gamma_1 = c_1$ and $\gamma_2 = c_2$ in the inequalities (9). Conseqeuntly, in this case the iteration count for the method (8) does not depend on the number of nodes of the grid: $n = O(\ln(2/\epsilon))$.

From the definition of the operators A and B it follows that, to find y_{k+1} from the preceding approximation y_k, it is necessary to solve the following difference problem:

$$\mathcal{R}y_{k+1} = -F_k, \quad x \in \omega, \quad F_k = \tau_{k+1}(\Lambda^- y_k + \varphi) - \mathcal{R}y_k,$$

$$y_{k+1} = g, \quad x \in \gamma.$$

In scalar form, this problem can be written in the form of the system

$$\sum_{\alpha=1}^{p} \left(y_{k+1}^{s}\right)_{\bar{x}_\alpha x_\alpha} = -F_k^s(x), \quad x \in \omega,$$
$$y_{k+1}^{s}(x) = g^s(x), \quad x \in \gamma, \quad s = 1, 2, \ldots, m_0. \tag{10}$$

Since each equation of the system (10) can be solved independently of the other equations, finding the approximation y_{k+1} leads to the solution of m_0 Dirichlet difference problems in a p-dimensional parallelipiped on the uniform rectangular grid $\bar{\omega}$.

If we solve the p-dimensional Dirichlet difference problem for Poisson's equation using the method of separation of variables in combination with the algorithm for the fast Fourier transform, then this requires $q \approx 4pN^p \log_2 N$ arithmetic operations ($N_1 = N_2 = \ldots = N_p = N = 2^{n_0}$). Consequently, to solve the system (10) requires $Q_{m_0} = m_0 q$ operations, and in total, finding the solution to the difference problem (4) to an accuracy ϵ requires $Q = nQ_{m_0} = nm_0q = O(m_0 pN^p \ln(2/\epsilon) \log_2 N)$ arithmetic operations.

We look now at the alternate-triangular iterative method. This iterative scheme has the form (8), where B is the factored operator $B = (E+\omega R_1)(E+\omega R_2)$, $R_1 = R_2^*$, $R_1 + R_2 = R$. The operators R_1 and R_2 are defined using the difference operators $\mathcal{R}_1$ and $\mathcal{R}_2$ as follows: $R_\alpha y = -\mathcal{R}_\alpha \mathring{y}$, $\alpha = 1, 2$, $y(x) = \mathring{y}(x)$ for $x \in \omega$ and $\mathring{y}(x) = 0$ for $x \in \gamma$, where

$$\mathcal{R}_1 y = -\sum_{\alpha=1}^{p} \frac{1}{h_\alpha} y_{\bar{x}_\alpha}, \quad \mathcal{R}_2 y = \sum_{\alpha=1}^{p} \frac{1}{h_\alpha} y_{x_\alpha}.$$

As in the scalar case, it can be proved that the inequalities $R \geq \delta E$, $R_1 R_2 \leq (\Delta/4)R$ are satisfied, where

$$\delta = \sum_{\alpha=1}^{p} \frac{4}{h_\alpha^2} \sin^2 \frac{\pi h_\alpha}{2l_\alpha}, \quad \Delta = \sum_{\alpha=1}^{p} \frac{4}{h_\alpha^2}.$$

From the general theory of the alternate-triangular method (cf. Section 10.1) it follows that, for the optimal value of the parameter $\omega = \omega_0 = 2/\sqrt{\delta\Delta}$, we have the operator inequalities

$$\mathring{\gamma}_1 B \leq R \leq \mathring{\gamma}_2 B, \quad \mathring{\gamma}_1 > 0, \tag{11}$$

where

$$\mathring{\gamma}_1 = \frac{\delta}{2(1+\sqrt{\eta})}, \quad \mathring{\gamma}_2 = \frac{\delta}{4\sqrt{\eta}}, \quad \eta = \frac{\delta}{\Delta}.$$

Comparing (7), (9), and (11), we find that the operators A and B satisfy the inequalities (9) with $\gamma_1 = c_1\mathring{\gamma}_1$ and $\gamma_2 = c_2\mathring{\gamma}_2$.

Using the Chebyshev set of parameters τ_k for the scheme (8), we obtain that this alternate-triangular method requires

$$n = O\left(\frac{1}{\sqrt{|h|}}\sqrt{\frac{c_2}{c_1}}\ln\frac{2}{\epsilon}\right)$$

iterations, where $|h|^2 = h_1^2 + h_2^2 + \ldots + h_p^2$. Since the move from y_k to y_{k+1} can be accomplished using the direct methods at a cost of $O(m_0 N_1 N_2 \ldots N_p)$ arithmetic operations, the total number of operations required to find the solution of problem (4) to an accuracy ϵ can be estimated using

$$Q = O\left(m_0 N^{p+0.5}\sqrt{\frac{c_2}{c_1}}\ln\frac{2}{\epsilon}\right),$$

if $l_1 = l_2 = \ldots = l_p$, $N_1 = N_2 = \ldots = N_p = N$.

In conclusion, we remark that the above iterative methods converge in the energy space H_D, where D can be taken to be one of the operators A, B, or $AB^{-1}A$.

14.3.2 A system of equations from elasticity theory. We now look at a system of equations from stationary elasticity theory (the Lamé equations)

$$Lu = \mu\Delta u + (\lambda+\mu)\,\mathrm{grad}\,\mathrm{div}\,u = -f(x), \tag{12}$$

where $u = (u_1, u_2, \ldots, u_p)^T$, $f = (f_1, f_2, \ldots, f_p)^T$, $x = (x_1, x_2, \ldots, x_p)$, and $\lambda > 0$ and $\mu > 0$ are the Lamé constants.

We write equation (12) in the form of the system

$$(Lu)^s = \mu\sum_{\alpha=1}^{p}\frac{\partial^2 u^s}{\partial x_\alpha^2} + (\lambda+\mu)\sum_{\beta=1}^{p}\frac{\partial^2 u^\beta}{\partial x_\beta \partial x_s} = -f^s, \quad s = 1, 2, \ldots, p. \tag{13}$$

For $p = 2$, the system (13) can be written in the form

$$(\lambda+2\mu)\frac{\partial^2 u^1}{\partial x_1^2} + \mu\frac{\partial^2 u^2}{\partial x_2^2} + (\lambda+\mu)\frac{\partial^2 u^2}{\partial x_1\partial x_2} = -f^1(x_1, x_2),$$

$$(\lambda+\mu)\frac{\partial^2 u^1}{\partial x_1\partial x_2} + \mu\frac{\partial^2 u^2}{\partial x_1^2} + (\lambda+2\mu)\frac{\partial^2 u^2}{\partial x_2^2} = -f^2(x_1, x_2).$$

This system describes the equilibrium of a homogeneous isotropic elastic solid body with a plane deformation. The unknown functions $u^1(x_1, x_2)$ and $u^2(x_1, x_2)$ represent the intersection points in the directions of the axes Ox_1 and Ox_2 respectively.

For the system (12), it is possible to pose the problem of finding the vector $u(x)$ that satisfies equations (12) in the region G and that takes on the given values

$$u(x) = g(x), \quad x \in \Gamma, \tag{14}$$

on the boundary Γ. Comparing (13) with (2), we find that the system (12), (14) can be written in the form (1), where $m_0 = p$,

$$k_{\alpha\beta}^{sm} = \mu\delta_{\alpha\beta}\delta_{sm} + (\lambda + \mu)[\theta\delta_{\alpha s}\delta_{\beta m} + (1 - \theta)\delta_{\alpha m}\delta_{\beta s}], \tag{15}$$

and θ is an arbitrary constant

$$\delta_{ij} = \begin{cases} 1, & i = j \\ 0, & i \neq j \end{cases}.$$

In fact, substituting (15) in (2), we will have

$$\begin{aligned}
(Lu)^s &= \sum_{\alpha,\beta=1}^{p} \sum_{m=1}^{p} \frac{\partial}{\partial x_\alpha}\left(k_{\alpha\beta}^{sm}\frac{\partial u^m}{\partial x_\beta}\right) \\
&= \mu \sum_{\alpha,\beta=1}^{p} \sum_{m=1}^{p} \delta_{\alpha\beta}\delta_{sm}\frac{\partial^2 u^m}{\partial x_\alpha \partial x_\beta} \\
&+ (\lambda + \mu)\left[\theta \sum_{\alpha,\beta=1}^{p} \sum_{m=1}^{p} \delta_{\alpha s}\delta_{\beta m}\frac{\partial^2 u^m}{\partial x_\alpha \partial x_\beta}\right. \\
&\left. + (1 - \theta) \sum_{\alpha,\beta=1}^{p} \sum_{m=1}^{p} \delta_{\alpha m}\delta_{\beta s}\frac{\partial^2 u^m}{\partial x_\alpha \partial x_\beta}\right] \\
&= \mu \sum_{\alpha=1}^{p} \frac{\partial^2 u^s}{\partial x_\alpha^2} + (\lambda + \mu)\left[\theta \sum_{\beta=1}^{p} \frac{\partial^2 u^\beta}{\partial x_s \partial x_\beta} + (1 - \theta)\sum_{\alpha=1}^{p} \frac{\partial^2 u^\alpha}{\partial x_\alpha \partial x_s}\right] \\
&= \mu \sum_{\alpha=1}^{p} \frac{\partial^2 u^s}{\partial x_\alpha^2} + (\lambda + \mu)\sum_{\beta=1}^{p} \frac{\partial^2 u^\beta}{\partial x_s \partial x_\beta}.
\end{aligned}$$

The assertion is proved.

We find now the constants c_1 and c_2 in the inequalities (3). We will show that $c_1 = \mu$. We have

$$\begin{aligned}\sum_{s,m=1}^{p} \sum_{\alpha,\beta=1}^{p} k_{\alpha\beta}^{sm} \xi_\alpha^s \xi_\beta^m &= \mu \sum_{\alpha,s=1}^{p} (\xi_\alpha^s)^2 \\ &+ (\lambda+\mu)\left[\theta \sum_{\alpha,s=1}^{p} \xi_\alpha^\alpha \xi_s^s + (1-\theta) \sum_{\alpha,s=1}^{p} \xi_\alpha^s \xi_s^\alpha\right] \\ &= \mu \sum_{\alpha,s=1}^{p} (\xi_\alpha^s)^2 + (\lambda+\mu)\left[\theta \left(\sum_{\alpha=1}^{p} \xi_\alpha^\alpha\right)^2 + (1-\theta)\sum_{\alpha,s=1}^{p} \xi_\alpha^s \xi_s^\alpha\right].\end{aligned} \tag{16}$$

Setting here $\theta = 1$, we find

$$\sum_{\alpha,\beta=1}^{p} (k_{\alpha\beta}\xi_\alpha, \xi_\beta) = \mu \sum_{\alpha=1}^{p} |\xi_\alpha|^2 + (\lambda+\mu)\left(\sum_{\alpha=1}^{p} \xi_\alpha^\alpha\right)^2 \geq \mu \sum_{\alpha=1}^{p} |\xi_\alpha|^2.$$

It is also not difficult to show that $c_2 = \lambda + 2\mu$. Setting $\theta = 0$ in (16) and using the Cauchy-Schwartz-Bunyakovskij inequality, we obtain

$$\begin{aligned}\sum_{\alpha,\beta=0}^{p} (k_{\alpha\beta}\xi_\alpha, \xi_\beta) &= \mu \sum_{\alpha=1}^{p} |\xi_\alpha|^2 + (\lambda+\mu) \sum_{\alpha,s=1}^{p} \xi_\alpha^s \xi_s^\alpha \\ &\leq \mu \sum_{\alpha=1}^{p} |\xi_\alpha|^2 + \frac{(\lambda+\mu)}{2}\left[\sum_{\alpha,s=1}^{p} (\xi_\alpha^s)^2 + \sum_{\alpha,s=1}^{p} (\xi_s^\alpha)^2\right] \\ &= \mu \sum_{\alpha=1}^{p} |\xi_\alpha|^2 + (\lambda+\mu) \sum_{\alpha,s=1}^{p} (\xi_\alpha^s)^2 = (\lambda+2\mu) \sum_{\alpha=1}^{p} |\xi_\alpha|^2.\end{aligned}$$

We now construct the difference scheme that approximates (12), (14). Substituting (15) in the difference scheme (4), we will have

$$\begin{aligned}(\Lambda^- y)^s &= \mu \sum_{\alpha=1}^{p} y^s_{\bar{x}_\alpha x_\alpha} + 0.5(\lambda+\mu) \sum_{\beta=1}^{p} \left(y^\beta_{\bar{x}_\beta x_s} + y^\beta_{x_\beta \bar{x}_s}\right) \\ &= -\varphi^s, \qquad x \in \omega, \\ y^s(x) &= g^s(x), \quad x \in \gamma, \quad s = 1, 2, \ldots, p,\end{aligned} \tag{17}$$

where $\bar{\omega} = \omega \cup \gamma$ is the grid introduced in Section 14.3.1.

It remains to define the operators A and R, as in Section 14.3.1. The symmetry condition (6) is satisfied; therefore, using the first Green formula, we find that the operators A and R are self-adjoint in H and satisfy the inequalities $c_1 R \le A \le c_2 R$, where $c_1 = \mu$, $c_2 = \lambda + 2\mu$.

Here, the remainder of the discussion is the same as that in Section 14.3.1. So the iterative method (8) with $B = R$ and with the Chebyshev parameters τ_k is characterized by the following estimate for the iteration count:

$$n \ge n_0(\epsilon) = \frac{\ln 0.5\epsilon}{\ln \rho_1}, \quad \rho_1 = \frac{1-\sqrt{\xi}}{1+\sqrt{\xi}}, \quad \xi = \frac{c_1}{c_2} = \frac{\mu}{\lambda + 2\mu},$$

and the alternate-triangular method, constructed on the basis of the regularizer R, is characterized by the same estimate, where

$$\xi = \frac{\mu}{\lambda + 2\mu} \frac{2\sqrt{\eta}}{1+\sqrt{\eta}}, \quad \eta = \frac{\delta}{\Delta},$$

$$\delta = \sum_{\alpha=1}^{p} \frac{4}{h_\alpha^2} \sin^2 \frac{\pi h_\alpha}{2 l_\alpha}, \quad \Delta = \sum_{\alpha=1}^{p} \frac{4}{h_\alpha^2}.$$

Thus, for the alternate-triangular method, the iteration count is proportional to

$$\sqrt{\frac{\lambda + 2\mu}{\mu}} = \sqrt{2 + \frac{\lambda}{\mu}},$$

thus:

$$n_0(\epsilon) = \sqrt{2 + \frac{\lambda}{\mu}} n_0^*(\epsilon),$$

where $n_0^*(\epsilon)$ is the iteration count for solving the p-dimensional Poisson difference equation using the alternate-triangular method.

14.4 Methods for solving elliptic equations in irregular regions

14.4.1 Difference problems in regions of complex form, and methods for their solution. In Chapters 3–4 effective direct methods were constructed for solving elliptic boundary-value difference problems — the cyclic reduction method, the separation of variables method using the fast Fourier transform, and also their combination. Application of these methods is limited to a class of problems with separable variables. In this section we consider certain aspects of using fast direct methods for solving elliptic boundary-value difference problems in cases when separation of variables is not available.

The first class of problems with nonseparable variables consists of boundary value problems for elliptic equations with variable coefficients in a rectangle. For example, suppose we need to solve

$$\Lambda y = \sum_{\alpha=1}^{2} (a_\alpha(x) y_{\bar{x}_\alpha})_{x_\alpha} = -f(x), \qquad x \in \omega,$$

$$y(x) = g(x), \qquad x \in \gamma,$$

where $\bar{\omega} = \omega \cup \gamma$ is a grid in the rectangle $\bar{G}$, and the coefficients $a_1(x)$ and $a_2(x)$ satisfy

$$0 < c_1 \le a_\alpha(x) \le c_2, \qquad , \alpha = 1, 2, \qquad w \in \bar{\omega}.$$

To approximately solve this problem it is possible to use, for example, the implicit Chebyshev iterative method, choosing as the operator B in the iterative scheme

$$B \frac{y_{k+1} - y_k}{\tau_{k+1}} + A y_k = f, \qquad k = 0, 1, \ldots.$$

the operator corresponding to the Laplace difference operator

$$Ry = \sum_{\alpha=1}^{2} y_{\bar{x}_\alpha \bar{x}_\alpha}.$$

In this case the number of iterations does not depend on the number of nodes of the grid:

$$n \ge n_0(\epsilon) = O\left(\sqrt{\frac{c_2}{c_1}} \ln \frac{2}{\epsilon}\right),$$

and the equation $By_{k+1} = F_k = (B - \tau_{k+1} A) y_k + \tau_{k+1} f$ can be solved using the direct methods indicated above. Thus this approach to solving the problem is appropriate in the case when the ratio c_1/c_2 is not small. For the case of strongly varying coefficients $a_\alpha(x)$ it is possible to use the alternate-triangle method.

The second class consists of boundary value problems for elliptic equations in an irregular region, different from a rectangle. Here several approaches are possible. If the corresponding difference problem is written in the form of a system of three-point vector equations, then it is possible to use the method of block elimination described in Section 2.4. The basic deficiency of this method is the necessity of computing and storing a large volume of temporary information. If a single problem is being solved, then finding these quantities constitutes a major part of the time spent to solve the entire problem.

Another situation arises if a series of problems is being solved where only the right-hand side changes. Here the basic volume of intermediate information is found only when solving the first problem, and this information can be used to solve each subsequent problem.

Moving to implicit iterative methods allows us to significantly reduce the volume of stored information, and also the number of arithmetic operations, and to reduce the solution of the given problem to the solution of a series of problems with an operator B having simple structure. Thus, in the case of the alternate-triangle method, the operator B is factored and allows the explicit solution of the equation $By_{k+1} = F_k$. Thus, we can restrict ourselves to the construction of a method for solving a series of problems given on a grid in a region of complex form, with operator B of simple structure. With this approach the necessary intermediate information, determined by the form of the region and the structure of the operator B and not depending on the right-hand side F_k, can be found once at the beginning of the iterative process at a preprocessing step of the algorithm.

In the methods described below, domain decomposition and augmentation of the region to a rectangle bring about the reduction of the problem $By_{k+1} = F_k$ to a problem, given on a grid in some rectangle (or rectangles), that allows the use of effective direct methods.

14.4.2 Decomposition of regions. Assume that, in a complex region composed of several regular regions, it is necessary to solve a boundary-value difference problem. We will assume that in each regular region the given difference equation with boundary conditions on the boundaries of the decomposition can be solved by one of the effective direct or iterative methods. Under the given assumptions, the desired problem in the complex region can be solved in two steps. Initially, the boundary conditions which the desired solution satisfies are determined on the boundaries of the decomposition of the region, then in each regular region the difference equations with the corresponding boundary conditions are independently solved.

We lay out this method on an example of a Dirichlet difference problem for Poisson's equation in a region consisting of two rectangles joined by a line segment. To solve the difference problem in a rectangle it is possible to use the direct methods outlined in Chapters 3–4.

Suppose that the region $\bar{G}$ consists of two rectangles

$$\bar{G} = \bar{G}^{(1)} \cup \bar{G}^{(2)} :$$

$$\bar{G}^{(\alpha)} = \left\{ l_1^{(\alpha)} \le x_1 \le L_1^{(\alpha)},\ l_2^{(\alpha)} \le x_2 \le L_2^{(\alpha)} \right\},$$

$$\alpha = 1, 2,$$

joined by a line segment $\bar{\Gamma}^{(0)} = \{\max(l_1^{(1)}, l_1^{(2)}) \le x_1 \le \min(L_1^{(1)}, L_2^{(2)}),$ $x_2 = l_2^{(2)} = L_2^{(1)}\}$ (see Figure 4)

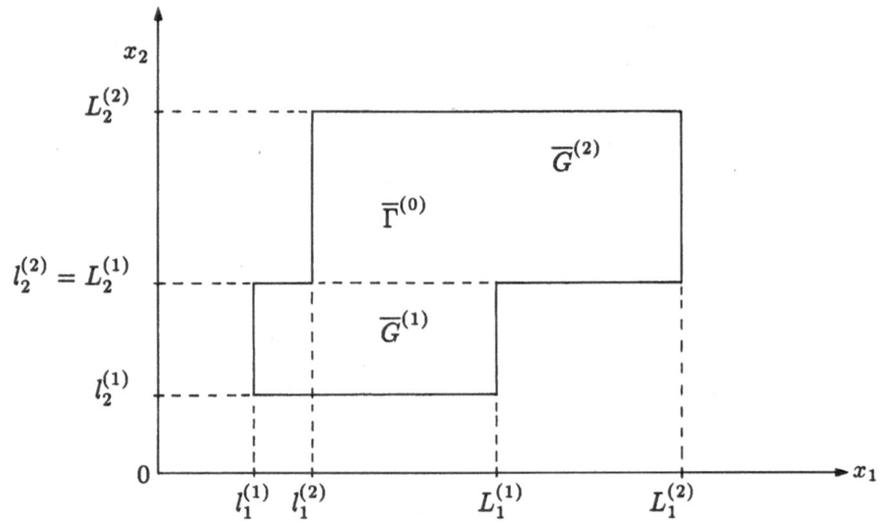

Figure 4.

We denote by Γ the boundary of the region $\bar{G}$. The part of the boundary of the rectangle $\bar{G}^{(\alpha)}$ without the interval $\Gamma^{(0)}$ we denote by $\Gamma^{(\alpha)}$. We pose the problem: find the solution of Poisson's equation in the region $\bar{G}$ that assumes on Γ the given values

$$\begin{aligned} Lu &= \frac{\partial^2 u}{\partial x_1^2} + \frac{\partial^2 u}{\partial x_2^2} = -f(x), \quad & x = (x_1, x_2) \in G, \\ u(x) &= g(x), & x \in \Gamma. \end{aligned} \tag{1}$$

We formulate initially the difference analog of (1). For this we introduce into the rectangle $\bar{G}^{(\alpha)}$ the grid $\bar{\omega}^{(\alpha)}$, uniform in the direction x_2:

$$\bar{\omega}^{(\alpha)}=\{x_{ij}=(x_1(i),x_2(j)), \quad x_1(i)=x_1(i-1)+h_1^{(\alpha)}(i), \quad 1\le i\le N_1^{(\alpha)},$$
$$x_1(0)=l_1^{(\alpha)}, \quad x_1(N_1^{(\alpha)})=L_1^{(\alpha)}, \quad x_2(j)=l_2^{(\alpha)}+jh_2^{(\alpha)},$$
$$0\le j\le N_2^{(\alpha)}, \quad N_2^{(\alpha)}h_2^{(\alpha)}=L_2^{(\alpha)}-l_2^{(\alpha)}\}, \quad \alpha=1,2.$$

The grids $\bar{\omega}^{(1)}$ and $\bar{\omega}^{(2)}$ must be consistent. This means that nodes of these grids lying on $\bar{\Gamma}^{(0)}$ coincide.

For the region illustrated in Figure 4, consistent grids can be made in the following way. On each interval $[l_1^{(1)}, l_1^{(2)}]$, $[l_1^{(2)}, L_1^{(1)}]$ and $[L_1^{(1)}, L_2^{(1)}]$ we construct a one-dimensional grid, uniform or non-uniform. Through the nodes construct lines parallel to the axis $0x_2$. Further, on the segments $[l_2^{(1)}, L_2^{(1)}]$ and $[l_2^{(2)}, L_2^{(2)}]$ introduce uniform one-dimensional grids with steps $h_2^{(1)}$ and $h_2^{(2)}$ respectively, through the nodes on lines parallel to the $0x_1$ axis. The points of intersection of the vertical and horizontal lines belonging to the rectangle $\bar{G}^{(\alpha)}$ form the grid $\bar{\omega}^{(\alpha)}$, $\alpha = 1, 2$.

The union of the grids $\bar{\omega}^{(1)}$ and $\bar{\omega}^{(2)}$ forms the grid $\bar{\omega}$ in the region $\bar{G}$. We denote by ω the interior and by γ the boundary nodes of the grid $\bar{\omega}$. By $\omega^{(\alpha)}$ we will denote the interior nodes of the grid $\bar{\omega}^{(\alpha)}$, and by $\gamma^{(\alpha)}$ the nodes of the grid $\bar{\omega}^{(\alpha)}$ lying on $\Gamma^{(\alpha)}$. By $\gamma^{(0)}$ we denote the common nodes of the grids $\bar{\omega}^{(1)}$ and $\bar{\omega}^{(2)}$ lying on the interval $\Gamma^{(0)}$. Then $\bar{\omega}^{(\alpha)} = \omega^{(\alpha)} \cup \gamma^{\alpha} \cup \gamma^{(0)}$, $\alpha = 1, 2$, and $\omega = \omega^{(1)} \cup \omega^{(2)} \cup \gamma^{(0)}$.

On the grid $\bar{\omega}$ we construct a difference scheme approximating (1):

$$\Lambda y = -\varphi(x), \ x \in \omega, \qquad y(x) = g(x), \ x \in \gamma, \tag{2}$$

where $\Lambda = \Lambda_1 + \Lambda_2$, and the difference operators Λ_1 and Λ_2 are defined by the formulas $\Lambda_1 y = y_{\bar{x}_1\hat{x}_1}$ and

$$\Lambda_2 y = \begin{cases} y_{\bar{x}_2 x_2}, & x \in \omega^{(\alpha)}, \\ \dfrac{2}{h_2^{(1)} + h_2^{(2)}} \left[\dfrac{y(x_1, x_2 + h_2^{(2)}) - y(x_1, x_2)}{h_2^{(2)}} \right. \\ \qquad \left. - \dfrac{y(x_1, x_2) - y(x_1, x_2 - h_2^{(1)})}{h_2^{(1)}} \right], & x \in \gamma^{(0)}. \end{cases}$$

We move on to the construction of a method for solving (2). We note that if the value of $y(x)$ is found for $x \in \gamma^{(0)}$, then (2) splits into two independent Dirichlet difference problems for Poisson's equation in the rectangles $\bar{G}^{(1)}$ and $\bar{G}^{(2)}$. Consequently, in this case a first-kind boundary condition must be determined on the boundary of the division of the regular regions (rectangles).

The desired values of $y(x)$ on $\gamma^{(0)}$ will be found in the following way. First we reduce (2) to a problem with homogeneous boundary conditions and right-hand side different from zero only for $x \in \gamma^{(0)}$. To do this we define the function $v(x)$ as the solution to

$$\begin{aligned} \Lambda v &= -\varphi(x), && x \in \omega - \gamma^{(0)}, \\ v(x) &= 0, && x \in \gamma^{(0)}, \\ v(x) &= g(x), && x \in \gamma. \end{aligned} \tag{3}$$

We note that (3) splits into two problems given on the grids $\bar{\omega}^{(1)}$ and $\bar{\omega}^{(2)}$. We have moved from (2) to (3), having changed the equation $\Lambda y = -\varphi$ to the condition $v(x) = 0$ at the nodes of the grid belonging to $\gamma^{(0)}$.

From (2), (3) we find that the function $z = y - v$ is the solution of

$$\begin{aligned} \Lambda z &= 0, && x \in \omega - \gamma^{(0)}, \\ \Lambda z &= -\psi(x), && x \in \gamma^{(0)}, \\ z(x) &= 0, && x \in \gamma, \end{aligned} \tag{4}$$

where $\psi(x) = \varphi(x) + \Lambda v(x)$. Since $v(x) = 0$ for $x \in \gamma^{(0)}$, then $y(x) = z(x)$ for $x \in \gamma^{(0)}$. Consequently, the desired solution of (2) can be found as the solution of the boundary-value problem

$$\begin{aligned} \Lambda y &= -\varphi(x), && x \in \omega - \gamma^{(0)}, \\ y(x) &= z^{(0)}(x), && x \in \gamma^{(0)}, \\ y(x) &= g(x), && x \in \gamma \end{aligned} \tag{5}$$

where $z^{(0)}(x)$ is the value of the solution $z(x)$ of (4) for $x \in \gamma^{(0)}$. The problem (5), and also (3), splits into two separate problems in $\bar{G}^{(1)}$ and $\bar{G}^{(2)}$.

We now find $z^{(0)}(x)$. To do this we study (4) for any $\psi(x)$. Since Λ is a linear difference operator and the function $z(x)$ satisfies homogeneous boundary conditions on γ, the solution of (4) is a linear function of the right-hand side of (4). Taking into account the structure of the right-hand side, we obtain that $z^{(0)}(x)$ and $\psi(x)$ are linked by a relation than can be written in the form

$$\sum_{\xi\in\gamma^{(0)}} a(x,\xi)z^{(0)}(\xi) = \psi(x), \ x \in \gamma^{(0)}, \tag{6}$$

where $a(x, \xi)$ are some coefficients.

We note that the nonsingularity of the matrix $\mathcal{A} = (a(x, \xi))$ in (6) follows from the uniqueness of the solution of (4). Besides, the coefficients $a(x, \xi)$ do not depend on $\psi(x)$, and are defined only by the operator Λ, the grid $\bar{\omega}$, and the geometry of the region $\bar{G}$. Therefore the problem of finding $a(x, \xi)$ is an independent part of this method. If these coefficients are determined, then $z^{(0)}(x)$ can be computed by solving (6) with $\psi(x)$ as indicated above.

We consider the method for finding $a(x, \xi)$. From (4), (6) it follows that for any $\psi(x)$ the solution of (4) satisfies

$$\sum_{\xi\in\gamma^{(0)}} a(x,\xi)z^{(0)}(\xi) = -\Lambda z(x), \ x \in \gamma^{(0)}. \tag{7}$$

Assume that $w(x)$ is the solution of

$$\begin{aligned} \Lambda w &= 0, & x &\in \omega - \gamma^{(0)}, \\ w(x) &= w^{(0)}(x), & x &\in \gamma^{(0)}, \\ w(x) &= 0, & x &\in \gamma, \end{aligned} \tag{8}$$

for an arbitrary given function $w^{(0)}(x)$. This solution corresponds to some function $\psi(x) = -\Lambda w(x)$, $x \in \gamma^{(0)}$. Consequently, $w(x)$ is the solution of

$$\begin{aligned} \Lambda w &= 0, & x &\in \omega - \gamma^{(0)}, \\ \Lambda w &= -\psi(x), & x &\in \gamma^{(0)}, \\ w(x) &= 0, & x &\in \gamma. \end{aligned}$$

Therefore by (7) we have

$$\sum_{\xi\in\gamma^{(0)}} a(x,\xi)w^{(0)}(\xi) = -\Lambda w(x), \ x \in \gamma^{(0)}, \tag{9}$$

allowing the coefficients $a(x, \xi)$ to be found.

Actually we set

$$w^{(0)}(x) = \delta(x-\eta) = \begin{cases} 0, & x \neq \eta, \\ 1, & x = \eta, \end{cases} \tag{10}$$

in (8), (9), where $x \in \gamma^{(0)}$ and η is an arbitrary node belonging to $\gamma^{(0)}$. From (9) we obtain

$$a(x,\eta) = -\Lambda w(x),\ x \in \gamma^{(0)} \tag{11}$$

where $w(x)$ is the solution of (8), (10). Sequentially choosing η as all nodes of the set $\gamma^{(0)}$, we find the values of $a(x,\xi)$ for all x and ξ belonging to $\gamma^{(0)}$.

Thus, to find the solution of (2) by the above method we need: 1) to compute the coefficients $a(x,\xi)$ for $x,\xi \in \gamma^{(0)}$; 2) to solve the auxiliary problem (3) and compute the function $\psi(x) = \varphi(x) + \Lambda v(x)$; 3) to solve the system of algebraic equations (6) and find the solution of (5).

Remark 1. The method of domain decomposition can also be applied when any combination of boundary conditions of first, second, or third kind are given on the sides $x_1 = l_1^{(\alpha)}$ and $x_1 = L_1^{(\alpha)}$, $\alpha = 1,2$, and also on the sides $x_2 = l_2^{(1)}$ and $x_2 = L_2^{(2)}$. For this the boundary condition of first kind given in (3) and (5) for $x \in \gamma$ should be changed to the corresponding non-homogeneous boundary condition on each part of the boundary γ. In (8) the homogeneous boundary condition of first kind on γ is changed to the corresponding homogeneous condition.

If a boundary condition not of first kind can be given on the part of the boundary Γ of the rectangle $\bar{G}$ lying on the line $x_2 = l_2^{(2)} = L_2^{(1)}$ (see Figure 4), then one should include those nodes of the grid $\bar{\omega}$ which belong to the indicated part of the boundary Γ in $\gamma^{(0)}$.

14.4.3 An algorithm for the domain decomposition method.

We now consider in more detail each step of this method.

14.4.3.1 *The preprocessing step.* The goal of the step is the computation of $a(x,\xi)$. If we solve a series of problems (2) with different $\varphi(x)$ and $g(x)$, then it is expedient to compute the coefficients $a(x,\xi)$ one time and store them. The coefficient $a(x,\eta)$ is found from (11), where $w(x)$ is the solution of (8), (10) for $\eta \in \gamma^{(0)}$.

The choice of method for solving this problem follows from the consideration that it is sufficient to find the function $w(x)$ only at the nodes positioned just below and above the $\gamma^{(0)}$ rows of the grid $\bar{\omega}$.

We denote $w^{(\alpha)}(x) = w(x)$, $x \in \omega^{(\alpha)}$, $\alpha = 1,2$. Using this notation, we write (8), (10) in the form of two problems

$$\begin{aligned} &\Lambda w^{(\alpha)} = 0, && x \in \omega^{(\alpha)}, \\ &w^{(\alpha)}(x) = 0, && x \in \gamma^{(\alpha)}, \\ &w^{(\alpha)}(x) = \delta(x-\eta), && x, \eta \in \gamma^{(0)},\ \alpha = 1,2. \end{aligned} \tag{12}$$

We introduce the vector $W_j^{(\alpha)}$, whose components are the values of $w^{(\alpha)}(x)$ at the nodes of the j-th row of the grid $\omega^{(\alpha)}$, and write (12) in the form of a system of three-point vector equations

$$-W_{j-1}^{(\alpha)} + C^{(\alpha)} W_j^{(\alpha)} - W_{j+1}^{(\alpha)} = 0,\ 1 \le j \le N_2^{(\alpha)} - 1, \alpha = 1,2, \tag{13}$$

the boundary conditions for which are given in the following form:

$$\begin{aligned} &W_0^{(1)} = 0, && W_{N_2^{(1)}}^{(1)} = F^{(1)}, && \text{if } \alpha = 1, \\ &W_0^{(2)} = F^{(2)}, && W_{N_2^{(2)}}^{(2)} = 0, && \text{if } \alpha = 2. \end{aligned} \tag{14}$$

Here $C^{(\alpha)}$ is a tridiagonal matrix corresponding to the difference operator $2E + [h_2^{(\alpha)}]^2 \Lambda_1$, and the vectors $F^{(1)}$ and $F^{(2)}$ have one non-zero component (E is the identity operator).

In Section 1.4.3, we obtained a formula for the general solution of a three-point vector equation with constant coefficients. Using this formula, we find the desired values for $W_{N_2^{(1)}-1}^{(1)}$ and $W_1^{(2)}$:

$$W_{N_2^{(1)}-1}^{(1)} = \left[U_{N_2^{(1)}-1}\left(\frac{C^{(1)}}{2}\right)\right]^{-1} U_{N_2^{(1)}-2}\left(\frac{C^{(1)}}{2}\right) F^{(1)},$$

$$W_1^{(2)} = \left[U_{N_2^{(2)}-1}\left(\frac{C^{(2)}}{2}\right)\right]^{-1} U_{N_2^{(2)}-2}\left(\frac{C^{(2)}}{2}\right) F^{(2)},$$

where $U_n(t)$ is the Chebyshev polynomial of the second kind of degree n.

Using the decomposition of the ratio of polynomials $U_{n-2}(t)/U_{n-1}(t)$ into the sum of elementary fractions, we obtain

$$W_{N_2^{(1)}-1}^{(1)} = \frac{2}{N_2^{(1)}} \sum_{k=1}^{N_2^{(1)}-1} \sin^2 \frac{k\pi}{N_2^{(1)}} \left(C^{(1)} - 2\cos\frac{k\pi}{N_2^{(1)}} E\right)^{-1} F^{(1)},$$

$$W_1^{(2)} = \frac{2}{N_2^{(2)}} \sum_{k=1}^{N_2^{(2)}-1} \sin^2 \frac{k\pi}{N_2^{(2)}} \left(C^{(2)} - 2\cos\frac{k\pi}{N_2^{(2)}} E\right)^{-1} F^{(2)}.$$

We will calculate these sums. Let $V_k^{(\alpha)}$ be the solution of

$$(C^{(\alpha)} - 2\cos\frac{k\pi}{N_2^{(\alpha)}}E)V_k^{(\alpha)} = \sin^2\frac{k\pi}{N_2^{(\alpha)}}F^{(\alpha)}, \; 1 \le k \le N_2^{(\alpha)} - 1, \tag{15}$$

then

$$W^{(1)}_{N_2^{(1)}-1} = \frac{2}{N_2^{(1)}} \sum_{k=1}^{N_2^{(1)}-1} V_k^{(1)}, \; W_1^{(2)} = \frac{2}{N_2^{(2)}} \sum_{k=1}^{N_2^{(2)}-1} V_k^{(2)}. \tag{16}$$

We transform (15) and (16) from vector to scalar form. We obtain that on the corresponding rows of the grid $\bar{\omega}^{(\alpha)}$ the functions $w^{(\alpha)}(x)$ are found from:

$$\begin{aligned} w^{(1)}(x_1,x_2) &= \frac{2}{N_2^{(1)}} \sum_{k=1}^{N_2^{(1)}-1} v_k^{(1)}(x_1), \; l_1^{(1)} \le x_1 \le L_1^{(1)}, \; x_2 = L_2^{(1)} - h_2^{(1)}, \\ w^{(2)}(x_1,x_2) &= \frac{2}{N_2^{(2)}} \sum_{k=1}^{N_2^{(2)}-1} v_k^{(2)}(x_2), \; l_1^{(2)} \le x_1 \le L_1^{(2)}, \; x_2 = L_2^{(2)} + h_2^{(2)}, \end{aligned} \tag{17}$$

where $v_k^{(\alpha)}(x_1)$ is the solution of the three-point boundary-value problem

$$\begin{aligned} \Lambda_1 v_k^{(\alpha)} - \lambda_k^{(\alpha)} v_k^{(\alpha)} &= -\frac{\delta(x-\eta)}{[h_2^{(\alpha)}]^2}\sin^2\frac{k\pi}{N_2^{(\alpha)}}, \; l_1^{(\alpha)} < x_1 < L_1^{(\alpha)}, \\ v_k^{(\alpha)}(l_1^{(\alpha)}) &= v_k^{(\alpha)}(L_1^{(\alpha)}) = 0, \; x,\eta \in \gamma^{(0)}, \\ \lambda_k^{(\alpha)} &= \frac{4}{[h_2^{(\alpha)}]^2}\sin^2\frac{k\pi}{2N_2^{(\alpha)}}, \; 1 \le k \le N_2^{(\alpha)} - 1. \end{aligned} \tag{18}$$

We compute the coefficient $a(x,\eta)$ using (11), which in this case has the form

$$\begin{aligned} a(x,\eta) &= \left(\frac{2}{h_2^{(1)}h_2^{(2)}}E - \Lambda_1\right) w^{(0)}(x) - \frac{2w^{(2)}(x_1, x_2 + h_2^{(2)})}{h_2^{(2)}[h_2^{(1)} + h_2^{(2)}]} \\ &+ \frac{2w^{(1)}(x_1, x_2 - h_2^{(1)})}{h_2^{(1)}[h_2^{(1)} + h_2^{(2)}]}, \; x \in \gamma^{(0)}, \; \eta \in \gamma^{(0)}, \end{aligned} \tag{19}$$

where $w^{(0)}(x) = \delta(x-\eta)$. Formulas (17)–(19) fully describe the preprocessing step.

If problem (18) is solved by the elimination method, then the computation of all the coefficients $a(x,\xi)$ requires $O(n(N_1^{(1)}N_2^{(1)} + N_1^{(2)}N_2^{(2)}))$ arithmetic operations, where n is the number of nodes belonging to $\gamma^{(0)}$.

14.4.3.2 *Auxiliary step.* The goal of this step is to find the solution of (3) at the nodes positioned just below and above the $\gamma^{(0)}$ rows of the grid $\bar{\omega}$, and to compute the function $\psi(x) = \varphi(x) + \Lambda v(x)$, where $x \in \gamma^{(0)}$.

We denote by $v^{(\alpha)}(x)$ the value of $v(x)$ for $x \in \bar{\omega}^{(\alpha)}$. From (3) we obtain two problems

$$\begin{aligned} &\Lambda v^{(\alpha)} = -\varphi(x), && x \in \omega^{(\alpha)}, \\ &v^{(\alpha)}(x) = g(x), && x \in \gamma^{(\alpha)}, \\ &v^{(\alpha)}(x) = 0, && x \in \gamma^{(0)}, \quad \alpha = 1, 2, \end{aligned} \tag{20}$$

each of which can be solved by the cyclic reduction method. Suppose that $N_2^{(1)}$ and $N_2^{(2)}$ are powers of 2 and use, for example, the first algorithm for the cyclic reduction method (cf. Section 3.2.3). We will show how to organize the computation on the example problem (20) for $\alpha = 2$. On the forward path of the reduction method we compute and store the vectors $\rho_j^{(k)}$. On the reverse path we sequentially compute but do not store the vectors $V_j^{(2)}$ for $j = N_2^{(2)}/2, N_2^{(2)}/4, \ldots, 1$. The components of the last computed vector give the desired values of $v^{(2)}(x)$ for $l_1^{(2)} \le x_1 \le L_1^{(2)}$, $x_2 = l_2^{(2)} + h_2^{(2)}$.

Analogously we find the values of $v^{(1)}(x)$ for $l_1^{(1)} \le x_1 \le L_1^{(1)}$, $x_2 = L_2^{(1)} - h_2^{(1)}$. Using these values, it is easy to compute the function $\psi(x)$ for $x \in \gamma^{(0)}$.

To realize this step it is necessary to expend $O(N_1^{(1)} N_2^{(1)} \log_2 N_2^{(1)} + N_1^{(2)} N_2^{(2)} \log_2 N_2^{(2)})$ arithmetic operations.

14.4.3.3 *Basic step.* At this step the solution of the desired problem is found. For this we solve (6) and compute $z^{(0)}(x)$. Then (5) is solved by the method of cyclic reduction, which can be written in the form of two problems

$$\begin{aligned} &\Lambda y^{(\alpha)} = -\varphi(x), && x \in \omega^{(\alpha)}, \\ &y^{(\alpha)}(x) = g(x), && x \in \gamma^{(\alpha)}, \\ &y^{(\alpha)}(x) = z^{(0)}(x), && x \in \gamma^{(0)}, \quad \alpha = 1, 2. \end{aligned} \tag{21}$$

The solution of these problems begins immediately with the reverse path of the cyclic reduction method, and uses the saved vectors $\rho_j^{(k)}$.

If (6) is solved by Gauss's method, then this requires $O(n^3)$ operations. Note, that if we solve a series of problems (2), then it is appropriate at the preprocessing step to factor the matrix A in (6) into the product $A = LU$ of lower-triangular and upper-triangular matrices. Then at the basic step of the solution of (6) only $O(n^2)$ operations are used. The solution of (21) by the method of cyclic reduction can be shown to require $O(N_1^{(1)}N_2^{(1)}\log_2 N_2^{(1)} + N_1^{(2)}N_2^{(2)}\log_2 N_2^{(2)})$ arithmetic operations.

14.4.4 The method of domain augmentation to a rectangle. Suppose that it is necessary to solve a Dirichlet problem for Poisson's equation (1) in the bounded region $\bar{G}$ with boundary Γ.

We construct a g bounded region $\bar{G}$ with boundary Γ.

We construct a grid $\bar{\omega}$ in the region $\bar{G}$. For this we introduce a family of straight lines $x_1 = x_1(i)$, $i = 0, \pm 1, \pm 2, \ldots$, and a family of equidistant straight lines $x_2 = x_2(j)$, $j = 0, \pm 1, \ldots$. The points $x_{ij} = (x_1(i), x_2(j))$ form a lattice on the plane. The points x_{ij} belonging to G form a family of interior nodes ω. The points x_{ij} belonging to Γ, and also the intersection points of Γ with some straight line from the constructed family of straight lines form a set of boundary nodes γ. The grid $\bar{\omega}$ is the union of ω and γ, $\bar{\omega} = \omega \cup \gamma$.

The interior nodes of the grid $\bar{\omega}$ will be divided into regular and irregular nodes. The nodes $x_{ij} \in \omega$ are called regular if all four neighboring points of the lattice $x_{i\pm 1,j}$, $x_{i,j\pm 1}$ belong to ω, otherwise they are irregular. We denote the set of regular nodes by ω°, and the irregular ones by ω^*.

The difference scheme approximating (1) is constructed in the usual way. The differential operator L is changed to a five-kind difference operator Λ at each node of ω, and first-kind boundary conditions are given at the boundary nodes γ. We note that the stencil of the operator Λ at a regular node consists of nodes of the lattice, while at an irregular node it may contain a node belonging to γ but not necessarily a node of the introduced lattice.

Our goal is to write the difference equations so that they are connected only by unknowns at nodes of the lattice. For this it is necessary to transform the difference equations at the irregular nodes, moving to the right-hand side the unknown values of the desired solution on γ. For this we change the structure of the difference operator Λ at each irregular node. We denote by Λ^* the operator Λ transformed at the points of ω^*. As a result we will have construced the difference scheme

$$\begin{aligned} \Lambda y &= -\varphi(x), \quad x \in \omega^\circ, \\ \Lambda^* y &= -\varphi^*(x), \quad x \in \omega^*, \end{aligned} \tag{22}$$

where $\Lambda y = y_{\bar{x}_1\hat{x}_1} + y_{\hat{x}_2 x_2}$.

We move on to the construction of a method for solving (22). Let $\bar{G}_0$ be a rectangle containing $\bar{G}$ whose sides are formed by straight lines from the family of straight lines constructed above. We denote by $\bar{\omega}_0$ the family of points x_{ij} of the lattice introduced above belonging to $\bar{G}_0$. Obviously $\omega = \omega^\circ \cup \omega^*$ is a part of the grid $\bar{\omega}_0$. We will denote by γ_0 the boundary of the grid $\bar{\omega}_0$. Thus, $\bar{\omega}_0$ is a uniform rectangular grid in the direction x_2 in the rectangle $\bar{G}_0$, with the difference equations (22) given at some of the nodes.

We note that the problem $\Lambda u = -f(x)$, $x \in \omega_0$, $u(x) = 0$ for $x \in \gamma_0$, where Λ is the difference operator defined above, can be effectively solved by one of the direct methods. Our goal is to reduce (22) to the indicated problem on the grid $\bar{\omega}_0$ with specially chosen right-hand side $f(x)$.

We pose the probem: choose the function $\psi(x)$ so that the solution of

$$\begin{aligned} \Lambda y &= -\varphi(x), && x \in \omega^\circ, \\ \Lambda y &= -\psi(x), && x \in \omega^*, \\ \Lambda y &= 0, && x \in \omega_0 - \omega, \\ y(x) &= 0, && x \in \gamma_0, \end{aligned} \tag{23}$$

satisfies

$$\Lambda^* y = -\varphi^*(x), \ x \in \omega^*. \tag{24}$$

If the function $\psi(x)$ is found, then the solutions of (22) and (23) coincide at the nodes ω.

The function $\psi(x)$ will be found as follows. First we reduce (23) to a problem whose right-hand side differs from zero only on ω^*. For this we define the auxiliary function $v(x)$ as the solution of

$$\begin{aligned} \Lambda v &= -\varphi(x), && x \in \omega^\circ, \\ \Lambda v &= 0, && x \in \omega_0 - \omega^\circ, \\ v(x) &= 0, && x \in \gamma_0. \end{aligned} \tag{25}$$

We note that the function $z = y - v$ is the solution of

$$\begin{aligned} \Lambda z &= 0, && x \in \omega^\circ, \\ \Lambda z &= -\psi(x), && x \in \omega^*, \\ \Lambda z &= 0, && x \in \omega_0 - \omega, \\ z(x) &= 0, && x \in \gamma_0. \end{aligned} \tag{26}$$

Besides, if $\psi(x)$ is the desired function, then from (23)–(25) we obtain

$$\psi^*(x) = -\Lambda^* z = \varphi^*(x) + \Lambda^* v(x), \ x \in \omega^*. \tag{27}$$

We note that, by the linearity of the operator Λ, the homogeneity of the boundary condition, and the structure of the right hand side of (26), the solution $z(x)$ is a linear function of $\psi(x)$. Since Λ^* is also a linear operator, it follows from (27) that $\psi(x)$ and $\psi^*(x)$ must be linearly related. Since the functions $\psi(x)$ and $\psi^*(x)$ are given on the one set ω^*, this relation has the form

$$\sum_{\xi \in \omega^*} a(x,\xi)\psi(\xi) = \psi^*(x), \ x \in \omega^*, \tag{28}$$

where $a(x,\xi)$ are some coefficients. If these coefficients are given, then solving (22) will consist of solving the auxiliary problem (25), computing the function $\psi^*(x)$ from (27), solving the system of algebraic equations (28) to determine $\psi(x)$, and solving (23) to find $y(x)$ at the nodes of the grid ω.

The coefficients $a(x,\xi)$ are found with the aid of a procedure analogous to that set out in Section 14.2.1. Suppose that $w(x)$ solves

$$\begin{aligned} \Lambda w &= 0, && x \in \omega^\circ, \\ \Lambda w &= -\delta(x-\eta), && x \in \omega^*, \\ \Lambda w &= 0, && x \in \omega_0 - \omega, \\ w(x) &= 0, && x \in \gamma_0, \end{aligned}$$

where η is a node belonging to ω^*. Then

$$a(x,\eta) = -\Lambda^* w(x), \ x \in \omega^*. \tag{29}$$

Sequentially choosing η as all points of ω^*, we obtain from (29) all the coefficients $a(x,\xi)$. This concludes our confrom (29) all the coefficients $a(x,\xi)$. This concludes our construction of the method for solving a Dirichlet difference problem for Poisson's equation in an arbitrary region.

In conclusion we make several comments. As was noted in the introduction, the methods outlined above are appropriate in the case when a series of problems are being solved with different right-hand sides. At the preprocessing step of the methods considered above we construct a matrix A which, in the method of domain decomposition, relates the right-hand side ψ and the solution $z^{(0)}$ on the boundary of the rectangles, and in the method of domain augmentation the given function ψ^* and the desired function ψ, the right-hand side at the irregular nodes ω^*. To solve a series of problems, the construction of the matrix A obviously need only be done once. Thus the volume of computation completed at the preprocessing step is an insignificant part of the total work.

In cases when the number of nodes belonging to $\gamma^{(0)}$ (or ω^*) is very great or a single problem is being solved, iterative methods should be used to solve the auxiliary problems (4). Since the right-hand side of (4) only differs from zero on $\gamma^{(0)}$, then it is natural to expect that iterative methods that do not take into account the specifics of the problem will not converge very rapidly. There arises the necessity of developing special iterative methods for finding projections of the solution of an equation onto some subspace for the case where the right-hand side belongs to this subspace.

Acceleration of convergence can be achieved at the expense of the choice of the operator B in the implicit iterative scheme, and also of the initial approximation from the subspace to which the right-hand side belongs.

Chapter 15

Methods for Solving Elliptic Equations in Curvilinear Orthogonal Coordinates

In this chapter we examine sample solutions to difference problems that approximate boundary-value problems for elliptic equations in curvilinear systems of coordinates. For problems in cylindrical and polar systems of coordinates we clarify the conditions for applying direct and iterative methods, in particular the alternating-directions method.

In Section 15.1, we pose boundary-value problems for the differential equations. Section 15.2 develops direct and iterative methods for solving difference problem in the (r, z) geometry, and also for problems on the surface of a cylinder. In Section 15.3, we look at methods for solving difference problems in a circle, a ring, and a ring sector.

15.1 Posing boundary-value problems for differential equations

15.1.1 Elliptic equations in a cylindrical system of coordinates. Suppose that we are given Poisson's equation

$$Lu = \frac{\partial^2 u}{\partial x_1^2} + \frac{\partial^2 u}{\partial x_2^2} + \frac{\partial^2 u}{\partial x_3^2} = -f(x), \quad x = (x_1, x_2, x_3). \tag{1}$$

If we are posed the problem of solving this equation in a finite circular cylinder or in an annular tube, then it is natural to work in cylindrical coordinates. In this system of coordinates, Poisson's equation (1) has the form

$$L_{r\varphi z} u = \frac{1}{r}\frac{\partial}{\partial r}\left(r\frac{\partial u}{\partial r}\right) + \frac{1}{r^2}\frac{\partial^2 u}{\partial \varphi^2} + \frac{\partial^2 u}{\partial z^2} = -f(r, \varphi, z), \tag{2}$$

where $r = \sqrt{x_1^2 + x_2^2}$, $\tan\varphi = x_2/x_1$, $z = x_3$.

Equation (1) describes, for example, a stationary temperature distribution $u = u(x_1, x_2, x_3)$ in a homogeneous medium. If the medium is not homogeneous, but isotropic, then in place of (1) we should consider the equation

$$Lu = \operatorname{div}(k \operatorname{grad} u) = \sum_{\alpha=1}^{3} \frac{\partial}{\partial x_\alpha}\left(k(x)\frac{\partial}{\partial x_\alpha}\right) = -f(x), \tag{3}$$

which in the (r, φ, z) system of coordinates corresponds to the equation

$$L_{r\varphi z} = \frac{1}{r}\frac{\partial}{\partial r}\left(rk\frac{\partial u}{\partial r}\right) + \frac{1}{r^2}\frac{\partial}{\partial \varphi}\left(k\frac{\partial u}{\partial \varphi}\right) + \frac{\partial}{\partial z}\left(k\frac{\partial u}{\partial z}\right) = -f. \tag{4}$$

If the medium is anisotropic, i.e. the coefficient of heat conductivity depends not only on a point but also on a direction, then in place of (3) we will have an equation with mixed derivatives

$$Lu = \sum_{\alpha,\beta=1}^{3} \frac{\partial}{\partial x_\alpha}\left(k_{\alpha\beta}\frac{\partial u}{\partial x_\beta}\right) = -f(x). \tag{5}$$

In a cylindrical system of coordinates, equation (5) corresponds to the equation

$$\begin{aligned} L_{r\varphi z} &= \frac{1}{r}\frac{\partial}{\partial r}\left[r\left(\bar{k}_{11}\frac{\partial u}{\partial r} + \frac{\bar{k}_{12}}{r}\frac{\partial u}{\partial \varphi} + \bar{k}_{13}\frac{\partial u}{\partial z}\right)\right] \\ &+ \frac{1}{r}\frac{\partial}{\partial \varphi}\left(\bar{k}_{21}\frac{\partial u}{\partial r} + \frac{\bar{k}_{22}}{r}\frac{\partial u}{\partial \varphi} + \bar{k}_{23}\frac{\partial u}{\partial z}\right) \\ &+ \frac{\partial}{\partial z}\left(\bar{k}_{31}\frac{\partial u}{\partial r} + \frac{\bar{k}_{32}}{r}\frac{\partial u}{\partial \varphi} + \bar{k}_{33}\frac{\partial u}{\partial z}\right) = -f(r, \varphi, z), \end{aligned} \tag{6}$$

where the cofficients $\bar{k}_{\alpha\beta}$ are expressed in terms of $k_{\alpha\beta}$ according to the formulas:

$$\begin{aligned} \bar{k}_{11} &= k_{11}\cos^2\varphi + (k_{12} + k_{21})\sin\varphi\cos\varphi + k_{22}\sin^2\varphi, \\ \bar{k}_{12} &= k_{12}\cos^2\varphi + (k_{22} - k_{11})\sin\varphi\cos\varphi - k_{21}\sin^2\varphi, \\ \bar{k}_{21} &= k_{21}\cos^2\varphi + (k_{22} - k_{11})\sin\varphi\cos\varphi - k_{12}\sin^2\varphi, \\ \bar{k}_{22} &= k_{11}\sin^2\varphi - (k_{12} + k_{21})\sin\varphi\cos\varphi + k_{22}\cos^2\varphi, \\ \bar{k}_{13} &= k_{13}\cos\varphi + k_{23}\sin\varphi, \quad \bar{k}_{23} = k_{23}\cos\varphi - k_{13}\sin\varphi, \\ \bar{k}_{31} &= k_{31}\cos\varphi + k_{32}\sin\varphi, \quad \bar{k}_{32} = k_{32}\cos\varphi - k_{31}\sin\varphi, \\ \bar{k}_{33} &= k_{33}. \end{aligned}$$

Equation (6) is called an *equation with mixed derivatives in a cylindrical system of coordinates*. If $\bar{k}_{\alpha\beta} = 0$ for $\alpha \neq \beta$, then (6) takes the form

$$L_{r\varphi z}u = \frac{1}{r}\frac{\partial}{\partial r}\left(r\bar{k}_1\frac{\partial u}{\partial r}\right) + \frac{1}{r^2}\frac{\partial}{\partial \varphi}\left(\bar{k}_2\frac{\partial u}{\partial \varphi}\right) + \frac{\partial}{\partial z}\left(\bar{k}_3\frac{\partial u}{\partial z}\right) = -f, \tag{7}$$

where $\bar{k}_\alpha = \bar{k}_{\alpha\alpha}$, $\alpha = 1, 2, 3$, and is called an equation without mixed derivatives.

Notice that if $k_{\alpha\beta} = k_{\beta\alpha}$, then $\bar{k}_{\alpha\beta} = \bar{k}_{\beta\alpha}$ and vice versa. The equations (2) and (4) given above are special cases of equation (7) corresponding to $\bar{k}_\alpha \equiv 1$ and $\bar{k}_\alpha \equiv k$.

Equation (5) will be strongly elliptic if there exists a constant $c_1 > 0$ such that for any ξ_1, ξ_2 and ξ_3 the following inequality is valid

$$\sum_{\alpha,\beta=1}^{3} k_{\alpha\beta}(x)\xi_\alpha\xi_\beta \geq c_1 \sum_{\alpha=1}^{3} \xi_\alpha^2. \tag{8}$$

If we make a change of variables in (8), setting

$$\xi_1 = \bar{\xi}_1\cos\varphi - \bar{\xi}_2\sin\varphi, \quad \xi_2 = \bar{\xi}_1\sin\varphi + \bar{\xi}_2\cos\varphi, \quad \xi_3 = \bar{\xi}_3,$$

then the inequality (8) takes the form

$$\sum_{\alpha,\beta=1}^{3} \bar{k}_{\alpha\beta}\bar{\xi}_\alpha\bar{\xi}_\beta \geq c_1 \sum_{\alpha=1}^{3} \bar{\xi}_\alpha^2. \tag{9}$$

In practise, we most often encounter two cases.

A) In the axially-symmetric case, the coefficients and right-hand side of the equation and also the solution do not depend on the angle φ. Here equation (6) can be simplified:

$$L_{rz}u = \frac{1}{r}\frac{\partial}{\partial r}\left[r\left(\bar{k}_{11}\frac{\partial u}{\partial r} + \bar{k}_{13}\frac{\partial u}{\partial z}\right)\right] + \frac{\partial}{\partial z}\left(\bar{k}_{13}\frac{\partial u}{\partial r} + \bar{k}_{33}\frac{\partial u}{\partial z}\right) = -f(r,z), \tag{10}$$

and in the absence of mixed derivatives the equation corresponding to (7) has the form

$$L_{rz}u = \frac{1}{r}\frac{\partial}{\partial r}\left(r\bar{k}_1\frac{\partial u}{\partial r}\right) + \frac{\partial}{\partial z}\left(\bar{k}_3\frac{\partial u}{\partial z}\right) = -f(r,z). \tag{11}$$

B) In the planar case the coefficients, right-hand side, and solution of equation (6) do not depend on z, and, consequently, equation (6) takes the form

$$L_{r\varphi}u = \frac{1}{r}\frac{\partial}{\partial r}\left[r\left(\bar{k}_{11}\frac{\partial u}{\partial r} + \frac{\bar{k}_{12}}{r}\frac{\partial u}{\partial \varphi}\right)\right] + \frac{1}{r}\frac{\partial}{\partial u}\left(\bar{k}_{21}\frac{\partial u}{\partial r} + \frac{\bar{k}_{22}}{r}\frac{\partial u}{\partial \varphi}\right) = -f(r,\varphi). \tag{12}$$

If mixed derivatives are absent, then the equation has the form

$$L_{r\varphi}u = \frac{1}{r}\frac{\partial}{\partial r}\left(r\bar{k}_1\frac{\partial u}{\partial r}\right) + \frac{1}{r^2}\frac{\partial}{\partial \varphi}\left(\bar{k}_2\frac{\partial u}{\partial \varphi}\right) = -f(r,\varphi). \tag{13}$$

In the planar case, it is said that (12) and (13) are elliptic equations in a polar coordinate system.

Notice that for $\bar{k}_\alpha \equiv 1$, $\alpha = 1,2,3$, the formulas (11) and (13) describe Poisson's equation in (r,z) and (r,φ) coordinate systems.

Sometimes it is necessary to solve Poisson's equation or a more general elliptic equation on the surface of a cylinder of radius R. In this case

$$L_{\varphi z} = \frac{1}{R}\frac{\partial}{\partial \varphi}\left(\frac{\bar{k}_{22}}{R}\frac{\partial u}{\partial \varphi} + \bar{k}_{23}\frac{\partial u}{\partial z}\right) + \frac{\partial}{\partial z}\left(\frac{\bar{k}_{32}}{R}\frac{\partial u}{\partial \varphi} + \bar{k}_{33}\frac{\partial u}{\partial z}\right) = -f(\varphi,z), \tag{14}$$

and, without mixed derivatives, equation (7) takes the form

$$L_{\varphi z} = \frac{1}{R^2}\frac{\partial}{\partial \varphi}\left(\bar{k}_2\frac{\partial u}{\partial \varphi}\right) + \frac{\partial}{\partial z}\left(\bar{k}_3\frac{\partial u}{\partial z}\right) = -f(\varphi,z). \tag{14'}$$

Notice that the change of variables $\varphi' = R\varphi$ enables us to reduce these equations to the usual elliptic equations with mixed derivatives.

15.1.2 Boundary-value problems for equations in a cylindrical coordinate system. We look first at the *axially symmetric case*. Since the solution does not depend on the angle φ, then in cylindrical coordinates (r,z), the region where we seek the solution is a rectangle $\bar{G} = \{l_1 \le r \le L_1,\ l_3 \le z \le L_3,\ l_1 \ge 0\}$. If the original region is an annular (hollow) cylinder, then $l_1 > 0$.

We now pose boundary-value problems for equation (10) in the rectangle $\bar{G}$. In the region G we are given equation (10), and on the sides $r = L_1$, $z = l_3$ and $z = L_3$ we are given boundary conditions of either first, second, or third kind. For example, third-kind boundary conditions have the form

$$
\begin{aligned}
-\bar{k}_{11}\frac{\partial u}{\partial r} - \bar{k}_{13}\frac{\partial u}{\partial z} &= \kappa_1^+ u - g_1^+(z), \quad r = L_1, \\
\bar{k}_{31}\frac{\partial u}{\partial r} + \bar{k}_{33}\frac{\partial u}{\partial z} &= \kappa_3^- u - g_3^-(r), \quad z = l_3, \\
-\bar{k}_{31}\frac{\partial u}{\partial r} - \bar{k}_{33}\frac{\partial u}{\partial z} &= \kappa_3^+ u - g_3^+(r), \quad z = L_3.
\end{aligned} \tag{15}
$$

For $l_1 = 0$, equation (10) has a singularity on the axis $r = 0$. In this case, we are usually interested in a bounded solution. If $l_1 > 0$, then on the side $r = l_1$ it is possible to give boundary conditions of first, second, or third kind. For example, third-kind boundary conditions have the form

$$
\bar{k}_{11}\frac{\partial u}{\partial r} + \bar{k}_{13}\frac{\partial u}{\partial z} = \kappa_1^- u - g_1^-(z), \quad r = l_1 > 0. \tag{16}
$$

If $l_1 = 0$, then the bounded solution is distinguished by the condition

$$
\lim_{r\to 0} r\left(\bar{k}_{11}\frac{\partial u}{\partial r} + \bar{k}_{13}\frac{\partial u}{\partial z}\right) = 0. \tag{17}
$$

In the conditions (15), (16), $\kappa_1^\pm(z)$ and $\kappa_3^\pm(r)$ are non-negative functions. If second-kind boundary conditions ($\kappa_\alpha^\pm \equiv 0$) are given on the boundary of the rectangle $\bar{G}$, then the problem (10), (15), (16) is soluble only if the following conditions is satisfied

$$
\begin{aligned}
\int_{l_1}^{L_1}\int_{l_3}^{L_3} r f(r,z)\,dr\,dz + \int_{l_3}^{L_3} \left[L_1 g_1^+(z) + l_1 g_1^-(z)\right] dz \\
+ \int_{l_1}^{L_1} r\left[g_3^+(r) + g_3^-(r)\right] dr = 0.
\end{aligned} \tag{18}
$$

In this case, the solution is not unique and is only defined up to a constant, i.e. $u(r,z) = u_0(r,z)$+constant, where $u_0(r,z)$ is some solution.

We look now at *equation* (14) *on the surface of a cylinder*. In the coordinates (φ, z), the region where we seek the solution is the rectangle $\bar{G} = \{l_2 \le \varphi \le L_2,\ l_3 \le z \le L_3,\ L_2 - l_2 \le 2\pi\}$.

On the sides $z = l_3$ and $z = L_3$ it is possible to give boundary conditions of first, second or third kind, for example

$$\begin{aligned} \frac{1}{R}\bar{k}_{32}\frac{\partial u}{\partial \varphi} + \bar{k}_{33}\frac{\partial u}{\partial z} &= \kappa_3^- u - g_3^-(\varphi), \quad z = l_3, \\ -\frac{1}{R}\bar{k}_{32}\frac{\partial u}{\partial \varphi} - \bar{k}_{33}\frac{\partial u}{\partial z} &= \kappa_3^+ u - g_3^+(\varphi), \quad z = L_3. \end{aligned} \tag{19}$$

Boundary conditions of the same kind can be given on the sides $\varphi = l_2$ and $\varphi = L_2$ if the surface is not closed ($L_2 - l_2 < 2\pi$). For example, third-kind boundary conditions have the form

$$\begin{aligned} \frac{1}{R^2}\bar{k}_{22}\frac{\partial u}{\partial \varphi} + \frac{1}{R}\bar{k}_{23}\frac{\partial u}{\partial z} &= \kappa_2^- u - g_2^-(z), \quad \varphi = l_2, \\ -\frac{1}{R^2}\bar{k}_{22}\frac{\partial u}{\partial \varphi} - \frac{1}{R}\bar{k}_{23}\frac{\partial u}{\partial z} &= \kappa_2^+ u - g_2^+(z), \quad \varphi = L_2. \end{aligned} \tag{20}$$

Here $\kappa_2^\pm(z) \geq 0$ and $\kappa_3^\pm(\varphi) \geq 0$.

For $\kappa_\alpha^\pm \equiv 0$, the solubility condition for problem (14), (19), (20) has the form

$$\int_{l_2}^{L_2}\int_{l_3}^{L_3} f(\varphi, z)d\varphi dz + \int_{l_3}^{L_3} \left[g_2^+(z) + g_2^-(z)\right] dz + \int_{l_2}^{L_2} \left[g_3^+(\varphi) + g_3^-(\varphi)\right] d\varphi = 0.$$

If the surface is closed ($L_2 - l_2 = 2\pi$), then the side $\varphi = l_2$ and $\varphi = L_2$ are identified and we are posed with the problem of finding a solution to equation (14) that is periodic with period 2π and that satisfies one of the conditions enumerated above on the sides $z = l_3$ and $z = L_3$. Here, if $\kappa_3^\pm \equiv 0$ in the conditions (19), then this problem can be solved (up to a constant) if

$$\int_{l_2}^{L_2}\int_{l_3}^{L_3} f(\varphi, z)d\varphi dz + \int_{l_2}^{L_2} \left[g_3^+(\varphi) + g_3^-(\varphi)\right] d\varphi = 0.$$

We formulate now *boundary-value problems for equation* (12) *in polar coordinates* for the case where the region under consideration is a circle, a ring, or a ring sector in Cartesian coordinates. In (r, φ) coordinates, this region corresponds to the rectangle $\bar{G} = \{l_1 \leq r \leq L_1,\ l_2 \leq \varphi \leq L_2,\ l_1 \geq 0,\ L_2 - l_2 \leq 2\pi\}$.

Suppose now that the ***original region is a circle.*** Equation (12) is given in G; for $r = L_1$ we are given boundary conditions of first, second, or third kind. For example, a third-kind boundary condition has the form

$$-\bar{k}_{11}\frac{\partial u}{\partial r} - \frac{\bar{k}_{12}}{r}\frac{\partial u}{\partial \varphi} = \kappa_1^+ u - g_1^+(\varphi), \quad r = L_1. \tag{21}$$

For the problem (12), (21) to be correct, it is necessary to pose an additional condition at the center of the circle. Usually we seek a bounded solution for $r = 0$. This solution satisfies the condition

$$\lim_{r \to 0} r\left(\bar{k}_{11}\frac{\partial u}{\partial r} + \frac{\bar{k}_{12}}{r}\frac{\partial u}{\partial \varphi}\right) = 0. \tag{22}$$

In view of the fact that the point $r = 0$ in the plane (x_1, x_2) has an arbitrary coordinate φ in polar coordinates, all the points on the side $r = 0$ of the rectangle $\bar{G}$ are identified. Here $u(0, \varphi) = u_0 =$ constant for $l_2 \le \varphi \le L_2$ by the continuity of the solution.

Further, the sides $\varphi = l_2$ and $\varphi = L_2$ are identified and we pose the problem of finding a solution of equation (12) that is periodic with period 2π and that satisfies the conditions enumerated above.

In the case where a second-kind boundary condition (21) with $\kappa_1^+(\varphi) \equiv 0$ is given for $r = L_1$, the solution of the problem exists if the condition

$$\int_0^{2\pi}\int_0^{L_1} r f(r, \varphi) dr d\varphi + L_1 \int_0^{2\pi} g_1^+(\varphi) d\varphi = 0 \tag{23}$$

is satisfied. The solution is not unique and is only defined up to a constant.

Suppose now that the ***original region is a ring,*** i.e. $l_1 > 0$. In this case, we seek a solution to equation (12) that is periodic with period 2π and that satisfies boundary conditions of first, second, or third kind on the sides $r = l_1$ and $r = L_1$. We give here the form of a third-kind boundary condition on the interior side of the ring

$$\bar{k}_{11}\frac{\partial u}{\partial r} + \frac{\bar{k}_{12}}{r}\frac{\partial u}{\partial \varphi} = \kappa_1^- u - g_1^-(\varphi), \quad r = l_1, \tag{24}$$

where $\kappa_1^-(\varphi) \ge 0$.

If we are given second-kind boundary conditions (21), (24) with $\kappa_1^{\pm}(\varphi) \equiv 0$, then the solution of this problem exists if the condition

$$\int_0^{2\pi}\int_{l_1}^{L_1} r f(r, \varphi) dr d\varphi + \int_0^{2\pi} \left[L_1 g_1^+(\varphi) + l_1 g_1^-(\varphi)\right] d\varphi = 0 \tag{25}$$

is satisfied. In this case the solution is determined up to a constant.

If the *region is a ring sector* ($l_1 > 0$, $L_2 - l_2 < 2\pi$), then we pose the problem of finding a solution of equation (12) that satisfies a boundary condition of first, second, or third kind on the sides of the rectangle $\bar{G}$, in particular with the conditions (21), (24) for $r = L_1$ and $r = l_1$ and with the third-kind boundary conditions

$$\begin{aligned} \bar{k}_{21}\frac{\partial u}{\partial r} + \frac{\bar{k}_{22}}{r}\frac{\partial u}{\partial \varphi} &= \kappa_2^- u - g_2^-(r), \quad \varphi = l_2, \\ -\bar{k}_{21}\frac{\partial u}{\partial r} - \frac{\bar{k}_{22}}{r}\frac{\partial u}{\partial \varphi} &= \kappa_2^+ u - g_2^+(r), \quad \varphi = L_2, \end{aligned} \tag{26}$$

for $\varphi = l_2$ and $\varphi = L_2$, $\kappa_2^{\pm} \geq 0$.

If we are given second-kind boundary conditions (21), (24), (26) with $\kappa_1^{\pm}(\varphi) \equiv 0$, $\kappa_2^{\pm}(r) \equiv 0$, then the solution exists if the condition

$$\int_{l_2}^{L_2}\int_{l_1}^{L_1} rf(r,\varphi)drd\varphi + \int_{l_2}^{L_2}(L_1 g_1^+ + l_1 g_1^-)d\varphi + \int_{l_1}^{L_1}(g_2^- + g_2^+)dr = 0 \tag{27}$$

is satisfied. Here the solution is not unique and is only determined up to a constant.

15.2 The solution of difference problems in cylindrical coordinates

15.2.1 Difference schemes without mixed derivatives in the axially-symmetric case. We shall look at boundary-value problems for elliptic equation not containing mixed derivatives in cylindrical coordinates for the axially-symmetric case.

In the rectangle $\bar{G} = \{l_1 \leq r \leq L_1,\ l_3 \leq z \leq L_3,\ l_1 \geq 0\}$ it is necessary to find a solution to the equation

$$\frac{1}{r}\frac{\partial}{\partial r}\left(rk_1\frac{\partial u}{\partial r}\right) + \frac{\partial}{\partial z}\left(k_3\frac{\partial u}{\partial z}\right) - qu = -f(r,z), \quad (r,z) \in G, \tag{1}$$

that satisfies the following boundary conditions on the boundary of the rectangle $\bar{G}$:

1) on the side $r = l_1$, $l_3 \le z \le L_3$,

$$u(r,z) = g_1^-(z), \quad \text{if} \quad l_1 > 0, \tag{2}$$

or

$$\begin{aligned} k_1 \frac{\partial u}{\partial r} &= \kappa_1^- - g_1^-(z), \quad \text{if} \quad l_1 > 0, \\ \lim_{r \to 0} r k_1 \frac{\partial u}{\partial r} &= 0, \quad \text{if} \quad l_1 = 0; \end{aligned} \tag{3}$$

2) on the side $r = L_1$, $l_3 \le z \le L_3$,

$$u(r,z) = g_1^+(z) \tag{4}$$

or

$$-k_1 \frac{\partial u}{\partial r} = \kappa_1^+ u - g_1^+(z); \tag{5}$$

3) on the side $z = l_3$, $l_1 \le r \le L_1$,

$$u(r,z) = g_3^-(r) \tag{6}$$

or

$$k_3 \frac{\partial u}{\partial z} = \kappa_3^- u - g_3^-(r); \tag{7}$$

4) on the side $z = L_3$, $l_1 \le r \le L_1$,

$$u(r,z) = g_3^+(r) \tag{8}$$

or

$$-k_3 \frac{\partial u}{\partial z} = \kappa_3^+ u - g_3^+(r). \tag{9}$$

It is assumed that the coefficients satisfy the conditions

$$\begin{gathered} k_1(r,z) \ge c_1 > 0, \quad k_3(r,z) \ge c_1 > 0, \quad g(r,z) \ge 0, \\ \kappa_1^\pm(z) \ge 0, \quad \kappa_3^\pm(r) \ge 0. \end{gathered}$$

In the case where $q \equiv 0$ and $\kappa_\alpha^\pm \equiv 0$ in the boundary conditions (3), (5), (7), (9) or where $l_1 = 0$ and we are given second-kind boundary conditions (5), (7), (9), we require that the solubility condition be satisfied (cf. (18) Section 15.1).

We will consider any combination of boundary conditions (2)–(9). We shall construct difference schemes corresponding to the indicated conditions.

In the rectangle $\bar{G}$, we introduce an arbitrary non-uniform rectangular grid

$$\bar{\omega} = \{(r_i, z_k) \in \bar{G},\ r_i = r_{i-1} + h_1(i),\ 1 \le i \le N_1,\ r_0 = l_1,\ r_{N_1} = L_1,$$
$$z_k = z_{k-1} + h_3(k),\ 1 \le k \le N_3,\ z_0 = l_3,\ z_{N_3} = L_3\},$$

and we define the average step

$$\hbar_\alpha(m) = \begin{cases} 0.5h_\alpha(1), & m = 0, \\ 0.5[h_\alpha(m) + h_\alpha(m+1)], & 1 \le m \le N_\alpha - 1, \\ 0.5h_\alpha(N_\alpha), & m = N_\alpha, \quad \alpha = 1, 3, \end{cases}$$

and the one-variable grid function

$$\rho(i) = r_i, \quad 1 \le i \le N_1, \quad \rho(0) = \begin{cases} \frac{1}{4}h_1(1), & l_1 = 0, \\ l_1, & l_1 > 0. \end{cases}$$

In the simplest case, where the coefficients k_1, k_3, q and f are continuous, the coefficients of the difference scheme will be defined using the formulas

$$a_1(i,k) = \bar{r}_i k_1(\bar{r}_i, z_k), \quad a_3(i,k) = k_3(r_i, \bar{z}_k),$$
$$d(i,k) = q(r_i, z_k), \qquad \varphi(i,k) = f(r_i, z_k),$$

where $\bar{r}_i = r_i - 0.5h_1(i)$, $\bar{z}_k = z_k - 0.5h_3(k)$.

Using this notation, we approximate (1) by the difference equation

$$\frac{1}{\rho}(a_1 y_{\bar{r}})_{\hat{r}} + (a_3 y_{\bar{z}})_{\hat{z}} - dy = -\varphi, \quad 1 \le i \le N_1 - 1, \qquad 1 \le k \le N_3 - 1. \tag{10}$$

Boundary conditions of first kind (2), (4), (6), (8) are approximated exactliy:

$$y(0,k) = g_1^-(z_k), \quad 0 \le k \le N_3, \tag{11}$$
$$y(N_1,k) = g_1^+(z_k), \quad 0 \le k \le N_3, \tag{12}$$
$$y(i,0) = g_3^-(r_i), \quad 0 \le i \le N_1, \tag{13}$$
$$y(i,N_3) = g_3^+(r_i), \quad 0 \le i \le N_1. \tag{14}$$

The difference analog of the boundary conditions (3) has the form

$$\frac{a_1^{+1}}{\rho\hbar_1}y_r + (a_3 y_{\bar{z}})_{\hat{z}} - \left(d + \frac{\kappa_1^-}{\hbar_1}\right)y = -\varphi - \frac{g_1^-}{\hbar_1}, \quad i = 0, \tag{15}$$

where $1 \le k \le N_3 - 1$ and $\kappa_1^- = g_1^- = 0$ if $l_1 = 0$. The boundary conditions (5), (7), (9) are approximated as follows:

$$-\frac{a_1}{\rho\hbar_1}y_{\bar{r}} + (a_3 y_{\bar{z}})_{\hat{z}} - \left(d + \frac{\kappa_1^+}{\hbar_1}\right)y = -\varphi - \frac{g_1^+}{\hbar_1}, \quad i = N_1, \tag{16}$$

where $1 \le k \le N_3 - 1$,

$$\frac{1}{\rho}(a_1 y_{\bar{r}})_{\hat{r}} + \frac{a_3^{+1}}{\hbar_3}y_z - \left(d + \frac{\kappa_3^-}{\hbar_3}\right)y = -\varphi - \frac{g_3^-}{\hbar_3}, \quad k = 0, \tag{17}$$

$$\frac{1}{\rho}(a_1 y_{\bar{r}})_{\hat{r}} - \frac{a_3}{\hbar_3}y_{\bar{z}} - \left(d + \frac{\kappa_3^+}{\hbar_3}\right)y = -\varphi - \frac{g_3^+}{\hbar_3}, \quad k = N_3, \tag{18}$$

where $1 \le i \le N_1 - 1$. Here we have used the notation $a_1^{+1} = a_1(i+1,k)$, $a_3^{+1} = a_3(i, k+1)$.

If we are given third-kind boundary conditions on intersecting sides of the rectangle $\bar{G}$, then we impose the following boundary conditions at the corner nodes of the grid $\bar{\omega}$

$$\frac{a_1^{+1}}{\rho\hbar_1}y_r + \frac{a_3^{+1}}{\hbar_3}y_z - \left(d + \frac{\kappa_1^-}{\hbar_1} + \frac{\kappa_3^-}{\hbar_3}\right)y = -\varphi - \frac{g_1^-}{\hbar_1} - \frac{g_3^-}{\hbar_3}, \quad i=k=0, \tag{19}$$

$$-\frac{a_1}{\rho\hbar_1}y_{\bar{r}} + \frac{a_3^{+1}}{\hbar_3}y_z - \left(d + \frac{\kappa_1^+}{\hbar_1} + \frac{\kappa_3^-}{\hbar_3}\right)y = -\varphi - \frac{g_1^+}{\hbar_1} - \frac{g_3^-}{\hbar_3}, \quad i=N_1, k=0, \tag{20}$$

$$\frac{a_1^{+1}}{\rho\hbar_1}y_r - \frac{a_3}{\hbar_3}y_{\bar{z}} - \left(d + \frac{\kappa_1^-}{\hbar_1} + \frac{\kappa_3^+}{\hbar_3}\right)y = -\varphi - \frac{g_1^-}{\hbar_1} - \frac{g_3^+}{\hbar_3}, \quad i=0, k=N_3, \tag{21}$$

$$-\frac{a_1}{\rho\hbar_1}y_{\bar{r}} - \frac{a_3}{\hbar_3}y_{\bar{z}} - \left(d + \frac{\kappa_1^+}{\hbar_1} + \frac{\kappa_3^+}{\hbar_3}\right)y = -\varphi - \frac{g_1^+}{\hbar_1} - \frac{g_3^+}{\hbar_3}, \quad i=N_1, k=N_3. \tag{22}$$

As before, if $l_1 = 0$, then in (19), and (21) we set $\kappa_1^- = g_1^- = 0$.

Notice that the difference problem (10), (15)–(22), with third-kind boundary conditions on each side of the rectangle $\bar{G}$, can be written in the compact form

$$\begin{gathered}\Lambda y = -f, \quad 0 \le i \le N_1, \quad 0 \le k \le N_3, \\ \Lambda = \Lambda_1 + \Lambda_3, \quad f = \varphi + \varphi_1/\hbar_1 + \varphi_3/\hbar_3,\end{gathered} \tag{23}$$

where

$$\varphi_1(i,k) = \begin{cases} g_1^-, & i = 0, \\ 0, & 1 \le i \le N_1 - 1, \\ g_1^+, & i = N_1, \end{cases} \quad \varphi_3(i,k) = \begin{cases} g_3^-, & k = 0, \\ 0, & 1 \le k \le N_3 - 1, \\ g_3^+, & k = N_3, \end{cases} \tag{24}$$

and the difference operators Λ_1 and Λ_3 are given by the formulas

$$\Lambda_1 y = \begin{cases} \dfrac{a_1^{+1}}{\rho\hbar_1} y_r - \left(d_1 + \dfrac{\kappa_1^-}{\hbar_1}\right) y, & i = 0, \\ \dfrac{1}{\rho}(a_1 y_{\bar r})_{\hat r} - d_1 y, & 1 \le i \le N_1 - 1, \\ -\dfrac{a_1}{\rho\hbar_1} y_{\bar r} - \left(d_1 + \dfrac{\kappa_1^+}{\hbar_1}\right) y, & i = N_1,\ 0 \le k \le N_3, \end{cases} \tag{25}$$

$$\Lambda_3 y = \begin{cases} \dfrac{a_3^{+1}}{\hbar_3} y_z - \left(d_3 + \dfrac{\kappa_3^-}{\hbar_3}\right) y, & k = 0, \\ (a_3 y_{\bar z})_{\hat z} - d_3 y, & 1 \le k \le N_3 - 1, \\ -\dfrac{a_3}{\hbar_3} y_{\bar z} - \left(d_3 + \dfrac{\kappa_3^+}{\hbar_3}\right) y, & k = N_3,\ 0 \le i \le N_1. \end{cases} \tag{26}$$

Here $d_1 + d_3 = d$, $d_1 \ge 0$ and $d_3 \ge 0$.

We now find the solubility condition for the difference scheme (23) in the case where $d \equiv 0$ and $\kappa_\alpha^\pm \equiv 0$, $\alpha = 1, 2$.

In the space H of grid functions defined on $\bar\omega$, we define the inner product using the formula

$$(u, v) = \sum_{i=0}^{N_1} \sum_{k=0}^{N_3} u(i,k) v(i,k) \rho(i) \hbar_1(i) \hbar_3(k). \tag{27}$$

We now define the operators A_1 and A_3, mapping into H, setting $A_\alpha = -\Lambda_\alpha$, $\alpha = 1, 3$. Then the difference scheme (23) can be written in the form of the operator equation

$$Au = f, \quad A = A_1 + A_3. \tag{28}$$

Using the first Green difference formula, we find that for the case $d \equiv 0$ and $\kappa_\alpha^\pm \equiv 0$

$$(Au, v) = \sum_{i=1}^{N_1}\sum_{k=0}^{N_3} h_1(i)\hbar_3(k)(a_1 u_{\bar r} v_{\bar r})_{ik}$$

$$+\sum_{i=0}^{N_1}\sum_{k=1}^{N_3} \hbar_1(i)\hbar_3(k)\rho(i)(a_3 u_{\bar z} v_{\bar z})_{ik} = (u, Av).$$

Consequently, the operator A is self-adjoint in H and is non-negative, where $(Au, u) = 0$ only in the case where $u(i,k) \equiv$ constant or $u(i,k) \equiv 0$. From this, using the Cauchy-Schwartz-Bunyakovskij inequality,

$$(Au, u)^2 \leq (Au, Au)(u, u)$$

it follows that $Au = 0$ for $u \neq 0$ if u is constant on $\bar\omega$. Thus, the kernel of the operator A consists of the grid functions that are equal to a constant on the grid $\bar\omega$. Therefore the problem (28) is soluble if the condition $(f, 1) = 0$ is satisfied or, by the definition of f, if the following condition is satisfied

$$\sum_{i=0}^{N_1}\sum_{k=0}^{N_3} \rho\varphi\hbar_1\hbar_2 + \sum_{k=0}^{N_3} \hbar_3(\rho(0)g_1^- + \rho(N_1)g_1^+) + \sum_{i=0}^{N_1} \hbar_1\rho[g_3^- + g_3^+] = 0. \quad (29)$$

The condition (29) is the difference analog of the solubility condition (18) for the differential problem corresponding to the difference problem (23).

If the condition (29) is satisfied, then, in the case where $d \equiv 0$ and $\kappa_\alpha^\pm \equiv 0$, the solution of problem (23) exists but is not unique: any two solutions can differ by a constant. Therefore, one of the possible solutions can be distinguished by fixing the value of $y(i,k)$ at some node of the grid $\bar\omega$.

15.2.2 Direct methods. We now examine the case where the difference problems (10)–(22) can be solved by one of the direct methods laid out in Chapters 3 and 4.

Suppose that the *coefficients* k_1, k_3, and q *in equation* (1) *do not depend on* z, i.e. $k = k_1(r)$, $k_3 = k_3(r)$, $q = q(r)$, that the coefficients κ_1^+ and κ_1^- are constants in the third-kind boundary conditions (3), (5), and that $\kappa_3^- = \kappa_3^+ \equiv 0$ in the conditions (7), (9).

We will allow any combination of the boundary conditions (2)–(9). It is assumed that the grid $\bar\omega$ is uniform in z, i.e. $h_3(k) \equiv h_3$, but it may be non-uniform in r. Under these assumptions, the difference problems (10)–(22) can be solved by either the cyclic reduction method or by a combination of the methods of incomplete reduction and separation of variables.

We shall illustrate one possible application of these direct methods on an example in which third- (second-) kind boundary conditions (3), (5) are given on the sides $r = l_1$ and $r = L_1$, and second-kind conditions are given for $z = l_3$ and $z = L_3$. Other combinations of boundary conditions can be examined analogously.

The difference scheme corresponding to this problem has the form (23). By the assumptions made above, the coefficients of the difference scheme are defined using the formulas (cf. Section 15.2.1) $a_1 = a_1(i) = \bar{r}_i k_1(\bar{r}_i)$, $a_3 = a_3(i) = k_3(r_i)$, $d = d(i) = q(r_i)$, so that $a_3^+ = a_3$. In the definition (25) of the difference operator Λ_1 we take $d_1 = d$, and in the formulas (26) for the operator Λ_2, we set $\kappa_3^- = \kappa_3^+ = 0$, $d_3 = 0$. Since the grid $\bar{\omega}$ is uniform in z, it follows that in (26) the difference expression $(a_3 y_{\bar{z}})_{\hat{z}}$ can be changed to $a_3 y_{\bar{z}z}$.

We now reduce the difference problem (23) to a system of three-point vector equations. To do this, we introduce the vector of unknowns

$$Y_k = (y(0,k), y(1,k), \dots, y(N_1,k))^T, \quad 0 \le k \le N_3,$$

containing the values of the desired grid function on the k-th row of the grid $\bar{\omega}$, and the right-hand-side vector

$$F_k = (\theta_0 f(0,k), \theta_1 f(1,k), \dots, \theta_{N_1} f(N_1,k))^T, \quad 0 \le k \le N_3,$$

where $\theta_i = h_3^2/a_3(i)$, $0 \le i \le N_1$. We define the square matrix C, setting

$$CY_k = ((2E - \theta_0\Lambda_1)y(0,k), \dots, (2E - \theta_{N_1}\Lambda_1)y(N_1,k))^T.$$

Using this notation, the difference problem (23) is written in the vector form

$$\begin{aligned} CY_0 - 2Y_1 &= F_0, & k &= 0, \\ -Y_{k-1} + CY_k - Y_{k+1} &= F_k, & 1 &\le k \le N_3 - 1, \\ 2Y_{N_3-1} + CY_{N_3} &= F_{N_3}, & k &= N_3. \end{aligned} \tag{30}$$

In order to convince oneself of this, it is sufficient to multiply each equation of the scheme (23) by $(-\theta_i)$ and transform to vector notation.

Recall that the cyclic-reduction method for the system (30) was constructed in Section 3.4.1. The combined method using incomplete reduction and separation of variables was examined in Section 4.3.2. Here, unlike in the discussions in Chapters 3 and 4, the examples contain a different definition of the operator Λ_1. But since the difference operator Λ_1 is, as before, a three-point operator, this difference does not affect the construction of these methods, nor the dependence of the operation count on the number of nodes of the grid $\bar{\omega}$. If $N_3 = 2^n$, then the operation count for these methods can be estimated as $O(N_1 N_3 \log_2 N_3)$.

In conclusion we remark that the application of the combined method was described in detail in Section 12.4.2 for a Cartesian coordinate system, where a particular solution was selected in the singular case ($d \equiv 0$, $\kappa_1^- = \kappa_2^+ \equiv 0$).

15.2.3 The alternating-directions method. We look now at a special case of the problem (1)–(9) for which $k_1 = k_1(r)$, $k_3 = k_3(z)$, $q =$ constant, $\kappa_\alpha^\pm =$ constant, $\alpha = 1, 3$, and where any combination of the boundary conditions (2)–(9) is given on the sides of the rectangle $\bar{G}$. In this case, the variables in the problem (1)–(9) are separated.

It is assumed that the grid $\bar{\omega}$ is arbitrary, non-uniform in each direction. Under these assumptions, the difference problems (10)–(22) can be solved by the alternating-directions method with the optimal set of iterative parameters given in Chapter 11 for the case of a Cartesian coordinate system.

We now illustrate the application of this method on an example in which third-kind boundary conditions (3), (5), (7), (9) are given on the sides of the rectangle G. The difference scheme corresponding to the problem (1), (3), (5), (7), (9) has the form (23), where the operators Λ_1 and Λ_3 are defined in (25), (26), and where the coefficients a_1, a_3, d_1, and d_3 are given by the formulas $a_1(i) = \bar{r}_i k_1(\bar{r}_i)$, $a_3(k) = k_3(\bar{z}_k)$, $d_1 = d_3 = 0.5d$, $d = q$.

In Section 15.2.1 it was proved that the difference problem (23) can be written in the form of the operator equation (28)

$$Au = f, \quad A = A_1 + A_3$$

in the Hilbert space of grid functions defined on $\bar{\omega}$. We indicate now the basic properties of the operators A_1 and A_3:

1) the operators A_1 and A_3 commute, $A_1 A_3 = A_3 A_1$;
2) A_1 and A_3 are self-adjoint operators, $(A_\alpha u, v) = (u, A_\alpha v)$;
3) the operators A_1 and A_3 are non-negative bounded operators, i.e. for any $u \in H$ the following inequalities are satisfied

$$\begin{gathered} \delta_\alpha(u,u) \le (A_\alpha u, u) \le \Delta_\alpha(u,u), \\ \delta_\alpha \ge 0, \quad \Delta_\alpha > 0, \quad \alpha = 1, 3. \end{gathered} \tag{31}$$

The commutativity of the operators A_1 and A_3 follows from the structure of the operators A_1 and A_3 and from the assumptions concerning the coefficients k_1, k_3, q, and $\kappa_\alpha^\pm$.

Further, using the definition (27) of the inner product in H and the Green difference formulas, we obtain the following equation for A_1 and for any $u, v \in H$

$$
\begin{aligned}
(A_1 u, v) = \sum_{i=1}^{N_1} \sum_{k=0}^{N_3} h_1(i)\hbar_3(k)(a_1 u_{\bar{r}} v_{\bar{r}})_{ik} + d_1(u,v) \\
+ \sum_{k=0}^{N_3} \hbar_3(k) \left[\kappa_1^- \rho uv \,|_{i=0} + \kappa_1^+ \rho uv|_{i=N_1}\right]
\end{aligned}
\tag{32}
$$

and the analogous equation for A_3

$$
\begin{aligned}
(A_3 u, v) = \sum_{k=1}^{N_3} \sum_{i=0}^{N_1} \rho(i)\hbar_i(i)h_3(k)(a_3 u_{\bar{z}} v_{\bar{z}})_{ik} \\
+ d_3(u,v) + \sum_{i=0}^{N_1} \rho(i)\hbar_1(i) \left[\kappa_3^- uv \,|_{k=0} + \kappa_3^+ uv|_{k=N_3}\right].
\end{aligned}
\tag{33}
$$

Interchanging u and v, we ascertain the self-adjointness of the operators A_1 and A_3.

If we set here $u = v$ and take into account the conditions $k_1 \geq c_1 > 0$, $k_3 \geq c_1 > 0$, $q \geq 0$, $\kappa_\alpha^\pm \geq 0$, $\alpha = 1, 3$, then we find that the operators A_1 and A_3 are non-negative, i.e. $(A_\alpha u, u) \geq 0$. If the condition

$$
d_\alpha^2 + (\kappa_\alpha^-)^2 + (\kappa_\alpha^+)^2 \neq 0, \quad \alpha = 1, 3, \tag{34}
$$

is satisfied, then the corresponding δ_α is positive. Assume that (34) is satisfied.

We give now a lower bound for δ_α.

From lemma 16 of Chapter 5 for fixed i, $0 \leq i \leq N_1$, we obtain the estimate

$$
\begin{aligned}
\delta_3 \sum_{k=0}^{N_3} \hbar_3(k) u^2(i,k) \leq \sum_{k=1}^{N_3} h_3(k) a_3(k) u_{\bar{z}}^2(i,k) \\
+ d_3 \sum_{k=0}^{N_3} \hbar_3(k) u^2(i,k) + \kappa_3^- u^2(i,0) + \kappa_3^+ u^2(i,N_3),
\end{aligned}
\tag{35}
$$

where $1/\delta_3 = \max\limits_{0\le k\le N_3} v(k)$, and $v(k)$ is the solution of the boundary-value problem

$$\begin{aligned} &(a_3 v_{\bar{z}})_{\hat{z}} - d_3 v = -1, \quad 1 \le k \le N_3 - 1, \\ &\frac{a_3^{+1}}{\hbar_3} v_z - \left(d_3 + \frac{\kappa_3^-}{\hbar_3}\right) v = -1, \quad k = 0, \\ &-\frac{a_3}{\hbar_3} v_{\bar{z}} - \left(d_3 + \frac{\kappa_3^+}{\hbar_3}\right) v = -1, \quad k = N_3. \end{aligned} \tag{36}$$

Since the condition (34) is satisfied, the solution of problem (36) exists and is unique. Now multiplying (35) by $\rho(i)\hbar_1(i)$ and summing for i between 0 and N_1, we obtain the inequality $\delta_3(u,u) \le (A_3 u, u)$. Solving problem (36) numerically, we determine δ_3. Thus, the constant δ_3 has been found. Analogously we estimate the constant δ_1: $1/\delta_1 = \max\limits_{0\le i\le N_1} \bar{v}(i)$, where $\bar{v}(i)$ is the solution of the boundary-value problem

$$\begin{aligned} &\frac{1}{\rho}(a_1 \bar{v}_{\bar{r}})_{\hat{r}} - d_1 \bar{v} = -1, && 1 \le i \le N_1 - 1, \\ &\frac{a_1^{+1}}{\rho\hbar_1} \bar{v}_r - \left(d_1 + \frac{\kappa_1^-}{\hbar_1}\right) \bar{v} = -1, && i = 0, \\ &-\frac{a_1}{\rho\hbar_1} \bar{v}_{\bar{r}} - \left(d_1 + \frac{\kappa_1^+}{\hbar_1}\right) \bar{v} = -1, && i = N_1. \end{aligned} \tag{37}$$

We obtain now estimates for Δ_1 and Δ_3. From (33) for $u = v$ we find

$$\begin{aligned} (A_3 u, u) = \sum_{i=0}^{N_1} \rho(i)\hbar_1(i) \Bigg[&\sum_{k=1}^{N_3} h_3(k) a_3(k) u_{\bar{z}}^2(i,k) \\ &+ d_3 \sum_{k=0}^{N_3} \hbar_3(k) u^2(i,k) + \kappa_3^- u^2(i,0) + \kappa_3^+ u^2(i,N_3) \Bigg]. \end{aligned}$$

We now estimate the expression standing in square brackets. From lemma 16 of Chapter 5 we obtain

$$\begin{aligned} &d_3 \sum_{k=0}^{N_3} u^2(i,k)\hbar_3(k) + \kappa_3^- u^2(i,0) + \kappa_3^+ u^2(i,N_3) \\ &\le m_1 \left[\sum_{k=1}^{N_3} a_3(k) u_{\bar{z}}^2(i,k) h_3(k) + \sum_{k=0}^{N_3} \hbar_3(k) u^2(i,k) \right], \end{aligned} \tag{38}$$

where $m_1 = \max\limits_{0\le k\le N_3} w(k)$, and $w(k)$ is the solution of the boundary-value problem

$$
\begin{aligned}
&(a_3 w_{\bar z})_{\hat z} - w = -d_3, && 1 \le k \le N_3 - 1,\\
&\frac{a_3^{+1}}{\hbar_3} w_z - w = -\left(d_3 + \frac{\kappa_3^-}{\hbar_3}\right), && k = 0,\\
&-\frac{a_3}{\hbar_3} w_{\bar z} - w = -\left(d_3 + \frac{\kappa_3^+}{\hbar_3}\right), && k = N_3.
\end{aligned}
\tag{39}
$$

Using lemma 17 of Chapter 5, we will have

$$\sum_{k=1}^{N_3} a_3(k) u_{\bar z}^2(i,k) h_3(k) \le m_2 \sum_{k=0}^{N_3} \hbar_3(k) u^2(i,k), \tag{40}$$

where

$$m_2 = \max\left(\frac{a_3(N_3)}{\hbar_3^2(N_3)}, \frac{a_3(1)}{\hbar_3^2(0)}, \max_{1\le k\le N_3-1} \frac{2}{\hbar_3(k)}\left[\frac{a_3(k)}{h_3(k)} + \frac{a_3(k+1)}{h_3(k+1)}\right]\right).$$

From (38) and (40) follows the estimate

$$
\begin{aligned}
&\sum_{k=1}^{N_3} h_3(k) a_3(k) u_{\bar z}^2(i,k) + d_3 \sum_{k=0}^{N_3} \hbar_3(k) u^2(i,k) + \kappa_3^- u^2(i,0)\\
&\quad + \kappa_3^+ u^2(i,N_3) \le \Delta_3 \sum_{k=0}^{N_3} \hbar_3(k) u^2(i,k), \quad \Delta_3 = m_1 + m_2(1+m_1).
\end{aligned}
$$

Multiplying this inequality by $\rho(i)\hbar_1(i)$ and summing it for i between 0 and N_1, we arrive at the estimate $(A_3 u, u) \le \Delta_3(u,u)$.

Analogously we find Δ_1: $\Delta_1 = \bar m_1 + \bar m_2(1+\bar m_1)$, where $\bar m_1 = \max\limits_{0\le i\le N_1} \bar w(i)$ and $\bar w(i)$ is the solution of the boundary-value problem

$$
\begin{aligned}
&\frac{1}{\rho}(a_1 \bar w_{\bar r})_{\hat r} - \bar w = -d_1, && 1 \le i \le N_1 - 1,\\
&\frac{a_1^{+1}}{\rho\hbar_1} \bar w_r - \bar w = -\left(d_1 + \frac{\kappa_1^-}{\hbar_1}\right), && i = 0,\\
&-\frac{a_1}{\rho\hbar_1} \bar w_{\bar r} - \bar w = -\left(d_1 + \frac{\kappa_1^+}{\hbar_1}\right), && i = N_1,
\end{aligned}
\tag{41}
$$

where

$$\bar{m}_2 = \max\left(\frac{a_1(N_1)}{\rho(N_1)\hbar_1^2(N_1)}, \frac{a_1(1)}{\rho(0)\hbar_1^2(0)},\right.$$
$$\left.\max_{1\le i\le N_1-1} \frac{2}{\rho(i)\hbar_1(i)}\left[\frac{a_1(i)}{h_1(i)} + \frac{a_1(i+1)}{h_1(i+1)}\right]\right).$$

Solving numerically problem (41) we determine $\bar{m}_1$ and, consequently, Δ_1. Thus, the constants δ_α and Δ_α, $\alpha = 1,3$ in the inequalities (31) have been found.

Recall that the iterative scheme of the alternating-directions method for the operator equation (28) has the form (cf. Chapter 11)

$$B_{k+1}\frac{y_{k+1}-y_k}{\tau_{k+1}} + Ay_k = f, \quad k = 0,1,\ldots, \quad y_0 \in H,$$
$$B_k = \left(\omega_k^{(1)}E + A_1\right)\left(\omega_k^{(3)}E + A_3\right), \quad \tau_k = \omega_k^{(1)} + \omega_k^{(3)}. \tag{42}$$

In Section 11.1.4, we constructed the optimal set of parameters $\omega_k^{(1)}$ and $\omega_k^{(3)}$, $k = 1,2,\ldots,n$ for the iterative scheme (42) where the operators A_1 and A_3 possess the properties 1)–3) listed above. With this set of parameters, a relative accuracy $\epsilon > 0$ can be attained ($\| y_n - u \|_D \le \epsilon \| y_0 - u \|_D$, $D = A, E$), if we perform $n \ge n_0(\epsilon)$ iterations, where

$$n_0(\epsilon) = \frac{1}{\pi^2}\ln\frac{4}{\eta}\ln\frac{4}{\epsilon}, \quad \eta = \frac{1-a}{1+a}, \quad a = \sqrt{\frac{(\Delta_1-\delta_1)(\Delta_3-\delta_3)}{(\Delta_1+\delta_3)(\Delta_3+\delta_1)}}.$$

The set of optimal parameters $\omega_k^{(1)}$ and $\omega_k^{(3)}$ for the case of second-kind boundary conditions ($d = 0$, $\kappa_\alpha^\pm \equiv 0$) was constructed in Section 12.4.1.

15.2.4 The solution of equations defined on the surface of a cylinder. We look now at a method for solving difference analogs of boundary-value problems for elliptic equations without mixed derivatives defined on the surface of a cylinder of radius R. We limit ourselves to considering a surface of a cylinder that is closed in φ, since the methods for solving problems in the case of a non-closed surface do not differ from the methods for solving planar problems in Cartesian coordinates.

Thus, in the region $\bar{G} = \{l_2 \le \varphi \le L_2,\ l_3 \le z \le L_3,\ L_2 - l_2 = 2\pi\}$ we seek a solution to the equation

$$\frac{1}{R^2}\frac{\partial}{\partial\varphi}\left(k_2\frac{\partial u}{\partial\varphi}\right) + \frac{\partial}{\partial z}\left(k_3\frac{\partial u}{\partial z}\right) - qu = -f(\varphi,z), \quad (\varphi,z) \in G, \tag{43}$$

that is periodic with period 2π, and that, on the sides $z = l_3$ and $z = L_3$, satisfies first-kind boundary conditions $u(\varphi, z) = g_3^-(\varphi)$ for $z = l_3$, $u(\varphi, z) = g_3^+(\varphi)$ for $z = L_3$, or second- or third-kind boundary conditions

$$\begin{aligned} k_3 \frac{\partial u}{\partial z} &= \kappa_3^- u - g_3^-(\varphi), \quad z = l_3, \\ -k_3 \frac{\partial u}{\partial z} &= \kappa_3^+ u - g_3^+(\varphi), \quad z = L_3, \end{aligned} \tag{44}$$

or any of their combinations. It is assumed that the coefficients satisfy the conditions

$$k_2(\varphi, z) \geq c_1 > 0, \quad k_3(\varphi, z) \geq c_1 > 0, \quad q(\varphi, z) \geq 0, \quad \kappa_3^{\pm}(\varphi) \geq 0.$$

In the region $\bar{G}$ we introduce the arbitrary non-uniform grid

$$\begin{aligned} \bar{\omega} = \{ &(\varphi_j, z_k) \in \bar{G}, \ \varphi_j = \varphi_{j-1} + h_2(j), \ 1 \leq j \leq N_2, \ \varphi_0 = l_2, \\ &\varphi_{N_2} = L_2, \ z_k = z_{k-1} + h_3(k), \ 1 \leq k \leq N_3, \ z_0 = l_3, \ z_{N_3} = L_3 \} \end{aligned}$$

and define the average step

$$\hbar_2(j) = \begin{cases} 0.5[h_2(1) + h_2(N_2)], & j = 0, \\ 0.5[h_2(j) + h_2(j+1)], & 1 \leq j \leq N_2 - 1. \end{cases} \tag{45}$$

The average step $\hbar_3(k)$ was defined above.

Taking into account the periodicity condition, we approximate equation (43) as follows:

$$(a_2 y_{\bar{\varphi}})_{\hat{\varphi}} + (a_3 y_{\bar{z}})_{\hat{z}} - dy = -\psi, \ 0 \leq j \leq N_2 - 1, \ 1 \leq k \leq N_3 - 1, \tag{46}$$

where we have used the relations $y(j, k) = y(N_2 + j, k)$, $j = 0, -1$, $a_2(0, k) = a_2(N_2, k)$, $h_2(0) = h_2(N_2)$ which follow from the periodicity. In the case of smooth coefficients k_2, k_3, q, and f, the coefficients in equation (46) can be selected, for example, as:

$$\begin{aligned} a_2(j, k) &= \frac{1}{R^2} k_2(\varphi_j - 0.5 h_2(j), z_k), \ d(j, k) = q(\varphi_j, z_k), \\ a_3(j, k) &= k_3(\varphi_j, z_k - 0.5 h_3(k)), \ \psi(j, k) = f(\varphi_j, z_k). \end{aligned}$$

The first-kind boundary conditions are approximated exactly

$$y(j, 0) = g_3^-(\varphi_j), \quad k = 0, \quad y(j, N_3) = g_3^+(\varphi_j), \quad k = N_3 \tag{47}$$

for $0 \le j \le N_2 - 1$, and the difference analog of the third-kind boundary conditions (44) for $0 \le j \le N_2 - 1$ has the form

$$\begin{aligned} &(a_2 y_{\bar{\varphi}})_{\hat{\varphi}} + \frac{a_3^{+1}}{\hbar_3} y_z - \left(d + \frac{\kappa_3^-}{\hbar_3}\right) y = -\psi - \frac{g_3^-}{\hbar_3}, \quad k = 0, \\ &(a_2 y_{\bar{\varphi}})_{\hat{\varphi}} - \frac{a_3}{\hbar_3} y_{\bar{z}} - \left(d + \frac{\kappa_3^+}{\hbar_3}\right) y = -\psi - \frac{g_3^+}{\hbar_3}, \quad k = N_3. \end{aligned} \tag{48}$$

In the problem (46), (47), the unknowns are the values $y(j,k)$ for $0 \le j \le N_2 - 1$, $1 \le k \le N_3 - 1$, and in the problem (46), (48), for the same values of j and for $0 \le k \le N_3$.

We now find the solubility conditions for the difference problem (46), (48) in the case where $d \equiv 0$, $\kappa_3^{\pm} \equiv 0$. First we write the scheme (46), (48) in the form

$$\begin{aligned} &\Lambda y = -f, \quad 0 \le j \le N_2 - 1, \quad 0 \le k \le N_3, \\ &\Lambda = \Lambda_2 + \Lambda_3, \quad f = \psi + \psi_3/\hbar_3, \end{aligned} \tag{49}$$

where the difference operator Λ_3 was defined in (26) with $d_3 = d$, and where the operator Λ_2 is given by the formula $\Lambda_2 y = (a_2 y_{\bar{\varphi}})_{\hat{\varphi}}$, $0 \le j \le N_2 - 1$,

$$\psi_3(j,k) = \begin{cases} g_3^-(\varphi_j), & k = 0, \\ 0, & 1 \le k \le N_3 - 1, \\ g_3^+(\varphi_j), & k = N_3. \end{cases}$$

Assume now that $d = 0$ and $\kappa_3^{\pm} = 0$. We denote by H the space of grid functions defined on $\bar{\omega}^* = \{(\varphi_j, z_k) \in \bar{\omega}, 0 \le j \le N_2 - 1, 0 \le k \le N_3\}$, where the inner product is defined by the formula

$$(u,v) = \sum_{j=0}^{N_2-1} \sum_{k=0}^{N_3} u(j,k) v(j,k) \hbar_2(j) \hbar_3(k).$$

We define the operators A_2 and A_3, mapping into H, by the equations: $A_3 = -\Lambda_3$, $A_2 y = -\Lambda_2 \bar{y}$, where $y(j,k) = \bar{y}(j,k)$ for $0 \le j \le N_2 - 1$, $0 \le k \le N_3$ and $\bar{y}$ satisfies the periodicity condition $\bar{y}(j,k) = \bar{y}(N_2 + j, k)$, $j = 0, -1$.

Using this notation, we write the difference scheme (49) in the form of the operator equation

$$Au = f, \qquad A = A_2 + A_3. \tag{50}$$

Taking into account the periodicity condition and using the Green difference formula, we obtain that

$$(Au,v) = -(\Lambda\bar{u},\bar{v}) = \sum_{k=0}^{N_3}\sum_{j=0}^{N_2-1} \hbar_3(k)h_2(j)(a_2\bar{u}_{\bar{\varphi}}\bar{v}_{\bar{\varphi}})_{jk}$$

$$+\sum_{k=1}^{N_3}\sum_{j=0}^{N_2-1} \hbar_2(j)h_3(k)(a_3\bar{u}_{\bar{z}}\bar{v}_{\bar{z}})_{jk} = (u,Av).$$

Consequently, the operator A is self-adjoint in H. In addition, examining the values of (Au,u), we find that the kernel of the operator A consists of the grid functions that are constant on ω^*. Therefore a solution of the difference problem (49) exists, if the condition $(f,1)=0$ is satisfied. Substituting here f from (49), we obtain

$$\sum_{j=0}^{N_2-1}\sum_{k=0}^{N_3} \hbar_2(j)\hbar_3(k)\psi(j,k) + \sum_{j=0}^{N_2-1} \hbar_2(j)[g_3^-(\varphi_j)+g_3^+(\varphi_j)] = 0.$$

If this condition is satisfied, a solution of the difference problem (46), (48) for $d=0$ and $\kappa_3^{\pm}=0$ exists, and any two solution of this equation differ by a constant.

We look now at the case where the solution of the difference problems (46)–(48) can be found by the direct methods outlined in Chapters 3 and 4.

The first case. The coefficients k_2, k_3, and q in equation (43) depend only on φ, $\kappa_3^{\pm}$ = constant, and the grid $\bar{\omega}$ is uniform in z. The difference problem (46), (48) can be written in the form of a system of three-point vector equations

$$\begin{aligned} (C+2\alpha E)Y_0 - 2Y_1 &= F_0, & k&=0,\\ -Y_{k-1}+CY_k-Y_{k+1} &= F_k, & 1&\le k\le N_3-1,\\ -2Y_{N_3-1}+(C+2\beta E)Y_{N_3} &= F_{N_3}, & k&=N_3, \end{aligned} \tag{51}$$

where $N_3 = 2^n$, $n>0$ is an integer, and

$$\begin{aligned} Y_k &= (y,(0,k),y(1,k),\ldots,y(N_2-1,k))^T,\\ F_k &= (\theta_0 f(0,k),\theta_1 f(1,k),\ldots,\theta_{N_2-1}f(N_2-1,k))^T,\\ CY_k &= ((2E-\theta_0\Lambda_2)y(0,k),\ldots,(2E-\theta_{N_2-1}\Lambda_2)y(N_2-1,k))^T \end{aligned}$$

for $0\le k\le N_3$. The operator Λ_2 was defined above, $f(j,k)$ is given in (49), and $\theta_j = h_3^2/a_3(j)$, $\alpha = h_3\kappa_3^-$, $\beta = h_3\kappa_3^+$.

Recall that in Section 3.4.3, we constructed the cyclic reduction method for solving problem (51) under the condition that $\alpha^2+\beta^2 \neq 0$. If $\alpha=\beta=0$ but $d \not\equiv 0$, then the algorithm for the method is laid out in Section 3.4.1. For the latter case we constructed, in Section 4.3.2, a combined method involving incomplete reduction and separation of variables.

The second case. The coefficients k_2, k_3, and q depend only on z, $\kappa_3^{\pm} =$ constant, and the grid $\bar{\omega}$ is uniform in φ. The difference problem (46), (48) can be written in the form of a system of three-point vector equations

$$\begin{aligned} -Y_{N_2-1}+CY_0-Y_1 &= F_0, & j&=0,\\ -Y_{j-1}+CY_j-Y_{j+1} &= F_j, & 1 &\le j \le N_2-2,\\ -Y_{N_2-2}+CY_{N_2-1}-Y_0 &= F_{N_2-1}, & j&=N_2-1. \end{aligned} \tag{52}$$

Here $N_2=2^n$, $n>0$ is an integer, and

$$\begin{aligned} Y_j &= (y(j,0), y(j,1),\dots,y(j,N_3))^T,\\ F_j &= (\theta_0 f(j,0), \theta_1 f(j,1),\dots,\theta_{N_3} f(j,N_3))^T,\\ CY_j &= ((2E-\theta_0\Lambda_3)y(j,0),\dots,(2E-\theta_{N_3}\Lambda_3)y(j,N_3))^T, \end{aligned}$$

where $0 \le j \le N_2-1$. The difference operator Λ_3 was defined in (26) with $d_3=d$ and $\theta_k = \hbar_3^2/a_2(k)$, $0\le k\le N_3$. The problem (52) can be solved by the cyclic-reduction method constructed in Section 3.4.2, or by a combined method using the algorithm for the discrete Fourier transform of a real periodic function. This algorithm was constructed in Section 4.1.4.

In conclusion we remark that, if the coefficients satisfy the conditions $k_2=k_2(\varphi)$ $k_3=k(z)$, $q=$ constant, $\kappa_3^{\pm}=$ constant, and the grid is non-uniform in each direction, then to find the solution of the problem (46), (48) it is possible to use the alternating-directions method with the optimal set of parameters

$$B_{k+1}\frac{y_{k+1}-y_k}{\tau_{k+1}}+Ay_k=f,\quad k=0,1,\dots,\quad y_0\in H,$$

$$B_k=\left(\omega_k^{(2)}E+A_2\right)\left(\omega_k^{(3)}E+A_3\right),\quad \tau_k=\omega_k^{(2)}+\omega_k^{(3)}.$$

Here we have $A_3=-\Lambda_3$, $A_2y=-\Lambda_2\bar{y}$, the difference operator Λ_3 is defined in (26) with $d_3=0.5d$, and $\Lambda_2 y=(a_2 y_{\bar{\varphi}})_{\hat{\varphi}}-0.5dy$. The constants δ_α and Δ_α, giving the bounds for the operator A_α, are estimated as follows: δ_3 and Δ_3 were found in Section 15.2, the constant δ_2 is found exactly: $\delta_2=0.5d$, and it is possible to take Δ_2 to be

$$\Delta_2=\max_{0\le j\le N_2-1}\left[\frac{2}{\hbar_2(j)}\left(\frac{a_2(j)}{h_2(j)}+\frac{a_2(j+1)}{h_2(j+1)}\right)+\frac{d}{2}\right].$$

15.3 Solution of difference problems in polar coordinate systems

15.3.1 Difference schemes for equations in a circle or a ring. We examine now methods for solving difference schemes for elliptic equations without mixed derivatives in polar coordinates. First of all, we study the case where the region in which we seek the solution is a circle or a ring in Cartesian coordinates. In polar coordinates, these regions correspond to a rectangle $\bar{G} = \{l_1 \le r \le L_1,\ l_2 \le \varphi \le L_2,\ l_1 \ge 0,\ L_2 - l_2 = 2\pi\}$. It is necessary to find a solution of the equation

$$\frac{1}{r}\frac{\partial}{\partial r}\left(rk_1\frac{\partial u}{\partial r}\right) + \frac{1}{r^2}\frac{\partial}{\partial \varphi}\left(k_2\frac{\partial u}{\partial \varphi}\right) - qu = -f, \quad (r,\varphi) \in G, \tag{1}$$

that is periodic in φ with period 2π and that satisfies the following conditions on the boundary of the rectangle $\bar{G}$:

1) for $r = L_1$, $l_2 \le \varphi \le L_2$, either the first-kind boundary conditions

$$u(r,\varphi) = g_1^+(\varphi), \tag{2}$$

or the second- or third-kind conditions

$$-k_1\frac{\partial u}{\partial r} = \kappa_1^+ u - g_1^+(\varphi); \tag{3}$$

2) for $r = l_1$, $l_2 \le \varphi \le L_2$, either the first-kind boundary conditions

$$u(r,\varphi) = g_1^-(\varphi), \tag{4}$$

or the second- or third-kind conditions

$$k_1\frac{\partial u}{\partial r} = \kappa_1^- u - g_1^-(\varphi); \tag{5}$$

for $r = l_1 = 0$ we pose the condition

$$\lim_{r\to 0} rk_1\frac{\partial u}{\partial r} = 0, \tag{6}$$

which singles out a bounded solution.

It is assumed that the coefficients satisfy the conditions $k_1(r,\varphi) \ge c_1 > 0$, $k_2(r,\varphi) \ge c_1 > 0$, $q_1(r,\varphi) \ge 0$, $\kappa_1^{\pm}(\varphi) \ge 0$.

We will consider any combination of the boundary conditions (2)–(5). We shall construct difference schemes corresponding to these boundary-value problems.

In the region $\bar{G}$, we introduce the arbitrary non-uniform rectangular grid

$$\bar{\omega} = \{(r_i, \varphi_j) \in \bar{G},\ r_i = r_{i-1} + h_1(i),\ 1 \le i \le N_1,\ r_0 = l_1,$$
$$r_{N_1} = L_1,\ \varphi_j = \varphi_{j-1} + h_2(j),\ 1 \le j \le N_2,\ \varphi_0 = l_2,\ \varphi_{N_2} = L_2\}.$$

The average step $\hbar_1(i)$ was defined in Section 15.2.1, and the step $\hbar_2(j)$ in Section 15.2.4 in formula (45). We define the grid function $\rho(i)$:

$$\rho(i) = \begin{cases} l_1 + \frac{1}{4} h_1(1), & i = 0, \\ r_i + \frac{1}{4}[h_1(i+1) - h_1(i)], & 1 \le i \le N_1 - 1, \\ L_1 - \frac{1}{4} h_1(N_1), & i = N_1. \end{cases} \tag{7}$$

In the simplest case of continuous coefficients k_1, k_2, q, and f, the coefficients of the difference scheme will be defined by the formulas

$$a_1(i,j) = \bar{r}_i k_1(\bar{r}_i, \varphi_j), \quad a_2(i,j) = k_2(r_i, \bar{\varphi}_j),$$
$$d(i,j) = q(r_i, \varphi_j), \qquad \psi(i,j) = f(r_i, \varphi_j),$$

where $\bar{r}_i = r_i - 0.5h_1(i)$, $\bar{\varphi}_j = \varphi_j - 0.5h_2(j)$.

Using this notation, we approximate (1) by the difference equation

$$\Lambda y = \frac{1}{\rho}(a_1 y_{\bar{r}})_{\hat{r}} + \frac{1}{\rho^2}(a_2 y_{\bar{\varphi}})_{\hat{\varphi}} - dy = -\psi,$$
$$1 \le i \le N_1 - 1, \quad 0 \le j \le N_2 - 1. \tag{8}$$

Here, for notational compactness, we have used the relations

$$y(i,j) = y(i, N_2 + j), \quad j = 0, -1, \quad a_2(i,0) = a_2(i, N_2),$$
$$h_2(0) = h_2(N_2), \tag{9}$$

which follow from the periodicity condition.

The boundary conditions (2), (4) are approximated exactly

$$y(N_1, j) = g_1^+(\varphi_j), \quad y(0,j) = g_1^-(\varphi_j), \quad 0 \le j \le N_2 - 1. \tag{10}$$

The difference analog of the third-kind boundary conditions (3), (5) has the form (for $0 \le j \le N_2 - 1$)

$$\Lambda y = -\frac{a_1}{\rho \hbar_1} y_{\bar{r}} + \frac{1}{\rho^2}(a_2 y_{\bar{\varphi}})_{\hat{\varphi}} - \left(d + \frac{r\kappa_1^+}{\rho\hbar_1}\right) y = -\psi - \frac{r g_1^+}{\rho \hbar_1}, \quad i = N_1, \quad (11)$$

$$\Lambda y = \frac{a_1^{+1}}{\rho \hbar_1} y_r + \frac{1}{\rho^2}(a_2 y_{\bar{\varphi}})_{\hat{\varphi}} - \left(d + \frac{r\kappa_1^-}{\rho\hbar_1}\right) y = -\psi - \frac{r g_1^-}{\rho \hbar_1}, \quad i = 0. \quad (12)$$

Here the relations (9) have been used.

It remains to construct the boundary-value difference problem on the side $r = l_1$ in the case where $l_1 = 0$. Since all the nodes lying on the side $r = 0$ are identified, then

$$y(0, j) = y_0, \quad 0 \le j \le N_2 - 1. \quad (13)$$

Further, since the origin is an interior point of the circle, then writing equation (1) in Cartesian coordinates and approximating it on a radially-annular grid using the condition (6), we obtain

$$\Lambda y = \frac{1}{2\pi\rho\hbar_1} \sum_{j=0}^{N_2-1} a_1^{+1} y_r \hbar_2 - dy = -\psi, \quad i = 0,$$
$$d(0,j) = d_0, \quad \psi(0,j) = \psi_0, \quad 0 \le j \le N_2 - 1. \quad (14)$$

Here y_0, d_0, and ψ_0 are the values of the corresponding grid functions at the center of the circle.

Thus, in the case of a circle, we have a non-local boundary condition (13), (14) on the side $r = 0$ of the rectangle $\bar{G}$. The difference schemes have been constructed.

Difference approximations to equation (1) often use a different grid in r in a neighborhood of the point $r = 0$, in which the point $r = 0$ is not included:

$$\bar{\omega} = \{(r_i, \varphi_j) \in \bar{G},\ r_i = (i + 0.5)h_1,\ 0 \le i \le N_1,\ r_{N_1} = L_1,$$
$$\varphi_j = \varphi_{j-1} + h_2(j),\ 1 \le j \le N_2,\ \varphi_0 = l_2,\ \varphi_{N_2} = L_2\}$$

(for simplicity, we assume that the grid in r is uniform).

Then $a_1(i,j) = \bar{r}_i k_1(\bar{r}_i, \varphi_j)$, $a_2(i,j) = k_2(r_i, \bar{\varphi}_j)$ and so forth, where $\bar{r}_i = ih_1$. Equation (8) remains unchanged, and for $i = 0$ we have the following difference equation:

$$\Lambda y = \frac{1}{r_0 h_1} a_1(1,j) y_r(1,j) + \frac{1}{r_0^2}(a_2 y_{\bar{\varphi}})_{\hat{\varphi}} - dy = -\psi$$

(here $r_0 = 0.5h_1$, $\bar{r}_1 = h_1$), which is analogous to a third-kind boundary condition.

A condition for $r = 0$ is absent; it is impossible to define the value of y at $r = 0$ from these difference equations.

15.3.2 The solubility of the boundary-value difference problems. In Section 15.3.1, we constructed difference schemes that approximate the problems (1)–(6). For a circle, the scheme is given by the formulas (8), (10), (11), (14), and for a ring by the formulas (8), (10), (12). We now investigate the solubility of these schemes.

We denote by $\bar{\omega}^*$ a part of the grid $\bar{\omega}$: $\bar{\omega} = \{(r_i, \varphi_j) \in \bar{\omega},\ 0 \le i \le N_1,\ 0 \le j \le N_2 - 1\}$. The space H consists of the grid functions defined on $\bar{\omega}^*$ that satisfy the auxiliary condition $y(0, j) =$ constant, $0 \le j \le N_2 - 1$, if $l_1 = 0$. The inner product in H is defined by the formula

$$(u, v) = \sum_{i=0}^{N_1} \sum_{j=0}^{N_2-1} u(i,j)v(i,j)\rho(i)\hbar_1(i)\hbar_2(j).$$

It is possible to show that, if the function $\rho(i)$ is defined by the formula (7), then the equation

$$(1,1) = 0.5(L_1^2 - l_1^2)(L_2 - l_2) = \pi(L_1^2 - l_1^2) \tag{15}$$

is valid, i.e. the square of the norm of the function that is identically equal to one on $\bar{\omega}^*$ is the same as the area of the circle ($l_1 = 0$) or the ring ($l_1 > 0$). In addition, if the region is a circle, then, using the fact that the grid functions in H are constant in j for $i = 0$, and also the equality

$$\sum_{j=0}^{N_2-1} \hbar_2(j) = L_2 - l_2 = 2\pi,$$

it is possible to obtain the following expression for the inner product introduced above

$$(u, v) = \rho(0)\hbar_1(0)2\pi u_0 v_0 + \sum_{i=1}^{N_1} \sum_{j=0}^{N_2-1} u(i,j)v(i,j)\rho(i)\hbar_1(i)\hbar_2(j), \tag{16}$$

where $u_0 = u(0, j)$, $v_0 = v(0, j)$.

We now investigate the solubility of the boundary-value difference problems (8), (11), (13), (14) for $l_1 = 0$ and (8), (11), (12) for $l_1 > 0$, if $d \equiv 0$, $\kappa_1^+ = \kappa_1^- \equiv 0$. We write the difference problems indicated above in the form of the operator equation

$$Au = f, \tag{17}$$

where the operator A is defined as follows: $Ay = -\Lambda\bar{y}$, $y(i,j) = \bar{y}(i,j)$ for $0 \le i \le N_1$, $0 \le j \le N_2 - 1$ and $\bar{y}$ satisfies the periodicity condition (9); in addition, $y(0,j) = \bar{y}(0,j) = \text{constant}$.

We look first at the operator A corresponding to the difference operator Λ for the problem (8), (11), (13), (14). Taking into account that Green's first difference formula for a function satisfying the periodicity condition takes the form

$$\sum_{j=0}^{N_2-1} (a_2 u_{\bar{\varphi}})_{\hat{\varphi}} v \hbar_2 = -\sum_{j=0}^{N_2-1} a_2 u_{\bar{\varphi}} v_{\bar{\varphi}} h_2,$$

we have, using (16), that

$$\begin{aligned}(Au, v) &= -(\Lambda\bar{u}, \bar{v}) \\ &= \sum_{j=0}^{N_2-1} \hbar_2 \left(\sum_{i=1}^{N_1} h_1 a_1 \bar{u}_{\bar{r}} \bar{v}_{\bar{r}} + \sum_{i=0}^{N_1} \rho \hbar_1 d\bar{u}\bar{v} + r\kappa_1^+ \bar{u}\bar{v}\big|_{i=N_1} \right) \\ &\quad + \sum_{i=1}^{N_1} \frac{\hbar_1}{\rho} \sum_{j=0}^{N_2-1} h_2 a_2 \bar{u}_{\bar{\varphi}} \bar{v}_{\bar{\varphi}} = -(\bar{u}, \Lambda\bar{v}) = (u, Av).\end{aligned}$$

Consequently, the operator A is self-adjoint in H.

For the operator A corresponding to the difference operator Λ from the problem (8), (11), (12), we obtain the analogous equation

$$\begin{aligned}&(Au, v) \\ &= \sum_{j=0}^{N_2-1} \hbar_2 \left(\sum_{i=1}^{N_1} h_1 a_1 \bar{u}_{\bar{r}} \bar{v}_{\bar{r}} + \sum_{i=0}^{N_1} \rho \hbar_1 d\bar{u}\bar{v} + r\kappa_1^- \bar{u}\bar{v}\,\big|_{i=0} + r\kappa_1^+ \bar{u}\bar{v}\big|_{i=N_1} \right) \\ &\quad + \sum_{i=0}^{N_1} \frac{\hbar_1}{\rho} \sum_{j=0}^{N_2-1} h_2 a_2 \bar{u}_{\bar{\varphi}} \bar{v}_{\bar{\varphi}} = (u, Av),\end{aligned}$$

from which follows the self-adjointness of the operator A.

If $d \equiv 0$, $\kappa_1^{\pm} \equiv 0$, then it follows from the self-adjointness of the operator A and from the Cauchy-Schwartz-Bunyakovskij inequality $(Au, u) \leq \| Au \| \| u \|$ that the kernel of the operator A consists of the grid functions which are equal to a constant on the grid $\bar{\omega}^*$. Therefore the existence condition for a solution to the operator equation (17) has the form $(f, 1) = 0$. For the problem (8), (11), (13), (14), it corresponds to the condition

$$\sum_{i=0}^{N_1} \sum_{j=0}^{N_2-1} \psi(i,j)\rho(i)\hbar_1(i)\hbar_2(j) + L_1 \sum_{j=0}^{N_2-1} \hbar_2(j) g_1^+(\varphi_j) = 0, \tag{18}$$

which is a difference analog to the condition (23) of Section 15.1. For the problem (8), (11), (12) the solubility condition has the form

$$\sum_{i=0}^{N_1} \sum_{j=0}^{N_2-1} \psi(i,j)\rho(i)\hbar_1(i)\hbar_2(j) + \sum_{j=0}^{N_2-1} \hbar_2(j)[L_1 g_1^+(\varphi_j) + l_1 g_1^-(\varphi_j)] = 0$$

and is an analog to the condition (25) of Section 15.1 that guarantees the solubility of the corresponding differential problem for a ring.

If these conditions are satisfied, then the solutions of these problems exist and any two solutions differ by a constant. The normal solution to these problems satisfies the condition $(\bar{y}, 1) = 0$.

Suppose that y is one of the solutions that can be found, for example, by fixing the desired solution at one node of the grid. Then, taking into account equation (15), we obtain that the function

$$\bar{y} = y - \frac{(y,1)}{\pi(L_1^2 - l_1^2)} = y - \frac{(y,1)}{(1,1)}$$

is the normal solution.

Remark. If we define the grid function $\rho(i)$ by the formulas

$$\rho(i) = r_i, \quad 1 \leq i \leq N_1, \quad \rho(0) = \begin{cases} h_1(0)/4, & l_1 = 0, \\ l_1, & l_1 > 0, \end{cases}$$

then we need only to change equation (15) in the case where $l_1 = 0$. In this case we will have

$$(1,1) = \pi L_1^2 + \frac{h_1^2(1)}{4}\pi = \pi\left(L_1^2 + \frac{h_1^2(1)}{4}\right).$$

15.3.3 The superposition principle for a problem in a circle. The solution of difference problems in a circle is complicated by the presence of the non-local boundary condition (14) given for $i = 0$. Notice that, if the problem is singular, and the solubility condition (18) is satisfied, then one of the solutions can be conveniently distinguished by fixing its value at the center of the circle, i.e. setting $y(0,j) = y_0$, $0 \le j \le N_2-1$. In this case, the condition (14) is not used, and the resulting problem with the given y_0 is analogous to the problem posed in the ring with a first-kind boundary condition on the interior circumference. Suppose now that the difference problem (8), (11), (13), (14) is non-singular. We shall show that its solution can be found by solving two auxiliary problems with local first-kind boundary conditions for $i = 0$, $0 \le j \le N_2 - 1$.

We will seek a solution to the problem (8), (11), (13), (14) in the form

$$y(i,j) = v(i,j) + y_0 w(i,j), \quad 0 \le i \le N_1, \quad 0 \le j \le N_2 - 1, \tag{19}$$

where y_0 is the value of the desired solution at the center of the circle, and $v(i,j)$ and $w(i,j)$ satisfy periodicity conditions

$$v(i,j) = v(i, N_2 + j), \quad w(i,j) = w(i, N_2 + j), \quad j = 0, -1$$

and are the solutions of the following boundary-value problems:

$$\left.\begin{aligned}
&\frac{1}{\rho}(a_1 v_{\bar r})_{\hat r} + \frac{1}{\rho^2}(a_2 v_{\bar\varphi})_{\hat\varphi} - dv = -\psi, && 1 \le i \le N_1 - 1,\\
&v(0,j) = 0, \quad i = 0, && 0 \le j \le N_2 - 1,\\
&-\frac{a_1}{\rho\hbar_1} v_{\bar r} + \frac{1}{\rho^2}(a_2 v_{\bar\varphi})_{\hat\varphi} - \left(d + \frac{r\kappa_1^+}{\rho\hbar_1}\right) v = -\psi - \frac{r g_1^+}{\rho\hbar_1}, && i = N_1,
\end{aligned}\right\} \tag{20}$$

$$\left.\begin{aligned}
&\Lambda w = \frac{1}{\rho}(a_1 w_{\bar r})_{\hat r} + \frac{1}{\rho^2}(a_2 w_{\bar\varphi})_{\hat\varphi} - dw = 0, && 1 \le i \le N_1 - 1,\\
&w(0,j) = 1, \quad i = 0, && 0 \le j \le N_2 - 1,\\
&-\frac{a_1}{\rho\hbar_1} w_{\bar r} + \frac{1}{\rho^2}(a_2 w_{\bar\varphi})_{\hat\varphi} - \left(d + \frac{r\kappa_1^+}{\rho\hbar_1}\right) w = 0, && i = N_1.
\end{aligned}\right\} \tag{21}$$

It is obvious that the function y defined according to (19) satisfies equation (8) and the conditions (11), (13). It remains to define y_0. Substituting (19) in the as yet unused condition (14), and taking into account the boundary conditions for v and w, we obtain

$$y_0 = \frac{\left[2\pi\rho\hbar_1\psi_0 + \sum_{j=0}^{N_2-1} a_1^{+1} v_r \hbar_2(j)\right]_{i=0}}{\left[2\pi\rho\hbar_1 d_0 - \sum_{j=0}^{N_2-1} a_1^{+1} w_r \hbar_2(j)\right]_{i=0}}. \tag{22}$$

We shall show that the denominator in (22) is non-zero. To do this, we take the inner product of equation (21) with w. Using the boundary conditions for w, the periodicity relations, and the Green difference formulas, we obtain

$$0 = \sum_{i=1}^{N_1-1}\sum_{j=0}^{N_2-1} (\Lambda w) w \rho \hbar_1 \hbar_2 = -\sum_{j=0}^{N_2-1} \hbar_2 \left(a_1^{+1} w_r \big|_{i=0} + L_1 \kappa_1^+ w^2\big|_{i=N_1}\right)$$
$$-\sum_{i=1}^{N_1}\sum_{j=0}^{N_2-1} a_1 w_{\bar{r}}^2 h_1 \hbar_2 - \sum_{i=1}^{N_1}\sum_{j=0}^{N_2-1} \hbar_1 \left[\frac{h_2}{\rho} a_2 w_{\bar{\varphi}}^2 + \hbar_2 \rho d w^2\right].$$

Since the function w is not constant, $d \geq 0$, $a_\alpha \geq c_1 > 0$, $\alpha = 1, 2$, and $\kappa_1^+ \geq 0$, where $d^2 + (\kappa_1^+)^2 \neq 0$, then from this we obtain that

$$\sum_{j=0}^{N_2-1} a_1^{+1} w_r \hbar_2 \big|_{i=0} < 0$$

and, consequently, the denominator in formula (22) is non-zero.

Thus, solving the original problem (8), (11), (13), (14) reduces to solving the two problems (20) and (21) with local boundary conditions and finding y_0 using the formula (22). The desired solution y is found from formula (19).

Notice that, if we are given a first-kind boundary condition $y(N_1, j) = g_1^+(\varphi_j)$ on the side $r = L_1$, then for the functions v and w we replace the third-kind boundary conditions in (20) and (21) by the conditions $v(N_1, j) = g_1^+(\varphi_j)$ and $w(N_1, j) = 0$ for $0 \leq j \leq N_2 - 1$. The formula (22) for y_0 is preserved. If the coefficients k_1, k_2, q, and κ_1^+ do not depend on φ, then the solution w of problem (21) also does not depend on φ. In this case, we have the following one-dimensional problem for the function w

$$\frac{1}{\rho}(a_1 w_{\bar{r}})_{\hat{r}} - dw = 0, \quad 1 \leq i \leq N_1 - 1,$$
$$w(0, j) = 1, \quad i = 0,$$
$$-\frac{a_1}{\rho\hbar_1} w_r - \left(d + \frac{r\kappa_1^+}{\rho\hbar_1}\right) w = 0, \quad i = N_1,$$

which can be solved by the elimination method.

15.3.4 Direct methods for solving equations in a circle or a ring. From the discussion above it follows that it is sufficient to consider methods for solving the difference problems (8), (10)–(12). First we study the cases for which these difference problems can be solved by one of the direct methods laid out in Chapters 3 and 4.

Suppose that the coefficients k_1, k_2, q of equation (1) do not depend on φ: $k_1 = k_1(r)$, $k_2 = k_2(r)$, $q = q(r)$. Such a situation occurs for Poisson's equation in polar coordinates. Assume in addition that k_1^- and k_1^+ are constants in the third-kind boundary conditions (11), (12). It is assumed that the grid $\bar{\omega}$ is uniform in φ, i.e. $h_2(j) \equiv h_2$, but it may be non-uniform in r. Under these assumptions, the difference equation (8) with any combination of the boundary conditions (10)–(12) can be solved by either the cyclic-reduction method, or by a combined method involving incomplete reduction and separation of variables.

We illustrate the possibility of applying direct methods on an example in which third-(second-) kind boundary conditions can be examined analogously.

In view of the assumptions made above, the coefficients of the difference scheme can be defined by the formulas

$$a_1(i) = \bar{r}_i k_1(\bar{r}_i), \quad a_2(i) = k_2(r_i), \quad d(i) = q(r_i),$$

and since the grid $\bar{\omega}$ is uniform in φ, the difference operator $(a_2 y_{\bar{\varphi}})_{\hat{\varphi}}$ can be changed to $a_2 y_{\bar{\varphi}\varphi}$.

We shall reduce the difference problem (8), (11), (12) to a system of three-point vector equations

$$\begin{aligned} -Y_{N_2-1} + CY_0 - Y_1 &= F_0, & j &= 0, \\ -Y_{j-1} + CY_j - Y_{j+1} &= F_j, & 1 &\le j \le N_2 - 2, \\ -Y_{N_2-2} + CY_{N_2-1} - Y_0 &= F_{N_2-1}, & j &= N_2 - 1. \end{aligned} \tag{23}$$

Here for $0 \le j \le N_2 - 1$ we have used the notation:

$$\begin{aligned} Y_j &= (y(0,j), y(1,j), \ldots, y(N_1,j))^T, \\ F_j &= (\theta_0 f(0,j), \theta_1(f(1,j), \ldots, \theta_{N_1} f(N_1,j))^T, \\ CY_j &= ((2E - \theta_0 \Lambda_1) y(0,j), \ldots, (2E - \theta_{N_1} \Lambda_1) y(N_1,j))^T, \end{aligned}$$

where

$$f(i,j) = \begin{cases} \psi(0,j) + \dfrac{l_1 g_1^-(\varphi_j)}{\rho(0)\hbar_1(0)}, & i = 0, \\ \psi(i,j), & 1 \le i \le N_1 - 1, \\ \psi(N_1,j) + \dfrac{L_1 g_1^+(\varphi_j)}{\rho(N_1)\hbar_1(N_1)}, & i = N_1, \end{cases} \tag{24}$$

the difference operator Λ_1 acts as follows:

$$\Lambda_1 y = \begin{cases} \frac{a_1^{+1}}{\rho \hbar_1} y_r - \left(d + \frac{r\kappa_1^-}{\rho \hbar_1}\right) y, & i = 0, \\ \frac{1}{\rho}(a_1 y_{\bar{r}})_{\hat{r}} - dy, & 1 \le i \le N_1 - 1, \\ -\frac{a_1}{\rho \hbar_1} y_{\bar{r}} - \left(d + \frac{r\kappa_1^+}{\rho \hbar_1}\right) y, & i = N_1, \end{cases} \tag{25}$$

and, finally, $\theta_i = \rho^2(i) h_2^2 / a_2(i)$, $0 \le i \le N_1$.

The system (23) is obtained from (8), (11) and (12) by multiplying each equation by the corresponding θ_i and transforming to vector notation.

Recall that the algorithm for the cyclic-reduction method applied to the system (23) was described in Section 3.4.2. In a combined method, we use the algorithm for the fast Fourier transform introduced in Section 4.1.4. These methods are characterized by the estimate $O(N_1 N_2 \log_2 N_2)$ for the operation count where $N_2 = 2^n$.

15.3.5 The alternating-directions method. Suppose now that the coefficients in equation (1) and the boundary conditions (3), (5) satisfy the conditions $k_1 = k_1(r)$, $k_2 = k_2(\varphi)$, $q =$ constant, $\kappa_1^\pm =$ constant, i.e. the alternating-directions method is applicable to the problem (1), (3), (5). It is assumed that the grid $\bar{\omega}$ is non-uniform in each direction. We consider the difference equation (8) with any combination of the boundary conditions (10)–(12). Under these assumptions, the variables in the difference scheme are separated, and an approximate solution can be found using the alternating-directions method with the optimal set of iterative parameters.

As an example, we consider the problem (8), (11), (12) with third-kind boundary conditions for $r = l_1$ and $r = L_1$. We write this problem in the form

$$\begin{aligned} \bar{\Lambda} y &= -\bar{f}, \quad 0 \le i \le N_1, \quad 0 \le j \le N_2 - 1, \\ \bar{\Lambda} &= \bar{\Lambda}_1 + \bar{\Lambda}_2, \quad \bar{f} = \rho^2 f, \end{aligned} \tag{26}$$

where $\bar{\Lambda}_1 = \rho^2 \Lambda_1$, the operator Λ_1 is defined in (25), the operator $\bar{\Lambda}_2$ is given by the equation $\bar{\Lambda}_2 y = (a_2 y_{\bar{\varphi}})_{\hat{\varphi}}$, where the relation (9) is satisfied, and the right-hand side f is defined in (24). The equation (26) was obtained from (8), (11), (12) by multiplying by ρ^2.

By the assumptions made above, the coefficients of the difference scheme (26) are chosen using the formulas $a_1(i) = \bar{r}_i k_1(\bar{r}_i)$, $a_2(j) = k_2(\bar{\varphi}_j)$, $d = q =$ constant.

In the space H of grid functions defined on $\bar{\omega}^*$, we define the inner product

$$(u,v) = \sum_{i=0}^{N_1} \sum_{j=0}^{N_2-1} \frac{\hbar_1(i)\hbar_2(j)}{\rho(i)} u(i,j)v(i,j). \tag{27}$$

The operators A_1 and A_2, mapping into H, are defined in the usual way: $A_\alpha y = -\Lambda_\alpha \bar{y}$, where $y(i,j) = \bar{y}(i,j)$ for $0 \le i \le N_1$, $0 \le j \le N_2 - 1$ and $\bar{y}$ satisfies the periodicity relation (9). Then the scheme (26) can be written in the form of the operator equation

$$Au = \bar{f}, \quad A = A_1 + A_2 \tag{28}$$

in the space H.

To solve equation (28), we use the alternating-directions method whose iterative scheme has the form

$$\begin{gathered} B_{k+1} \frac{y_{k+1} - y_k}{\tau_{k+1}} + Ay_k = \bar{f}, \quad k = 0,1,\ldots, \quad y_0 \in H, \\ B_k = \left(\omega_k^{(1)} E + A_1\right)\left(\omega_k^{(2)} E + A_2\right), \quad \tau_k = \omega_k^{(1)} + \omega_k^{(2)}. \end{gathered} \tag{29}$$

The self-adjointness of the operators A_1 and A_2 in the space H is established using the Green difference formulas, and their commutativity can be easily verified.

We now find bounds for the operators A_1 and A_2, i.e. the constants δ_α and Δ_α, $\alpha = 1,2$, in the inequalities

$$\delta_\alpha(u,u) \le (A_\alpha u, u) \le \Delta_\alpha(u,u).$$

We first find δ_2 and Δ_2. Since the periodicity condition (9) is satisfied for the function $\bar{u}(i,j)$, we have

$$(A_2 u, u) = -(\Lambda_2 \bar{u}, \bar{u}) = \sum_{i=0}^{N_1} \sum_{j=0}^{N_2-1} \frac{\hbar_1(i)h_2(j)}{\rho(i)} \left(a_2 \bar{u}_{\bar{\varphi}}^2\right)_{ij},$$

and then

$$\delta_2 = 0, \quad \Delta_2 = \max_{0 \le j \le N_2-1} \left[\frac{a_2(j+1)}{h_2(j+1)} + \frac{a_2(j)}{h_2(j)}\right] \frac{2}{\hbar_2(j)}.$$

Here we have used the relation (9) for a_2 and h_2.

Further, using an analog of lemma 16 in Chapter 5, we find that δ_1 can be estimated as follows: $1/\delta_1 = \max_{0\le i\le N_1} v(i)$, where $v(i)$ is the solution of the boundary-value problem

$$\begin{aligned} &\rho(a_1 v_{\bar{r}})_{\hat{r}} - d\rho^2 v = -1, \quad 1 \le i \le N_1 - 1, \\ &\frac{\rho a_1^{+1}}{\hbar_1} v_r - \left(d + \frac{r\kappa_1^-}{\rho\hbar_1}\right)\rho^2 v = -1, \quad i = 0, \\ &-\frac{\rho a_1}{\hbar_1} v_{\bar{r}} - \left(d + \frac{r\kappa_1^+}{\rho\hbar_1}\right)\rho^2 v = -1, \quad i = N_1. \end{aligned} \tag{30}$$

The problem (30) can be solved by the elimination method.

We obtain now an estimate for Δ_1. Using Green's first difference formula and the definition (27) of the inner product, we find

$$\begin{aligned} (A_1 u, u) &= -(\bar{\Lambda}_1 \bar{u}, \bar{u}) \\ &= \sum_{j=0}^{N_2-1} \hbar_2(j)\left[\sum_{i=1}^{N_2-1} h_1(i) a_1(i) \bar{u}_{\bar{r}}^2(i,j) + d\sum_{i=0}^{N_1} \hbar_1(i)\rho(i)\bar{u}^2(i,j)\right. \\ &\qquad \left. + l_1\kappa_1^- \bar{u}^2(0,j) + L_1\kappa_1^+\bar{u}^2(N_1,j)\right] \end{aligned}$$

We now estimate the expression in square brackets. From an analog of lemma 16 in Chapter 5 we obtain the estimate

$$\begin{aligned} &d\sum_{i=0}^{N_1} \hbar_1(i)\rho(i)\bar{u}^2(i,j) + l_1\kappa_1^-\bar{u}^2(0,j) + L_1\kappa_1^+\bar{u}^2(N_1,j) \\ &\qquad \le m_1\left[\sum_{i=1}^{N_1}\left(a_1\bar{u}_{\bar{r}}^2\right)_{ij} h_1(i) + \sum_{i=0}^{N_1}\frac{\hbar_1(i)}{\rho(i)}\bar{u}^2(i,j)\right] \end{aligned} \tag{31}$$

where $m_1 = \max_{0\le i\le N_1} w(i)$, and $w(i)$ is the solution of the problem

$$\begin{aligned} &\rho(a_1 w_{\bar{r}})_{\hat{r}} - w = d\rho^2, \quad 1 \le i \le N_1 - 1, \\ &\frac{\rho a_1^{+1}}{\hbar_1} w_r - w = -\left(d + \frac{r\kappa_1^-}{\rho\hbar_1}\right)\rho^2, \quad i = 0, \\ &-\frac{\rho a_1}{\hbar_1} w_{\bar{r}} - w = -\left(d + \frac{r\kappa_1^+}{\rho\hbar_1}\right)\rho^2, \quad i = N_1. \end{aligned} \tag{32}$$

Further, from an analog of lemma 17 of Chapter 5, we obtain the estimate

$$\sum_{i=1}^{N_1} a_1(i)\bar{u}_{\bar{r}}^2(i,j)h_1(i) \le m_2 \sum_{i=0}^{N_1} \frac{\hbar_1(i)}{\rho(i)}\bar{u}^2(i,j), \tag{33}$$

where

$$m_2 = \max\left(\frac{a_1(N_1)\rho(N_1)}{\hbar_1^2(N_1)}, \frac{a_1(1)\rho(0)}{\hbar_1^2(0)}, \max_{1\le i\le N_1-1} \frac{2\rho(i)}{\hbar(i)}\left[\frac{a_1(i)}{h_1(i)} + \frac{a_1(i+1)}{h_1(i+1)}\right]\right).$$

From (31) and (33) follows the estimate

$$\sum_{i=1}^{N_1} h_1 a_1 \bar{u}_{\bar{r}}^2 + d\sum_{i=0}^{N} \hbar_1 \rho \bar{u}^2 + l_1 \kappa_1^- \bar{u}^2\,\big|_{i=0} + L_1 \kappa_1^+ \bar{u}^2\big|_{i=N_1}$$

$$\le \Delta_1 \sum_{i=0}^{N_1} \frac{\hbar_1}{\rho}\bar{u}^2, \qquad \Delta_1 = m_1 + m_2(1+m_1).$$

Multiplying this inequality by $\hbar_2(j)$ and summing for j between 0 and N_2-1, we obtain $(A_1 u, u) \le \Delta_1(u,u)$.

Thus, the constants δ_α and Δ_α, $\alpha = 1, 2$, have been found. Recall that the formulas for the iterative parameters $\omega_k^{(1)}$ and $\omega_k^{(2)}$ were obtained in Section 11.1.4.

In an analogous manner, we construct the alternating-directions method for the difference problem (8), (10) with first-kind boundary conditions. The constants δ_α and Δ_α are estimated in the same way as in the earlier case, only in (30) and (32), the third-kind boundary conditions should be changed to the conditions $v(0) = 0$, $v(N_1) = 0$ and $w(0) = 0$, $w(N_1) = 0$.

In conclusion we remark that for $d = 0$, $\kappa_1^\pm = 0$, the problem (8), (11), (12) is singular, and if the solubility condition

$$\sum_{i=0}^{N_1}\sum_{j=0}^{N_2-1} \psi\rho\hbar_1\hbar_2 + \sum_{j=0}^{N_2-1} \hbar_2[L_1 g_1^+ + l_1 g_1^-] = 0,$$

is satisfied, then the problem has a non-unique solution. For this case, the set of parameters $\omega_k^{(1)}$ and $\omega_k^{(2)}$ for the alternating-directions method was constructed in Section 12.4.1.

15.3.6 Solution of difference problems in a ring sector. We look now at methods for solving boundary-value difference problems for an ellitptic equation without mixed derivatives in a ring sector.

In the region $\bar{G} = \{l_1 \le r \le L_1,\ l_2 \le \varphi \le L_2,\ l_1 > 0,\ L_2 - l_1 < 2\pi\}$, it is necessary to find a solution to equation (1) that satisfies one of the boundary conditions (2)–(5) on the sides $r = l_1$ and $r = L_1$, and that satisfies on the sides $\varphi = l_2$ and $\varphi = L_2$ one of the conditions

$$u(r,\varphi) = g_2^-(r), \quad \varphi = l_2 \tag{34}$$

or

$$\frac{k_2}{r}\frac{\partial u}{\partial \varphi} = \kappa_2^- u - g_2^+(r), \quad \varphi = l_2, \tag{35}$$

$$u(r,\varphi) = g_2^+(r), \qquad \varphi = L_2, \tag{36}$$

or

$$-\frac{k_2}{r}\frac{\partial u}{\partial \varphi} = \kappa_2^+ u - g_2^+(r), \quad \varphi = L_2. \tag{37}$$

It is assumed that the coefficients satisfy the conditions $k_1(r,\varphi) \ge c_1 > 0$, $k_2(r,\varphi) \ge c_1 > 0$, $q(r,\varphi) \ge 0$, $\kappa_1^\pm(\varphi) \ge 0$, $\kappa_2^\pm(r) \ge 0$.

In the region $\bar{G}$, we introduce an arbitrary non-uniform rectangular grid $\bar{\omega}$ (cf. Section 15.3.1):

$$\bar{\omega} = \{(r_i,\varphi_j) \in \bar{G}, \quad r_i = r_{i-1} + h_1(i), \quad 1 \le i \le N_1, \quad r_0 = l_1,$$
$$r_{N_1} = L_1, \quad \varphi_j = \varphi_{j-1} + h_2(j), \quad 1 \le j \le N_2, \quad \varphi_0 = l_2, \quad \varphi_{N_2} = L_2\}$$

and define the average steps $\hbar_1(i)$ and $\hbar_2(j)$:

$$\hbar_\alpha(m) = \begin{cases} 0.5h_\alpha(1), & m = 0, \\ 0.5[h_\alpha(m) + h_\alpha(m+1)], & 1 \le m \le N_\alpha - 1, \\ 0.5h_\alpha(N_\alpha), & m = N_\alpha, \quad \alpha = 1,2. \end{cases}$$

Equation (1) can be approximated by the difference equation

$$\frac{1}{\rho}(a_1 y_{\bar{r}})_{\hat{r}} + \frac{1}{\rho^2}(a_2 y_{\bar{\varphi}})_{\hat{\varphi}} - dy = -\psi, \tag{38}$$
$$1 \le i \le N_1 - 1, \quad 1 \le j \le N_2 - 1.$$

The first-kind boundary conditions (2), (4), (34), (36) are approximated exactly:

$$y(N_1,j) = g_1^+(\varphi_j), \quad y(0,j) = g_1^-(\varphi_j), \quad 0 \le j \le N_2, \tag{39}$$

$$y(i,N_2) = g_2^+(r_i), \quad y(i,0) = g_2^-(r_i), \quad 0 \le i \le N_1. \tag{40}$$

The third-kind conditions (3) and (5) that were given for $r = L_1$ and $r = l_1$, are changed to the conditions (11) and (12) for $1 \le j \le N_2 - 1$.

The difference analog of the boundary conditions (35) and (37) has the form

$$\frac{1}{\rho}(a_1 y_{\bar{r}})_{\hat{r}} + \frac{a_2^{+1}}{\rho^2 \hbar_2} y_\varphi - \left(d + \frac{\kappa_2^-}{\rho \hbar_2}\right) y = -\psi - \frac{g_2^-}{\rho \hbar_2}, \quad j = 0, \tag{41}$$

$$\frac{1}{\rho}(a_1 y_{\bar{r}})_{\hat{r}} - \frac{a_2}{\rho^2 \hbar_2} y_{\bar{\varphi}} - \left(d + \frac{\kappa_2^+}{\rho \hbar_2}\right) y = -\psi - \frac{g_2^+}{\rho \hbar_2}, \quad j = N_2. \tag{42}$$

If we are given third-kind boundary conditions on intersecting sides of the rectangle, then we pose the following boundary conditions at the corner nodes of the grid $\bar{\omega}$:

$$\frac{a_1^{+1}}{\rho \hbar_1} y_r + \frac{a_2^{+1}}{\rho^2 \hbar_2} y_\varphi - \left(d + \frac{r\kappa_1^-}{\rho \hbar_1} + \frac{\kappa_2^-}{\rho \hbar_2}\right) y = -\psi - \frac{r g_1^-}{\rho \hbar_1} - \frac{g_2^-}{\rho \hbar_2}, \tag{43}$$

if $i = j = 0$;

$$-\frac{a_1}{\rho \hbar_1} y_{\bar{r}} + \frac{a_2^{+1}}{\rho^2 \hbar_2} y_\varphi - \left(d + \frac{r\kappa_1^+}{\rho \hbar_1} + \frac{\kappa_2^-}{\rho \hbar_2}\right) y = -\psi - \frac{r g_1^+}{\rho \hbar_1} - \frac{g_2^-}{\rho \hbar_2}, \tag{44}$$

if $i = N_1$, $j = 0$;

$$\frac{a_1^{+1}}{\rho \hbar_1} y_r - \frac{a_2}{\rho^2 \hbar_2} y_{\bar{\varphi}} - \left(d + \frac{r\kappa_1^-}{\rho \hbar_1} + \frac{\kappa_2^+}{\rho \hbar_2}\right) y = -\psi - \frac{r g_1^-}{\rho \hbar_1} - \frac{g_2^+}{\rho \hbar_2}, \tag{45}$$

if $i = 0$, $j = N_2$; and, finally,

$$-\frac{a_1}{\rho \hbar_1} y_{\bar{r}} - \frac{a_2}{\rho^2 \hbar_2} y_{\bar{\varphi}} - \left(d + \frac{r\kappa_1^+}{\rho \hbar_1} + \frac{\kappa_2^+}{\rho \hbar_2}\right) y = -\psi - \frac{r g_1^+}{\rho \hbar_1} - \frac{g_2^+}{\rho \hbar_2}, \tag{46}$$

if $i = N_1$, $j = N_2$.

If we consider the difference problem (38), (11), (12), (41)–(46) with $d \equiv 0$ and $\kappa_\alpha^\pm \equiv 0$, $\alpha = 1, 2$, then a solution exists if the following condition is satisfied

$$\sum_{j=0}^{N_2}\sum_{i=0}^{N_1} \rho \hbar_1 \hbar_2 \psi + \sum_{j=0}^{N_2} \hbar_2 (L_1 g_1^+ + l_1 g_1^-) + \sum_{i=0}^{N_1} \hbar_1 (g_2^- + g_2^+) = 0;$$

this is the difference analog of the solubility condition (27) of Section 15.1 for the corresponding differential equation problem. Here, any two solutions of this equation differ by a constant.

This assertion is proved in almost the same way as in Section 15.3.2 for the case of a circle and a ring. Here the inner product in the space H of grid functions defined on $\bar{\omega}$ is given by the formula

$$(u, v) = \sum_{i=0}^{N_1}\sum_{j=0}^{N_2} u(i,j) v(i,j) \rho(i) \hbar_1(i) \hbar_2(j). \tag{47}$$

Notice that the coefficients a_1, a_2, q, and the function $\rho(i)$ are defined in this sub-section in the same way as in Section 15.3.1.

We now comment on methods for solving difference problems we have constructed. If the coefficients k_1, k_2, and q depend only on r, $\kappa_1^\pm$ are constants, and $\kappa_2^\pm = 0$, if we are given the boundary conditions (3), (5), (35), (37), and if the grid $\bar{\omega}$ is uniform in φ, then the corresponding difference problems can be solved by the direct methods constructed in Chapters 3 and 4.

If the conditions $k_1 = k_1(r)$, $k_2 = k_2(\varphi)$, $q =$ constant, $\kappa_\alpha^\pm =$ constant are satisfied, and if the grid $\bar{\omega}$ is non-uniform in each direction, then we can solve the difference problems using the alternating-directions method with the optimal set of parameters. In this case, as was done in the previous sub-section, the difference equations are first multiplied by $\rho^2(i)$.

15.3.7 The general variable-coefficients case. We look now at the case where the variables are not separated and where it is necessary to use an iterative method to find a solution to the boundary-value problem.

Suppose, for example, that it is necessary to find a solution to a Dirichlet problem for equation (1) on the grid $\bar{\omega}$ under the assumptions that the grid $\bar{\omega}$ is uniform in φ ($h_2(j) \equiv h_2$), $q = 0$, and the coefficients k_1 and k_2 satisfy the conditions

$$0 < c_1 \le k_\alpha(r, \varphi) \le c_2, \quad \alpha = 1, 2. \tag{48}$$

Under these assumptions, the difference problem can be written in the form

$$\Lambda y = \frac{1}{\rho}(a_1 y_{\bar{r}})_{\hat{r}} + \frac{1}{\rho^2}(a_2 y_{\bar{\varphi}})_{\varphi} = -\psi, \quad (r,\varphi) \in \omega, \tag{49}$$

where

$$\begin{aligned} a_1(i,j) &= \bar{r}_i k_1(\bar{r}_i, \varphi_j), & a_2(i,j) &= k_2(r_i, \bar{\varphi}_j), \\ \bar{r}_i &= r_i - 0.5h_1(i), & \bar{\varphi}_j &= \varphi_j - 0.5h_2. \end{aligned} \tag{50}$$

In the space H of grid functions defined on ω, we define the inner product using

$$(u,v) = \sum_{i=1}^{N_1-1} \sum_{j=1}^{N_2-1} u(i,j)v(i,j)\rho(i)\hbar_1(i)h_2$$

and the operators A and R, mapping in H, by $Ay = -\Lambda \mathring{y}$, $Ry = -\mathcal{R}\mathring{y}$, where $y(r,\varphi) = \mathring{y}(r,\varphi)$ for $(r,\varphi) \in \omega$ and $\mathring{y}(r,\varphi) = 0$ for $(r,\varphi) \in \gamma$. Here the difference operator $\mathcal{R}$ is defined by the relation

$$\mathcal{R}y = \frac{1}{\rho}(\bar{r} y_{\bar{r}})_{\hat{r}} + \frac{1}{\rho^2} y_{\bar{\varphi}\varphi}, \quad (r,\varphi) \in \omega.$$

Using the Green difference formulas, it is possible to prove that the operators A and R are self-adjoint in H and, in addition, for any $y \in H$ we have the equations

$$(Ay,y) = \sum_{i=1}^{N_1} \sum_{j=1}^{N_2-1} a_1 \mathring{y}_{\bar{r}}^2 h_1 h_2 + \sum_{j=1}^{N_2} \sum_{i=1}^{N_1-1} \frac{a_2}{\rho} \mathring{y}_{\bar{\varphi}}^2 \hbar_1 h_2,$$

$$(Ry,y) = \sum_{i=1}^{N_1} \sum_{j=1}^{N_2-1} \bar{r} \mathring{y}_{\bar{r}}^2 h_1 h_2 + \sum_{j=1}^{N_2} \sum_{i=1}^{N_1-1} \frac{1}{\rho} \mathring{y}_{\bar{\varphi}}^2 \hbar_1 h_2.$$

From this and from (48), (50), it follows that the operators A and R are energy equivalent with constants $\gamma_1 = c_1$ and $\gamma_2 = c_2$:

$$\gamma_1(Ry,y) \le (Ay,y) \le \gamma_2(Ry,y), \quad \gamma_1 > 0. \tag{51}$$

The difference problem (49) can be written in the form of the operator equation

$$Au = f$$

with the operator A defined above. To solve this problem, we use an implicit iterative scheme

$$B\frac{y_{k+1} - y_k}{\tau_{k+1}} + Ay_k = f, \quad k = 0,1,\ldots, \quad y_0 \in H, \tag{52}$$

where $B = R$.

From the general theory of iterative methods outlined in Chapter 6, it follows that, if the parameters τ_{k+1} in the scheme (52) are selected using the formulas for the Chebyshev method

$$\tau_k = \frac{\tau_0}{1+\rho_0\mu_k},$$

$$\mu_k \in \mathcal{M}_n^* = \left\{-\cos\frac{(2i-1)\pi}{2n},\ 1 \le i \le n\right\}, \quad k = 1,2,\ldots,n,$$

then we have the following estimate for the error $z_n = y_n - u$

$$\| y_n - u \|_D \le \epsilon \| y_0 - u \|_D,$$

where $D = A$, $D = B$, or $D = AB^{-1}A$; the iteration count can be estimated using

$$n \ge n_0(\epsilon) = \ln(0.5\epsilon)/\ln\rho_1.$$

Here

$$\tau_0 = \frac{2}{\gamma_1+\gamma_2}, \quad \rho_0 = \frac{1-\xi}{1+\xi}, \quad \rho_1 = \frac{1-\sqrt{\xi}}{1+\sqrt{\xi}}, \quad \xi = \frac{\gamma_1}{\gamma_2}.$$

Since γ_1 and γ_2 do not depend on the step-size of the grid $\bar{\omega}$, the iteration count is proportional to $|\ln 0.5\epsilon|$ and does not change as the grid is refined.

To find y_{k+1}, we have the difference problem

$$\mathcal{R}y_{k+1} = -F, \quad (r,\varphi) \in \omega, \quad y_{k+1} = g, \quad (r,\varphi) \in \omega$$

with a known right-hand side $F = -\mathcal{R}y_k + \tau_{k+1}(\Lambda y_k + \psi)$. Notice that this problem satisfies all the conditions necessary for application of direct solution methods, for example the cyclic-reduction method with its operation count of $O(N_1N_2\log_2 N_2)$, where $N_2 = 2^n$. Thus, the total number of operations required to find the solution of this difference problem to an accuracy ϵ can be estimated by the quantity $O(N_1N_2\log_2 N_2\ln(2/\epsilon))$.

Analogously, it is possible to construct, under corresponding assumptions, iterative methods for solving boundary-value difference problems posed in previous sub-sections involving cylindrical and polar coordinates.

Appendix A

Construction of the Minimax Polynomial

1. In Section 6.2, when examining two-level iterative schemes, we formulated the following problem: construct the polynomial of degree n whose value at zero is 1, and whose maximum modulus on the interval $[\gamma_1, \gamma_2]$ is minimal.

We now solve this problem. It will be convenient for us to carry out our investigation, not on the interval $[\gamma_1, \gamma_2]$, but on the interval $[-1, 1]$. To do this, we make a linear change of variable, mapping the interval $\gamma_1 \leq t \leq \gamma_2$ onto the interval $-1 \leq x \leq 1$, and the point γ_1 to the point 1. This change of variable has the form

$$t = \frac{1 - \rho_0 x}{\tau_0}, \quad \tau_0 = \frac{2}{\gamma_1 + \gamma_2}, \quad \rho_0 = \frac{1 - \xi}{1 + \xi}, \quad \xi = \frac{\gamma_1}{\gamma_2}.$$

Under this transformation, the point $t = 0$ corresponds to the point $x = 1/\rho_0 > 0$.

Thus, the problem formulated above is equivalent to the problem: among all polynomials of degree n that take on the value 1 at the point $x = 1/\rho_0 > 0$, find the one that deviates least from zero on the interval $[-1, 1]$.

This is the classical Chebyshev problem from the theory of function approximation; its solution is well known, but it will be useful for us to derive this solution here. To do this, we will require

Theorem 1. *For any two continuous functions $g(x) > 0$ and $f(x)$ on the interval $[-1, 1]$, there exists a unique polynomial $P_n(x)$ of degree at most n such that*

$$q_n = \max_{-1 \leq x \leq 1} g(x)|f(x) - P_n(x)| = \min_{\left\{ \substack{R_k(x) \\ k \leq n} \right\}} \max_{-1 \leq x \leq 1} g(x)|f(x) - R_k(x)|. \square$$

This polynomial is fully characterized by the following property: there are at most $n+2$ sequential points on the interval $[-1,1]$ at which the function $g(x)(f(x) - P_n(x))$ takes on the value q_n with alternating sign.

We transform this problem to the problem described in theorem 1. Taking into account that the desired polynomial takes on the value 1 at the point $x = 1/\rho_0$, we represent it in the form

$$P_n(x) = 1 - \left(\frac{1}{\rho_0} - x\right) R_{n-1}(x) = \frac{1-\rho_0 x}{\rho_0}\left[\frac{\rho_0}{1-\rho_0 x} - R_{n-1}(x)\right],$$

where $R_{n-1}(x)$ is a polynomial of degree at most $n-1$.

From this it follows that our problem reduces to the problem of finding a polynomial $R_{n-1}(x)$ of degree at most $n-1$ that gives the best uniform approximation (with weight function $g(x) = (1-\rho_0 x)/\rho_0 > 0$) to the function $f(x) = \rho_0/(1-\rho_0 x)$ on the interval $[-1,1]$.

This is the problem given in theorem 1.

Therefore, on the basis of theorem 1, there exist at least $n+1$ points $x_1, x_2, \ldots, x_{n+1}$ in the interval $[-1,1]$ at which the desired polynomial $P_n(x)$ takes on the value q_n with alternating sign.

We show first that here must be exactly $n+1$ such points. If a continuous function takes on the value q_n with alternating sign at more than $n+1$ points on the interval $[-1,1]$, it must vanish for at least n points on this interval.

Since the polynomial $P_n(x)$ is not identically zero, it must vanish at no more than n points on the interval $[-1,1]$. Therefore, the desired polynomial $P_n(x)$ takes on the value q_n with alternating sign precisely $n+1$ times on the interval $[-1,1]$.

We now characterize these points. If the polynomial $P_n(x)$ takes on its maximal value at an interior point of the interval $[-1,1]$, then the derivative $P_n'(x)$ vanishes at this point. But the degree of $P_n'(x)$ is equal to $n-1$ and, consequently, the derivative of the desired polynomial can vanish at only $n-1$ points. Therefore, the polynomial has $n-1$ interior extremal points on $[-1,1]$ and, consequently, two boundary extrema, i.e.

$$|P_n(-1)| = |P_n(1)| = q_n.$$

Thus we have

$$P_n(\omega_j) = 0, \quad j = 1,2,\ldots,n \quad |P_n(x_j)| = q_n, \quad j = 1,2,\ldots,n+1,$$

where ω_j are the roots of the polynomial, and x_j are the extremal points

$$-1 = x_{n+1} < \omega_n < x_n < \cdots < \omega_2 < x_2 < \omega_1 < x_1 = 1.$$

In addition, since $P_n(1/\rho_0) = 1$ and all the roots of the polynomial $P_n(x)$ lie on the interval $[-1, 1]$, then $P_n(1) = q_n$ and, consequently, we have the equation

$$P_n(x_j) = (-1)^{j-1} q_n, \quad j = 1, 2, \ldots, n+1. \tag{1}$$

Thus we obtain

Lemma 1. *Among all polynomials $P_n(x)$ taking on the value* 1 *at $x = 1/\rho_0$ the one that deviates least from zero on the interval $[-1, 1]$ satisfies the differential equation*

$$(1 - x^2)(P')^2 = n^2(q_n^2 - p^2). \tag{2}$$

Proof. The points $x_2, x_3, \ldots, x_n$ are simple roots of the polynomial $P_n'(x)$. Obviously, these points are double roots of the polynomial $q_n^2 - P_n^2(x)$, and the points $x_{n+1} = -1$ and $x_1 = 1$ are simple roots of this polynomial. Therefore the polynomials $(1 - x^2)(P_n'(x))^2$ and $q_n^2 - P_n^2(x)$ have the same roots. Consequently, they are proportional, i.e.

$$(1 - x^2)(P_n')^2 = c(q_n^2 - P_n^2(x)).$$

Equating the coefficients of highest degree in the two polynomials, we find that $c = n^2$. The lemma is proved. □

2. We move on now to the construction of the polynomial $P_n(x)$ using equation (2). This equation, in addition to the unknown function $P_n(x)$, contains the as yet unknown parameter q_n. We will not separately fix auxiliary conditions which will uniquely determine the solution to equation (2), but we will instead use all the known information concerning $P_n(x)$.

We look first at equation (2) on the interval $[-1, 1]$. In this case $|P_n(x)| \le q_n$, and, consequently, on the left- and right-hand sides of equation (2), it is possible to take the square root

$$\pm \frac{dP}{\sqrt{q_n^2 - P^2}} = n \frac{dx}{\sqrt{1 - x^2}}, \quad 0 \le x \le 1. \tag{3}$$

We shall investigate the left-hand side of (3). If $P_n(x_{j+1}) = q_n$, then, as the variable x varies between x_{j+1} and x_j, the function $P_n(x)$ decreases from q_n to $-q_n$. Here the differential dP is negative, and therefore it follows that we should choose the minus sign in the left-hand side of equation (3). Analogously we find that, if $P_n(x_{j+1}) = -q_n$, then we should choose the plus sign. Taking into account (1), we obtain that, on the interval $[x_{j+1}, x_j]$, equation (3) should be written in the form

$$(-1)^{j-1}\frac{dP}{\sqrt{q_n^2 - P^2}} = n\frac{dx}{\sqrt{1-x^2}}, \quad x \in [x_{j+1}, x_j], \quad j = 1, 2, \ldots, n. \tag{4}$$

We obtain now an expression for $P_n(x)$ on the interval $[-1, 1]$. Suppose that x is any point of the interval $[-1, 1]$, and for definiteness assume that x belongs, for example, to the interval $[x_{j+1}, x_j]$.

When integrating the right-hand side of equation (4) for x between x and 1. We obtain

$$n\int_x^1 \frac{dx}{\sqrt{1-x^2}} = n \arcsin x\Big|_x^1 = n \arccos x.$$

We then integrate the left-hand side of equation (4). As x changes from x_{j+1} to x_j, the function $P(x)$ changes from $P(x_{j+1}) = (-1)^j q_n$ to $P(x_j) = (-1)^{j-1} q_n$. Therefore

$$(-1)^{j-1}\int_{P(x_{j+1})}^{P(x_j)} \frac{dP}{\sqrt{q_n^2 - P^2}} = \int_{-q_n}^{q_n} \frac{dP}{\sqrt{q_n^2 - P^2}} = \arcsin \frac{P}{q_n}\Big|_{-q_n}^{q_n} = \pi.$$

Further, when we integrate the left-hand side of (4) from $P(x)$ to $P(x_k)$ we obtain

$$(-1)^{k-1}\int_{P(x)}^{P(x_k)} \frac{dP}{\sqrt{q_n^2 - P^2}} = \int_{(-1)^{k-1}P(x)}^{q_n} \frac{dP}{\sqrt{q_n^2 - P^2}} = \arccos(-1)^{k-1}\frac{P(x)}{q_n}.$$

Since

$$\int_x^1 \frac{dx}{\sqrt{1-x^2}} = \int_x^{x_k} \frac{dx}{\sqrt{1-x^2}} + \sum_{j=1}^{k-1}\int_{x_{j+1}}^{x_j} \frac{dx}{\sqrt{1-x^2}},$$

then finally we obtain

$$n \arccos x = (k-1)\pi + \arccos(-1)^{k-1}\frac{P(x)}{q_n}. \tag{5}$$

From this we find

$$P_n(x) = q_n \cos(n \arccos x), \quad |x| \le 1. \tag{6}$$

Setting $x = \omega_k \in [x_{k+1}, x_k]$ in (5), we find the roots of the polynomial $P_n(x)$

$$\omega_k = \cos\frac{(2k-1)\pi}{2n}, \quad k = 1, 2, \ldots, n.$$

The formula (6) defines the polynomial $P_n(x)$ for $x \in [-1, 1]$. We now find the form of the polynomial $P_n(x)$ for $|x| \geq 1$ and determine q_n. To do this we notice that

$$\omega_{n-k+1} = \cos\left(\pi - \frac{2k-1}{2n}\pi\right) = -\omega_k, \quad k = 1, 2, \ldots, n.$$

Therefore $P_n(-x) = (-1)^n P_n(x)$ and, consequently, it is sufficient to determine $P_n(x)$ for $x \geq 1$.

We now investigate equation (2) for $x \geq 1$. In this case, we write it in the following form:

$$(x^2 - 1)(P')^2 = n^2(P^2 - q_n^2), \quad x \geq 1.$$

Since $x \geq 1$, then $P(x) \geq q_n$ and the function is increasing. Therefore, extracting a root, we obtain

$$\frac{dP}{\sqrt{P^2 - q_n^2}} = n\frac{dx}{\sqrt{x^2 - 1}}.$$

To integrate the right-hand side of this equation from 1 to x, we integrate the left-hand side between q_n and $P_n(x)$. Therefore

$$\begin{aligned}\int_{q_n}^{P_n(x)} \frac{dP}{\sqrt{P^2 - q_n^2}} &= \ln\left(\frac{P_n(x)}{q_n} + \sqrt{\frac{P_n^2(x)}{q_n^2} - 1}\right) = \text{arc }\cosh\frac{P_n(x)}{q_n} \\ &= n\int_1^x \frac{dx}{\sqrt{x^2 - 1}} = n\ln\left(x + \sqrt{x^2 - 1}\right) = n \operatorname{arc\,cosh} x.\end{aligned} \tag{7}$$

From this we obtain

$$P_n(x) = q_n \cosh(n \operatorname{arc\,cosh} x), \quad x \geq 1.$$

Since $P_n(x) = (-1)^n P_n(-x)$, then for $x \leq 1$ we find

$$P_n(x) = (-1)^n q_n \cosh(n \operatorname{arc\,cosh}(-x)) = q_n \cosh(n \operatorname{arc\,cosh} x) \quad x \leq -1.$$

Thus, for $|x| \geq 1$ we obtain the following expression for the polynomial $P_n(x)$:

$$P_n(x) = q_n \cosh(n \operatorname{arc\,cosh} x), \quad |x| \geq 1. \tag{8}$$

We now find q_n. Setting $x = 1/\rho_0$ in (8), and taking into account that $P_n(1/\rho_0) = 1$, we obtain

$$q_n = 1/\cosh(n \operatorname{arc\,cosh}(1/\rho_0)).$$

On the other hand, setting $x = 1/\rho_0$ in (7), we find

$$\ln \frac{1+\sqrt{1-q_n^2}}{q_n} = n \ln \frac{1+\sqrt{1-\rho_0^2}}{\rho_0} = n \ln \frac{1}{\rho_1},$$

where

$$\rho_1 = \frac{\rho_0}{1+\sqrt{1-\rho_0^2}} = \frac{1-\sqrt{\xi}}{1+\sqrt{\xi}}, \quad \xi = \frac{\gamma_1}{\gamma_2}, \quad \rho_0 = \frac{2\rho_1}{1+\rho_1^2}.$$

Consequently,

$$q_n = \frac{1}{\cosh\left(n \operatorname{arc\,cosh} \frac{1}{\rho_0}\right)} = \frac{2\rho_1^n}{1+\rho_1^{2n}} < 1. \tag{9}$$

Combining (6) and (8), we obtain

$$P_n(x) = q_n T_n(x) = T_n(x)/T_n(1/\rho_0), \tag{10}$$

where

$$T_n(x) = \begin{cases} \cos(n \arccos x), & |x| \leq 1, \\ \cosh(n \operatorname{arc\,cosh} x), & |x| \geq 1. \end{cases}$$

The polynomial $T_n(x)$ is called the Chebyshev polynomial of the first kind of degree n.

Thus, the problem has been completely solved. Its solution is given by formulas (9) and (10). Transforming to the variable t, we obtain the desired polynomial

$$Q_n(t) = P_n\left(\frac{1-\tau_0 t}{\rho_0}\right) = q_n T_n\left(\frac{1-\tau_0 t}{\rho_0}\right),$$

which deviates least from zero on the interval $[\gamma_1, \gamma_2]$.

Appendix B

Bibliography

1. Wasow W.R., Forsythe, G. *Raznostnye metody resheniya differentsial'nykh uravneniy v chastnykh proivodnykh.* Moscow: Foreign Literature Publishers, 1963. Translated from *Finite-difference methods for partial differential equations.* New York: Wiley, 1960.

2. Gelfond A.O. *Ischisleniye konechnykh raznostnye.* Moscow: Nauka, 1967. Translated as *Calculus of finite differences.* India: Hindustan Publishing Company, 1972.

3. Karchevskiy M.M., Lyashko A.D. *Raznostnye skhemy dlya nelineynykh zadach matematicheskoy fiziki* [*Difference schemes for non-linear problems in mathematical physics*]. Kazan: Kazan State University Press, 1976.

4. Krasnoselskij M.A., Vainikko G.M., et al. *Priblizhennoye resheniye operatornykh uravneniy.* Moscow: Nauka, 1969. Translated as *Approximate solution of operator equations.* Groningen: Wolters-Noordhoff Publishing, 1977.

5. Marchuk G.I. *Metody vychislitel'noy matematiki.* Novosibirsk: Nauka, 1973. Translated as *Methods of numerical mathematics.* New York: Springer-Verlag, 1975.

6. Oganesyan L.A., Rivkind V.Y. Rukhovets L.A. *Variatsionno-raznostnye metody resheniya ellipticheskikh uravneniy* [*Variational-difference methods for solving elliptic equations*], parts 1 and 2. *Differential Equations and their Applications* 5 (1973) and 8 (1974).

7. Ortega J.M., Rheinbolt W.C. *Iteratsionnye metody resheniya nelineynykh sistem uravneniy so mnogimi neizvestnymi.* Moscow: Mir, 1975. Translated from *Iterative solution of non-linear equations in several variables.* New York: Academic Press, 1970.

8. Samarskii A.A. *Vvedenie v teoriyu raznostnykh skhem* [*An introduction to the theory of difference schemes*]. Moscow: Nauka, 1971.
9. Samarskii A.A. *Teoriya raznostnykh skhem* [*The theory of difference schemes*]. Moscow: Nauka, 1977.
10. Samarskii A.A., Gulin A.V. *Ustoichivost' raznostnykh skhem* [*The stability of difference schemes*]. Moscow: Nauka, 1973.
11. Samarskii A.A., Andreev V.B. *Raznostnye metody dlya ellipticheskikh uravneniy* [*Difference methods for elliptic equations*]. Moscow: Nauka, 1976.
12. Samarskii A.A., Karamzin Y.N. *Raznostnye uravneniya* [*Difference equations*]. Moscow: Znanie, 1978.
13. Faddeev D.K., Faddeeva V.N. *Vychislitel'nye metody lineynoy algebry.* Moscow: State publishing house for physico-mathematical literature, 1963. Translated as *Computation methods of linear algebra.* San Francisco: W.H. Freeman and Company, 1963.
14. Young D.M. *Iterative solution of large linear systems.* New York: Academic Press, 1971.

Appendix C

Translator's Note

A number of the methods described in this book are also discussed in the Western scientific literature, sometimes from a quite different perspective. In this note, I would like to indicate some of these sources.

The early chapters discuss direct methods for solving large sparse linear systems of special form. The major ideas are: Gaussian elimination, cyclic reduction, and application of the fast Fourier transform (FFT). An extensive discussion of Gaussian elimination, along with a detailed error analysis, can be found in Wilkinson [1965]; special adaptations of this technique, as well as cyclic reduction and its generalizations, are discussed in Buzbee, Golub, and Nielson [1970]. The FFT is developed in Hamming [1973]. The bibliographies of these works indicate the development of these topics in the west.

The remaining chapters are devoted to iterative methods. As the bibliography of this book implies, the work of Young [1971] has been fundamental to the field of iterative methods based on matrix splittings. Further results can be found in Varga [1962]. Conjugate-gradient and related methods are treated in Concus, Golub, and O'Leary [1976] and Paige and Saunders [1975].

As in any translation, there are dilemmas imposed by terminology. Some of the methods discussed in this book, even though they have similar names and formulas, are not identical to their western counterparts. In these cases, I have tried to use different terminology to distinguish them. Whenever practical, terms common to English-language mathematical literature have been used, even when they differ considerably from their Russian names. In most respects, the translation is styled closely after the original.

Stephen Nash

References:

- Buzbee B.L., Golub G.H., Nielson C.W. *On direct methods for solving Poisson's equation.* SIAM Journal of Numerical Analysis 7 (1970): 627–656.
- Concus P., Golub G.H., O'Leary D.P. *A generalized conjugate-gradient method for the numerical solution of elliptic partial differential equations.* In *Sparse Matrix Computations*, edited by J. Bunch and D. Rose, pp. 309–332. New York: Academic Press, 1976.
- Hamming R.W. *Numerical methods for scientists and engineers.* New York: McGraw-Hill, 1973.
- Paige C.C., Saunders M.A. *Solution of sparse indefinite systems of linear equations.* SIAM Journal of Numerical Analysis 12 (1975): 617–629.
- Varga R.S. *Matrix iterative analysis.* Englewood Cliffs, New Jersey: Prentice-Hall, 1962.
- Wilkinson J.H. *The Algebraic Eigenvalue Problem.* Oxford: Oxford University Press, 1965.
- Young D.M. *Iterative solution of large linear systems.* New York: Academic Press, 1971.

Index